DANA G. MARSH
W9-BYV-695

TOLERANCE DESIGN

TOLERANCE DESIGN

A HANDBOOK FOR DEVELOPING OPTIMAL SPECIFICATIONS

C. M. Creveling

Addison-Wesley

An Imprint of Addison Wesley Longman, Inc.
Reading, Massachusetts Harlow, England Menlo Park, California
Berkeley, California Don Mills, Ontario Sydney
Bonn Amsterdam Tokyo Mexico City

For more information, please contact:
Corporate & Professional Publishing Group
Addison Wesley Longman, Inc.
One Jacob Way
Reading, Massachusetts 01867

Library of Congress Cataloging-in-Publication Data

Creveling, Clyde M., 1956–
Tolerance design : a handbook for developing optimal specifications / C. M. Creveling
p. cm.
Includes bibliographical references and index.
ISBN 0-201-63473-2 (alk. paper)
1. Tolerance (Engineering) 2. Design, Industrial. I. Title.
TS172.C74 1997
620′.0045—DC21 96-37196′
CIP

Text printed on recycled and acid-free paper.

0-201-63473-2
123456789–MA–00999897

First printing, December 1996

This book is dedicated in loving memory of

Katherine Jane Creveling,
my mother. Her love, encouragement, and support were universal constants in my life.

And

Robert Byron Dome,
my father-in-law. Bob was the most prolific engineer (115 U.S. patents) and finest gentleman I have had the pleasure of knowing. His "tolerance" for the diversity of people in this world was and continues to be an inspiration to me.

These wonderful people passed away during the writing of this book. I shall never forget the contributions they made to my life.

CONTENTS

Preface *xv*
Acknowledgments *xix*

Section I: Setting the Stage for Understanding the Science of Tolerancing *1*

1 Introduction to Tolerancing and Tolerance Design *3*

The Historical Roots of Tolerancing *3*
The State of the Art in Tolerancing Techniques *6*
What Should a Modern Tolerance Development Process Look Like? *6*
The Relationship between Traditional Tolerancing Methods and the Taguchi Approach *6*
Developing Tolerances: The Role of Engineers and Designers *9*
Concepts, Definitions, and Relationships *10*
Matching Design Tolerances with Appropriate Manufacturing Processes *12*
Introduction to the Taguchi Approach to Tolerance Analysis *12*
Review of the Three Phases of a Quality Engineering-Based Product-Development Process *13*
Taguchi's Approach to Tolerancing *20*
Summary *23*
References *24*

2 The Relationship of Quality Engineering and Tolerancing to Reliability Growth *25*

The Three Initial Phases of Product Development *26*
Phase 1: Subsystem Concept Selection and Robustness Optimization *28*
Gate 1 *39*
Phase 2: System Robustness Optimization *39*

Gate 2 42
Phase 3: Product Design 42
Gate 3 47
The Reliability Bathtub Curve and Tolerancing 48
Summary 49
References 50

3 Introductory Statistics and Data Analysis for Tolerancing and Tolerance Design *51*

The Role of Data in Tolerance Analysis 51
Graphical Methods of Data Analysis 52
The Histogram 54
Quantitative Methods of Data Analysis 54
Introduction to the Fundamentals of Descriptive Statistics 55
The First Moment of the Data about the Mean–The Arithmetic Average 55
The Second Moment about the Mean: The Variance 56
The Standard Deviation 57
The Third Moment about the Mean: The Skew 58
The Fourth Moment about the Mean: The Kurtosis 59
The Coefficient of Variation 61
The Use of Distributions 62
The Use of Distributions in Taguchi's Parameter and Tolerance Design 62
The Use of Distributions in Traditional Tolerance Analysis 63
Introduction to the Fundamentals of Inferential Statistics 63
The Z Transformation Process 63
The Student-t Transformation Process 66
The Adjusted Z Transformation Process for Working with Nonnormal Distributions 67
Manufacturing Process Capability Metrics 72
Six Sigma Process Metrics 76
The Relationship between the Quality-Loss Function, C_p, and C_{pk} 79
Summary 81
References 83

Section II: Traditional Tolerance Analysis *85*

4 Using Standard Tolerance Publications and Manufacturer's Process Capability Recommendations *87*

Starting the Tolerance Design Process 87
The Three Sigma Paradigm 88
Processes for Establishing Initial Tolerances 90
Establishing Process Capability and Process Control for Identifying Initial Tolerances 96
Creating a Library of Process Capabilities as a Data Base 98

Summary 99
References 99

5 Linear and Nonlinear Worst-Case Tolerance Analysis 101

Standard Worst-Case Methods 101
Six Sigma Worst-Case Tolerance Analysis 102
Analysis for Nonlinear Tolerance Stacks Using the Worst-Case Method 104
Process Diagrams for Worst-Case Analysis for the Stackup of Tolerances in Assemblies 105
A Case Study for a Linear Worst-Case Tolerance Stackup Analysis 109
A Process for Modeling Worst-Case Nonlinear Tolerance Stackup Problems 111
Summary 122
References 123

6 Linear and Nonlinear Statistical Tolerance Analysis 124

The Root Sum of Squares (RSS) Appoach 126
The Z Transformation Process 127
One-Dimensional Tolerance Stacks 127
Motorola's Dynamic Root Sum of Squares Approach 130
Motorola's Static Root Sum of Squares Approach 131
The Nonlinear RSS Case Method 132
Process Diagrams for the Statistical Methods of Tolerance Stackup Analysis 133
The Linear RSS Method 133
The Dynamic RSS Method 138
The Static RSS Method 139
The Nonlinear RSS Example 145
Summary 147
References 148

7 Sensitivity Analysis and Related Topics 149

The Various Approaches to Performing Sensitivity Analysis 149
Mathematical Sensitivity Analysis 149
Tolerance-Range Sensitivity Analysis 150
Component Manufacturing Process Output Distribution Sensitivity Analysis 151
Nonlinear Tolerance Sensitivity Analysis 151
The Steps for Basic Nonlinear Sensitivity Analysis 152
Empirical Sensitivity Analysis 153
ANOVA Sensitivity Analysis 153
One-Factor-at-a-Time Sensitivity Analysis 153
Using Sensitivity Analysis in Concept Design 154
An Example of the Use of Sensitivity Analysis in Concept Development 156

Summary *158*
References *159*

8 Computer Aided Tolerancing Techniques *160*

Various Software and Platform Options to Support CAT Analysis *161*
Commercially Available Pro/ENGINEER-Based Tolerance Analysis Software *162*
Monte Carlo Simulations in Tolerance Analysis *163*
Characterizing Probability Distributions for Tolerance Analysis Applications *167*
The Common Probability Distributions Available in *Crystal Ball* *167*
The Less Common Probability Distributions Available in *Crystal Ball* *168*
Step-by-Step Process Diagrams for a *Crystal Ball* Monte Carlo Analysis *170*
Sensitivity Analysis Using *Crystal Ball* *172*
How to Use *Crystal Ball* *175*
Running the Monte Carlo Simulation *181*
Preparing Engineering Analysis Reports *184*
Another Computer-Aided Tolerance Approach *185*
Summary *185*
References *186*

9 Introduction to Cost-Based Optimal Tolerancing Analysis *187*

Skills Required for Cost-Based Optimal Tolerance Analysis *188*
The Various Approaches to Cost-Based Optimal Tolerance Analysis *188*
Cost versus Tolerance Plots *190*
Summary *190*
References *190*

10 Strengths and Weaknesses of the Traditional Tolerance Approaches *191*

Using Standard Tolerance Publications and Manufacturer's Process Capability Recommendations *192*
Worst-Case Tolerance Analysis *193*
The Statistical Methods of Tolerance Analysis *193*
The Root Sum Square Approach *193*
The Dynamic Root Sum of Squares Approach *193*
The Static Root Sum of Squares Approach *194*
The Nonlinear Tolerance Approaches *194*
Sensitivity Analysis *194*
Computer-Aided Tolerancing *194*
Cost-Based Optimal Tolerance Analysis *195*
How the Six Processes Relate to the Overall Product Tolerancing Process *195*
Summary *196*
References *197*

Section III: Taguchi's Approach to Tolerancing and Tolerance Design *199*

11 The Quality-Loss Function in Tolerancing and Tolerance Design *201*

Linking Cost and Functional Performance *201*
An Example of the Cost of Quality *202*
The Step Function: An Inadequate Description of Quality *206*
The Customer Tolerance *207*
The Quality-Loss Function: A Better Description of Quality *208*
The Quality-Loss Coefficient *209*
An Example of the Quality-Loss Function *210*
  The Types of Quality-Loss Functions *210*
Developing Quality-Loss Functions in a Customer's Environment *219*
Constructing the Quality-Loss Economic Coefficient $(A_0/(\Delta_0)^2)$ *220*
Summary *222*
References *223*

12 The Application of the Quadratic Loss Function to Tolerancing *224*

The Difference between Customer, Design, and Manufacturing Tolerances *224*
  Customer Tolerances *224*
  Design and Manufacturing Tolerances *225*
The Taguchi Tolerancing Equations *225*
  Taguchi's Economical Safety Factor *225*
  The Derivation of Taguchi's Tolerance Equation and the Factor of Safety *226*
Relating Customer Tolerances to Engineering Tolerances *229*
An Example of Tolerancing Using the Loss Function (Nominal-the-Best Case) *229*
Relating Customer Tolerances to Subsystem and Component Tolerances *230*
The Linear Sensitivity Factor, β *231*
Using the Loss Function for Multiple-Component Tolerance Analysis *232*
An Example of Applying the Quality-Loss Function to a Multicomponent Problem *233*
Setting Up the Problem *234*
Identifying Critical Parameters *234*
  Mapping the Critical Parameters and Their Sensitivities *235*
Converting the Traditional Tolerance Problem into a Quality-Loss Tolerance Problem *237*
How to Evaluate Aggregated Low Level Tolerances *238*
  Case 1: Ratio of Losses are Much Less than Unity *239*
  Case 2: Ratio of Losses are Much Greater than Unity *239*
  Case 3: Ratio of Loss Approximately Equal Unity *239*
Using the Loss Function Nonlinear Relationships *240*
Developing Tolerances for Deterioration Characteristics in the Design *241*
Tolerancing the Deterioration Rate of a Higher Level Product Characteristic *242*
Determining Initial and Deterioration Tolerances for a Product Characteristic *243*

Summary *244*
References *245*

13 General Review of Orthogonal Array Experimentation for Tolerance Design Applications *246*

Developing Tolerances Using a Designed Experiment *246*
Use of Orthogonal Arrays in Tolerance Design *247*
The Build-Test-Fix Approach *248*
Introduction to Full Factorial Experiments *248*
Designed Experiments Based on Fractional Factorial Orthogonal Arrays *249*
Degrees of Freedom: The Capacity to do Experimental Analysis *251*
Degrees of Freedom for the Full Factorial *251*
The DOF for the Fractional Factorial Orthogonal Array *251*
DOF for Two-Level Full Factorial Arrays *252*
Methods to Account for Interactions within Tolerance Design Experiments *255*
Interactions Defined *255*
Quantifying the Effect of an Interaction *257*
Defining the Degrees of Freedom Required for Evaluating Interactions *259*
Summary *260*
References *261*

14 Introducing Noise into a Tolerance Experiment *262*

Defining Noises and Creating Noise Diagrams and Maps *262*
Selecting Noise Factors for Inclusion in a Tolerance Experiment *264*
Noise Diagrams and System Noise Maps *265*
Experimental Error and Induced Noise *267*
The Noise Factor Experiment *271*
Creating Compound Noise Factors *272*
Setting Up and Running the Noise Experiment *273*
Analysis of Means for the Noise Experiment *274*
Verification of the Predicted Response *276*
Summary *277*
References *277*

15 Setting Up a Designed Experiment for Variance and Tolerance Analysis *278*

Preparing to Run a Statistical Variance Experiment *279*
Selecting the Right Orthogonal Array for the Experiment *279*
Special Instructions for Running Three-Level Statistical Variance Experiments *281*
Using the $\sqrt{3}\ \sigma$ Transformation *292*
A Comparison of Output Statistics *296*
Conducting a Tolerance Experiment for Worst-Case Conditions *300*

An L_9 Experiment and Monte Carlo Simulation Using $\mu \pm 3\sigma$ Levels Assuming Uniform Distributions 301
Metrology and Experimental Technique 304
Summary 304
References 305

16 **The ANOVA Method** 306

Accounting for Variation Using Experimental Data 307
A Note on Computer-Aided ANOVA 308
An Example of the ANOVA Process 308
Degrees of Freedom in ANOVA 311
Error Variance and Pooling 312
Error Variance and Replication 312
Error Variance and Utilizing Empty Columns 313
The *F*-Test 313
A WinRobust ANOVA Example 314
An ANOVA-TM Example 318
Summary 323
References 324

17 **The Tolerance Design Process: A Detailed Case Study** 325

The Steps for Performing the Tolerance Design Process 325
Option 1: A Company Cost-Driven Process 327
Option 2: A Manufacturing Capability and Cost-Driven Process 328
The ASI Circuit Case Study 330
Identifying the Parameters for Tolerancing 330
Improving Quality in Cases where Parameter Design was Not Done 331
Constructing the Loss Function 332
Setting Up and Running the Experiment 334
Two- versus Three-Level Experiments 337
Techniques for Putting Noise into the Tolerance Experiment 338
Running the Experiment 338
Data Entry 340
Measuring the Proper Response Characteristic 343
Interactions in Tolerance Experiments 343
Analyzing the Data 344
Applying ANOVA 344
Relating the ANOVA Data to the Loss Function and Process Capability (Cp) 349
Defining the Critical to Function (CTF) Factors 350
Defining the Cost Improvement Parameters (CIP) 351
Identifying and Quantifying the Costs Associated with Improving Quality 352
Working with Suppliers to Lower Customer Losses through Reducing the Component Parameter Standard Deviations 352

Calculating New Variances and MSD Values Using the Variance Equation 353
Quantifying the Cost of Reducing the Parameter Standard Deviations 354
Identifying and Quantifying the Opportunities for Lowering Costs 356
Relaxing Tolerances and Material Specifications of CIPs to Balance Cost and Quality 356
New Loss and Cp after Upgrading the Critical To Function Factors and Downgrading the Cost Improvement Parameters 357
Using Tolerance Design to Help Attain Six Sigma Quality Goals 357
Using Tolerance Design to Improve System Reliability 357
Summary 358
References 359

Section IV: Industrial Case Studies 361

18 Drive System Case Studies 363

The Drive System 364
The Drive Module 365
Case 1: Defining Tolerances for Standard Drive Module Components 367
Tolerancing Custom Parts to Assemble with Standard Parts 368
Case 2: Drive Module for Worst-Case Assembly Analysis 369
Worst-Case Linear Stack Analysis 369
Case 3: Drive Module for RSS (Statistical) Assembly Analysis 371
RSS Case Linear Stack Analysis 371
Case 4: Drive Module for Computer-Aided Assembly Analysis 373
Case 5: Drive System Aided by the Use of a Designed Experiment 380
Preparing for the Drive System Tolerance Development Experiment 381
Setting Up the Experiment 382
Data Acquisition 385
Results from the L_9 Experiment 386
Summary 388

Appendix A: The Z Transformation Tables and the t Transformation Table 389

Appendix B: The Adjusted Z Transformation Tables 393

Appendix C: The F Tables 403

Appendix D: Additional References for Tolerance Design 409

Appendix E: Suppliers for Tolerance Design 411

Index 413

PREFACE

This is the first book in the history of engineering science to comprehensively address the analytical and experimental methods available for the development of tolerances. This indicates, on one hand, an unfortunate legacy of neglecting a very important engineering topic. On the other hand, it provides an opportunity for engineering practitioners, students, and educators to fill a void in their skill sets.

This book is written for a fairly diverse audience. It is intended to be used as a handbook for industrial applications as well as a course text in technology and engineering science programs at the college and university level. It is structured so that undergraduate technology and engineering students at the associate *and* baccalaureate levels can equally share in its informational value. The author views the tolerance development process as the responsibility of the design engineering *team.* This requires that the book be suitable for engineers, designer/drafters, and technicians. Each of these engineering team members has a valuable role to play in the development of product specifications. *Tolerance Design* contains detailed information for each of these team members.

This book is broad in scope. It reviews the three initial phases of the product-development process where tolerance development resides as the final cost-versus-quality balancing process. It explains, in detail, how tolerance design relates to concept design and parameter design. It also relates the tolerance design process to many other engineering and product development tools and tasks, including reliability growth activities. It covers all the basic analytical methods that exist in modern and traditional tolerancing techniques. It also rigorously covers the latest experimental methods that can be employed in the tolerance design process.

This text details step-by-step the analytical and experimental tolerance-development methods so the reader can more easily and properly develop tolerances. Many examples and case studies are used to convey the process of defining the right tolerances so they can be properly communicated to the manufacturing, assembly, and service communities. The text illustrates how to balance product cost and quality through the tolerance development process. Great emphasis is placed on how the engineering and manufacturing teams must work harmoniously and concurrently to develop optimal tolerances.

Tolerance Design instills a philosophy that tolerance development is as much about cost optimization as it is about functional integrity.

This book has been heavily critiqued and scrutinized, on your behalf, by many of the top experts in the U.S. engineering, tolerance analysis, and tolerance communication communities. Much of their work has been included in this text. This was done to make the depth and breadth of the book rock solid. Since the book is the first of its kind, the author and publisher believed it had to be thoroughly evaluated, perhaps even more rigorously than usual, so that it could stand as a major contribution to engineering literature for many years to come. The book was "engineered" to meet your needs so you can meet your customer's needs.

The world is so competitive now that there is simply no way to keep our industries economically viable by using the same ad hoc engineering processes that have evolved over the past 50 years. With this in mind, a major focus in this text is given to the disciplined methods of quality engineering as defined by Dr. Genichi Taguchi. Engineering *processes* are now being taught in universities and applied in industries all over the United States. This book is very process oriented. It strives to teach discipline in the development of tolerances. In fact, you will have difficulty succeeding in tolerance optimization if the engineering processes that follow the tolerancing process are void of rigor and discipline. Taguchi's approach to product development provides an excellent engineering process context for our discussions.

The literature of off-line quality engineering is primarily focused on the parameter design process. There are numerous books and papers available on this topic, which is frequently referred to as robust design. The other two phases of off-line quality engineering—concept design and tolerance design—have received much less attention in published material. This is particularly true of tolerance design. Until now there has been no single book dedicated to the engineering process of designing tolerances using the quality-loss function. Furthermore, there has not been a single text on how to establish six-sigma process capability in the context of the quality-loss function and upper and lower specification limits that have been produced from the output from orthogonal array matrix experiments.

The question of why so little has been written on this practical subject is relatively easy to answer. Taguchi and his colleagues (myself included) have focused the majority of their writing, teaching, and consulting on developing robustness (insensitivity to sources of variability) as early as possible in the product development process. Consequently, the time and energy spent on promoting tolerance design is relatively small in comparison to the work done to promote the proper application of parameter design. Taguchi has written several interesting chapters that show the power of continuing to improve the quality of a product or process through the methods of tolerance design.

His unique contributions to tolerance literature include the application of orthogonal array matrix experiments—complete with stressful noises, ANOVA data decomposition, and the quality-loss function. He has provided a way to bring the design into production with a deliberate, balancing process to be sure both the customer and the corporation have the right quality and cost designed into the product. Dr. Taguchi's work is the basis for the third section of this book.

In the United States, the traditional approach to tolerancing has been primarily in the domain of designers and drafters as opposed to that of engineers. The engineering focus has been on defining the nominal set points at which the part or design functions correctly. The designers are frequently left to finish the job by defining the tolerances and then properly communicate them to the appropriate manufacturing personnel through drawings or specification documents. Tolerancing occurs naturally as a two-step process: the determination of the tolerance and the communication of the tolerance. In his book *Total Quality Development,* Don Clausing states:

> *In parameter design, the best nominal (target) values are selected for the critical design parameters. In tolerance design, we select the economical precision around the nominal values in two steps: (1) selecting the right production precision, and (2) putting tolerance values on drawings . . . The intrinsically more important part of tolerance design is the selection of the best production precision. Strictly speaking, this has nothing to do with tolerances at all. Tolerances are the numbers that go on the drawing (pp. 249–250).*

The point is that regardless of whether the engineer or the designer or both establish the tolerances, the tolerances are critical to the successful manufacture and performance of the product over its intended life cycle. This book is all about the process of determining production tolerances. The process of communicating the tolerances is left to other fine texts, such as *Geometrics III,* by Lowell Foster.

This book includes a focus on the economics of the tolerance development process. As Taguchi points out in his book *Taguchi on Robust Technology Development,* the average loss N products will incur over time, T, is derived from the following general relationship:

$$L(y) = \frac{1}{N}\sum_{i=1}^{N}\int_{0}^{T} L_i(t, y)\,dt$$

This integral equation captures the central concept of how, on average, money ($L(y)$) is lost over time (T, total life and t, any point in time) as a measurable parameter (y) in a design deviates from its intended target value. It is clear that tolerances can play a significant role in the loss ($L_i(t, y)$) of dollars incurred over the life (T) of the product. Thus, tolerance design can also have an impact on the economics associated with product reliability. This book is designed to make this particular concept of accounting for loss and improving reliability understandable to any technical person with a basic understanding of algebra. The reader should not be alarmed that an integral equation is used to communicate the nature of loss. It is my intention to make the loss function clear to *all* readers—not just those with a background in advanced mathematics. One of Taguchi's finest professional attributes is that he simplifies complex optimization processes to the point where practicing engineers and technical personnel can rapidly learn and apply them. I took great pains to continue this approach in this text. This is not a book to dazzle the theoreticians; rather, it is a practical guide to solving real engineering problems using a disciplined process. It is also meant as a follow-up text to *Engineering Methods for Robust Product Design,* by William Y. Fowlkes and myself to help engineers define the optimum nominal (target) set points for a design. *Tolerance Design* will help you see your design through to the end of the design process by defining the tolerances that support the nominal set points.

As in all engineering tasks, the need for a disciplined process is essential to assure that the life cycle cost goals, including the customer's cost goals, are actually met. To this end, the overall tolerancing process is important. It is too important to be left as a detail that receives little or no serious thought until a project is nearly ready to go into production. It is a deliberate set of steps that are every bit as important as concept design and parameter design. The detailed procedures that are necessary to perform tolerance design on components that have first undergone parameter design is thoroughly documented and flow-charted, as is the process of performing tolerance design on components that have not been made robust and are in need of improvement.

This book is also about matching the designed tolerances with manufacturing processes that are capable of economically producing parts or components that maximize quality and minimize loss over the life of the product. Considerable material is presented to provide a strategy for attaining six-sigma-level part quality. The best tolerance design projects can fall far short of their intended goals if *manufacturing* process variability is not understood and optimized.

Optimizing both the numerator *and* the denominator of the process capability index (Cp) are considered in this text:

$$C_p = \frac{USL - LSL}{6\sigma} \quad \begin{array}{l} \leftarrow \text{ Allowable design variability} \\ \longleftarrow \text{ Real manufacturing variability} \end{array}$$

Three software packages are illustrated in this book. For personal-computer applications, I recommend the use of *Crystal Ball*—a flexible Monte Carlo simulation package that is readily applicable to statistical tolerance analysis. For analysis of variance and computer-aided experimental design, ANOVA software and WinRobust software are recommended. *Tolerance Design* places a strong emphasis on the use of personal computers in both the analytical and experimental forms of tolerance development, and provides an overview of workstation-based computer applications.

The book concludes with an entire chapter dedicated to a related series of five case studies from Eastman Kodak Company. These case studies draw upon many of the topics and processes discussed in this text—including how and when to transition from analytical to experimental tolerance development methods.

It has been my pleasure to have been accorded the freedom and time to develop a complete course on tolerancing and tolerance design, and to apply the techniques at Eastman Kodak Company. It has been through these experiences that this text has taken form.

C. M. Creveling
June 1996

ACKNOWLEDGMENTS

I would like to thank the following people for their help and encouragement in making this book a reality: Jennifer Joss, Tara Herries, David Ullman, Bill Greenwood, Ken Chase, Lowell Foster, Peter Newman, Bill Fowlkes, Genichi Taguchi, Shin Taguchi, Alan Wu, Don Clausing, John Gallenberger, Tom Jacoby, Tomas Roztocil, David Mabee, and David Bettiol (and any other managers I work for), Tom Barker, Juliann Nelson, Margaret Hackert-Kroot, and Tom Albrecht (Kodak's tolerance experts), and James Minor.

Special thanks to my wife, Kathy, and my son, Neil, for their generous support of my efforts in writing a second book.

SECTION I

Setting the Stage for Understanding the Science of Tolerancing

The book is written in three sections. Section 1 includes Chapters 1 through 3 and is an introduction to tolerance design in a modern concurrent engineering context. Section 2 contains Chapters 4 through 10 and focuses on the fundamental principles of traditional methods of analytical tolerance development. Section 3 contains Chapters 11 through 17 and focuses on the Taguchi approach to tolerancing and on his experimentally based process of tolerance design. Section 4 contains Chapter 18 and concludes the book with five industrial case studies.

Section 1 has been structured to help introduce the general topics associated with tolerance design in a modern concurrent engineering context. It contains a broad range of material that may at first seem odd to present in this text. This book is unique in that it is about a very old engineering topic that requires a very modern view of applying traditional engineering tools in a quality engineering context. Analytical and experimental tolerance development approaches require engineers to work in teams that are constrained by the discipline and rigor of well-thought-out engineering processes. Because of this focus, the first section of the book has been designed to give all the readers a common vocabulary and process orientation.

Section 1 sets the stage for the need to design tolerances through a disciplined process. It does this in the light of historical practices, the lack of current teaching on the subject, and the excellent options that exist for practicing tolerance design in harmony with modern product development processes.

The text assumes that not all readers are familiar with the following topics:

- Descriptive and inferential statistics,
- Reliability engineering,

- Taguchi's approach to quality engineering (Taguchi methods),
- Six-sigma methods, and
- Product commercialization and development engineering processes.

Many readers may have read the author's previous text, *Engineering Methods for Robust Product Design;* W. Y. Fowlkes & C. M. Creveling; Addison-Wesley (1995). For many readers, some of Section 1 will be a review; however, much of it will hold unique and helpful engineering process insights for even the most seasoned veteran of Taguchi's methods of quality engineering.

To tolerance well, one must understand statistics. To tolerance well in a modern industrial product development context, one must understand engineering processes and how they relate to the entire commercialization process. To tolerance well in the context of satisfying customers, one must understand six-sigma methods, reliability engineering, and Taguchi's approach to quality engineering. Without Section 1, the reader would be at a disadvantage when these topics appear throughout the book.

Section 1 can serve as a reference as the reader learns to relate the specific tools of tolerance development to the broad context of product development. Many of the decisions and actions taken prior to tolerance development can and will have profound effects on the ease of tolerance development.

Having a "mini-text" on basic statistics within the book will facilitate the use of statistics and help the reader become proficient in their application. Since Six Sigma methods are used later in the book, an introduction to this topic is presented to get the reader used to its symbology and mechanics.

The goal of Section 1 is to prepare and stimulate the reader to sufficiently understand the culture and vocabulary (both verbal and mathematical) of tolerance design.

CHAPTER 1

Introduction to Tolerancing and Tolerance Design

The Historical Roots of Tolerancing

Tolerancing techniques have evolved over the years and have reached their current state through an ever increasing quest for efficiency in producing products that satisfy customer requirements. As society has progressed to its current standard of living, manufacturing technology has had a lot to do with how expectations have developed. Conversely, the ever increasing demand for higher quality products and services has stimulated remarkable changes in manufacturing processes and in the upstream product development and design processes. Parallels may be drawn between the development of tolerancing methods and the development of manufacturing technology. Tolerancing methods have two general historical categories. The concept of tolerancing evolved in support of a design's basic functionality, somewhat independent of how many items were being made. Later, tolerances evolved in support of a design's interchangeability and mass production. The former sets the early historical context (prenineteenth century), while the latter sets the modern historical context (nineteenth and twentieth centuries).

The earliest uses of design specifications, including tolerances, were to record dimensional and material requirements that preserved the "recipe" for the successful replication of a design that had been proved to work. The issue was the preservation of functionality through craftsmanship. As time and technical sophistication progressed, the means for recording human endeavors became an important element in the design process. Humankind evolved from using natural tools incorporated from the immediate environment to actually crafting tools and simple devices that supported basic needs. The need for products had not yet arisen because human beings had yet to develop the interactive social systems that segregated individuals by occupation. It was still every person for him- or herself.

As people began to form the social groups that eventually gave rise to villages, rudimentary monetary systems began to evolve. Bartering and other forms of exchange converted work into a tradable commodity. People learned to exchange their skills for objects that would elevate their capacity to survive and thrive. The need for replication of valued objects that made life easier began to grow. The human condition went from a survival mode to a broader mode of existence that included leisure time and pursuits that joined evolving artistic capabilities with the growing capacity to perform work more efficiently.

Tools used for everyday tasks were embellished with graphical designs and pictorial representations that today provide fascinating accounts of the lives of our ancestors (we still engrave sporting weapons with intricate patterns and scenes of the hunt, or overlay cooking utensils with pictures of common herbs and seasonings). As people developed craftsmanship in their ability to make things, they began to keep oral or written accounts of how they did it. Many considered their knowledge of how to make things work as a secret to be guarded over a lifetime. Probably the most famous "recipe" that was taken into eternity was the process and specifications used by Stradivarius to make the most prized of all musical instruments.

Design specifications can be traced back to the biblical account of Noah's Ark, circa 2350 B.C. From there, we can trace a progression of using specifications to generate replicas of useful items that were in demand for commercial, medicinal, and artistic purposes. By the beginning of the nineteenth century a transition was to forever change the meaning and purpose of specifications: The creation of a formal geometrical dimensioning process occurred in Europe around 1795 [R1.1].

The industrial revolution brought mass production to the forefront of humankind's commercial processes. Henry Ford, Thomas Edison, George Eastman, Eli Whitney, and countless others of that period in the United States and Europe invented or developed products that greatly changed how society lived and played. These events gave component and manufacturing process replication a new paradigm. Probably the most notable change was the advent of the universal interchangeability of components and assemblies. Interchangeability existed prior to the Industrial Revolution; the difference really was in the scale and scope of its application. Tolerances became very important to the assembly and production processes being developed at that time. One thousand parts could be made independent of another thousand mating parts, and any one part could be expected to mate with any other and still function properly.

Engineering, as a science and profession, was taking on new dimensions. By the 1920s and 1930s the early forms of modern quality control were forming in the minds of men like Sir Ronald Fisher, Frank Yates, and Walter Shewhart. Tolerances played a major role in the quality control processes of this period. Later, in the 1940s and 1950s, Genichi Taguchi and W. Edwards Deming began to educate the world (mainly from Japan) that quality was a matter best dealt with long before the product went into production. The role of tolerances continued to increase in importance as the fledgling mass production economy became the gigantic industrial engine that characterized Western industry in the '40s and '50s.

The birth of the quality focus that is currently sweeping Western industry took its roots in the '60s. The great space race and the cold war had a profound effect on modern engineering educational processes. Engineering academicians shifted their focus from a design-based paradigm, at the university level, to a science-based paradigm—literally with the launching of Sputnik. This set the stage for a trend in engineering education in the United States away from the balance that kept practical design-oriented curricula active across the country and toward a more theoretical and mathematical approach. This country has been the birthplace of much of the world's technology but has often left the manufacturing—and the money that follows it—to other nations. The lack of visibility of a standardized, modern tolerance development process in academia is but one indicator of the reasons why the West is struggling to catch up to Japan and other Pacific Rim countries that excel at product design and manufacturing.

It would take the United States until the 1980s to begin learning and practicing many of the processes the Japanese have been employing since the '50s to drive quality into their products upstream of the manufacturing processes. To the U.S. engineering community, tolerances were mainly an afterthought to the product design process of build-test-fix.

The postwar Japanese have focused on designing quality into the embryonic product, using many of the best statistical methods and quality processes invented in the United States and Europe. The strongest focus in Japan is on the manufacturing processes and the shrinking of variability within those processes. Thus, an economical tolerance in Japan in terms of standard deviations is a whole world apart, literally and figuratively, from what an economical tolerance is, based on standard deviations from industries in the West.

It was not until the early 1980s, when the advent of the highly praised methods of Six Sigma emerged from an aggressive and forward thinking U.S. company (Motorola), that U.S. industrial thinking really began to change. Motorola made global news by successfully competing with the Japanese, in their own markets, in the area of electronic telecommunications equipment. The Six Sigma Institute and Motorola University were formed to focus the company's corporate vision around *variance reduction.* This led to significant gains in product-defect reduction. The Six Sigma methods were built around the statistical sciences and their use in a context of excellence in engineering process execution. This book will make heavy use of the methods of Six Sigma. Mikel J. Harry and Reigle Stewart wrote an exceptionally practical manual titled *Six Sigma Mechanical Design Tolerancing* (1988) [R1.2]. Much of Harry's writing parallels the work Taguchi has done in the sense that both authors are completely focused on variance reduction and the disciplined pursuit of robust design in the development of products. In Section II, frequent reference will be made to the Six Sigma tolerancing manual, as well as to numerous other sources that will construct a clear and comprehensive capability around traditional tolerancing processes.

The ad hoc use of build-test-fix methods during product development dominated the historical process of specification development. While this process did work, and still does to a lesser extent today, trial-and-error engineering left behind a trail of accrued knowledge. The knowledge of how to make things work predominately included the relative precision of specific product-interface requirements. This accrued knowledge became the standard for all trained engineers. Mentoring of young engineers by older, experienced engineers was the primary mode of education in the science and art of establishing tolerances; it is still considered good practice. To this day the dominant method of learning how to develop tolerances is not done by following structured engineering processes but through a loosely defined on-the-job training process that is not standardized or organized in any recognized fashion by academia or industry. The material in this book is offered as a significant step toward bringing the important process of tolerancing into a single source as a part of the engineering sciences.

Currently, few engineering schools offer a single course on how to both develop and communicate tolerances. This book is meant to be used in a comprehensive course in tolerance development and communication, including the topic of geometric dimensioning and tolerancing (GD&T). GD&T is a *specification communication process focusing on the development of a graphical model of the design after the analytical and physical model development is completed through tolerance develop-*

ment. Knowing how to design a product or process implies that one knows how to *fully* develop specifications. The material in this text will provide most of the information needed to complete a design by providing insight on how to develop a complete set of specifications for simple to highly complex components, assemblies, and products. The tolerance communication process, geometric dimensioning and tolerancing, is not covered in this text. The author recommends the GD&T text *Geometrics III,* Lowell Foster [R1.3] as a good reference in this area.

The State of the Art in Tolerancing Techniques

The topic of establishing tolerances in modern product development is discussed every day in businesses all over the world. Every product made is toleranced. Every manufacturing process used to make a product is controlled by some form of specification limit. There is a distinction to be made between the *engineering process* involved in developing tolerances and the *design process* involved in communicating tolerances. The current state of the art provides many options that make it possible to develop tolerances that have a strong role in the quality, cost, and cycle time resident in the development of a product. One significant problem that limits the use of the current methods of tolerancing is that most schools and industries do not view tolerancing as a major part of the engineering process. Most schools and industries struggle with the task of aligning engineering tools within an engineering process to efficiently carry out the development of products and manufacturing processes. This book compiles the major tolerance analysis tools that currently exist into a rational tolerance engineering process.

What Should a Modern Tolerance Development Process Look Like?

This book will use many highlighted sections of text to illustrate the steps required to carry out a specific tolerancing process. Figures 1.1 and 1.2 illustrate the overall tolerance development process covered in this book; each process element will be explained in detail.

Figure 1.1 illustrates that tolerance development takes a path that begins and ends with customer expectations and requirements. Design and manufacturing tolerances must derive from customer tolerances. Design and manufacturing process capabilities (Cp) must be optimized to *minimize customer loss.*

Figure 1.2 describes a general process for developing tolerances.

The Relationship between Traditional Tolerancing Methods and the Taguchi Approach

There are a number of books and published papers that discuss tolerancing as a minor topic in a broader design or process control context. Many of them are listed as references at the end of each chapter. There are numerous published papers that treat tolerancing as a topic unto itself. Much of the current tolerancing literature that might be considered novel is typically expressed in a format that is

Figure 1.1 The path of tolerance development

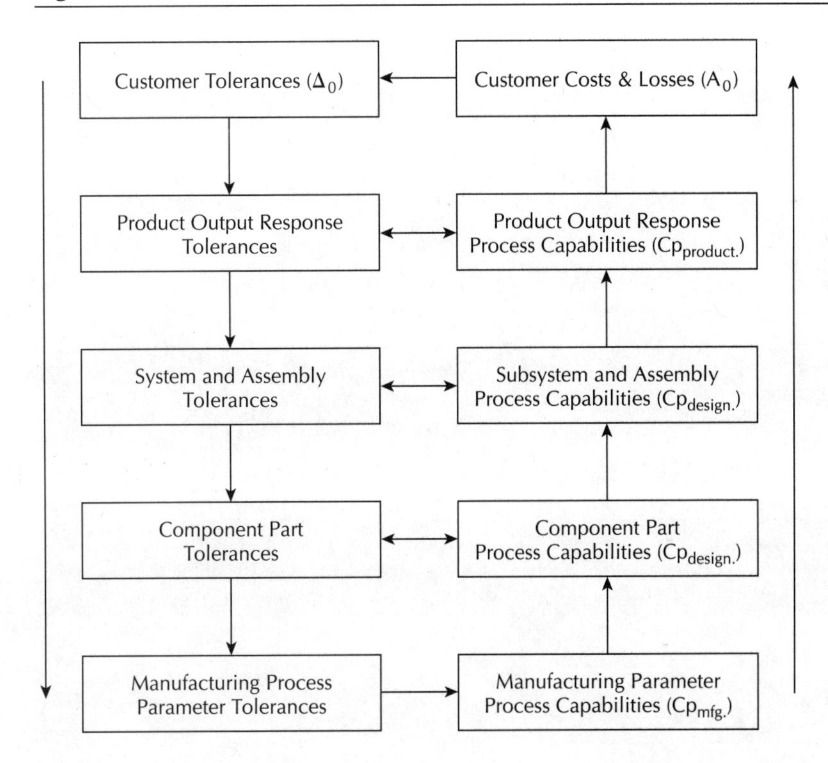

difficult for most practicing engineers to obtain, decipher, and apply. The current methods of tolerancing are often not linked very well with the overall product development process. Within the global collection of tolerancing techniques a definite link can be made between traditional analytical tolerancing methods and the empirical approach to tolerance design, including the use of the quality-loss function developed by Taguchi. This book will lay out the connection between various tolerance development techniques. The book also places all the tolerance engineering methods into a practical engineering process that will define a step-by-step approach to developing optimal tolerances while balancing costs for any given design, and illustrates many of these using shaded step-by-step procedures we will call *process flow diagramming.*

There is no single tolerancing method that does it all, from component parts up through the subsystem and finally to the system level. There is a tolerance engineering *process* that draws on the capabilities of the world's best practices for tolerancing that will provide a comprehensive approach to optimizing tolerances in all the levels of the architecture for the complete product design. That is why this book includes an introduction to all the mainstream traditional tolerance methods in addition to the empirically based Taguchi tolerance design methods.

Figure 1.2 The general tolerance development process

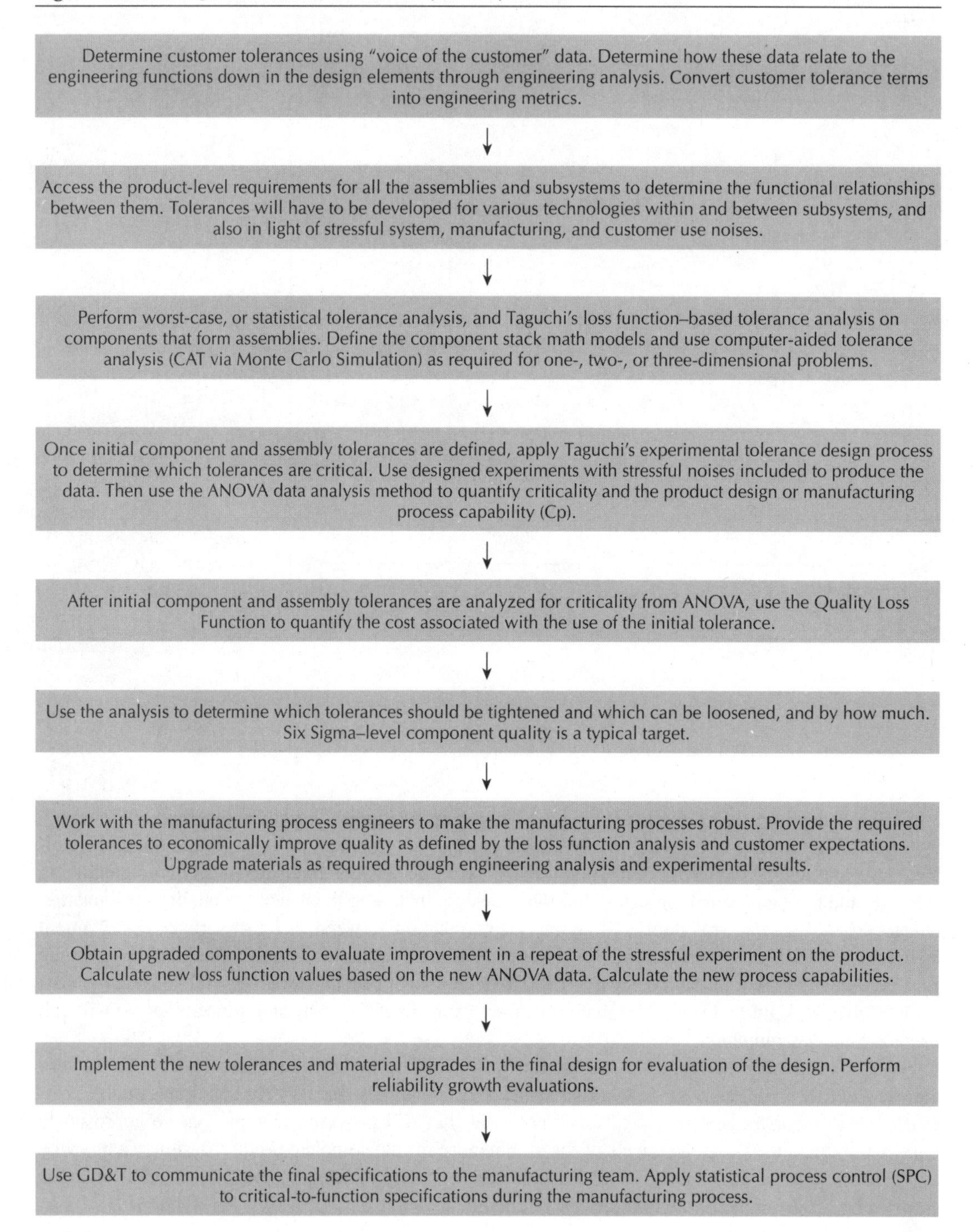

Developing Tolerances: The Role of Engineers and Designers

We previously mentioned that this book is about the *engineering* process for developing tolerances. There is a distinction to be made about the sequence of events within the tolerance engineering process. A tolerance must be developed before it can be communicated. Many high schools, colleges, and universities train students on how to communicate tolerances. Today the standard means of tolerance communication is through the methods of GD&T. Most technicians, designers, and engineers who have been exposed to a modern technical curriculum have been exposed to GD&T, if not comprehensively trained in its use. The same cannot be said for the methods of *developing* tolerances.

The process an engineering team must follow to develop the tolerances for component parts, subassemblies, and a total system is rarely discussed, either formally or informally, in any detail in our universities or industrial training processes. Every design engineer has an ad hoc tolerance design process. Some of these processes are methodically thought out and executed approaches that consider internal and customer costs, functional quality, and overall system reliability growth, while others are undisciplined and leave the tolerances as little more than educated guesses just prior to beginning production. World-class companies no longer leave the engineering process open to every individual's personal whim; instead, they focus on disciplined, methodical, integrated engineering processes that are designed and optimized much like the products they are intended to produce. *The process by which a product is toleranced is every bit as important as the tolerances themselves.*

This book lays out the physics of the tolerancing process, and takes the view that tolerancing and tolerance design fall within the domain of the engineering sciences. Tolerances must derive from our engineering knowledge and analysis of the functional capabilities and limitations of our designs and their requisite manufacturing processes. This is true particularly with respect to the design's sensitivity to sources of variability within the manufacturing and customer-use environments. Tolerance communication is the follow-up process that permits the design team to convey all the necessary design information to the manufacturing team that will bring the design into physical being. A miscommunicated tolerance is a problem that can be assessed and rectified relatively quickly, but a poorly developed tolerance can take months to figure out and correct.

There is a big difference between these two maladies. Fixing a drawing error is much to be preferred over having to figure out whether a tolerance is in need of total redefinition. The former often requires little more than a phone call and an eraser, while the latter can take up weeks or months of an entire engineering team's analytical and experimental efforts.

Because this book constructs an engineering process approach to tolerancing and tolerance design, we will draw on the tools of mathematics, descriptive and inferential statistics, probability analysis, the engineering sciences and physics, personal computers, design of experiments, the method of analysis of variance, reliability engineering methods, cost analysis, process capability analysis, statistical process control, Six Sigma methods, and the quality-loss function. Clearly, there are a lot of tools available to aid in the tolerancing process. This text assumes that the tools of mathematics, basic statistics, engineering science, physics, and personal computer skills are already in the reader's command. The other tools will be defined and developed, at a practical level, for the engineering team's use in tolerance analysis and design.

Concepts, Definitions, and Relationships

The following section provides additional concepts, definitions, and relationships for the terms used in tolerancing. These terms will be used extensively throughout this book. Here we will establish their meaning; later we will define detailed processes to put these concepts to work in the tolerance development phase of the commercialization of products.

The term *tolerance* exists because we live in a probabilistic universe where energy is always conserved. Entropy (loosely defined as the conversion of energy from a useful form into nonuseful forms) is a very real part of every energy-transformation process involving random events. Ironically, the most constant thing happening in our universe is change. Thus, tolerances must exist in our world *because things change.* Tolerances play a key role in defining the constraints necessary to promote consistent transformations of energy. Functional quality, in one way or another, depends on the efficient and consistent transformation of energy. Tolerances must be viewed in the context of how they help constrain the efficient transformation of energy during the function of the design. How well a manufacturing process can produce on-target performance is also a function of the efficiency and stability of energy transformations. This becomes important when studying the output variability associated with a manufacturing process.

Tolerances are defined, fundamentally, as limits or boundaries. In the realm of mechanical-design engineering and product manufacturing they are defined by the ANSI Y14.5M-1982 (R1988) Standard [R1.4] as "The total amount by which a given dimension may vary, or the difference between the limits." This definition has slightly different meaning for manufacturing disciplines in other industries such as chemical processing or consumer electronics; for our purposes the differences are not important. The concepts of tolerance, variability, and limitation beg a few questions: What can change? By how much? And who or what sets the limits?

Tolerances should follow a sequential path of origination. *Who* determines tolerances is at the beginning of the tolerance design process. The customer must be foremost in the design engineer's mind at the beginning of the tolerancing process. Then comes the question of *what* determines the tolerances. That is to say, now that we know what a customer can and will tolerate, what will the physics of the manufacturing processes and the functional capability of the product design allow? Can the customer's tolerances be accommodated by the product? Can the product's tolerances be accommodated by the manufacturing process? One cannot hope that the customer will tolerate what has been created—one must design (manipulate what the laws of physics and probability will allow) to the customer's standards. In today's lean and mean business environment it is very risky to anticipate or assume customer tolerances of performance (even around new features that customers did not know they wanted or needed). This book provides detailed processes to engineer tolerances in the context of customer satisfaction and low cost of ownership. Later, in Chapter 11, we will discuss how to obtain customer information to aid in the construction of quality-loss functions, which will in turn govern the development of tolerance values.

In Taguchi's world of quality engineering, tolerances are *economically established operating windows of functional variability for optimized control factor set points to limit customer loss.* To many engineers and designers in the United States, tolerances are simply the production or manufacturing limits that can be held within economic boundaries based on the unit manufacturing cost of the component. Typically the limits are based on performance that provides three sigma (3σ) quality. This means that 2,700 out of every 1,000,000 parts made will be outside of the tolerance limits. This text seeks to shift the engineer's focus from unit-to-unit manufacturing cost to total quality loss.

The standard definition of a tolerance is also concerned with geometric and locational sufficiency—including the three key areas of form, fit, and function. Another common paradigm associated with tolerancing is that geometric dimensioning and tolerancing (GD&T) is the birthplace of tolerances. This may be so in a lot of cases in current industry, but the true genesis of a tolerance should be in the marriage of physics, engineering, and economics. Tolerances must be *engineered.* They are to be developed through the tools of mathematical analysis (engineering and statistics), empirical analysis (design of experiments including stressful noise factors), computer-aided analysis (various Monte Carlo or geometric simulations), and economic analysis of manufacturing, assembly, service, and maintenance and customer usage (life-cycle cost models and the quality-loss function). The methods of GD&T are profoundly useful in *communicating* tolerances, and play a critical role in how tolerances are ultimately relayed to the manufacturing community. The methods of traditional tolerance analysis, tolerancing, and tolerance design covered in this book will develop a tolerance that is ready to be communicated through the GD&T process. The discipline and clarity that GD&T brings to the tolerancing task is indisputable. No less discipline and clarity ought to go into the engineering of the tolerances.

Some tolerances can be found in various technical reference sources (see Chapter 4). The practice of design engineering requires the knowledge of how to assign a tolerance when employing *standard* components that are being purchased or manufactured. There are many items in the engineering of products or processes that have been used for years and have a universal standard that has been developed regarding the proper application of the component with respect to functional capabilities and tolerancing. These tolerances have been engineered for general cases of applications and are one of the topics of discussion in Chapter 4. The text will point out when it is advisable to look up a tolerance as opposed to when it is wise to engineer custom tolerances.

How do tolerances affect product cost and customer loss? How do they affect part, subsystem, and system quality and customer satisfaction? Tolerances have far more impact on cost, quality, and customer satisfaction than they have traditionally been accorded. Often, during product design, the focus is on the nominal specification and its effect on functional quality. Typically, the tolerance becomes the focus of serious scrutiny only when manufacturing costs are being established or parts are coming in above or below the tolerance limits and the assembly process is disrupted. Tolerances are best known for the strong impact they have on part cost.

Generally, if the engineer requires tight tolerances, costs are going to be at a premium. If tolerances can be relaxed the parts will drop in price. Many engineering design books have graphs and diagrams that illustrate the relative cost of standard types of manufacturing processes and the grades of precision within them. This will also be discussed in Chapter 4.

If an engineer possesses a metric that provides accurate and precise information concerning the impact each and every design parameter has on the functional variability of the product, then he or she can discriminate between which tolerances have to be tightened and which can be loosened. Not only could the parameters be examined for their effect on functional variation, but if the *right* metric existed, the cost of the part tolerancing could be evaluated against the losses incurred by the customer when the product's performance is not on target. Such a metric exists. It is called the quality-loss function. It provides a unifying metric between deviations from functional performance and the consequential loss incurred by the business and the customer. This will be the focus of Chapter 11.

For now, we shall define the quality-loss function as the loss in dollars when a customer tolerance is surpassed by a design parameter deviation away from the target set point.

Matching Design Tolerances with Appropriate Manufacturing Processes

Traditional tolerancing and Taguchi's tolerancing and tolerance design approaches are *dependent* in their ability to add value to the product design; that is, they are processes that are applied to the results of concept and parameter design. If poor engineering choices were made during these phases of product development, then tolerancing and tolerance design will add only as much value as the design and its manufacturing processes are inherently capable of supporting. For instance, if the choice of a design concept dictates that a well characterized and optimized manufacturing process be employed, and the parameter design process develops a specific level of insensitivity to the unit-to-unit variation coming from the optimized manufacturing process, then the tolerancing processes defined in this book will work well in conjunction with the *engineered* capability of the design. If a poor decision was made with respect to the selection of the manufacturing process at the concept phase, or if no manufacturing process was considered at that early point in the commercialization process, then difficulties can be encountered in deploying the tolerancing process.

Trying to fit a manufacturing process to the design tolerances just prior to production is a suboptimal approach to developing design and manufacturing process capability. There should be strong manufacturing engineering expertise applied at the earliest stages of product conceptualization and development. Good design process capability is developed right from the start of technology development and is strengthened by consistent manufacturing, engineering, planning, and analysis through concept design and parameter design. This is a good reason to have production suppliers serve as partners early in product development.

A tool that manufacturing engineers frequently bring into the product development process is a catalogue of numerous manufacturing process variances. These can estimate expected levels of quality and process capability from historical data from the current state of the art in manufacturing. If their data is not directly applicable, it gives the manufacturing engineer the time to develop a manufacturing process that is relevant and capable for the case at hand. This information is typically not foremost in the minds of scientists, physicists, and development engineers active in the earliest stages of technology and product development. However, considering the manufacturing processes' variances and ultimate impact on product functionality can no longer be an optional or occasional development activity. World-class companies consider, very early in the development process, how to deploy the product functionality through capable manufacturing processes so that the loss to the customer is minimized over the life of the product. Developing the need for tolerances that no one has the process capability (Cp) to maintain in the four-to-six sigma range is a sign that a business has fundamental flaws in its product and manufacturing engineering processes.

Introduction to the Taguchi Approach to Tolerance Analysis

While this text has a strong focus on traditional tolerancing techniques, we will blend these with the methods developed by Taguchi to build a full, modern portfolio of tolerancing methods. This portfolio will then be used to define a comprehensive tolerancing process that can be applied within any general product or manufacturing process development engineering process (see Chapter 2).

The *process* of engineering a new or improved product is at the heart of Taguchi's system of quality engineering. How an engineer steps through the day-to-day activities to assure that an optimum balance exists between product cost, product quality, and the product development cycle time is central to his thinking. Taguchi's approach has benefits for the process of engineering as well as for the products that come out of the engineering process. For companies that engineer products, the cost and quality of the product development cycle is fundamental to their ability to compete in the global marketplace. In the final analysis, the product any company offers to its customers is a direct reflection of the engineering process employed to create it. Since tolerancing and tolerance design are engineering processes, they will strongly influence the nature of the final product.

Review of the Three Phases of a Quality Engineering–Based Product-Development Process

The topic of tolerancing and tolerance design is the final phase in the quality engineering process Taguchi has referred to as off-line quality control. This chapter is designed to introduce this final balancing process before the design is released into production. The first two phases of off-line QC are concept design and parameter design. Once the design is in production, on-line quality control processes are employed to sustain, maintain, and stabilize the quality that has been designed into the product. Thus, there are four major components to the overall process of quality engineering, as illustrated by Figure 1.3.

Figure 1.3 Quality engineering processes for product commercialization

Off-Line Q.C.	
Processes for Developing Generically Robust Technology	Research & Technology Development
Concept Design	
Processes for Developing Insensitivity to Specific Product Noise	Product Development
Parameter Design	Manufacturing Process Development
Processes for Balancing Cost & Quality	Product Design
Tolerance Design	Manufacturing Process Design
On-Line Q.C.	Production
Processes for Stability & On-Target Performance	Service & Maintenance

Before the topics of tolerancing and tolerance design are discussed, a brief review of the first two off-line quality control processes is in order. Developing tolerances is dependent on the upstream engineering processes, as we shall presently see. The following material will illustrate the sequential nature of the three phases of off-line QC.

Concept design: Selecting inherent robustness and concept superiority

Concept design is the first phase of product development. It is a *selection* process. It is here that the concepts to be included in a product are developed. Product development is different from technology development. Technology development is the process of conducting research and development around fundamental phenomenological building blocks. This is where the laws of physics are uncovered and/or explored for potential commercial applications. It is where what *can* happen is discovered, documented, and certified as a viable technology that can be made to produce repeatable results. Once a technology is certified for its stability of behavior and general susceptibility to various sources of noise (variation), it is then ready to be used in the first phase of product commercialization—concept design.

Concept design is used to compare various competing technologies that have been proved to be mature enough to safely enter into commercialization. Concept design is a form of benchmarking, which usually compares one or more existing products against other competing products from the marketplace. In *Total Quality Development* [R1.5], Clausing recommends employing an enhanced version of quality function deployment (EQFD) using the Pugh concept selection process to adequately perform concept selection. EQFD is a process that is used to bring products to market based on the disciplined gathering, translation, and deployment of the voice of the customer (VOC) into the product concept. EQFD is used to construct comparison matrices called "houses of quality." These houses graphically depict the competing requirements in a developing product so that the multifunctional engineering team can make rational trade-offs and ultimately align the engineering, manufacturing, and commercialization skills of the corporation with the product and service needs of the customer [R1.6].

Another powerful tool that is capable of aiding in the identification of superior design concepts is TRIZ. TRIZ is a Russian acronym for the theory of inventive problem solving. This process involves the vast technical concept resources available in the world's patent literature. It also draws upon the history of technological trends in a manner that leads the engineering team to conceive of logical future states in which the concepts for product or process inventions are likely to reside. TRIZ is an excellent tool to aid in the speed and creativity of the invention process. The computer tools that TRIZ employs are easy to use on any PC platform. To explore the capabilities of TRIZ, contact the American Supplier Institute (see Appendix E).

Concept design includes the best of existing product concepts, but can also include new ideas that may not have been commercialized yet. This can present a conflict to the development team since the existing products have a track record while the new technologies do not. The Pugh concept selection process was developed by Stuart Pugh at the University of Strathclyde, Scotland, in the 1980s to deal with this specific problem. This process is fully developed in his book, *Total Design* [R1.7].

The engineering team can perform sensitivity analysis on various concepts to isolate their *inherent robustness.* This approach uses classical math-model-based optimization methods or experimental optimization techniques in conjunction with known process capability and tolerance data to determine which concepts hold the most promise for further robustness optimization and ease of tolerance design. These screening steps are excellent means beyond simply defining candidate concepts that have low associated noise susceptibility and confirmed means of measuring functional performance. In this process, defined by David Ullman (in an unpublished paper [R1.8]), the design concepts can be prescreened for their amenability to later stages of optimization. These analytical methods are the first crucial steps required to assure an optimum design. See Chapter 7 for more on this form of sensitivity analysis.

During the Pugh process there may need to be some prototyping and experimentation conducted on the new concepts to fairly assess their capability against the concept selection criteria—particularly with regard to noise. A particularly helpful experimental approach is the use of Taguchi's dynamic methods for technology robustness optimization [R1.9]. The selection criteria is defined by a multifunctional product development team that identifies rigorous criteria that will substantiate a concept's superiority in engineering terms that are based on customer needs, referred to in quality function deployment terms as the *voice of the customer.*

One of the key metrics to be used in defining superiority is inherent robustness, which is the measure of how many significant noises to which a design concept is sensitive. The magnitude of the sensitivity to specific noises is important, as is the sheer number of noises that can assail a particular design. Certainly a design concept that has mild sensitivity to a relatively few noises is going to inherently out-perform a concept that is highly sensitive to many noises. The goal of the selection process (see Figure 1.4) is to establish one or two concepts that demonstrate clear superiority over the others.

Initially, in the Pugh process, a datum concept is picked to represent the best existing concept presently on the market. The concept being developed may be a new component, subsystem, or a complete product. Then the iterative process, which Pugh called a converging-diverging process, is used to isolate the superior concept to take into parameter design. The process is shown in Figure 1.5.

In summary, concept design uses the following tools, typically referred to as best practices, to develop a superior, inherently robust concept:

- Enhanced quality function deployment
- TRIZ process
- Sensitivity analysis on concepts to isolate inherent robustness
- The Pugh concept selection process
- Product benchmarking experimentation and evaluation
- Dynamic signal-to-noise ratio experimentation to evaluate general noise susceptibility

A key thing to remember about concept design is that it is intended to reject any concepts based on *immature* technologies that are trying to creep into the commercialization process. Attempting to develop immature technology during the product commercialization process is a certain recipe for a quality, cost, and cycle-time disaster. The sooner this is stopped, the sooner industries will see their next leap in product development productivity and efficiency.

Figure 1.4 The comparison of two concepts for inherent robustness

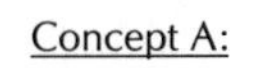

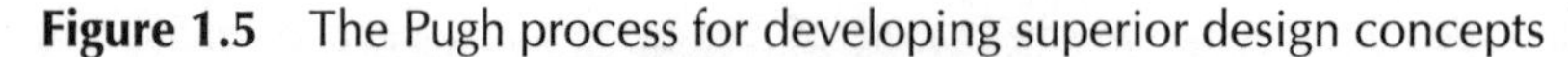

Figure 1.5 The Pugh process for developing superior design concepts

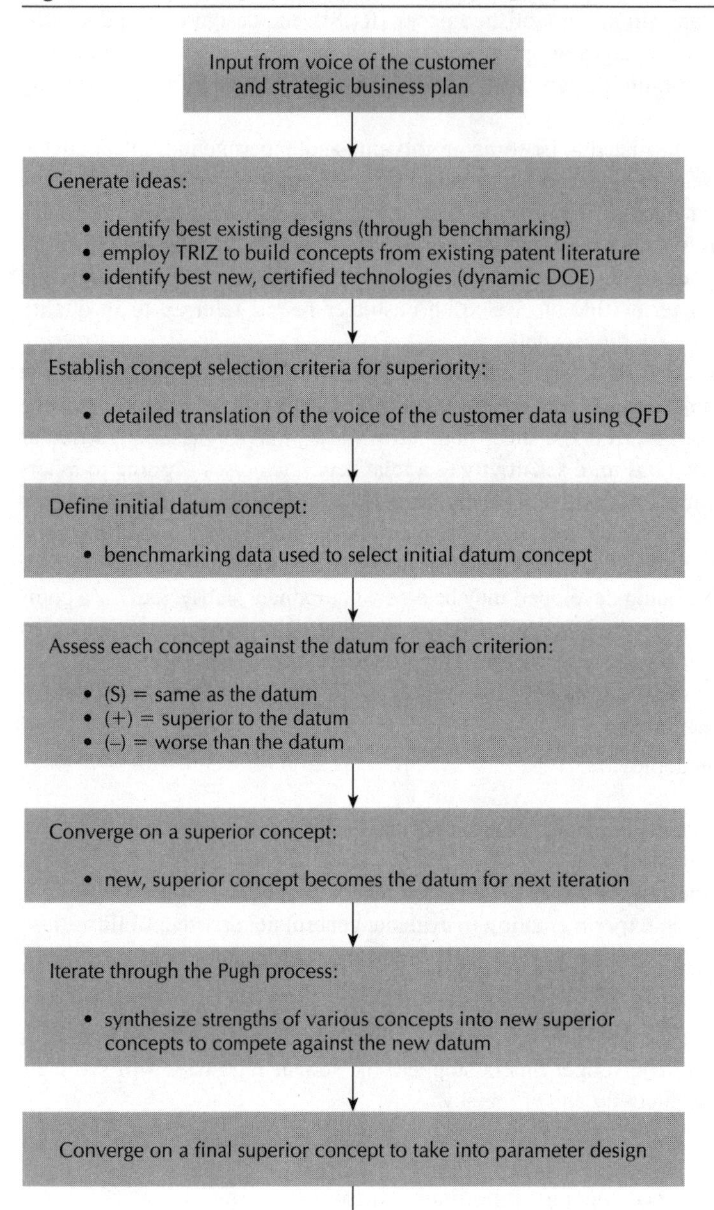

One way to check on how well concept design has been done is to look for evidence that the superior design has clearly defined response characteristics that make it easy to evaluate with known and trusted transducers, meters, and data acquisition systems. If the engineering team can't define how it is going to measure the concept's development progress—that is, design for testability—in engineering units, then the superiority of the concept should come into question. That which cannot be physically measured cannot be easily optimized. Concept design is also an excellent process to prevent products from being developed outside of the context of a rigorous customer focus. The final deliverable that comes out of concept design is a "real" design that is shown to be superior to all others. This design must be inherently and readily robust for *specific product application* in the next phase of quality engineering—parameter design.

Parameter design: Developing and optimizing insensitivity to noise

Parameter design is the process that will be employed by the multifunctional engineering team to optimize the superior concept against the noises inherent in the fabrication and use of the design. Parameter design is, therefore, an *optimization* process. The value in doing concept design first is that one gets to pick the noises that are going to have to be accommodated during the rest of the product development process. It is much easier to perform parameter design on a new, superior concept than it is to try to improve an existing design that is laced with sensitivities to noises that may not be readily apparent to the current engineering team. When this is the case, extra effort must be expended by the team to identify the noises and evaluate their severity. Even then, the existing design may have only minimal optimization capability—it all depends on the physics underlying the design and the strength of the interactions between the engineering parameters and the noise factors.

Parameter design is typically 80 percent engineering analysis and planning and 20 percent experimentation and data analysis. It is the process by which the engineering team uses a series of designed experiments (Chapter 13) to assist in the optimization of the design so that it becomes relatively insensitive to the application-specific noises that will be assailing it during manufacture and customer use. Remember that in concept design the specific application noises may not yet be totally apparent. This is why during concept design we focus on general or generic noises associated with the concept and wait until parameter design to do the detailed and rigorous product specific noise evaluations (see Chapter 14).

During the analysis and planning stage of parameter design the *ideal function* of the design is defined; signal and control factors (parameters the engineer *can* control) are identified; noise factors are further defined, compounded, and evaluated; and the all-important functional responses are established. The deliverable from the parameter design process is a critical parameter drawing or document (significant, evolutionary geometric model) that contains the optimum control factor set points that define the nominal or target values where the design is least susceptible to the effects of noise. In parameter design we try to improve the design's quality without adding a lot of cost. It is like trying to make a silk purse out of a sow's ear. If we can make the design minimally sensitive to noise before moving on to tolerancing and tolerance design, where it is required that we spend additional money on improving product quality, then we have gone a long way toward minimizing total product cost, including the cost of ownership and maintenance. We would ideally like to use tolerance design to add cost only where it is absolutely necessary to provide the very highest customer satisfaction that is economically feasible.

Traditionally in the United States, most companies have not used structured product development processes that include a major phase of product robustness optimization. The traditional method employed by many engineers is the build-test-fix method. This iterative approach to product development is slow, costly, and often leaves the product in a suboptimal condition when it is delivered to the market. The engineering team is often left to perform round after round of reliability improvement to complete the optimization of the product. Changing a product after it has been tooled and is in mass production is extremely expensive.

Another focus in parameter design is to engineer the design to be free of strong control factor interactions (yet another form of susceptibility to noise; see Chapter 13). When this is not done the design is left in a vulnerable position where noise, acting over time, can cause numerous control factor levels (design set points) to change, thus cascading performance variability into the functional response of the product. It is not unusual for a design to be modified for any number of reasons prior to its introduction to the market. If the control factors are all interdependent on one another's optimal set points, then any alteration to one factor could cause the effect of another control factor to disrupt the desired performance of the product. When this happens the engineering team has to iteratively rebalance the performance of the design, which puts undue restrictions on the tolerances, since they become the main countermeasure against the sensitivity of the interactive factors.

The parameter design phase of off-line quality control should remedy this debilitating cycle of iterative improvement activity. Once robustness is developed into the product and large control factor interactions are engineered out of the design (well upstream from production implementation), the engineering team can focus its attention on balancing cost and performance with little risk of compromising basic functional performance—even in the presence of degrading noises.

Figure 1.6 reviews the various steps in the parameter design process. Each step of the process is briefly outlined. This flow chart is adapted from *Engineering Methods for Robust Product Design* (Fowlkes and Creveling) [R1.9]. The book's focus is on the engineering process and tools (including WinRobust Lite data-analysis software) that will help enable the engineering team to develop robustness in a technology, product, or manufacturing process well in advance of production implementation. As previously mentioned, this leaves time for the engineering team to primarily focus its attention on *refining* the design by balancing cost and quality. The parameter design process positions the product to be in a robust condition where it is optimized against key sources of variability, including the desensitization of strong interactions between control factors. The design can still fall short of the quality level required by highly discriminating consumers, and will also need to be prepared for economic production and use. Tolerance design is the final stage of off-line QC, where additional quality can be designed into the product under the discipline of a data-driven balancing process wherein money is strategically spent to further improve the final production design.

Tolerancing and tolerance design: Balancing product cost and quality

Tolerance design is a *balancing* process. It is designed to balance product or process functional quality with overall cost. The term tolerance has no meaningful definition apart from *the ability to limit the variation of the nominal design parameter (target) set points.* The nominal or target set points are the direct output from the parameter design process.

We shall further define these nominal set points by recognizing them as the core of what is known as a *design specification,* which is the specific elements (targets and tolerances) that define the functional capability of the design. A complete design specification includes both the target set point and the tolerance associated with the target. Every product is technically defined as an accumulated set of design specifications.

Ideally, every product would provide the ultimate in customer satisfaction if the target specifications could be met perfectly every time the product was manufactured. This almost never happens. It is not uncommon in the manufacture of consumer products such as automobiles and office copiers, products with high degrees of complexity, to find a very small number of the products working almost perfectly over extended periods of time. Engineers are both perplexed and encouraged by this occurrence:

Figure 1.6 The parameter design process

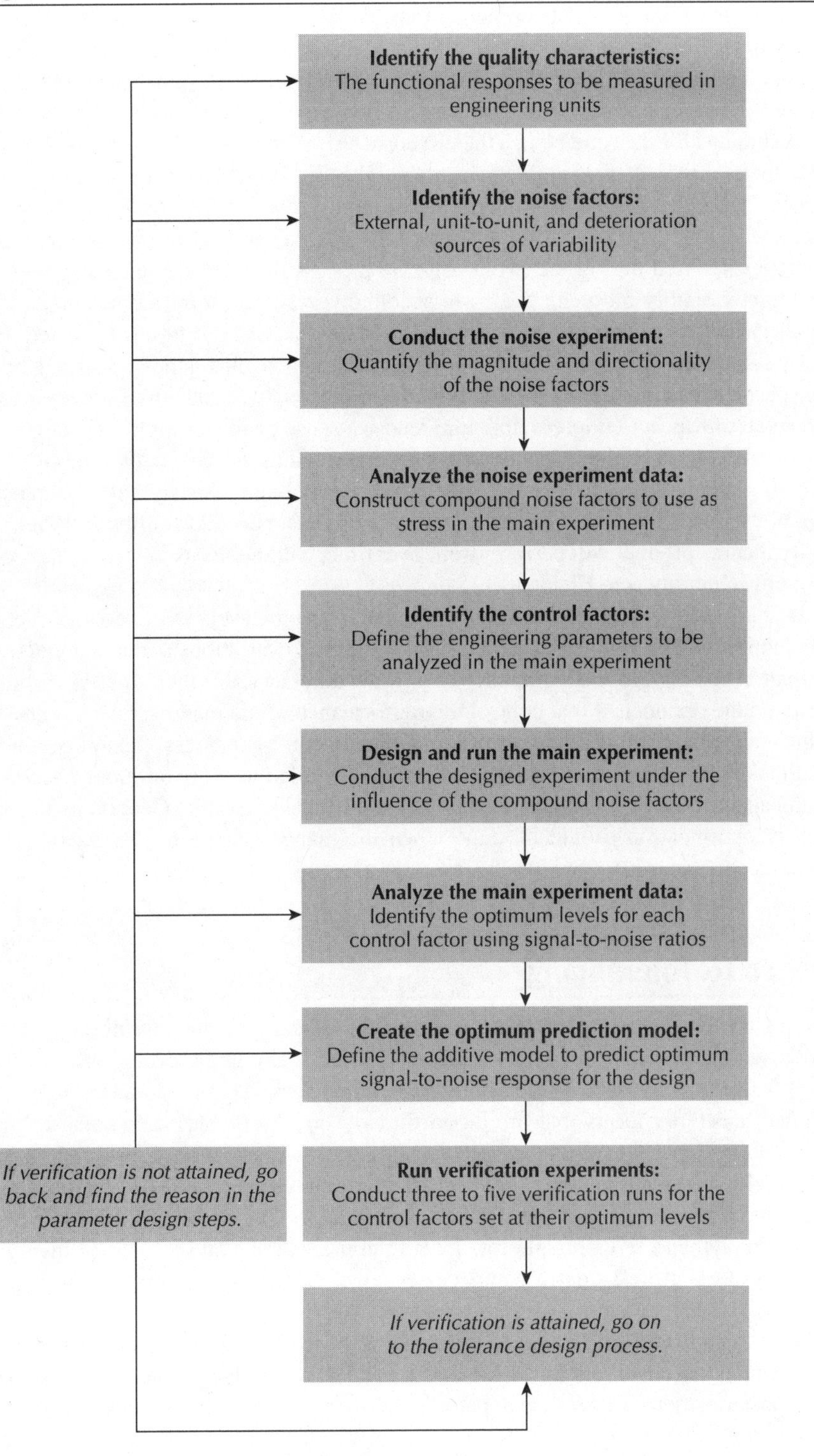

If a few products can exhibit this level of exceptional performance, why can't they all? What combination of specifications allows this remarkable phenomenon to occur? A common engineering practice in some companies, when there is uncertainty about the effects of manufacturing variability, is to build the "perfect design" embodiment of a subsystem or component. If the "perfect design" is defined as all specifications (optimized design set points) being *on target,* then the expected high level of performance should occur, excluding (for the time being) the effects of use over time in the customer's environment. Blocking out the real effects of manufacturing variability like this provides a starting point from which the engineering evaluation of manufacturing variability can begin. The point is that a design might be fully capable of excellent performance when the right set points (targets) are in place. If the tolerance limits are surpassed during the manufacturing process then the customer is going to experience an out-of-the-box quality problem; these are absolutely deadly to a business's reputation and are to be avoided at just about any cost. If the tolerance limits are passed during the expected life of the product, typically some time after the initial *break-in* period, then the customer will sense a reliability problem. These problems are not quite as bad as out-of-the-box quality problems, but they aren't far behind. The intent in all of this is to impress upon the reader a sense of responsibility for defining these subtle tolerance bands and to engineer the design to operate within them over time in an economic fashion. Optimizing the design process (functional) capability simultaneously with the manufacturing process capability, assembly process capability, and service/support capability will yield a much higher probability that the product will perform well, over time, within the customer's tolerances. This is the development of reliability (see Chapter 2).

It is common, during product parameter design, to evaluate various levels of the control factor nominal set points to identify the optimum level at which insensitivity to variations in the manufacturing process exist. In manufacturing process parameter design we do essentially the same thing, only now the optimization is on the parameters that control the performance of the manufacturing process. Ideally, the development of design process capability and manufacturing process capability, for a specific product, should take place simultaneously. In this context the tolerances are developed according to inputs from both capabilities. This is a major goal in concurrent engineering. Generic manufacturing process capability optimization should be done when the manufacturing process is initially invented and developed (see Chapter 3).

Taguchi's Approach to Tolerancing

Tolerancing, in Taguchi's context, is defined as the process of establishing manufacturing tolerances (low-level tolerances) based on customer tolerances (high-level tolerances). High-level tolerances are always expressed as a value that is directly determined by customers. These tolerances may be expressed in nontechnical customer terms that are based directly on observable phenomena or are simple perceptions of subtle departures from expected (on-target) performance. Low-level tolerances are expressed in highly technical engineering terms that are direct expressions of the physical measures that lie behind the performance the customer sees and expects. One could say that high-level tolerances are expressed in layman's terms, while low-level tolerances are always expressed in engineering terms that are part of the professional vocabulary developed in college, university, and industry environments.

It is very important to properly translate the true meaning of the high-level tolerance into the world of engineering at which low-level tolerances are developed. Many products have fallen short of customer expectations because there was no clear path established between the engineering terms being used in product development and the terms customers used to express their perception and expectation of the product's features and stability of performance. Voice-of-the-customer gathering processes and quality-function deployment (QFD) processes are designed to provide detailed insight into the customer's target expectations and tolerances for product performance. Taguchi's tolerancing process (Chapter 12) is built around the relationships that can be established between customer tolerances and engineering tolerances through the mathematics of the quality-loss function.

Figure 1.7 illustrates the steps for Taguchi's tolerancing process. This process is a reasonable alternative to the standard selection of ± 3 sigma tolerances (standard deviations of the output response coming from a manufacturing process). Careful attention must be paid to close communications and partnership with the component manufacturer to work out a common alignment around the customer-driven tolerances as they relate to the realistic limits of actual production capability. There will be times when the customer's tolerances are going to demand more than the current state of the art in one or more component manufacturing process can deliver. Often an alternative component can be used to

Figure 1.7 Overview of Taguchi's tolerancing process

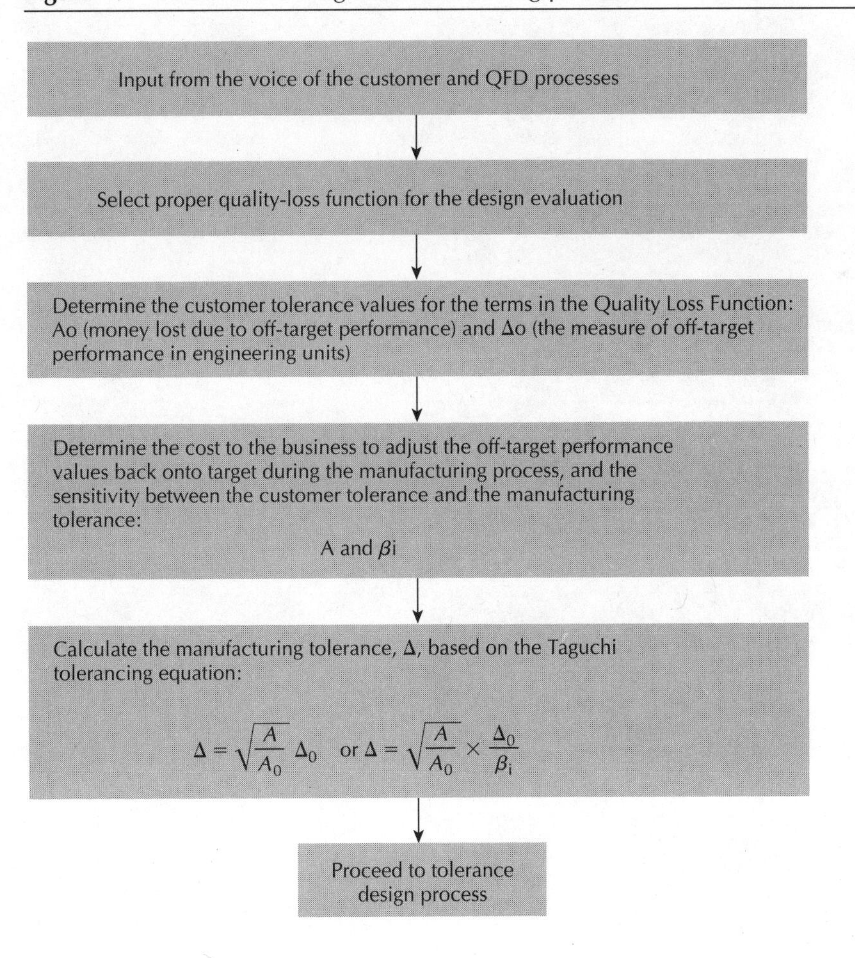

Figure 1.8 Overview of Taguchi's tolerance design process

Input from parameter design:

- optimum control-factor set points defined
- tolerance estimates determined from engineering analysis
- initial material grades selected

↓

Generate quality-loss function:

- identify customer costs for intolerable performance

↓

Select and load an orthogonal array for the tolerance experiment:

- select control factors and their initial tolerances
- reuse the noise factors from the parameter design experiments in the tolerance experiments

↓

Conduct the experiments and collect the data:

- take replicates to quantify experimental error

↓

Analyze the data using ANOVA:

- determine the percent contribution to variability from each parameter in the tolerance experiment

↓

Determine the mean square deviation (MSD) for each control factor, calculate the quality loss due to each factor, and determine the process capability index for the existing design

↓

Determine new, upgraded tolerances, their increase in cost, the new MSD created by tighter tolerances, the new quality loss values, and the quality gain in the improved design

↓

Determine which factors are economically feasible to upgrade:

- recalculate new MSD, quality loss, and Cp values
- define critical-to-function specifications

↓

Work with manufacturing engineers to make the manufacturing process robust and then attain lower variance components.

↓

Run verification experiments and make comparisons between old quality loss and Cp to the new quality loss and Cp

↓

Communicate tolerances via GD&T, and create quality plans with suppliers for SPC and other on-line quality processes make up the difference in performance by tightening a tolerance that *can* be maintained with the alternative component production process.

Tolerance design, in this book, refers to the process of using designed experiments, ANOVA data processing, and the quality-loss function to simultaneously evaluate the effect of various levels of tolerance boundaries on the cost and performance of the product when exposed to realistically stressful noise factors. Tolerance design follows the process illustrated in the flow diagram shown in Figure 1.8.

The tolerance design process is a unique tool available to balance the cost and quality resident in a design just prior to manufacturing. This process is anchored in the customer's expectations and tolerances as defined in the quality-loss function. No other tolerancing process I know of has the capability to account for the loss to both the customer and the business as a function of performance variability in the design or a specific component. As in all engineering processes, the quality of the effort and analysis of data used to develop the loss function, the control and noise factors, and the cost of upgrading the design will greatly affect the outcome of the tolerance design process.

Chapters 4 through 10 are devoted to the portfolio of existing analytical tolerancing methods that are common today. The strengths and weaknesses of these traditional approaches will be discussed in some detail. The bottom line on Taguchi's tolerance design process is that there simply is no other process available that is capable of identifying and applying the specific sources of variability in a system of parameters that are undergoing the stress of external, unit-to-unit, and deterioration noises simultaneously. All the other methods deal mainly with unit-to-unit variability in the pristine environment of probabilistic math models that are highly susceptible to the "noise" of human judgment and assumption-making skills. For certain subassemblies, assemblies, and products, where various mixes of technologies exist, there is no comparison between the quality of information that comes from experimental data developed in the presence of induced noise versus the aseptic world of computer-simulated variability based on probability density functions (statistical models).

Summary

This introductory chapter has covered a broad range of topics relative to tolerance design. The following outline summarizes the major topics that have been discussed. We have reviewed many of the general concepts and terms required to understand how the tolerance development process fits into the overall product development process. There is much that can be done prior to tolerance design to help make the development of tolerances a smooth and efficient process. Habitual performance of concept design and parameter design prior to tolerance design can remove many difficulties and extra work in the development of tolerances.

Chapter 1: Introduction to Tolerancing and Tolerance Design

- The Historical Roots of Tolerancing
- The State of the Art in Tolerancing Techniques
  - What Should a Modern Tolerance Development Process Look Like?
  - The Relationship between Traditional Tolerancing Methods and the Taguchi Approach

- Developing Tolerances: The Role of Engineers and Designers
- Concepts, Definitions, and Relationships
- Matching Design Tolerances with Appropriate Manufacturing Processes
- Introduction to the Taguchi Approach to Tolerance Analysis
  - Review of the Three Phases of a Quality Engineering–Based Product-Development Process
    - Concept design: Selecting inherent robustness and concept superiority
    - Parameter design: Developing and optimizing insensitivity to noise
    - Tolerancing and tolerance design: Balancing product cost and quality
- Taguchi's Approach to Tolerancing

In general terms, the process of tolerance development for the purpose of reducing variability in a product has been developed, as shown in Figures 1.1 and 1.2. Chapter 2 will detail how specifications are derived from the customer's statements and requirements all the way to the final component, assembly, and product level specifications that are used to define the final product design. It also provides a comprehensive process for developing reliability in a product. The reader may be curious about why a book on tolerance design has an entire chapter dedicated to the topic of reliability growth. Tolerance design, when properly conducted, can have a very profound effect on product reliability. Chapter 2 defines a complete process, based on a comprehensive portfolio of engineering best practices, to systematically predict and grow reliability simultaneously. The specific role of tolerance design relative to reliability growth will be explained in detail.

References

R1.1: Cardwell, D., ed., *The Norton History of Technology* (New York: W. W. Norton & Co., 1995).

R1.2: Harry, M., and R. Stewart, *Six Sigma Mechanical Design Tolerancing* (Motorola Inc., Publication No. 6s-2-10/88, 1988).

R1.3: Foster, Lowell, *Geometrics III* (Addison-Wesley Publishing Co., 1995).

R1.4: *Dimensioning and Tolerancing* ANSI Y14.5 M1982, ASME (R1988) pg. 14.

R1.5: Clausing, D., *Total Quality Development* (ASME Press, 1994).

R1.6: Cohen, L., *Quality Function Deployment* (Addison-Wesley Publishing Co., 1995).

R1.7: Pugh, S., *Total Design* (Addison-Wesley Publishing Co., 1990).

R1.8: Ullman, D., and Y. Zhang, *A Computer-Aided Design Technique Based on Taguchi's Philosophy,* unpublished; 1994, pg. 25.

R1.9: Fowlkes, W. Y., and C. M. Creveling, *Engineering Methods for Robust Product Design* (Addison-Wesley Publishing Co., 1995).

Note: All page numbers at the end of the references are referencing where this material was used in this book—not the pages from quoted work.

CHAPTER 2

The Relationship of Quality Engineering and Tolerancing to Reliability Growth

This chapter is designed to introduce the reader to a topic rarely discussed in modern texts on product development. Experts in the field of quality engineering are often asked how the methods and metrics of reliability engineering can be related to the methods and metrics of robust design. The question can be further posed about selecting superior product concepts, optimizing nominal parameter set points, balancing costs and quality through the development of parameter tolerances, and the effect these three processes have on reliability growth as a product is brought through a commercialization process. To properly understand how concept, parameter, and tolerance design play a major role in product reliability growth, one must study the larger product development context in which they reside.

A responsible design engineer is a lifelong student of the processes and best practices through which competent design is executed. The following discussion outlines the major steps that are required to relate a number of product development best practices, including Taguchi's methods of concept design, parameter design, and tolerance design, to the process of reliability growth. The following reliability growth process is benchmarked against a widely accepted set of engineering design evaluation practices. This is done to illustrate the strong ties the proposed reliability growth process has to long-standing and accepted engineering tools. The reader may be wondering why such a broad set of processes is being considered when the purpose of this book is to teach the methods of tolerance design. The answer lies partly in the need to understand how upstream product development processes effect the tolerance development process and how *all* these processes effect reliability growth.

When tolerances are developed properly, they can have a profound effect on design reliability. This will be specifically illustrated through a detailed case study in Chapter 17. When concepts are not superior and parameter nominal set points are over-sensitive to noise, the burden of keeping the design under some degree of control falls on its tolerances. If the tolerances become the primary countermeasure against the effects of noise, their task is likely to be well beyond their capacity to deliver reliability over the long run. The following material is proposed to help the reader develop the appropriate contributions that tolerances can make in conjunction with the many other design parameters and specifications at one's disposal to make a design as reliable as the laws of physics will allow.

The Three Initial Phases of Product Development

Figure 2.1 is representative of the processes that are to be followed in the first three phases of a product commercialization process. It is in these three phases that most of the product's inherent reliability will be established. The discussion will enter the product commercialization process in Phase 1 and will conclude at the end of Phase 3. The total product commercialization process being discussed here actually is made up of numerous phases. The additional phases, after Phase 3, pertain to design certification where an engineering model of the product is evaluated and tuned; preproduction readiness where preproduction models of the product are built to production process standards and evaluated for the optimization of the production assembly process and the maturing product design; production approval, where shipping approval for the product is granted; steady state production, where on-line quality control is applied; and the final phase, discontinuing the product, where an economical ramp down of production is engineered and implemented.

Figure 2.2 shows a system reliability growth flow diagram, which plays a key role in laying the foundation for the material presented in the rest of this chapter, including an explanation of each block within the process, and references for understanding the appropriate best practices that enable the process to be completed.

Before getting into the explanation of what is to occur in each process block, we need to discuss *The Mechanical Design Process* by David Ullman [R2.1], which concerns the evaluation of a product for function and performance. The only way to control and monitor the physics of reliability growth is through the disciplined application of engineering principles and metrics within the product development process steps. Ullman's book contains information about the overall engineering design process

Figure 2.1 The three initial phases of product development

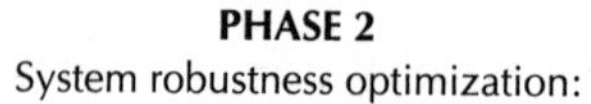

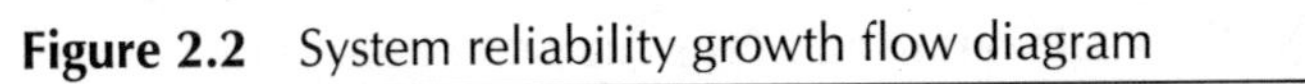
Figure 2.2 System reliability growth flow diagram

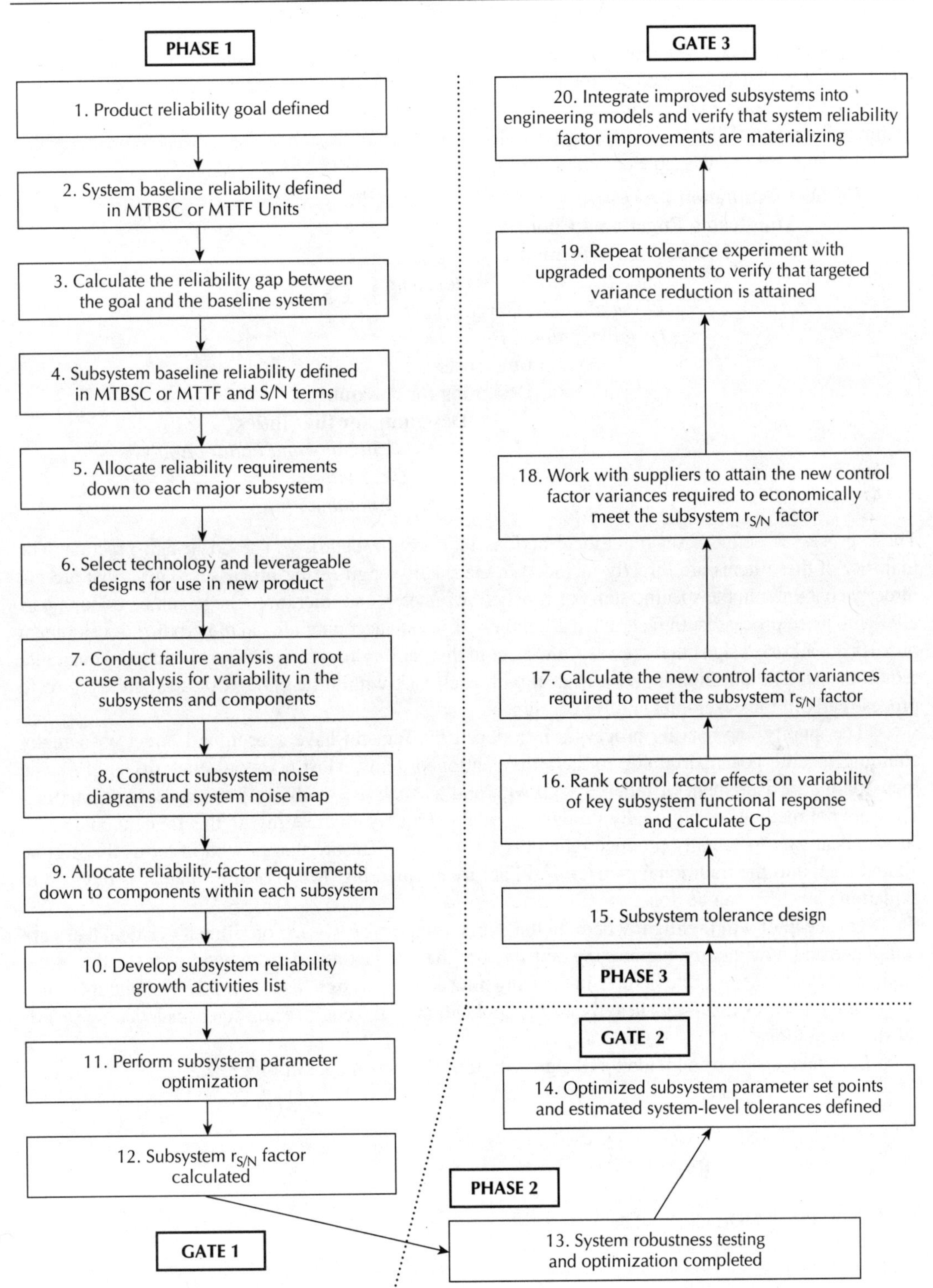

that will provide a good benchmark for the effectiveness of the processes being discussed in this chapter. Without a disciplined and thorough effort to quantitatively evaluate a product's development progress with respect to function and performance, there is no way to credibly manage the product's commercialization. Ullman has captured a well-accepted array of standard engineering practices that should be integral to any viable commercialization effort.

Product Evaluation Processes:

- **Monitoring Functional Change**
    - **Evaluating Performance**
        - *Analytical Model Development*
        - *Physical Model Development*
        - *Graphical Model Development*
            - **Evaluating Costs**
                - **Designing for Assembly**
                    - **Designing for the "ilities"**
                        - *Reliability and Failure Analysis*
                        - *Testability*
                        - *Maintainability*

These processes define a set of rational actions that are the bedrock of good design practice. The majority of these items are directly included in the quality engineering processes. Those that are not directly contained in the specific steps of concept, parameter, and tolerance design can be done in parallel with these processes. In fact, when all of these tasks are accounted for in the product development process it leads to designs that are very complete in their ability to satisfy both corporate and customer requirements. The question is *when* to deal with each task within the context of a reliability growth process early in the development of the design.

The quality engineering processes introduced by Taguchi have a profound effect on actually attaining specific goals pertaining to the aforementioned items. Most texts on reliability engineering focus on the measurement of reliability growth, and as such are really only methods of accounting. They are not methods of physically producing growth. Quality engineering, in the Taguchi context, is an excellent way to actually produce reliability growth and measure that growth in terms that can be related back into the traditional metrics of reliability engineering. Much of this chapter is devoted to explaining how this can be done.

This chapter will highlight where, in the engineering process, each of Ullman's critical tasks are completed and how quality engineering helps assure that the lasting effects extend over the life of the product. Ullman's list is quite helpful in showing how concept design, parameter design, and tolerance design are processes that focus heavily on the application of the engineering sciences aided by practical statistical tools.

Let's take a step-by-step look at the growth flow diagram shown in Figure 2.2.

Phase 1: *Subsystem Concept Selection and Robustness Optimization*

Block 1.

Product reliability goal defined *—using VOC & business case*

Determining the overall product (system) reliability goal is the starting point for establishing the reliability growth process. It is extremely rare that a product can go right from conception to production without a period of improvement of the functional performance of the components as they are integrated into the system we ultimately call the product. Whether one is designing a new thimble or a new pickup truck, the need to quantify the mean time to failure (MTTF) or mean time between service calls (MTBSC) is a must. MTBSC is used in industries where the product has enough value to be repaired, maintained, or adjusted. These time spans, referred to as reliability metrics, should be based on the VOC data gathered during pre-Phase 1 activities, when the overall product concept is being defined. If the new product is very closely related to a previous product then a data base of reliability performance is typically available to help clarify how much better the new product must be.

The new product reliability goal should be built on information that comes from experience within the business and engineering teams, the capabilities of new technologies and manufacturing processes being employed, an assessment of competitive products (benchmarking), the business case for the product, and the specific requirements that customers have communicated to the product-development team. As for Ullman's focus on product evaluation, this is where design for reliability has its roots. The reliability targets will set the stage for many of the activities to follow. It is here that the business team must make its initial assessment as to whether the corporation possesses the technical and financial capability to attain the reliability goals. It is beyond the scope of this discussion to define the specific best practices that are to be used to properly construct a competitive new product concept. The following books are recommended for help in establishing initial reliability goals:

- *Practical Reliability Engineering, 3rd Ed.;* Patrick O'Conner; J. Wiley, 1991
- *Reliability for the Technologies;* Doty L. A.; ASQC Quality Press, 1989
- *Quality Function Deployment;* Lou Cohen; Addison-Wesley, 1995
- *Total Quality Development;* Don Clausing; ASME Press, 1994
- *Benchmarking: The Search for Industry Practices that Lead to Superior Performance;* ASQC Quality Press, 1989
- *Reliability Based Design,* S. S. Rao, McGraw Hill, 1992

Block 2.

System baseline reliability defined in MTBSC or MTTF units *—from service support data base &* *competitive benchmarking*

Once the overall system reliability goal is defined, it is then necessary to find out just where the current product and its closest competition's baseline reliability capability resides in reliability metrics. Depending on how records are kept within a company and how much competitive data is obtained and assessed, defining current reliability numbers can be challenging.

System reliability data is frequently obtained from service and warranty records. These records help establish the time intervals between component failures, adjustments, consumable-item depletion, and other events associated with the perception of reliability. Customer returns of defective products either right out of the box or after some period of time are another way of tracing product reliability data. Sources for helping develop this information are listed below:

- *Practical Reliability Engineering, 3rd Ed.;* Patrick O'Conner; J. Wiley, 1991
- *Reliability for the Technologies;* Doty L. A.; ASQC Quality Press, 1989

- *Mechanical Applications in Reliability Engineering;* R. Sadlon; Reliability Analysis Center; ITT Research Institute, 1993

Block 3.

Calculate the reliability gap between the goal & the baseline system —*expressed in MTBSC or MTTF Units*

The reliability gap, or difference between the goal and the baseline system performance, can now be calculated. This system-level gap now becomes the focus of the reliability growth effort. The engineering team, the manufacturing community, and the service and customer support groups must all pull together to develop, design, manufacture, and support the product reliability growth in a seamless manner. The absolute accuracy of the reliability gap is not critical in most cases. The number must be credible and able to be substantiated with data. In this context, the entire development team will align behind the common goal and work together to meet the challenge of closing the gap. Everyone in the business and technical community must do their share of the work. Reliability growth is a total product team effort, not just an engineering activity.

It is important to clearly document the specific components of the gap. What is the nature of the problems that create the gap? Are they rooted in weak technologies present in the design? Are they due in part to weak service processes? Are the component suppliers sources of discrepant parts that are failing prematurely or off target as delivered? Have the engineering teams released drawings that are improperly toleranced? Does the design possess internal features that force the customer to use the product in a way that limits reliability? The list can go on and on. Which leads us to our next topic—allocating specific amounts of reliability growth to the subsystems that make up the product. While this is being done, the nonengineering teams within the business can be forming strategies to deliver their best efforts to help the product with respect to reliability growth. Their contributions will become important later in the process, after the engineering team has developed the functional framework for the product.

Block 4.

Subsystem baseline reliability defined in MTBSC or MTTF & S/N terms —*from service support data base, in-house noise experiments, and competitive benchmarking*

The *system* reliability growth requirement can be broken down into specific improvement factors that each subsystem must attain so that the cumulative effect can promote the system reliability to meet or exceed the goal. For example, let's say the product is made up of four distinct subsystems. Each subsystem could be "taxed" with a uniform reliability improvement factor, which we will call r, for reliability. In this case we might see each of the four subsystems required to deliver a 3× improvement in MTBSC or MTTF. Another common approach is to scale the tax to the *criticality* of the subsystem. This case might produce a need for two of the subsystems to attain 5× improvement, while the third subsystem is required to attain a 2× improvement and the fourth to deliver only a 1.5× improvement. Sometimes, due to design requirements, certain subsystems actually deliver negative reliability growth. This is expressed with an r factor of less than one (for example, 0.80, expressing a 20 percent reduction in reliability). The r factor will be thoroughly discussed in the next section. For now, it is expressed in reliability units (MTBSC or MTTF).

At the end of Phase 1 the engineering team is going to have to offer proof that the subsystem is better than the benchmark and is delivering its allocated portion of reliability growth. Each existing subsystem must be benchmarked against a general set of stressful noises to determine the baseline signal-to-noise (S/N) performance of the design. Although some readers may be familiar with Taguchi's S/N ratio, we will introduce its use at this point in the reliability growth process.

The S/N ratio is a measure of a design's insensitivity to various sources of noise. It is used to transform experimental data that has been processed into a mean and standard deviation. These are sample statistics gathered in the presence of stressful noises that are intentionally introduced during the experimentation. When a specific target is to be attained, the S/N ratio takes on the following form (called the *nominal the best S/N ratio*):

$$\frac{S}{N} = 10\log\left(\frac{\bar{y}^2}{s^2}\right)$$

The baseline S/N ratio will be used to compare with the new, improved S/N ratio from the parameter optimization process done during subsystem parameter optimization. This will be the proof that real performance improvement is occurring relative to the goal. A new form of the r factor will come from this comparison of S/N ratios ($r_{S/N}$). Chapter 1 discussed the parameter design process. This is where the S/N ratio is quantified through experimentation and analysis.

Running a credible benchmark, or baseline, S/N experiment requires good metrology and test capabilities. Thus, the engineering team is given its first opportunity to design and build a core capability in measuring functional performance that will be repeated over and over throughout the development of their particular subsystem. Benchmarking is the beginning of a long series of data-acquisition tasks. The transducers, meters, and data-acquisition equipment must be flexible, precise, and accurate.

- *Benchmarking: The Search for Industry Practices that Lead to Superior Performance;* ASQC Quality Press, 1989
- *Engineering Methods for Robust Product Design;* W. Y. Fowlkes & C. M. Creveling, Addison-Wesley, 1995
- *Experimental Methods for Engineers 4th Ed.;* J. P. Holman; McGraw Hill, 1984
- *Handbook of Transducers,* Norton, Prentice Hall, 1989
- *Measurement Systems, Application and Design 4th Ed.;* Doebelin, McGraw Hill, 1990
- *Mechanical Measurements;* Bechwith, Marangoni & Lienhard 5th Ed.; Addison-Wesley, 1993
- *LabView Graphical Programming;* G. W. Johnson; McGraw Hill, 1994

Block 5.

Allocate reliability requirements down to each major subsystem *—in MTTF terms that are approximately equated to the S/N-based reliability growth factor:*

$$r_{reliability} \approx r_{S/N} = 10^{\frac{S/N_{new} - S/N_{baseline}}{10}}$$

where "$r_{reliability}$" = (growth #) × MTTF or MTBSC (baseline)

At this point we can and must begin to build a relationship between reliability metrics and engineering metrics. Now that we are interested in allocating system goals down to the subsystems we must refine our metrics to get closer to our fundamental engineering objectives. This means making a transition from reliability metrics to *engineering metrics.* The bridging of these metrics involves an assumption that will leave some doubt as to the accuracy and precision of the correlation between reliability units and engineering units. The strength of the correlation is very much dependent on the team's engineering skills, instrumentation skills, experimental technique, selection of noise factors, understanding of the customer use habits, service processes, and craftsmanship.

The use of stressful noise experiments on existing products is an excellent way to begin to build a common understanding between reliability metrics and engineering metrics (see Chapter 14 for information on conducting noise experiments). The data created to build this understanding is very dependent on using the appropriate deterioration noise factors in the experiment.

Reliability, at its most basic meaning, is on-target performance over the designed life of the product. By engineering metrics we mean scalar and vector units such as time, temperature, force, acceleration, pressure, and so on. Engineering metrics also include metrics that take the laws of probability into account. These include the mean and standard deviation of a sample of product-performance data. In this way we can account for trends in data, the central tendency of a set of data, and the dispersion of the data about the point of central tendency (see Chapter 3). The assumed correlation between reliability metrics based on statistical distribution parameters and engineering metrics based on statistical distribution parameters is approximate, but is not unreasonable.

Robustness metrics are built from probabilistic (statistical) metrics and are considered a subset of engineering metrics. Engineering units account for the specific physical activities that underlie the failure events the reliability metrics quantify. Engineering units are a more direct measure of physical performance, while the reliability metrics are a measure of after-the-fact performance that is sometimes more perceptible to the customer. Some reliability metrics, like MTBSC, have an inherent human element that has little to do with physics and everything to do with emotion. This must be taken into account and gives rise to our caution of the approximate nature of the correlation between reliability and robustness metrics. Because reliability metrics are typically more easily perceived by customers, they are also convenient for business management to use in determining the health of an existing or developing product. As engineers, technicians, and designers we must use both reliability and engineering metrics to properly communicate the progress and capability of our evolving designs. It is a mistake to focus on reliability metrics alone during product development. In fact, the engineering team must work in engineering units as a priority and then convert information into reliability units to properly communicate with team members in management and other business units.

Madhve Phadke, in his book *Quality Engineering Using Robust Design* [R2.2], discusses a S/N ratio-based reliability improvement factor designed to convert engineering units into an approximation of reliability units. This reliability improvement factor, $r_{S/N}$, is decomposed from the difference of the baseline S/N ratio (obtained in Block 4) of a component, subsystem, or product and the new, improved S/N ratio that results from the parameter optimization process (obtained in Block 11). This factor provides a bridge between reliability and robustness metrics, and is defined as follows:

$$r_{S/N} \approx 10^{\frac{(S/N_{new} - S/N_{old})}{10}}$$

The S/N-based reliability factor, $r_{S/N}$, is actually a measure of the change in the variance of the measured response of the component or product due to the improvements that have been made from the parameter robustness optimization process (see Figure 1.6 in Chapter 1). *The growth in reliability is assumed to approximately follow the behavior due to variance reduction in the functional response, as measured by the S/N ratio.* Thus, if variance is reduced by a factor of two (3 dB in S/N units), then reliability can be expected to improve by a similar factor. The assumption is based on the concept that reliability is directly affected by changes, which variance numerically quantifies, in the response variable. As long as appropriate attention is paid to properly stressing the design, with noises that are time-dependent and are representative of true deterioration, the correlation will be reasonable. The S/N ratio is an excellent measure of the insensitivity a design has to the various noises that ultimately affect the real and perceived reliability of a product. A good measure of reliability is how well a product produces on-target performance over time. The proven way to get a product to provide on-target performance is to design control factors to support this performance even though sources of variability are active around and within these parameters. In this way, parameter design plays a significant role in reliability growth. The $r_{S/N}$ factor is an indicator of the trend being established for reliability growth.

An insightful paper has been written by Gary Wasserman of Wayne State University to establish an analytical basis to substantiate the $r_{S/N}$ factor proposition [R2.3]. It is highly recommended reading for an introduction to the topic of analytically relating S/N to traditional reliability metrics.

A critical requirement for trusting the improvement factor $r_{S/N}$ is the integrity of the noises that the engineering team uses to quantify the S/N ratio. This is true for assessing the *baseline* product's sensitivity to sources of variability and the *new* product's sensitivity to the same sources of variability. If critical noises, such as deterioration of performance due to the effects of usage over time and initial off-target performance due to unit-to-unit (manufacturing) noise, are not deployed in the benchmarking experiments, then the credibility of the $r_{S/N}$ factor may be in question.

We will use the r factor in two ways. First, it is *estimated* by simply substituting the difference between the baseline MTBSC or MTTF values for the components or subsystems and the reliability goals of the new product ("$r_{\text{reliability}}$"). This provides a target for the engineering team to hit and a means to back-calculate the required difference in subsystem S/N ratios to meet the allocated goal. This is what is to be done at this point in the reliability growth process. The second way it is used is to directly calculate $r_{S/N}$ as the baseline S/N ratio subtracted from the new S/N ratio (dividing this difference by 10 and then taking the proper base-10 antilog) that comes from the subsystem parameter optimization experiments, which we will discuss shortly in Block 11.

Block 6.

Select technology & leverageable designs for use in new product *—using the Pugh concept selection process, TRIZ, QFD, benchmarking, dynamic S/N characterization—all based on the VOC*

All our efforts up to this point have focused on understanding the reliability goals of the product, the baseline reliability of existing products similar to the new product, and the allocation of goals downward into the subsystems that will make up the product. It is now time to define exactly what new or leveraged technologies or existing designs are mature enough to be included in the architecture of the new product. This is the point at which concept design (explained in Chapter 1) is employed to judge new technologies and existing designs against engineering criteria that have been translated from the VOC data. It is at this point that the noise sensitivity of the design is first selected, evaluated, and characterized (typically through failure-mode analysis and noise experimentation). It is also the point that the ability of the candidate concept to be physically tested for performance is designed into the product. If a design can't be evaluated by measuring its physical functions in engineering units then it will be difficult to optimize and control. The ability of the design to be maintained by service and repair personnel is also largely determined during concept selection. The concept selection criteria must include all the dimensions of Ullman's product-evaluation items.

The superior concepts will be allowed to enter the commercialization process only after the rigors of enhanced QFD, TRIZ, competitive benchmarking, engineering analysis, and appropriate dynamic S/N characterization have been completed. This is the engineering team's earliest and best chance of selecting inherently *robust and tunable* technology to build upon throughout the rest of the development process. As mentioned previously, these choices will saddle the team with a portfolio of components and subsystems that carry the "baggage" of noise susceptibility with them. Noise is unavoidable in the design, but the *degree* to which it is present can be selected during concept design. The $r_{S/N}$ factor assumption is based on the concept of reduced variability directly having a bearing on product reliability, and will thus be greatly affected by the level of noise sensitivity that carries on through the optimization steps within the off-line quality engineering process. This is one of the most critical points in the development process because it is here that the *capacity* of a design to be receptive to the reliability growth process is determined. Remember, the $r_{S/N}$ factor is as good as our engineering skills, and thus will be sensitive to the outcome of applying those skills and how well they are used to execute the recommended best practices here in the concept selection process.

Referring back to the Ullman product-evaluation process, the engineering team will begin forming the *analytical* models needed to assess the subsystems and components. Physical models will begin to be built and assessed to certify concepts for generic robustness capacity and stability of performance as commercializable technologies. Virtually every item on the evaluation list will, in one way or another, appear as a criterion for use in the Pugh concept selection process. For further information on this topic the following references are recommended.

- *Quality Function Deployment;* Lou Cohen; Addison-Wesley, 1995
- *Total Quality Development;* Don Clausing; ASME Press, 1994
- *Total Design;* Stuart Pugh; Addison-Wesley, 1991
- *Engineering Methods for Robust Product Design,* Fowlkes & Creveling, Addison-Wesley, 1995
- *Benchmarking: The Search for Industry Practices that Lead to Superior Performance;* ASQC Quality Press, 1989

Block 7.

Conduct failure analysis and root cause analysis for variability in the subsystem & components —*using FMECA, FTA, Ishikawa diagramming, and appropriate fundamental engineering analysis*

The next step in the reliability growth process is to identify the fundamental reasons why performance is capable of degrading. Another grouping of best practices can be employed in the area of failure analysis. Three methods are typically used by engineering teams to aid in the root-cause analysis of why designs fail to perform as expected. They are:

- *Failure modes, effects, and criticality analysis* (FMECA) is a detailed process designed to identify the various ways in which a product, subsystem, or component can possibly fail to perform. This method classifies the effect of the failures on the system at large. FMECA also adds the quantification of the probability of the failure modes to the analysis. This is called *criticality analysis.* FMECA is typically viewed as a technique that provides analysis beginning at the component level and progressing up into the subsystem level. Once the component failure modes and effects are identified, the process works out to the effects on other subsystems and finally out to the entire system. From this *"Failure rates can be estimated and calculations can be made to determine the overall probability that the product will operate for a certain period of time, or the probable operating time between failures."*

- *Fault tree analysis* (FTA) is similar to FMECA. It differs in the sense that it takes a top-down approach to failure analysis. Again, according to Doty, "Fault tree analysis uses the concepts of Boolean algebra to determine the probability of a predetermined event occurring." Fault trees are built from binary events that are expressed as either/or, 0/1, or operational/failed situations. FTA can be performed bidirectionally from the top effects down to fundamental causes or from the fundamental causes up to higher level effects. Each subevent can be assigned a probability that is linked to the AND/OR nature of the Boolean analysis of the subsystem interactions. The AND probabilities can then be *multiplicitively* related to calculate a high-level probability of a failure; the OR probabilities can be *added* to estimate a high-level probability of a failure.
- *Ishikawa diagrams* can be used to graphically relate root causes of specific failures up to a system-level failure. These diagrams are also known as cause-and-effect diagrams or fishbone diagrams. They provide structure and order to identify the many possible failure paths within a system.

Each of these failure-analysis techniques ultimately rely on the corporate wisdom and engineering analysis of the development team. Here again, the analytical and physical models of the design will be called upon to aid in the evaluations. Without sound, fundamental engineering analysis underlying these methods they are no more than speculation. The real sources of failures are often subtle but are always rooted in physical changes that are related to the noises present in the manufacturing or operating environment of the product. Good failure analysis will lead back to the sources of variability (noise), and therefore are excellent ways to promote the identification of the critical noise factors that will be used to stress the designs as they progress through the optimization process. The following references are suitable for preparing the team to employ the methods of failure analysis:

- *Mechanical Applications in Reliability Engineering;* R. Sadlon; Reliability Analysis Center; ITT Research Institute, 1993
- *Practical Reliability Engineering 3rd Ed.;* Patrick O'Conner; J. Wiley, 1991
- *Reliability for the Technologies;* Doty, L. A.; ASQC Quality Press, 1989
- *Failure Mode, Effects, and Criticality Analysis (FMECA);* R. Borgovini, S. Pemberton and M. Rossi; Reliability Analysis Center; ITT Research Institute, 1993
- *Fault Tree Analysis Application Guide;* D. Mahar and J. W. Wilbur; Reliability Analysis Center; ITT Research Institute, 1990
- *Guide to Quality Control;* K. Ishikawa; Asian Productivity Organization, 1982

Block 8.

Construct subsystem noise diagrams & system noise map *—with input from service data & failure analysis*

The system, subsystems, and components are now quite well known to the engineering team with respect to their failure modes, effects, criticality, and general susceptibility to sources of variation. We can now specifically focus on the detailed physical sources of variation commonly referred to as noise. The three types of noise that we will use in constructing subsystem noise diagrams and the system noise map are:

- *External* sources of variability,
- *Unit-to-unit* sources of variability, and
- *Deterioration* sources of variability.

Creating detailed diagrams of these forms of noise is a primary means of preparing for the appropriate stress testing of the designs prior to system robustness optimization (see Chapter 14 for examples). Robust subsystems that have included system-level noises in their optimization are helpful in smoothing the transition during the integration of the entire system. These diagrams are also highly useful in constructing system stress tests, as well as the proper stressing in the tolerancing experiments that follow system optimization and finally balance product cost and quality. Without properly introducing noise into the optimization experiments and forcing what can happen to happen, the reliability growth process will not be anchored in reality. The $r_{S/N}$ factor and the S/N ratios will be overstated (understressed), and thus meaningless for the assessment of reliability growth. This is another pivotal point in building integrity into the generation of the $r_{S/N}$ factor. The quality that goes into the identification of critical noise factors is a major milestone in the success of any reliability growth activity, since these factors are what lie beneath the true quantification of the variance associated with the product's functional performance. Remember that we are relying heavily on the assumed relationship between reliability and the variance of functional performance characteristics. Inducing noise properly is our link back to reality in assessing reliability growth through variance reduction over time. References for constructing noise diagrams and maps include:

- *Engineering Methods for Robust Product Design;* Fowlkes & Creveling; Addison-Wesley, 1995
- *Taguchi Methods;* G. Peace; Addison-Wesley, 1993
  (Also see Chapter 14 of this text)

Block 9.

Allocate "$r_{reliability}$" factor requirements down to components within each subsystem —*the % contribution to "$r_{reliability}$" is divided among the components with help from service data & failure analysis*

Now that we have assessed the subsystems and components for failure modes and their effects on the system and each other, and have assigned criticality factors, we are ready to take the assignment of reliability growth down into the component level. Each subsystem has been allocated an r factor for improvement relative to the baseline performance. This r factor can now be further broken down and allocated to the specific critical components that make up the subsystem.

Each component will be independently assessed for the percentage of reliability growth it is expected to provide so that the roll-up of improvements equals the improvement "$r_{reliability}$" expected from the subsystem. This assessment comes from reviewing existing service data and data from returned products, customer complaints, and engineering analysis from a failure perspective. This allocation process is very similar to the one previously discussed to determine subsystem requirements for reliability growth. For example, a subsystem may be made up of 25 component parts. Each part will be assigned a percentage of improvement contribution depending on its criticality (from FMECA) and capacity for improvement based on engineering analysis. Some parts will fall out of the process because they are not critical contributors to reliability, or are so far down on the scale as to be inconsequential in comparison to other parts.

Block 10.

Develop subsystem reliability growth activities list *—with input from service data, noise diagrams/map, & failure analysis*

The subsystem engineering teams should now know a great deal about their respective reliability gap closure requirements, their failure issues, and the specific noises to which their subsystems are susceptible. They must now focus their energies around specific engineering activities that are targeted to deliver results in the reliability growth of the subsystems for which they are responsible using input from service data, noise diagrams or map, and failure analysis. The activities lists should comprehensively link each failure mode with a corrective engineering action. Each subsystem should be represented by a parameter diagram that relates the control factors, the noise factors, and the functional response variables that reflect the quality of the subsystem design. These diagrams will be the foundation of the subsystem optimization efforts that are the next step in reliability growth. The activities lists are excellent structures for planning resource requirements and constructing timelines for completion dates; they also help supplement the FMECA process for predicting the reliability growth potential at this point in the development process. This is the team's first actionable glimpse into the real results they are going to have at their disposal to retire their reliability deficit. This is also a good point in the process to hold a status review for the management team.

Block 11.

Perform subsystem parameter optimization *—with input from subsystem noise diagrams & system noise map* *—improved S/N ratio defined*

This step is the second opportunity to really affect reliability. Our first opportunity was in the concept-selection process, during which we attempted to choose subsystems that were superior, inherently robust, and amenable to further optimization. This includes the product evaluation issues of testability, maintainability, and reliability. Predicting reliability requirements is very helpful for knowing where the design is and where it needs to be with respect to reliability. Actually *producing* reliability lies in the domain of concept design, parameter design, and tolerance design. These best practices are at the heart of the engineering processes used to grow reliability and meet targeted goals, because of their focus on variance reduction and on on-target functional performance in the presence of noise.

At this point the engineering team has a great deal of knowledge about the subsystem, focusing in particular on failure analysis and noise susceptibility. The specific areas of reliability growth opportunity have been identified. There is much knowledge about the limitations of the design. Now it is time to delve into the control factors of the subsystem and explore the experimental design space to see if there are control-factor set points that will leave the design minimally susceptible to the critical noise factors and thus reduce the variance of the design's output. It is time to define just how well the design can be made to perform. A key step in this endeavor is to include system-level noises, along with the subsystem noises already found. This prepares the subsystem to enter the system-integration tests with a first cut at making it insensitive to system noises that are known at this point. This helps minimize the occurrence of negative subsystem integration interactions during system optimization.

The outputs of the subsystem parameter optimization experiments are a new set of improved signal-to-noise ratios (S/N_{new}) that reflect the growth in reliability from all the Phase 1 activities and the optimized parameter set points. A critical parameter drawing is created from these set points as the first formal graphic model of the design. Resources to aid in the parameter design process are as follows:

- *Engineering Methods for Robust Product Design;* Fowlkes & Creveling; Addison-Wesley, 1995
- *Taguchi Methods;* G. Peace; Addison-Wesley, 1993
- *Quality Engineering Using Robust Design;* M. Phadke; Prentice Hall, 1989

Block 12.

Subsystem "$r_{S/N}$" factor is calculated *—with input from baseline subsystem S/N ratio & improved S/N ratio after parameter design*

With the new S/N ratio in hand the team can return to the original baseline S/N ratio (obtained in block 4) and perform the $r_{S/N}$ factor calculation.

This equation represents the reliability improvement factor due to the improved S/N ratios in relation to the baseline S/N ratios.

$$r_{S/N} = 10^{\frac{(S/N_{new} - S/N_{baseline})}{10}} \quad \text{where } r_{reliability} \approx r_{S/N}$$

This $r_{S/N}$ factor, which is entirely based on actual performance data from laboratory experiments, can now be compared to the predicted r factor that was estimated in Block 5. This comparison will indicate whether the rate of reliability improvement is on an appropriate growth trend that is commensurate with the overall system goals. This is the value of the r factor: It is a helpful tool in the management of technical risk at the various gates in the development process where key decisions must be made. In this case we are approaching Gate 1, where we must provide credible evidence that the subsystems have been made robust and are showing strong evidence that they can deliver their portion of the allocated reliability growth. More growth is to be developed in Phases 2 and 3, though these phases are not likely to play as strong a role in reliability growth as Phase 1 activities did. Every product is different, but in general if Phase 1 does not provide approximately 60 to 80 percent of the reliability growth by means of the $r_{S/N}$ factor metric then the likelihood of Phase 2 system optimization processes and Phase 3 tolerancing processes making up the differences in a timely and economical fashion is relatively low. Concept design and parameter design are the two best practices that are designed to provide early reliability growth for a minimum cost in terms of engineering expenses, product inception costs (costs associated with getting a design to work and ready to launch in a stable form late in the design process), and unit manufacturing costs. They are also designed to let the business team know, early enough in the product-development process, whether the product should be modified or abandoned due to an inability to meet cost, quality, and cycle time goals.

Gate 1

At Gate 1 the product-development management team must assess many issues to decide whether the subsystems are mature enough to meet the goals of the specific product application. The following is a review of the product-evaluation processes suggested by Ullman to verify these items are present in the process blocks that represent the Phase 1 activities. We will illustrate which blocks have activities that align with the product evaluation processes.

Product Evaluation Processes:

Monitoring Functional Change:	Blocks 4, 11, and 12
Evaluating Performance	
Analytical Model Development:	Blocks 6, 7, and 11
Physical Model Development:	Blocks 6, 7, and 11
Graphical Model Development:	Blocks 6 and 11
Evaluating Costs:	Block 6
Designing for Assembly:	Block 6 and 11
Designing for the "ilities"	
Reliability and Failure Analysis:	Blocks 1, 6, 7, 8, 10, and 11
Testability:	Blocks 6, 7, 8, and 11
Maintainability:	Blocks 6, 7, and 11

Each of the key product-evaluation processes is addressed in one or more of the process blocks used to represent the product-reliability growth activities. In this particular discussion about the development of a product, we are primarily viewing things from a reliability perspective. It is interesting to see how profound the effect of performing the reliability growth process is on virtually every performance issue facing the product engineering team. Of course there is a lot more to comprehensively developing a product than what the reliability growth process implies. The intent of this chapter is to address the technical side of product development. A series of marketing and sales processes are taking place simultaneous to the reliability growth process. This text discusses only the crossover required between these teams in a context of multifunctional teamwork. Let's move on to Phase 2 in our endeavor to define how *system* robustness evaluations fit into the reliability growth process.

Phase 2: *System Robustness Optimization*

Block 13.

System robustness testing & optimization completed *—subsystem interactions defined & problems corrected* *—component life testing performed on critical parts*

The optimized subsystems are now ready to enter Phase 2 of the product development process. It begins with Block 13, in which system robustness testing and optimization are completed; subsystem interactions are defined and problems are corrected; and component life testing is performed on critical parts. Subsystems have now earned the right to be assembled into a system configuration that is highly representative of the final product. The engineering team has shown strong evidence, through the r factor and $r_{S/N}$ factor comparisons and other engineering metrics, that the individual subsystems have continued to show the capability to meet the final reliability goal.

There are still some risks associated with the ongoing optimization of the product. The component parts are to be made from materials and processes that are as close to the final production materials and processes as possible. This leaves some concern over the differences that exist between *real* subsystems built from production parts and *practice* subsystems made by other means. This is a product-development "noise" and is one of the reasons the product-development process requires iterative system optimization through the construction of system models, engineering models, and preproduction models. Developing system-functional integrity can be quite difficult depending on the number of component parts and subsystems, and the interactivity between the subsystems as they are joined into the system architecture.

The Phase 2 system optimization process is designed to provide the structure, discipline, and appropriate best practices to develop a substantive level of maturity in the product design. The goal of the system optimization phase is to demonstrate that the system is robust against subsystem and system-level noises. It is to prove that all the technology in the product is safe to deploy into the design and production phases of the product commercialization process. The evaluation of the system models is designed to smoke out any mistakes, hidden interactions, barriers to full functionality on a spatial level, and safety, regulatory, and environmental problems. The system evaluations also provide the manufacturing, assembly, human factors, service, and product-support teams an opportunity to assess functionality, maintainability, and other issues that come from assembling and operating the complete product.

System evaluations are complex series of tests that require critical noise factors, identified in Block 9 of Figure 2.2, to be realistically applied during the system tests. Only in this context can the system be proved to be robust. Many products never experience the full force of noise until they are actually in production and are operating in the customer's use environment. This situation gives rise to postshipping approval reliability improvement activities that are wildly expensive and embarrassing to the firm's reputation with customers. If this is to be largely avoided, the engineering team must apply the stressful noises to the developing physical model of the product long before the manufacturing community and customers have a chance to do it.

It is imperative that two sets of performance metrics be obtained during the system tests. The optimization task and the Gate 2 decision-making process are too complex to be fueled by reliability metrics alone. Counting failures and service calls is too remote, from the physics of variation that cause these events, to be helpful in rapid system optimization. The system-performance analysts must oversee the comprehensive instrumentation of the system models for acquisition of fundamental functional performance in engineering units (scalars and vectors). Sample data must be gathered in sufficient quantity, on critical functional parameters, to provide good statistical measures of average performance and the variance associated with that performance when real noise is applied to the system. This noise will be fairly random in the real world but the development team cannot wait for random variation to occur; they must force what can happen to happen during the system-level evaluations.

A heavy emphasis on data acquisition and instrumentation is critical to the engineering mission of detecting the root causes of internal subsystem problems as well as interactive problems between subsystems. The team cannot afford to waste time trying to figure out what happened when variability occurs – they must place instrumentation at strategic interfaces between and within subsystems to measure it as it happens. This is particularly helpful in rapidly defining hardware versus software problems. Obviously one cannot instrument the machine to the point where the process of data acquisition interferes with system performance or bankrupts the development budget. The engineering teams must join forces and judiciously construct a system-instrumentation architecture that is designed to explore the most sensitive and critical areas of the design. The instrumentation plan must include flexibility in anticipation of new sources of variability coming to light during the evaluations. While gathering this engineering data, the system analysts should also be monitoring reliability metrics such as mean time to failure or product stoppages, mean time between service calls, and other time-based performance measures. In this way, two sets of "books" can be kept to account for product performance during the system optimization phase.

Typically the reliability metrics will be used mainly by the management community to assess risk, allocate resources, and make business decisions at a relatively high level. The engineering metrics will mostly be used by the technical community to do root-cause analysis, trouble shooting, and optimization, and to make other technical decisions.

An important side note: All critical component and assembly dimensions and set points must be inspected prior to inclusion in the system evaluations. Only in this way can the unit-to-unit noises be quantified before their effects are intentionally varied later in the system evaluations.

Life testing must also be initiated at this point. The engineering team must develop a knowledge of *long-term* robustness, which is often called reliability or durability, about the critical components within the product. This is particularly true for fatigue-sensitive items or components that must withstand high mechanical, thermal, or electrical loads along with repetitive use in particularly noisy environments. Use of accelerated life testing (see *Accelerated Testing* by W. Nelson) may be appropriate for some components that are amenable to that process. Certain components will need to be pre-aged to simulate the effects of time and deterioration. This is a form of stressful life testing that adds to the understanding of how the system reacts to wear and other effects that move set points off target. In this way, the subsystems and system can be prepared for the onslaught of sources of variability that is to come so that no surprises occur in the later phases of the product's development, manufacture, and use. It is often appealing to construct the system with all the components' set points right on target and everything as perfect as can be attained. This is fine if the team wants to know the particular point of performance capability that some small percentage of the final product will exhibit. With this in mind, it is arguable that the system tests should be run, after a brief study of the nominal set points, in two scenarios: first under a reasonably extreme set of noise conditions to represent actual conditions that could occur, and second under an average set of noises that are moderate in nature and more generalized across the spectrum of manufacturing and use conditions. These two scenarios provide general estimates of performance that will force the engineering team to go the extra mile in upgrading performance in light of the reliability that customers expect.

When problems are uncovered during the stressful system evaluations, the subsystem teams have an excellent opportunity to correct the problems before Gate 2. Gate 2 is a very serious checkpoint. It is here that the product either proceeds with the full support of the business's decision makers or is halted. Without stressful robustness evaluations, the decision makers have a very biased data base to draw upon. It is at this point of the product-development process where the deepest scrutiny and most honest assessment of system performance are required. The next phase, product design, is where sizable sums of capital begin to be spent on tooling, additional engineering resources, and the support structure for taking the product into the manufacturing phases. Taking an unproven, lightly stressed product into today's business climate is unthinkable.

Block 14.

Optimized subsystem & system parameter set points & estimated system tolerances defined —*input from system testing*

The outcome of the system robustness tests is a new or at least reevaluated and proven set of optimized control factor set points for the components, the subsystems, and the system-level interfaces between the subsystems. Major interactions will have been identified. Many mistakes will have been found and corrected. An understanding of subsystem-interface sensitivities will have been evaluated such that the system-level tolerances are initially defined. Subsystem tolerances also should be coming into a higher state of clarity. In general, the business and technical teams should have a very good understanding of the capability of the product with respect to functional system performance.

Gate 2 requires proof of system-level robustness and further evidence of reliability growth. It is at this point that the engineering team must prove they can truly meet the reliability goals of the product. The $r_{S/N}$ factors generated in Phase 1 can be compared to the Phase 2 $r_{S/N}$ factors as they are calculated from system-model data. Here we would like to see evidence that the $r_{S/N}$ factor has grown such that, at a minimum, the total reliability growth should be approaching 70 to 80 percent of the intended goal. Phase 3 will be used to develop the remaining reliability and close the remaining gap. A great deal of accountability is assumed when the engineering team states their position on reliability. The output from the system-optimization experiments provides the data base upon which major funding and resourcing decisions will be made. There is little room for compromise on the integrity and validity of the system test data. Once Gate 2 is passed it is very expensive to turn back.

The ensuing phases of the product-development process are very much dependent on the set points that are brought out of Phase 2. It is this set of target (nominal) set points that will be taken into the tolerance design phase of the commercialization of the product. Since it is difficult to perform tolerance design on a moving target, the Gate 2 parameter set points must, for the sake of cycle time and cost control, be stable.

Gate 2

Phase 3: *Product Design*

Block 15.

Subsystem tolerance design —*input from system noise map, subsystem noise diagrams, & estimated tolerances*

Tolerance design is the third major opportunity for the engineering team to grow reliability in a significant way. It is done in the fifteenth phase of production shown in Figure 2.2, using input from the system noise map, the subsystem noise diagrams, and estimated tolerances. Now the reliability growth is definitely going to increase the product manufacturing costs. The process of tolerance design is probably the least understood and most loosely practiced of all the engineering sciences. Prior to publishing this text, there was no single, modern source of information solely dedicated to the topic of engineering tolerances in preparation for communicating (GD&T) them to the manufacturing community. The tolerancing phase of the product begins with a clear understanding of the nominal set points that have been developed in the preceding phases. If there is ambiguity in these target values that promote optimal performance, the process of developing tolerances to complete the definition of the specifications cannot proceed easily.

With the nominal set points in hand the engineering team must develop the tolerances in a serial process as follows:

1. For mechanical and electrical "stack-up" problems one can employ traditional analytical tolerancing techniques starting with component parts and working up to subassemblies and finally to the system interfaces. Sometimes the subsystem-interface tolerances will force the allocation of tolerances down to the component level (see Chapters 5 to 10).

For a more customer-focused approach, Taguchi's loss function-based tolerancing methods are recommended (see Chapters 11 and 12).

Component-part tolerancing will typically be initiated by using standardized tables and charts that have been developed over the years for the multitude of manufacturing processes that currently exist. This literally means the engineers must look up or ask the component manufacturer for the recommended tolerances for the geometry, materials, and manufacturing process being employed. Some companies have developed their own catalogue of process capabilities for the specific types of manufacturing processes they use. Note that this process is totally focused on what the *standard* tolerancing should be—based on the capability of the manufacturing process. Hopefully the engineering team selected the design concept, back in Phase 1, based on the careful consideration of this particular manufacturing process. These recommended tolerances are only the starting point for the rest of the tolerancing process. The traditional standard for component tolerances is based on $\pm$ 3 standard deviations of the output response of the production manufacturing process. This can be a significant issue in developing reliability growth in a six-sigma quality goal context. The three-sigma case will be discussed in Chapter 4.

For assemblies that do lend themselves to tolerance stack analysis we will begin with the recommended tolerances for the components. The best practice that must now be applied is any one of the many forms of statistical tolerancing. Most engineers know this by the name *root sum of squares tolerancing* (see Chapter 6). This technique applies the laws of probability to the tolerancing process. This approach is more realistic than the worst-case approach to tolerancing (see Chapter 5). These tolerances can be hand-calculated for simple tolerance stacks of mechanical elements. For complex subassemblies (2- and 3-dimensional cases) and electronic applications the engineer will need to apply computer-aided tolerancing techniques (see Chapter 8). CAT is typically based on Monte Carlo simulations of the probability of the component assuming a specific dimension or value. With modern software any number of discrete or continuous probability distributions can be employed in this process. CAT programs are usually able to handle most tolerance cases that can be expressed as a math model.

With the component and subassembly tolerance baselines established, the engineering team can then move on to apply Taguchi's tolerance design process to complete the cost versus quality balancing (see Chapters 11 to 17). Taguchi's empirical approach is especially useful when tolerancing mixtures of various technologies represented in components or process set points in a single subassembly or an entire system.

The Taguchi tolerancing process allows the team to balance cost and functional quality through the experimental development of optimal tolerances in the face of the various noises defined in Block 9. Statistical tolerancing and CAT cannot include all the noises that a tolerancing process based on a designed experiment can (see Chapter 17), nor can they handle multitechnological cases that are not amenable to math modeling. Taguchi's approach facilitates the development of tolerances in subsystems made up of many different technological elements. For example, a subsystem that employs mechanical, electrical, electronic, chemical, magnetic, electrostatic, thermal, and optical elements can readily be optimized for target set points and tolerances using Taguchi's empirical approach. It provides data that can be applied to the quality loss function (see Chapters 11 and 12), which in turn will relate the functional quality of the design, as the tolerance limits are approached, to the cost implications to the business and customer. Now a tolerance can be linked to the cost of ownership as well as to the perception of reliability.

Tolerance design typically requires that the subsystem control factors be set at their high and low standard deviations from supplier data bases and statistical tolerance studies (see Chapter 15). This two-level control factor is then loaded into an orthogonal array. The very same noises that were used to stress the product during parameter design are now employed to stress the tolerance experiment (see Chapter 14). From this activity the data can be processed using the analysis of variance technique (see Chapter 16).

Block 16.

Rank control factor effects on variability of
key subsystem functional responses
& calculate Cp
—*ANOVA data decomposition & process capability analysis on subsystem tolerance limits and* s *from experimental data*

Analysis of variance (ANOVA) is a data-decomposition process that is applied to the results obtained from a designed experiment. It will unveil the specific contribution each control factor makes relative to the measured response of the design. In a tolerance experiment this tells the engineer which tolerances are having the biggest impact on performance in the face of induced noise. Chapter 16 provides a practical introduction to the mechanics and application of the ANOVA process.

Control factors that have high levels of effect on the response can be retoleranced as necessary, at a specific cost, to reduce the overall variance of the targeted functional response. This is where the engineer can control the physics behind the reliability as embodied in the variance of the response. The power to alter the variance is the power to improve reliability. Again, the correlation is approximate but quite real.

The engineering team should, from enhanced QFD and VOC data, have an upper and lower tolerance limit assigned to the subsystem functional response. From the ANOVA data we can determine the overall standard deviation of the functional response. With the product design process capability index:

$$Cp = \frac{USL - LSL}{6\sigma}$$

we can calculate the process capability of the subsystem in the light of stressful noises. This is an excellent indicator of the current performance of the subsystem with initial tolerances from standard tolerancing processes. We are now equipped with enough information to begin the *upgrading process* to increase the process capability of the subsystem and to use tolerances as a means of controlling costs while improving reliability growth.

Block 17.

> Calculate the new control factor variances required to meet the subsystem "$r_{reliability}$" factor
> —*reduce critical parameter variances by their % contribution to overall subsystem "$r_{reliability}$" allocation*

The ANOVA data is quite helpful to the engineering team at this point. The ANOVA data contains the specific variance information about each control factor as well as the variance for the overall subsystem response. We know from the allocated portion of the r factor just how much growth in reliability is required from each component. Block 8 produced the initial estimate of these values. This can be related to the control factors of the design. Each control factor variance can be targeted for an appropriate reduction that can be attained through tighter tolerances or material upgrades. Remember, the key assumption in this process is that reliability goes up approximately as variance goes down. We must be circumspect in our attempt to engineer variances down in such a manner that they stay low over time. Just tightening a tolerance may not last in the face of deterioration noises, which is why tolerance design must be done in the noisy environments that are representative of the types found in the manufacture and customer use of the product.

Block 18.

> Work with suppliers to attain the new control factor variances required to meet the subsystem "$r_{reliability}$" factor
> —*tighten critical parameter tolerances to attain proper % contribution to reduce overall subsystem response variance to meet "$r_{reliability}$" factor allocation*

Lower variances in control-factor contributions to the overall subsystem response must somehow be attained and paid for. This is why we say tolerance design is a *balancing* process between cost and quality. This is the point at which the relationship between the supplier of the production parts and the engineering team may come under some stress. The team is likely to be using standard tolerances generated by traditional economic means (probably $\pm 3\sigma$). To improve the quality of the parts, typically by holding the tolerances to closer limits around the optimum nominal set points, the supplier must find an economical and technically feasible way to reduce part variation. It is desirable to do this on a routine basis that can be monitored and controlled by statistical process control techniques. An excellent way to enable the supplier to react favorably to requests for tighter tolerance control is to put the manufacturing processes through the robustness optimization process prior to this point (preferably before or at least during Phase 2). If the manufacturing technology is both robust and tunable, then, for an economical price, the manufacturer can accommodate the request for a quality upgrade. This upgrade process is to be applied to all the control factors that show statistical significance, through the F ratio in the ANOVA data (see Chapter 16). Then the engineering team can apply the variance reduction provided by the supplier, as well as the upgrade cost to the quality-loss function and the process capability equation. Decisions can then be made to determine which control factors should be upgraded permanently. The new variances can be used to recalculate the Cp so it can be compared to the original Cp based on the initial tolerances (typically $\pm 3\sigma$). This process promotes the balancing of cost with quality by allowing accurately targeted expenditures on *only* those control factor tolerances that really have strong quality (variance reduction) payback. The quality payback is quantified through the quality-loss function. This process is fully explained through an example in Chapter 17.

There is another benefit from this process that can help control part costs. Those factors in the design that showed little effect on the variability of the functional response may be considered for material downgrades or *tolerance loosening*. This can potentially have a minor impact on the functional quality of the design. Of course this all has to be verified, along with the changes made by tightening critical control factor tolerances. Every case is unique but some of the cost increases may be offset by identifying tolerances that may initially have been set too tight. Before these downgrades are made, engineering analysis should be done with respect to component life. In some cases, life tests will have to be run. As in all these optimization cases, sound engineering judgment, backed by data, must prevail.

We have just outlined a process to justify increasing the cost of the design in order to reduce variability, for the sake of increasing reliability. As each targeted control factor is upgraded with respect to tolerance tightening and material upgrading, we can trace the benefit back to the subsystem r factor. This is done through the percent improvement made in the overall subsystem response variance by the individual control-factor variance reductions. The process is to be repeated, in parallel, across all the subsystems in the product. The accrued benefit of all the variance reductions is the physics behind the reliability growth process. Once the new parts are available from the upgraded and improved manufacturing processes, the engineering team is ready to verify the improvements.

Block 19.

Repeat tolerance experiment with upgraded components to verify targeted variance reduction is attained
—*Redo ANOVA & new Cp*

The engineering team is ready to rerun the initial tolerance experiment, but now will include the tighter tolerances defined by the upgrade process and the looser tolerances defined by the downgrade process. This is a verification experiment and as such may need to be replicated enough times to assure the engineering team that the results are durable. This is up to the judgment of the engineering team. The standard recommendation is three to five replications using the appropriate noise factors to stress the design just as was done earlier.

The ANOVA process on the new data will be repeated as well. The data can be used to confirm that the targeted variance reductions are taking place in the control-factor tolerance contributions to the overall subsystem response. The overall subsystem variance will produce a new standard deviation that can be plugged into the subsystem Cp equation to define the new process capability for the upgraded subsystem. Again, this process is demonstrated in Chapter 17.

Block 20.

Integrate improved subsystems into engineering models & verify system r factor improvements are materializing
—*compare to subsystem & system goals*

The final step in the tolerance design process and its contribution to reliability growth is to verify the improvements at the system level, comparing them to subsystem and system goals. This can be done by integrating the improved subsystems into the next iteration of system product models typically called engineering models. Another way to evaluate the improvements is to subject the subsystem to the original robustness verification experiment from Block 11 and then compare the S/N from the current experiment back to the original optimum S/N from Block 11. A new $r_{S/N}$ factor can be calculated by taking the difference between the Block 20 S/N and the baseline S/N from Block 4. Hopefully the comparisons will indicate the subsystem $r_{S/N}$ factors are growing as necessary to meet the system goals. The engineering-model testing will provide the data to verify the growth in both engineering and reliability metrics. The engineering models (the Phase 3 version of the product) should be evaluated using standard reliability tests. These tests are used to generate Duane Plots (MTBSC level plotted over time), which are useful in communicating reliability growth in reliability units. When to spend the time and resources to generate this information is up to the engineering team. It would not be unreasonable to create such plots at the end of Phase 2. The system reliability performance should be in sight of the original Block 1 goal at this point.

By now the other teams mentioned in Block 3—the product assembly community and the service/customer support groups—will have defined how their processes can contribute to overall system reliability. Their contributions can now be fully integrated to the overall product-development effort. These contributions should be the final gap closure that drives the system reliability numbers to meet or exceed the goals. As in all developmental efforts, the engineering contributions to reliability growth will have to be debugged as the engineering models mature and verify the validity of the Phase 3 design efforts.

Gate 3

This concludes the overview of the reliability growth process between phases 1 and 3. While it would be great if the goals for system reliability were met at this point, it is not yet time to celebrate. The following phases and gates are designed to transition the engineering models into preproduction models and finally into fully functional production models that will bear revenue. Many errors, disturbances, and sources of variation can creep into the final stages of the product-development process, particularly when all components are coming from real production processes and tools. This is when the engineering team must remain focused on maintaining and stabilizing the quality they have engineered into the system by employing the best practices within the realm of on-line quality control. The other teams must focus on holding up their end of the reliability growth process to be sure the total product reliability stabilizes and is translated into the production phases of the project. This is particularly true for the product-assembly teams.

Chapters 4 to 17 discuss, in great detail, how to deploy tolerance design in the commercialization of a product. The preceding process shows exactly where tolerancing and tolerance design reside in the development of a product. It also shows how the tolerancing process is capable of playing a meaningful role in reliability growth. It should be clear to the reader at this point that the parameter design process must precede the tolerance design process. In this sequence, tolerancing is performed on optimized nominal set points rather than on nominal set points that have not been scrutinized for their sensitivity to noise. Noise is the cause of reliability degradation. We will conclude this discussion on that theme, with a brief discussion on the effect of particular noises on reliability in the context of tolerancing.

The Reliability Bathtub Curve and Tolerancing

The bathtub curve (Figure 2.3), commonly used to express the frequency of component or product failures relative to time in service, can be used to illustrate where (in the product) and when (in service life) the outcome of a proper tolerancing process can help cut the cost of quality.

There are three periods of time in a product's life where noise plays a role in the perception of reliability. These periods have significant overlap. We employ the bathtub curve to illustrate what these noises are and why they are important in tolerancing. The two portions of the curve where failure rates are at a maximum are during the early life and end of life portions. At these two stages of life, the unit-to-unit noises and deterioration noises are having their most significant effect. How we establish and maintain tolerances will have a definite impact on the sensitivity to unit-to-unit noises. Defining the critical-to-function parameters (CTF) and defining proper tolerances for each of them, to minimize variance, will have a settling effect on the occurrence of early life failures, particularly if the process is kept on target and not allowed to drift. It is critical to understand the nature of the particular process capabilities pertaining to the components before the beginning of product life so that it can be engineered to support the reliability goals for the product.

If tolerances are developed outside of the effects of the wear and deterioration of the components, they will have no capacity to support the ideal function of the design as its service life increases in the sea of variability found in the customer's environment. This text focuses on actively developing data-driven tolerances—not just passively assigning them through engineering judgment alone. The process developed in these pages is meant to encourage the use of traditional analytical tolerancing techniques and Taguchi's tolerancing process as a basis for conducting designed experiments to produce data to evaluate the effects of noise upon critical tolerance limits. During this experimental process, one must force what variations can happen, to happen. In this context the effects of time (random events) and use will have to be intentionally induced and applied to the critical optimum control factor set points placed at their ± 1 sigma limits. Tolerancing, in this context, will assure a healthy start for the product and a minimized impact on performance as the product nears the end of its useful life. We would like to see the product leave the factory insensitive to noise and designed to maintain that insensitivity for as long as is required by the customer's specifications. The tolerances can be set to help lengthen part life and make the design more predictable with regard to preventative maintenance planning. Parameter design provides an economically feasible method to design in robustness. Tolerance design provides the next stage, of sustained robustness, by allowing financial resources to be expended wisely to balance the cost and performance over the life of the product.

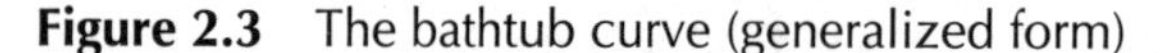
Figure 2.3 The bathtub curve (generalized form)

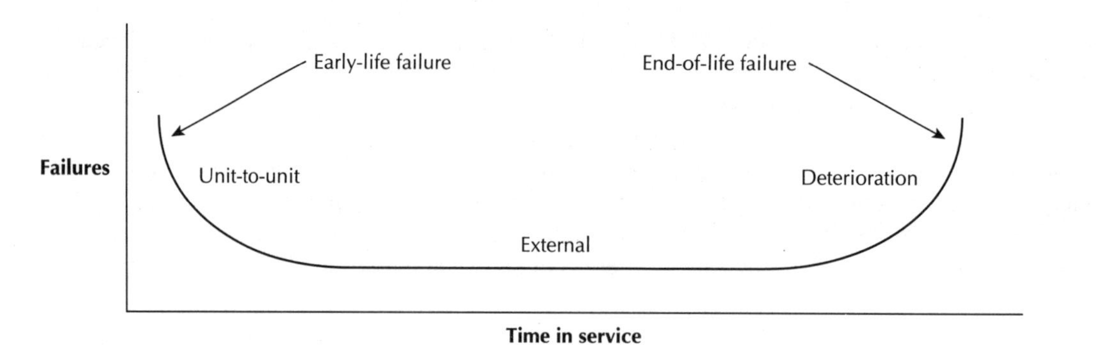

Summary

A case has been made for the use of a disciplined engineering process to generate reliability growth early in the development of a product. The chapter is outlined as follows:

Chapter 2: The Relationship of Quality Engineering and Tolerancing to Reliability Growth

- The Three Initial Phases of Product Development
  - Phase 1: Subsystem Concept Selection and Robustness Optimization
    Blocks 1–12
  - Phase 2: System Robustness Optimization
    Blocks 13 & 14
  - Phase 3: Product Design
    Blocks 15–20
- The Reliability Bathtub Curve and Tolerancing

Chapter 2 provides a detailed set of 20 "blocks" that contain specific activities to promote reliability growth. These activities are set in the context of the initial three phases of a product development process. The main focus is on the synthesis of a widely accepted set of engineering best practices to actively develop reliability while simultaneously quantifying the rate of growth through a metric called the r factor. The r factor is developed through the following general process:

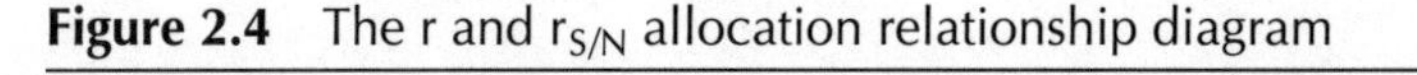

Figure 2.4 The r and $r_{S/N}$ allocation relationship diagram

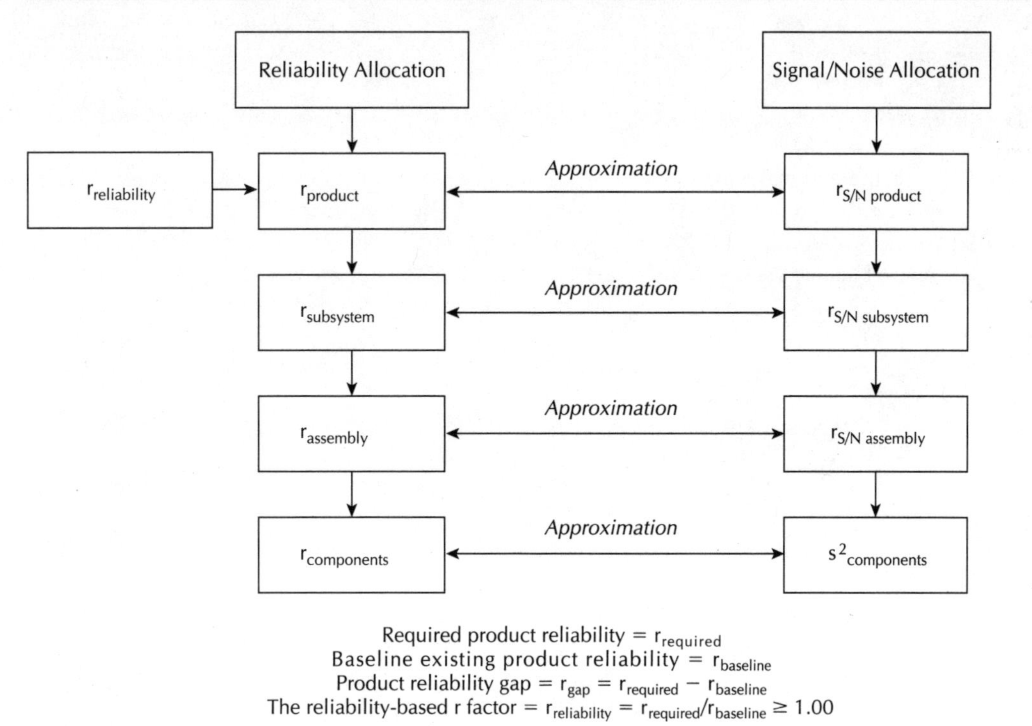

The concept of the approximate relationship between reliability metrics and robustness metrics is documented in the text and summarized in Figure 2.4.

The general context for engineering concepts, nominal parameter set points, and tolerances in support of reliability growth has been developed. The serial nature of performing concept design, parameter design, and tolerance design is fundamental to adequately controlling costs, meeting product evaluation goals, reducing cycle time, and promoting reliability growth. The application of the r factor has been discussed and its various stages of allocation illustrated. The need to develop both engineering metrics and reliability metrics has been explained. The overarching lesson that is to be learned from this discussion is that the early, methodical application of quality engineering best practices within a disciplined product-development process will yield improved reliability along with many other economic and quality-related benefits.

References

R2.1: *The Mechanical Design Process;* David Ullman; (pg. 227, McGraw-Hill, 1992) pg. 52.
R2.2: *Quality Engineering Using Robust Design;* Madhav Phadke; (Prentice Hall, 1989) pg. 62.
R2.3: *A Modeling Framework for Relating Robustness Measures with Reliability;* Gary Wasserman (Quality Engineering, 1995) pg. 63.

All remaining references are listed at the end of each of the blocks that define the 20 steps in the reliability growth process. This is done to aid in aligning the proper reference resources with the particular process being discussed.

CHAPTER 3

Introductory Statistics and Data Analysis for Tolerancing and Tolerance Design

The Role of Data in Tolerance Analysis

This chapter is designed to prepare the reader to work with data, apply statistical methods, and understand the fundamentals of six-sigma methods. The practice of applying analytical or experimental methods to optimize the tolerancing of a product or process always requires the collection of data. The statistics used in tolerance development will typically be produced from *sample* data, which must be properly gathered and processed. In the case of analytical or computer-aided tolerance analysis the output of the math model will need to be confirmed through the collection of data from actual hardware. In the case of experimentally based tolerance analysis, the process Taguchi calls tolerance design, data is the key mechanism for enabling tolerance optimization. This data will come directly from a manufacturing process, from an experiment on a product, or from an element within a product.

As mentioned in Chapter 1, we are always dealing with two issues in tolerancing: the optimization of manufacturing process performance and the optimization of product-design performance. It is considered good practice for an engineering team to optimize these concurrently for a specific application. Collecting viable data and properly analyzing it is essential for truly optimizing cost and quality.

The ease of working with data to optimize some form of functional performance depends on the type of data. In general, it is difficult to make the necessary determinations to improve a product's performance by looking at the raw data collected during an experiment. The data must be organized into a form that is useful for building knowledge and ultimately for drawing conclusions that lead to improved product or process performance.

Data can consist of numbers that represent *discrete* variables such as whole integers (1, 2, 3, 4) or *continuous* variables such as rational numbers (1.245, 2.657, 3.912, 4.003, etc.). While discrete

Figure 3.1 Illustration of data—sample versus population

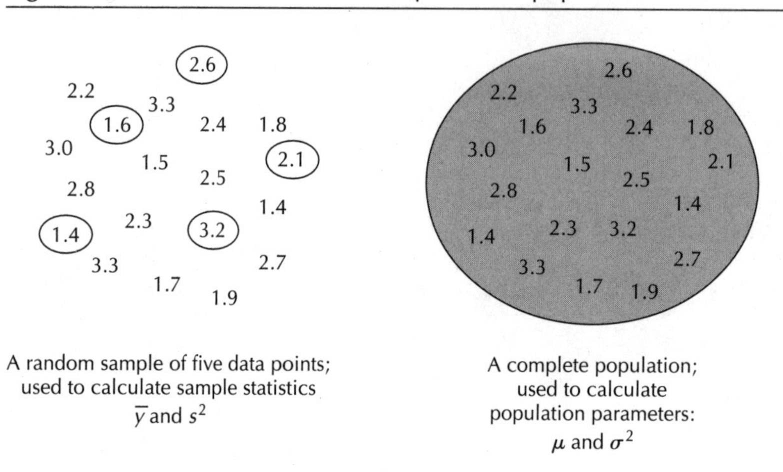

values such as counting events, gradation and classification, binary values (zeroes and ones), and pass/fail can be used in tolerance design activities, data having continuous values such as measured deviations from a target, engineering units (scalars and vectors), and direct measures of physical performance are preferred. In statistical terms, data may be classified by two features: A data set is either a *sample* from an unknown population or is made up of *all possible values* from an entire population. Examples of a sample of data and a population of data are displayed in Figure 3.1.

Data that are a sample from a population can be mathematically processed into descriptive values called *sample statistics.* Data that comprise an entire population can be mathematically processed into descriptive values called *population statistics.* Some population statistics can never be quantified because they are so large as to be uncountable or are ongoing—as in a manufacturing process—so that it is not possible to gather the entire data set. We are almost always going to be working with samples from populations.

Graphical Methods of Data Analysis

The Frequency Distribution Table [R3.1, R3.2]

Whenever a number of measurements are taken as data, some amount of variation will exist among the values. It is also typical to find the experimental measurements tending to fall around a value that suggests a *central tendency* in the data. One way to describe the nature of data is to graphically display the shape or distribution of the data. The distribution displays the variability in the data as well as the propensity of the data to be spread around a central value. A data set can be organized and displayed according to the *frequency* of the occurrence of specific values. From such a display three observations can be made:

1. The variation that exists between the data points,
2. The tendency of the data to fall around a specific central value, and
3. The shape of the data as it is distributed around the point of central tendency.

To illustrate these three items let's look at a sample, shown in Table 3.1, of 100 data points—in this case, diameter measurements taken from a batch of several hundred one-inch-nominal-diameter tubes that were extruded through an aging extrusion die. In this section we will describe the process we'll use to determine the type of distribution that describes the probabilistic nature of a manufacturing process output variable (see Chapter 8).

In order to visualize the data it is necessary to process it from its raw form into a summary form called a frequency distribution table. A frequency distribution table is used to generate a plot of the frequency of sample or population data values by placing them within ranges of numbers in a graphical format. The distribution is established by counting how many data points fall into each range. In Table 3.2, the 100 data points are counted in 11 equally sized ranges of 0.005 inches starting at 0.975 inch and ending at 1.030 inches.

Table 3.1 Table of data for one-inch-nominal-diameter tubes

1.013	1.010	1.014	1.009	0.996	1.003	0.997	1.000	1.007	0.996
1.007	0.984	1.012	0.996	0.991	0.995	1.006	0.988	1.005	0.992
0.993	1.002	1.005	1.008	0.982	1.015	1.010	1.004	0.987	1.014
1.000	0.985	1.006	0.994	1.001	1.012	1.006	0.983	1.004	0.994
0.998	1.008	0.985	0.995	1.009	0.992	1.002	0.986	0.995	1.030
1.019	1.001	1.021	0.993	1.015	0.997	0.993	0.994	1.008	0.990
0.994	1.007	0.998	0.994	0.996	1.005	0.986	1.018	1.003	1.013
1.009	0.990	0.990	0.993	0.995	1.017	1.000	1.009	1.006	1.005
1.020	1.005	1.003	1.005	0.998	0.999	1.000	0.997	1.000	0.995
1.007	1.005	1.015	0.985	0.989	1.015	1.005	1.011	0.992	0.984

Table 3.2 The frequency distribution table[1]

Range number	*Range boundaries*	*Frequency*
1	$0.975 \le y < 0.980$	0
2	$0.980 \le y < 0.985$	4
3	$0.985 \le y < 0.990$	8
4	$0.990 \le y < 0.995$	17
5	$0.995 \le y < 1.000$	15
6	$1.000 \le y < 1.005$	16
7	$1.005 \le y < 1.010$	21
8	$1.010 \le y < 1.015$	9
9	$1.015 \le y < 1.020$	7
10	$1.020 \le y < 1.025$	2
11	$1.025 \le y < 1.030$	1

1. Users of the spreadsheet Excel can easily use it to produce histograms. Consult your user's manual.

Figure 3.2 Histogram of extrusion data

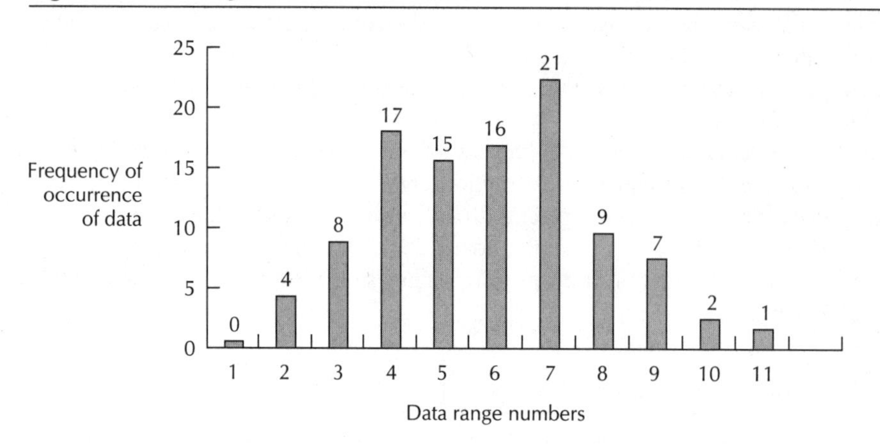

The Histogram

The frequency distribution table can be shown using a graphical display called a *histogram.* Histograms display the shape, dispersion, and central tendency of the distribution of a population, or sample, of data. The extrusion histogram shown in Figure 3.2 has the same 11 ranges shown in Table 3.2; they have been numbered from 1 to 11 to act as placeholders for the data points that fall within each range. The vertical axis represents the frequency of data occurrence. This form of distribution is reasonably representative of a *normal* distribution (defined later in this chapter).

Quantitative Methods of Data Analysis

A graphical display of the central tendency, distribution width, and dispersion geometry of a data set is a useful analysis tool. But there is much more to be gained by *quantifying* the nature of the data—specifically its central tendency and variability, and measures of the geometric shape of its distribution. The shape of the distribution of a data set is of interest to us particularly with respect to flatness (or peakedness) and skew to one side or another. A quantitative description of the data using just a few summary statistics allows us to develop a number of key quality measures, including the quality-loss function and the signal-to-noise ratio. Knowing the detailed nature of how manufacturing processes behave over time in the presence of noise is very helpful in the tolerancing process. Characterizing the behavior of a component, a subassembly, or an entire product in relation to the variation present in these items due to manufacturing processes and the numerous other noises in the customer-use environment is also crucial to proper tolerance design. We need to take a few moments to review some useful methods of quantifying the presence of variability in product designs and manufacturing processes. In fact, we need to take *four* moments—that is, measures related to the mean loosely defined as mathematical summations of values raised to a specific power in numerical or physical space—to quantify the key effects that are so critical to beginning proper tolerance analysis.

Introduction to the Fundamentals of Descriptive Statistics

The First Moment of the Data about the Mean: The Arithmetic Average

When we assess the data from a measured-response variable associated with a manufacturing process or the performance of a product design, the first characteristic we seek to quantify is the specific value about which the data is dispersed. We have previously referred to this value as a measure of central tendency. This means that when we assess the first moment about the mean by using the following equation:

$$\mu_1 = \frac{1}{n}\sum_{i=1}^{n} (y_i - \bar{y})^1 \quad \textbf{(3.1)}$$

where

$$\bar{y} = \frac{1}{n}\sum_{i=1}^{n} y_i \quad \textbf{(3.2)}$$

we find that the value of this summation is zero. This happens because we first used all the data values (y_i) in a summation process (Σ) to obtain a specific overall magnitude of the aggregated sample of data. We then took this pool of summed data and divided it by the total number of data points (n). This defines the arithmetic mean ($\bar{y}$). *All* of the data has been used to define the arithmetic mean. We can now run all of the individual data through a summation and averaging process, as shown in Equation 3.1, where the mean is subtracted from each data point and raised to the power of one. This is the mechanics of taking the first moment, μ_1—raising the differences between the data points and the arithmetic mean to the first power. Once this process is completed for each data point, the result is that the first moment about the mean must always equal zero. Obtaining any other result means there has been an error in the calculations.

Figure 3.3 The mean of a sample of data

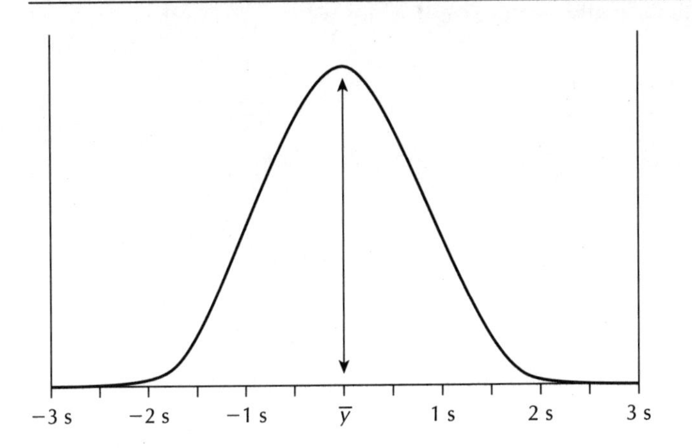

Again, the central tendency of the sample data is given by the sample mean, $\bar{y}$, shown in equation 3.2. This expression states that a measure of the sample mean or average of all the sample data can be calculated by adding (Σ) all the individual data points (y_i) and then dividing by the total number of individual data points (n).

For the 100 random samples shown in Table 3.1:

$$\bar{y} = \frac{1}{100}(1.013 + 1.007 + .993 + 1.000 + .998 + \ldots) = 1.001 \ldots \textit{ the sample mean}$$

Figure 3.3 illustrates the location of the mean in a normal distribution.

The Second Moment about the Mean: The Variance

When we perform the same summation and division (averaging) process on the *square* of the individual data deviations from the arithmetic mean we obtain the *second* moment, formally called the *variance.* It is literally the measure of the average deviation the data has with respect to the mean. To convert the variance into what is commonly known as the standard deviation we must take the square root of the variance.

The second moment about the mean:

$$\mu_2 = S^2$$

The variability of sample data is given by the *sample variance,* S^2, shown in Equation 3.3.

$$S^2 = \frac{1}{n-1}\sum_{i=1}^{n}(y_i - \bar{y})^2 \tag{3.3}$$

This expression states that a measure of the dispersion, or scatter, of the data is based on the average of the squared deviations of the individual data points from the sample mean. A sample deviation, $(y_i - \bar{y})$, is given by the difference between the individual data points, y_i, and the sample mean $\bar{y}$. The value $n - 1$ has been shown to provide an estimation of a more accurate value of the sample variance (that is, closer to the actual value of the variance for the complete population). If this is not used, the variance tends to understate what would actually be the case if many samples (more than 30) were taken [R3.1 pg. 73].

For the 100 samples from Figure 2.1:

$$S^2 = \frac{1}{100-1}\sum_{i=1}^{100}[(1.013 - 1.003)^2 + (1.007 - 1.003)^2 + (.993 - 1.003)^2 + \ldots]$$

$S^2 = 0.001$: the sample variance

The definition of the population variance is:

$$\sigma^2 = \frac{1}{N}\sum(y - \bar{y})^2 \tag{3.4}$$

Note that we use (N), the total population, for this calculation. If you see the Greek symbol σ used in practice, be wary of it—few data sets ever truly contain N. Remember (n) is a sample of data that is a representation of the population. Depending on how it is obtained, it can be a *biased* sample. Inducing noise in an experiment creates a biased set of data. This is essential in Taguchi's approach to robust design and tolerance design, but quite undesirable in a designed experiment being used to generate data for constructing a math model that is representative of the fundamental functional contributions of control factors to the design's performance. In tolerance analysis, we need the effects of manufacturing and customer-use noises present in the sample data so that the tolerances can be optimized in the presence of this variation. If we do not include this noise, the tolerance may be too wide for real situations. As a wise professor I had in college used to say, *true* truth is required to make enlightened decisions. Truth is how a design or process functions when all dimensions and set points are perfectly set at their optimum nominal conditions. *True* truth is what you get when the design parameters are set at their manufacturing tolerance limits and then exposed to debilitating noise factors (see Chapter 15, on worst-case tolerance experiments). One set of data expresses how the design or process functions when the world is nearly perfect; the other expresses what the design or process is really capable of in a worst-case scenario as it progresses through its life cycle.

The Standard Deviation

Another measure of variability in the data is commonly used. The square root of the sample variance is a statistic called the *sample standard deviation,* shown in Equation 3.5 as S.

$$S = \sqrt{S^2} \quad \textbf{(3.5)}$$

The sample standard deviation is the positive *root* mean square deviation of the data from the sample mean. For the 100 samples from Table 3.1, $S = 0.010$.

Figure 3.4 illustrates the dispersion in the data as measured in the units of the standard deviation (plus or minus 1, 2, and 3 standard deviations away from the mean):

Figure 3.4 The standard deviation

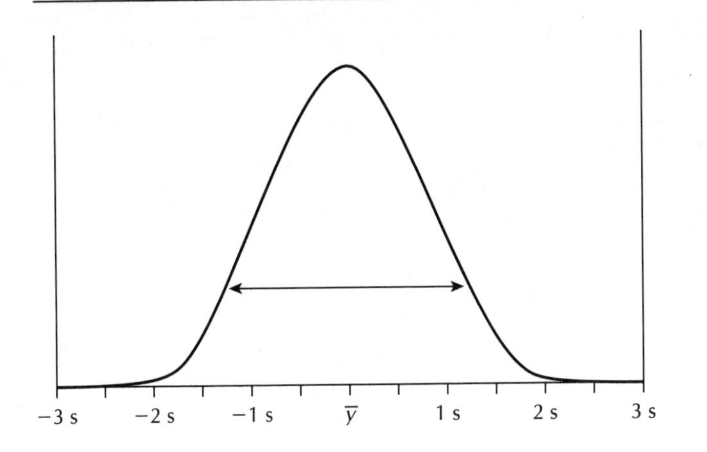

The Third Moment about the Mean: The Skew

The measure of the symmetry of a distribution of data about the mean is referred to as the *skew*. When we take the third moment about the mean we obtain a value indicative of the skew:

$$\mu_3 = \frac{1}{n}\sum_{i=1}^{n}(y_i - \bar{y})^3 \tag{3.6}$$

Skew is then represented by the equation:

$$\text{Skew} = \frac{\mu_3}{s^3} \tag{3.7}$$

where s^3 is the sample standard deviation raised to the third power.

The following equation is commonly used in statistical-analysis computer programs:

$$\text{Skew} = \frac{1}{n}\sum_{i=1}^{n}\left(\frac{y_i - \bar{y}}{s}\right)^3 \tag{3.8}$$

When the distribution of the data is symmetrical about the mean, then the skew is equal to zero. A continuous positively skewed distribution will appear as shown in Figure 3.5.

The skew will assume positive values when the data is biased to the right, and negative values when the data is biased to the left. Skew is also commonly calculated by a simple equation relating the difference between the mean and the median of the data in proportion to the standard deviation:

$$\text{Skew} = \frac{3(\bar{y} - \tilde{y})}{S} \tag{3.9}$$

where $\tilde{y}$ is the median (the physical center value of the data set).

For the 100 random samples the median is:

$$\tilde{y} = 1.001$$

Figure 3.5 Skewed distribution

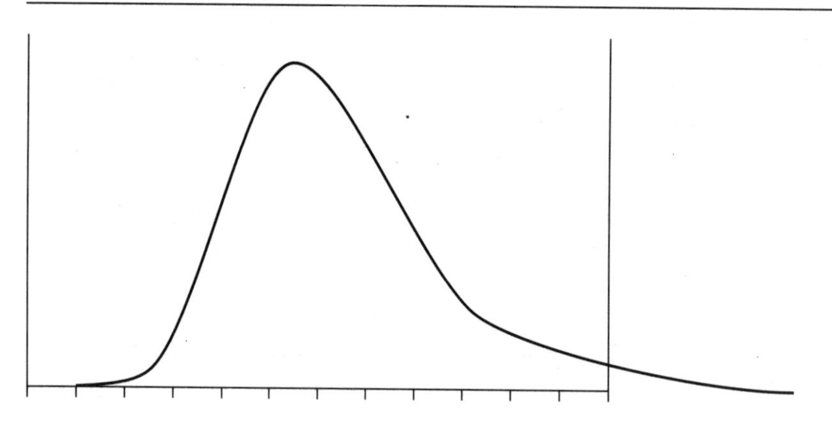

The skew value will take on meaning when the mean and median from a data set are not the same number. In our case the mean = 1.001, the median = 1.001, and $S = 0.010$.

$$\text{Skew} = \frac{3(1.001 - 1.001)}{0.010} = 0$$

So, we see that our data has no measurable skew when the median method is used. When we use the method many statistical packages in the PC world use, the formula does produce a measurable skew:

$$\text{Skew} = \frac{1}{100}\sum_{i=1}^{n}\left(\frac{y_i - \bar{y}}{s}\right)^3 = +0.181$$

This method is more sensitive than the median method and therefore may be preferred when assessing data from critical technologies or processes. The skew is still quite small and suggests we are looking at data from a normal distribution.

The Fourth Moment about the Mean: The Kurtosis

The fourth moment will provide a numerical value associated with the peakedness or flatness of the data as it is distributed about the mean, also known as kurtosis. The following equation incorporates the fourth moment about the mean and the fourth power of the sample standard deviation to measure kurtosis.

$$\mu_4 = \frac{1}{n}\sum_{i=1}^{n}(y_i - \bar{y})^4 \tag{3.10}$$

$$\text{Kurtosis} = \frac{\mu_4}{s^4} \tag{3.11}$$

The following equation is commonly used to calculate a zero-based kurtosis in statistical-analysis computer programs:

$$\text{zero-based kurtosis} = \frac{1}{n}\sum_{i=1}^{n}\left(\frac{y_i - \bar{y}}{s}\right)^4 - 3 \tag{3.12}$$

Note that the value of 3 is subtracted from the kurtosis value. This forces the value to be zero-based, as opposed to being centered around the number 3. The common approach to quantifying kurtosis is that a normally peaked distribution is centered about the value of 3. As the kurtosis deviates above or below 3, the peakedness or flatness begins to take on numerical significance as described below.

There are three general distribution types used to define the nature of kurtosis. The first is *mesokurtic distribution,* shown in Figure 3.6. In it the data is normally distributed about the mean. The kurtosis will equal 3.0.

The second is *platykurtic distribution,* shown in Figure 3.7. In it the data is dispersed about the mean in a manner that is flat in nature; the kurtosis will be less than 3.0.

The third is *leptokurtic distribution,* shown in Figure 3.8. In it the data is dispersed about the mean in a manner that is very peaked in nature; the kurtosis will be greater than 3.0.

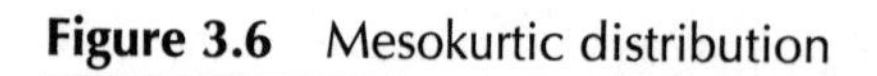

Figure 3.6 Mesokurtic distribution

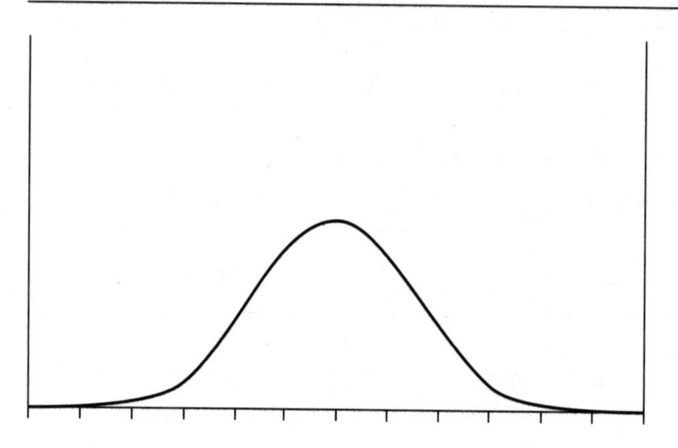

Figure 3.7 Platykurtic distribution

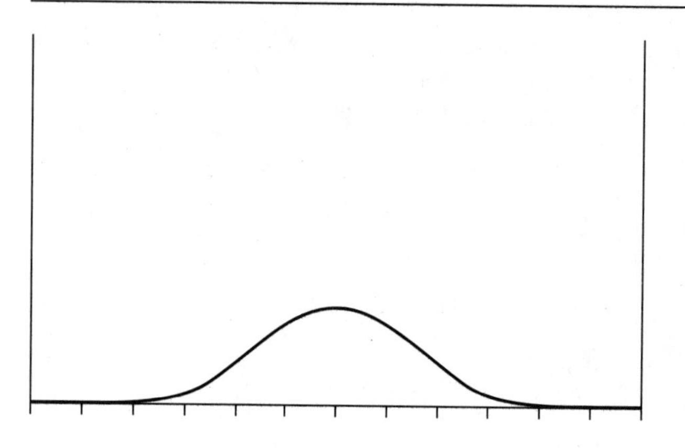

Figure 3.8 Leptokurtic distribution

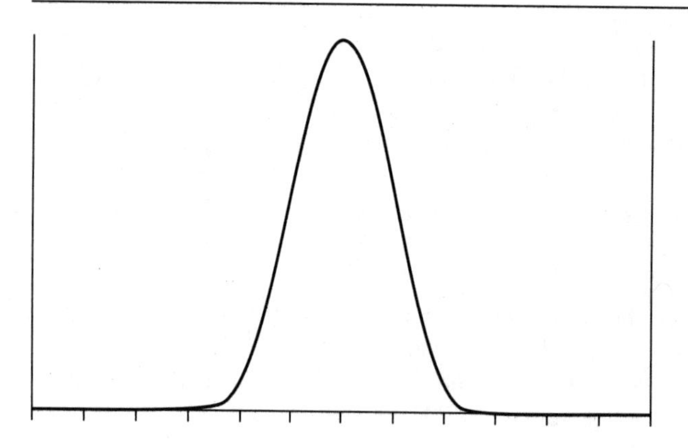

The standard value of kurtosis using Equation 3.11 would be +2.554, indicating the sample is reasonably representative of a normal, or mesokurtic, distribution. There is some slight flattening or extending of the tails of the distribution evident in the kurtosis. For the 100 samples, the zero-based computer value for kurtosis is: −0.446. Please watch out for adjustments like this in common statistical software packages you may be using for your data analysis.

Often a simple rule of thumb is applied to measures of kurtosis falling between 2.5 and 3.5. When a distribution falls in this range it is generally appropriate to assume it is reasonably normal. Values outside of this range are likely to be indicative of a *nonnormal* distribution. A word of caution is in order: small samples (less than 30) can be a problem in measuring kurtosis since they are limited in their ability to be truly representative of the population distribution. When it is necessary to be quite sure the distribution is reasonably normal, try to obtain 30 or more samples when calculating the kurtosis.

The Coefficient of Variation

When the nature of the design requires a response that can have variability on either side of a target value, the mean and the standard deviation can have a meaningful relationship to each other that can be expressed numerically as a ratio called the coefficient of variation (COV):

$$\text{COV} = \frac{S}{\bar{y}}(100) \tag{3.13}$$

The COV expresses the variation in the data as a percentage of the mean value. This permits one to say, for example, that the variation may be 1, 10, or 50 percent of the mean response being measured. This highlights the relative size of the variation with respect to the mean. Knowing this relationship enhances the engineer's understanding of the nature of the variability taking place in the experiments. In this case, the COV is

$$\text{COV} = \frac{.010}{1.001}(100) = .999$$

Which of the two measures of variation, S^2 and COV, should be used is an engineering decision that depends upon the nature of the system being optimized. It is usually helpful to examine the nature of the relationship between variability and the mean.

If the COV is *constant* as the mean rises and falls in value for a particular factor within the experimental design being studied, the standard deviation is proportional to the mean. This means that optimizing the system based on only the variance may be in conflict with putting the mean value onto the desired target for a particular factor. This can greatly complicate the analysis and can be avoided by optimizing the COV, which is independent for certain factors that can be used for adjustment of the mean. This is a key strategy in parameter design.

If the COV is *changing* as the mean increases or decreases in value during design-space experimentation, then another quality metric, such as S^2, should be used to evaluate the system's quality. The point here is that what type of data should be measured and how that data should be analyzed is entirely an engineering decision. Typically in tolerance design and tolerance analysis the variance is sufficient to enable the analysis process.

The Use of Distributions

The Use of Distributions in Taguchi's Parameter and Tolerance Design

Just the few specific statistical quantities, given in equations 3.1 through 3.4, are needed to define the metrics used for analyzing the robustness of a product or process. Understanding these fundamental statistical measures is a requirement for properly applying the methods of parameter design. Much the same is true for applications of orthogonal-array-based designed experiments with stressful noises included for the study of tolerance effects on overall design performance.

Most things have a definite pattern to their variability. The specific shape of the distribution of a data sample dictates how the data can and should be analyzed. Many products are made up of components that vary randomly. Their distribution approximates what is called the *normal* or *Gaussian* distribution, shown in Figure 3.3.

The distribution shown in Figure 3.9 is the *standardized* normal distribution. This simply means the average is located at zero and the data are distributed about the mean in standard deviation units. We will make use of this standardized normal distribution later, when estimates of component quality levels are being made. This pattern can represent the nature of how manufacturing variability affects certain part dimensions. It can also represent how external sources of customer use and external environmental conditions can affect the variation in a product's performance. There are numerous other types of distributions found in nature. It is important for you to know they exist; though their description is not directly relevant to parameter and experimental tolerance design, they can be quite important to analytical methods for tolerance analysis [R3.4].

The two basic sample statistics ($\bar{y}$ and S^2) used to describe the distribution's central location and width are fundamental to quantifying variability. In robust design it is not necessary to assume that the data are coming from a normal distribution, or even from a reasonable approximation of a normal distribution. The simple statistics discussed here are particularly useful because they are good measures regardless of the exact distribution shape. In fact, through the application of noise factors to purposefully induce variation, it is common to apply these statistics to nonnormal distributions. Care must be taken in this area of representative distribution characteristics, especially for the topic of computer-

Figure 3.9 The standardized normal distribution

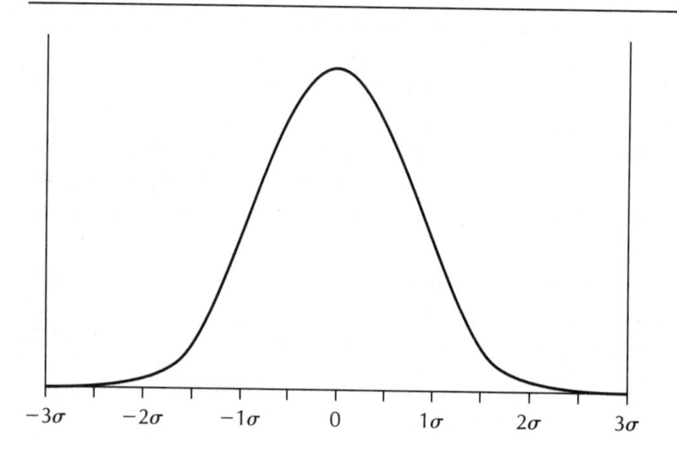

aided statistical tolerance analysis. Monte Carlo simulations may be very dependent on the distribution characteristics of an input variable (see Chapter 8).

The Use of Distributions in Traditional Tolerance Analysis

When a distribution is normal or approximately normal many forms of analysis are available for routine application in an engineering context. When distributions are not normal, special methods are often required to properly analyze the data and apply the results in a traditional tolerance-analysis context. It is important to assess the relative normality of a set of data. We can do this using the mean, variance, skew, and kurtosis, along with a few simple process steps. A distribution's descriptors become important when estimates are made to predict how many parts per million (ppm) will likely be beyond the tolerance limits from a manufacturing process.

For a more complete discussion of the issues of the risk associated with assuming the distribution of data approximates a normal distribution, see *Statistics for Experimenters,* by G. E. P. Box, et al. [R3.5]. The chapter on computer-aided tolerance analysis contains a detailed section on the use of many types of distributions in the context of Monte Carlo simulation.

When a data set is *properly* assumed or proven to be normal (consult a basic statistics text for additional information on these procedures), we can perform six-sigma-quality (or any sigma we desire) analysis using a mathematical transform called the Z transform. When data is not from a normal distribution, a different process called the *adjusted Z transform* can be used to allow the estimation to be made. We must remember that small samples of data (less than 30) probably are not sufficient to properly assess normality, and in fact may not be from a standard normal distribution. When this is the case we must use a different sampling distribution, still assuming a degree of normality in the data that is capable of quantifying the risk assessment being made in working with small sample sizes. This distribution is called the student-t distribution.

Introduction to the Fundamentals of Inferential Statistics

The Z Transformation Process

The Z transformation is a mathematical calculation that is performed on large samples ($n > 30$) of data. It is theoretically based on a well-known statistical theorem called the central limit theorem, which states that for large samples, the sampling distribution of the mean can be approximated closely with a normal distribution [R3.6].

In the case of the central limit theorem we are taking samples of size n from a population and calculating their mean. This sampling and averaging is repeated numerous times to produce a set of averages, which are in turn averaged to estimate μ. The standard deviation of that average of the averages is called the standard error of the mean

$$\left(\frac{\sigma}{\sqrt{n}}\right).$$

The standard expression of the Z transform is:

$$Z = \frac{(x - \mu)}{\frac{\sigma}{\sqrt{n}}} \tag{3.14}$$

In Equation 3.14, x is a sample mean being subtracted from the estimated population mean μ. This difference is then divided, or *normalized,* by the standard error of the mean; n is the number of sample averages taken. Analysts use the Z transform to convert into standard units estimates of population data that have been summarized into population statistics μ and σ. This means that for a data set that provides a distribution with an average and a standard deviation that fit within the context of a normal distribution, we can convert the mean to equal zero ($\mu = 0$) and the standard deviation to equal one ($\sigma = 1$).

In most cases the assumption is made that samples greater than 30 are sufficient to allow the Z transform to be used. Obviously 50 or 100 would be better because one is working with more of the actual population. The closer one gets to including the entire population in the data set the closer one is to knowing the *true* mean and standard deviation. Doing this transformation allows one to use the area under the curve for a standard Z or normal distribution to define the *likelihood of the occurrence,* or the probability, of an event. In our case, the event is whether a part is within or outside the tolerances. The total area under the Z curve is equal to one, or 100 percent. Technically, it is called a *probability density function,* and is defined mathematically as:

$$f(x) = \frac{1}{\sigma\sqrt{2\pi}} \exp\left[-\frac{1}{2}\left(\frac{x-\mu}{\sigma}\right)^2\right], \quad -\infty < x < \infty \tag{3.15}$$

where x is a continuous random variable such as a part dimension from a manufacturing process. The population mean is μ and the population standard deviation is σ; μ is often estimated by applying the central limit theorem through the gathering of many samples and working with the average of the sample averages. Figure 3.10 is the graphical representation of the normal distribution defined by Equation 3.15.

Figure 3.10 The normal distribution

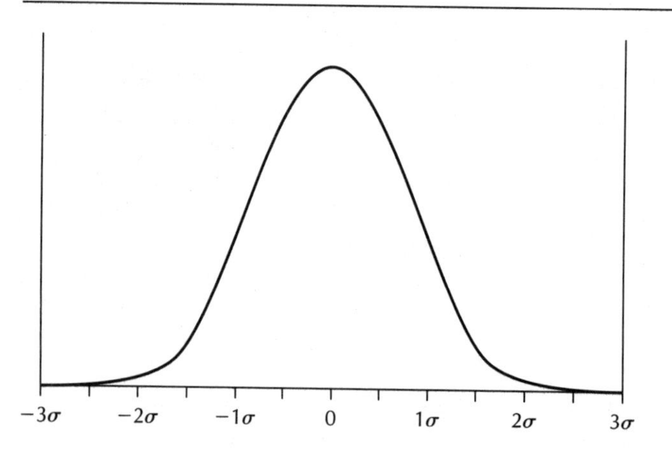

In this symmetrical distribution the variation is easy to quantify in terms of the percentage of the area that will occur between one, two, and three standard deviations from the mean, μ. In Z transform units, $Z = k\sigma$. Thus, when $Z = 1\sigma$, the data point is one standard deviation away from the mean. The amount of area contained between plus and minus one standard deviation ($Z = \pm 1$) is equal to 68.26 percent of the total area under the standard normal curve. The area contained between plus and minus two standard deviations ($k = \pm 2$) is equal to 95.44 percent of the total area under the normal curve. For plus and minus three standard deviations ($k = \pm 3$) about the mean the area encompasses 99.73 percent of the total area under the Z curve. We will relate these percentages of areas to actual parts. That is to say, out of one million parts, if the manufacturing process is running at a three-sigma level of quality, we can expect 99.73 percent of the parts to be within plus and minus three standard deviations of the total population; that is, 997,300 parts will be within the tolerance limits if they are set at the three-sigma level. There will be 2,700 parts out of the specification boundaries (half will occur on each side of the tolerance boundaries). Remember, much of modern Western industry works to three-sigma tolerance standards today. Customer expectations are forcing this to change, as Motorola's six-sigma thrust has shown.

The Z transformation is very useful in allowing us to apply the normal distribution to any set of data, assuming it does follow a normal distribution, to determine the number of parts that are going to fall out of the tolerance boundaries. In this use it can tell us only how good or bad the news really is. We must use engineering judgment, analysis, and a disciplined process to react to the poor sigma performance when necessary to improve quality.

For tolerance analysis, the Z transform works as follows. Let's say we have a sample value, or estimate of μ, of an output dimension from a manufacturing process that is calculated to be 1.500 inches in diameter. What percentage of parts will be unacceptable if we use an *upper tolerance limit* of 1.506 inches? The standard deviation, σ, about μ is computed to be .002. The Z transform tells us that 1.506 inches is three standard deviations away from the mean of 1.500. We are using a modified version of the Z transform that recognizes that we are working with a single large sample of data ($n = 1$). Thus, we use a specific large sample value to estimate μ and the population standard deviation (σ).

$$Z = \frac{(x - \mu)}{\sigma}$$

$$Z = \frac{(1.506 - 1.500)}{.002}$$

$$Z = 3\sigma \tag{3.16}$$

We can rewrite Equation 3.16 to reflect either the upper or lower tolerance limit:

$$Z = \frac{\text{USL or LSL} - \mu)}{\sigma} \tag{3.17}$$

This allows us to relate the location of the mean to the tolerances to determine just how many parts are going to fall outside of the tolerances.

For the lower spec limit the Z values will be negative numbers. This is fine since the standard table of Z values (see Appendix A) is set up to accommodate values of $Z = -4.90$ to $+4.90$, and provide a corresponding range of probabilities from .0002 to .9997. For positive or negative numbers, we use the symmetry of the normal distribution directly to calculate the number of parts that will fall beyond the tolerance limits. If a negative Z number is calculated, its corresponding probability value, from the table, can easily be located to identify the percent out of specification parts that are beyond the tolerance limit (see the instructions in Appendix A).

The Student-t Transformation Process

When one is working with small samples of data (less than 30) it is usually considered good practice to shift from using a standard Z table to the standard-t table (see Appendix A). Here, one will find that it is necessary to select a level of confidence or statistical precision around the estimated t value (remember: t is representative of the number of sample standard deviations a value is away from the sample mean).

This selected confidence level is the value that represents how many components will fall within the tolerance limit. Thus, the $1 - \alpha$ values in a t table must be understood to mean the values shown in Table 3.3 (Appendix A).

Now, consider three-sigma quality. We have shown that we have three-sigma quality when 99.73 percent of the components are within specifications (0.27 percent of components are outside of the tolerance limits). Small sample sizes greatly limit the predictability of quality levels. This is one reason why the student-t distribution is not used in tolerance analysis: It is not very useful when the starting point for most analyses begins at numerical levels that are beyond the t distribution's reasonable range of applicability.

The following example is used to illustrate this point so the reader understands its limitations. The low confidence-level constraints (99.5 percent) are necessary because there are so few data samples. This process works exactly the same as a Z transform from a calculational point of view.

The standard t transform is as shown in Figure 3.11, which illustrates the relative difference between a standard normal distribution (solid line), which represents the true nature of the population distribution, and the student-t distribution (dotted line), based on just a few samples (approximately five in this case). The student-t distribution takes into account the fact that the distribution will depart from the norm as the sample size diminishes from 30 toward 2. The sample size is used to help define a metric called the *degrees of freedom* (d.f.) for using the t distribution. The degrees of freedom (DOF) are calculated by subtracting 1 from the sample size (DOF $= n - 1$). Notice how a statistical confi-

Table 3.3 Confidence table

$1-\alpha$	α	*Percentage in specification*	*Percentage out of specification*
0.60	0.40	60	40
0.70	0.30	70	30
0.80	0.20	80	20
0.90	0.10	90	10
0.95	0.05	95	5
0.975	0.025	97.5	2.5
0.99	0.010	99	1
0.995	0.005	99.5	0.5

Figure 3.11 Comparison of the Z and t distributions

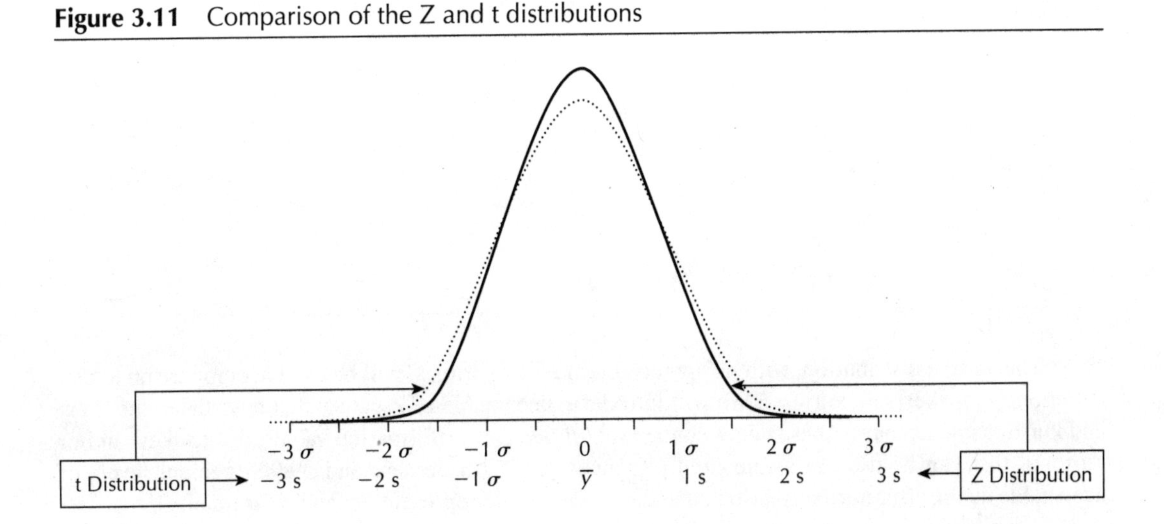

dence level ($t_{.100}$, $t_{.050}$, $t_{.025}$, $t_{.010}$, $t_{.005}$) must also be selected when working with a t table. This is required because one is working with a distribution that is only estimating, with limited probability, that the small sample is actually representing the true population parameters. Remember that a statistic is an estimation of a number, while a parameter is a true expression of what a population is quantitatively doing.

The Adjusted Z-Transformation Process for Working with Nonnormal Distributions

There is a process for working with nonnormally distributed data. That is to say, if the manufacturing process being used to produce components has an output distribution with a large skew and/or a significant kurtosis, then the process shown in figures 3.16 and 3.17 can allow one to determine where to shift the location of the tolerance limits to maintain a certain level of quality. When the distribution of output is nonnormal, the percentage of response data can be shifted in the following ways.

The distribution of response data can be biased to the right or the left due to excessive skew, as shown in Figure 3.12. This means the tolerances are no longer going to be symmetrically located about the mean, and will have to be shifted accordingly.

The distribution of response data can be biased to be pushed out or pulled in due to excessive kurtosis, as shown in Figure 3.13.

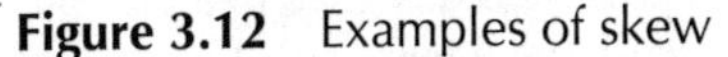

Figure 3.12 Examples of skew

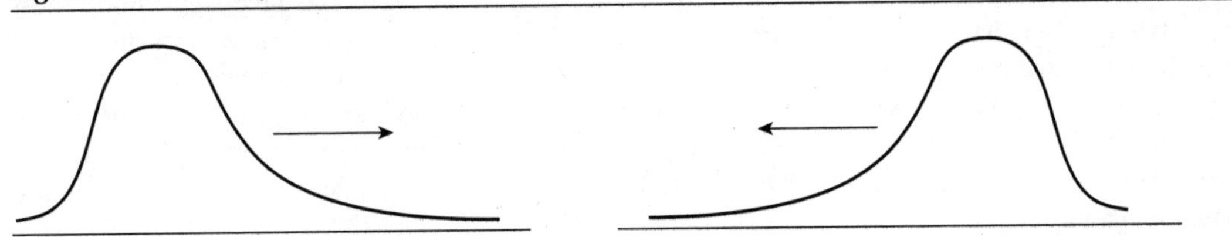

Figure 3.13 Examples of kurtosis

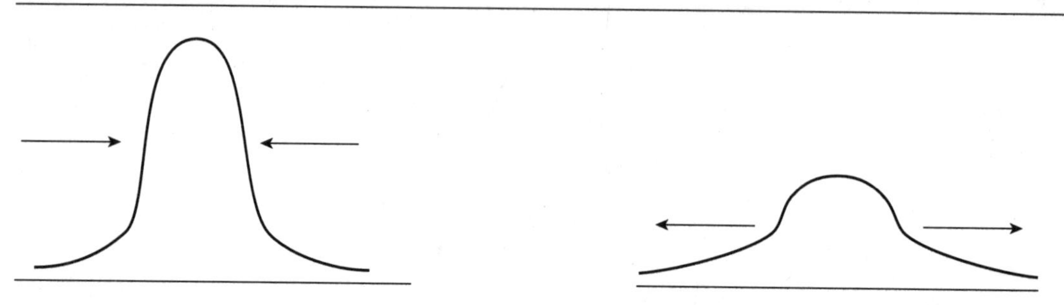

The normal distribution, with no appreciable skew or kurtosis, will be used in comparison to distributions with skew and kurtosis. We will introduce a new table of Z values that have taken the skew and kurtosis into account. These tables contain the *shifted* Z-transformation values, due to skew and/or kurtosis, that can be used to locate limiting values for 99.73 percent and 99.994 percent levels of acceptable quality (for normally distributed data these would represent 3σ and 4.5σ quality).

The tables included in Appendix B [R3.7] contain values for 99.73 percent quality and 99.994 percent quality scenarios.

The adjusted-Z transformation process is useful for aiding in the estimation of tolerance limits when there is clear evidence that the distribution is nonnormal. The process uses the zero-based form of kurtosis. This means if the distribution statistics are computed using the kurtosis calculation that bases no measurable kurtosis at the value of 3, then one must subtract 3 from that value to use the tables included in this text (see Appendix B). Once this is done the process proceeds.

The adjusted Z tables contain two axes: the horizontal axis contains a row of values for increasing levels (−1.0 to 0, and 0 to 1.0) for skew. The vertical axis contains a series of increasing values (−1.2 to 4.0) for kurtosis. There are two tables available for use. The first is calculated for a 99.73 percent quality scenario, where the table values represent the Z values (number of standard deviations from the mean) that represent where the probability values (tolerance limits) of 0.00135 (for the left tail of the distribution) and .99865 (for the right tail) should be *shifted* to due to the effects of skew and/or kurtosis. Figure 3.14 shows a normal distribution with the aforementioned limits prior to shifting due to skew and/or kurtosis.

Remember, Figure 3.14 shows a normal distribution *before* the effects of skew and/or kurtosis. The three sigma limits are located at the symmetrical probability values that state that there is a 0.135 percent probability for a value occurring beyond the lower limit and a 0.135 percent probability for a value occurring beyond the upper limit for a total of .27 percent probability that a value will fall beyond these limits (100 − 0.27 = 99.73). Recall that this simply repeats what was previously described as three-sigma quality for a normally distributed set of performance data from a manufacturing process or design output response where 2,700 parts per million are expected to fall beyond the tolerance limits. This is the situation that the normal Z-transformation process is designed to represent.

When skew is present, the symmetry of the distribution is disrupted. The three-sigma limits must be shifted to new locations to keep the probabilities balanced in proportion to the new, nonsymmetrical distribution of the area under the curve. The adjusted Z table provides new, shifted values that will replace the three sigma values that are present when the skew and kurtosis are very low (normality case).

We will demonstrate how to use the adjusted Z tables by exploring three examples. The first example will illustrate how the adjusted Z table works for zero skew and zero kurtosis (a normal distribution case).

Figure 3.14 Normal distribution labeled with three-sigma limits to be shifted

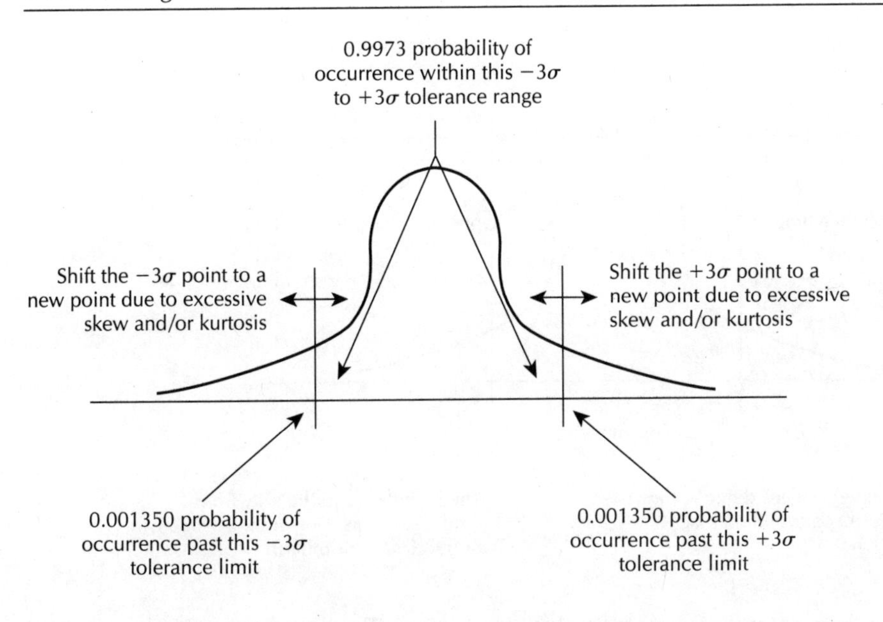

This case will use the adjusted Z table for a 99.73 percent quality level. Thus, we are going to use the table to guide us in estimating where to place tolerances so that no more than 2,700 parts per million will exceed the estimated tolerance limits. This also means that 99.73 percent of the parts will fall within the tolerance limits.

A sample of data is collected and it is found to have the following statistics:

- mean = 3.0
- standard deviation = 1.0
- skew = 0
- kurtosis = 0 (remember we are using zero-based kurtosis to represent normality)

When we look at the adjusted Z tables (must use *both* 99.73 percent tables: one for each tail of the distribution) for skew = 0 and kurtosis = 0, we find the value of −3.0 for the left-tail table and +3.0 for the right-tail table. These are the tolerance limits for this normally distributed case, which means there is a probability of 0.135 percent for a value falling below the lower limit of the tolerance, and a probability of 99.73 percent that the output will fall between the −3.0 and +3.0 limits. Thus, the tolerance range is six units wide (−3.0 to +3.0, with the mean located at 3.0). With 99.865 percent of the probability used up so far, we find that there remains a probability of 100 to 99.865 percent that there will be a value falling beyond the upper-tail tolerance limit. Thus there is a remaining probability of 0.135 percent that a value will surpass the upper tolerance limit. Figure 3.15 illustrates this case.

Now, we can see that the specifications for this case are:

$$\text{lower tolerance} = \text{mean} + (\text{left-tail Z value} \times \sigma) = 3.0 + (-3.0 \times 1) = 0$$

$$\text{upper tolerance} = \text{mean} + (\text{right-tail Z value} \times \sigma) = 3.0 + (+3.0 \times 1) = 6$$

Figure 3.15 The three-sigma quality case for normally distributed data

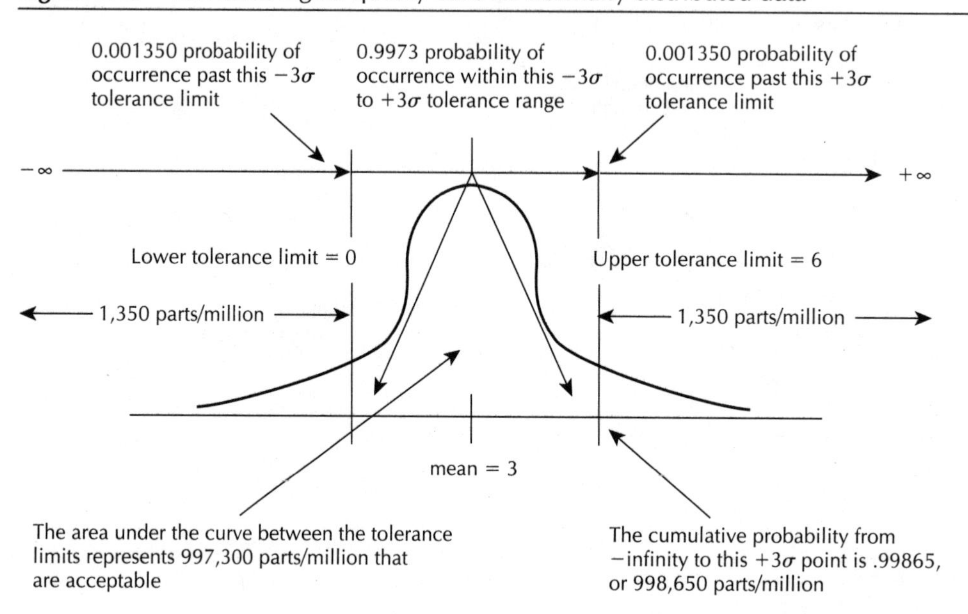

Let's explore a case where there is positive skew and kurtosis. For a case where the skew is +.40 and the kurtosis is +1.4 we find that shifting the tolerance limits takes on the form shown in Figure 3.16.

This shift is accomplished by looking up the values for the positive skew and the positive kurtosis from the 99.73 percent adjusted Z tables (remember to use the tables for the left and right tails of the distribution). In this case the table values for the lower tail with a *positive* skew = 0.40 and a *positive* kurtosis = 1.4 must come from the lower-tail table, which provides the value of −3.100. The same process must be done to find the shifted upper-tail Z value. Again, using a skew of 0.40 and a kurtosis of 1.4 we find the upper-tail value equal to +4.060. This is the shifted Z value that has been moved from 3 to account for the skew and kurtosis.

For the case where the mean is equal to 3, the standard deviation is equal to 1, and now with skew equal to 0.40 and kurtosis equal to 1.4, we find the following new specifications:

$$\text{lower tolerance} = \text{mean} + \text{left-tail Z value} = 3.0 + (-3.1 \times 1) = 0.10$$

$$\text{upper tolerance} = \text{mean} + \text{right-tail Z value} = 3.0 + (+4.06 \times 1) = 7.06$$

As the reader can see, the tolerances have shifted to account for the fact that the distribution was skewed to the right and broadened by the kurtosis. Because both skew and kurtosis effects were present in the new distribution, the tolerance range moved to acquire a higher positive limit. The entire tolerance range broadened, primarily because of the effect of the kurtosis.

The use of the adjusted Z tables for negative skew can be a little confusing. We will carefully demonstrate how to use the 99.73 percent table that accounts for negative skew. For a case where the skew is −.40 and the kurtosis is +1.4 we find the shifting of the limits takes on the form shown in Figure 3.17. This shift is accomplished by looking up the values for the negative skew and the positive kurtosis from the adjusted Z tables that are provided for negative skew (Appendix B).

Figure 3.16 Nonnormal distribution with shifted limits

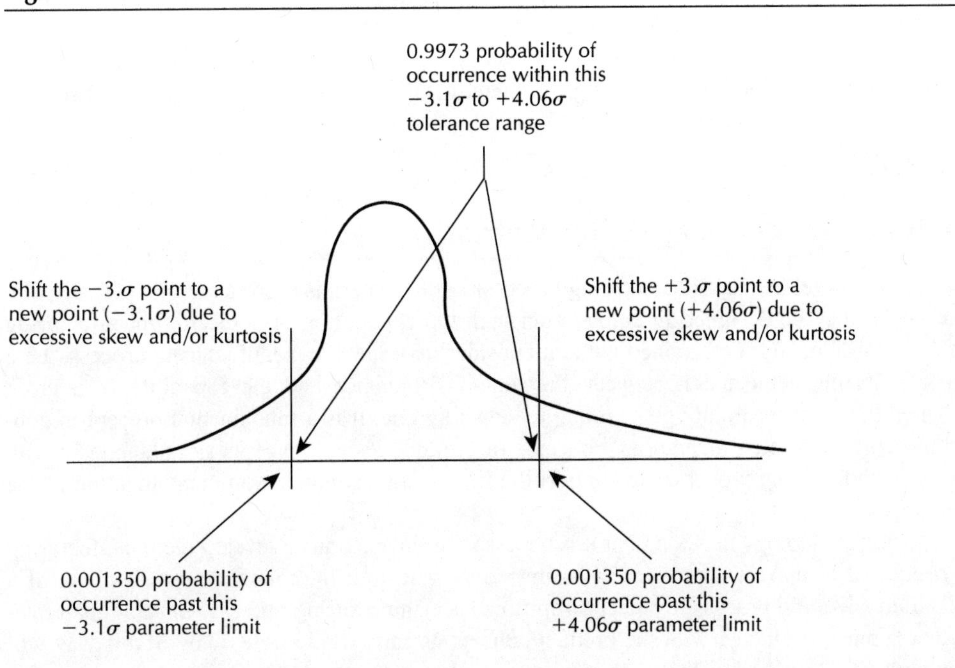

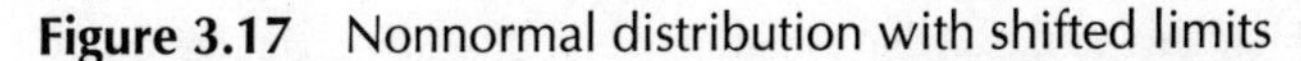

Figure 3.17 Nonnormal distribution with shifted limits

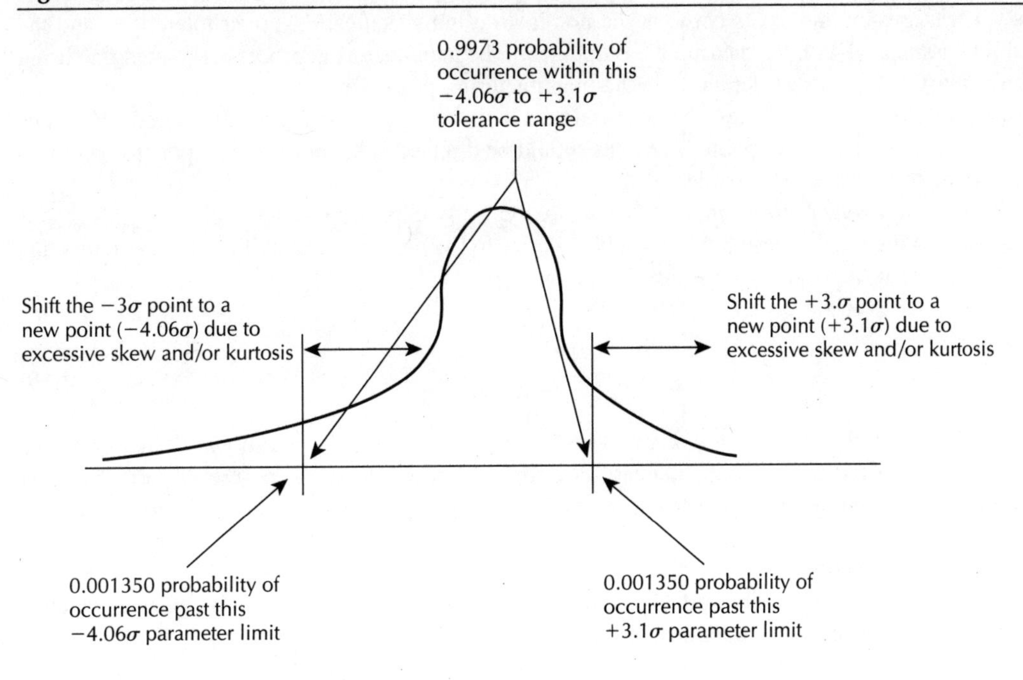

Using the 99.994 percent table is exactly the same as using the 99.73 percent table. We have demonstrated how to handle nonnormal distributions in the context of defining tolerance limits at several levels of quality. Now we must build a knowledge of how to relate process capability, including the measure of the distribution from any given manufacturing process, to the metrics and science of tolerancing.

Manufacturing Process Capability Metrics

Tolerances are always related to manufacturing processes or to materials used in the manufacture of a product, and they must be designed in conjunction with the application of a specific manufacturing process. If a tolerance band is determined without considering a specific manufacturing process, there is great risk in having a mismatch between the required tolerance and the capability of a given process—when the engineer finally gets around to selecting one. It is a fundamental precept in concurrent engineering to develop technology concepts or product-design concepts simultaneously with the necessary manufacturing processes to support the timely and economic commercialization of the desired product.

Often during technology development it is necessary to invent and codevelop the manufacturing technology required to make the product. It is unwise to wait until the tolerance design phase of a product-commercialization process to select or optimize a manufacturing process. Capable manufacturing processes must be aligned with the product concept as early as possible. Only in this way will there be enough time to develop the necessary relationship between tolerances and manufacturing processes. It is also essential to perform manufacturing-process parameter optimization just as one would for design-component parameters.

The design engineer and the manufacturing engineer have to define a common metric that quantifies the relationship that exists between the nominal design specifications, their tolerances, and the variability associated with the measurable output from the manufacturing process. The manufacturing engineer must also provide tolerances on the manufacturing process parameters and raw materials to help the team stay well within the tolerances assigned to the component being manufactured. The focus on manufacturing process set-point tolerances should be directed at keeping the component specifications as close to on-target as possible.

The *manufacturing process capability index,* typically expressed as Cp, Cpk, $Cp_{(\text{upper limit})}$, or $Cp_{(\text{lower limit})}$, is the ratio of design tolerance boundaries to the measured variability of the manufacturing process output response.

Cp is defined arithmetically as follows:

$$\text{Cp} = \frac{\text{USL} - \text{LSL}}{6\sigma} \tag{3.18}$$

USL stands for *upper specification limit,* and LSL stands for *lower specification limit.* The 6σ stands for six times the short-term sample standard deviation of the production measure of part quality in engineering units; the use of σ is really a misapplication of the population parameter for a standard deviation. The true measure of variability most often used in the calculation of Cp is $6s$ (where σ is the sample standard deviation).

Three-sigma quality is achieved when Cp = 1. Figure 3.18 shows a normal distribution, which is on target, in relationship to the tolerance limits for Cp = 1.

Figure 3.18 Distribution for Cp = 1

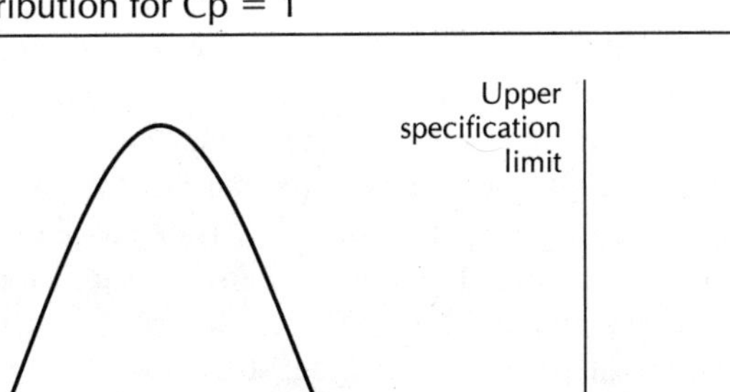

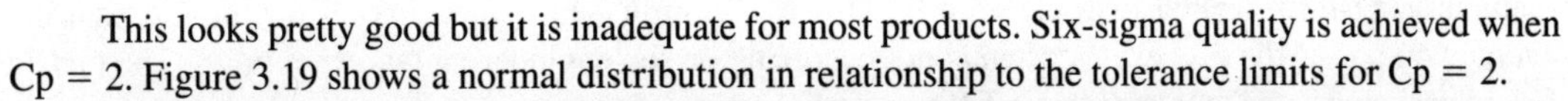

This looks pretty good but it is inadequate for most products. Six-sigma quality is achieved when Cp = 2. Figure 3.19 shows a normal distribution in relationship to the tolerance limits for Cp = 2.

This form of manufacturing process capability (Cp) expresses the unique synergy that must exist between the design-specification limits and the actual short-term variability measured from the output of the manufacturing process. Cp is said to be a short-term metric of process capability because it displays capability from *recent statistical samples* of data. In this sense it is technically wrong to use the population standard deviation (σ) as the measure of variability in the denominator. The sample standard deviation (s) is the proper metric to use in this case. If one has more than 30 samples (50 to 100 is good) from the process then it is reasonable to use σ instead of s, but remember that using σ is still just an *estimate* of the true population standard deviation. An example of establishing Cp for a one-inch target-diameter shaft that is to be held between 1.002 and .998 inches would involve collecting at least 30 samples of the manufactured shafts, measuring the diameters, and then calculating the short-term sample standard deviation.

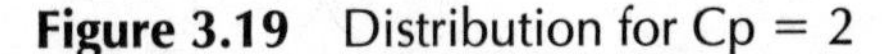

Figure 3.19 Distribution for Cp = 2

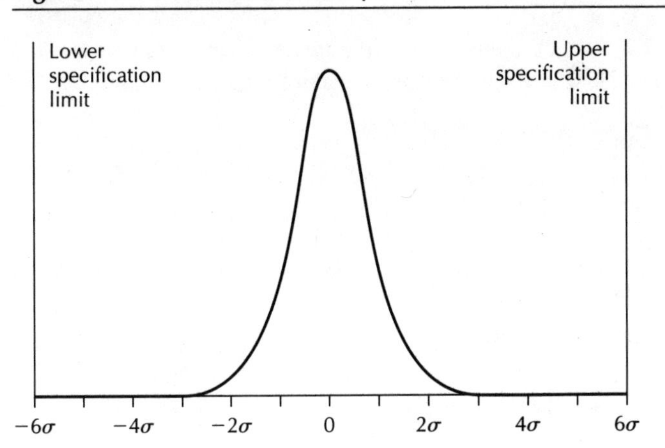

The measure of Cp is heavily influenced by what is required for the product to properly function in comparison to what is possible for the manufacturing process to deliver. Many manufacturing engineers have been forced to tell disappointed designers that they should have consulted with them before designing a concept that simply could not be made by existing manufacturing processes. The rule of good design practice is simple: Do not design in a vacuum—you must first know what is *economically possible* to be made, and then design within those constraints. This is one of the essential practices for designing for six-sigma component performance. To aid in attaining six-sigma quality, the team must use an engineering process that includes manufacturing-process sample standard deviations in the parameter optimization experiments as unit to unit noise. Better still is to know the standard deviations of numerous manufacturing processes when selecting concepts during concept design. Another helpful technique is a predesign analysis, performed during concept design, to minimize tolerance sensitivity prior to parameter design (see example at the end of Chapter 7). In this way one can design to the manufacturing processes that are in existence and that are proved to be economically applicable to the design's intent.

There are many instances in existing manufacturing processes where the process average is not centered on the specified target value. Manufacturing process capability can be expressed in a more realistic, long-term format than Cp affords. This is accomplished by accounting for the fact that the average output of the process, $\bar{y}$, often is not occurring on the engineering target set point. The metric C_{pk} is designed to account for this off-target performance of the average output of the manufacturing process. It is defined as follows:

$$C_{pk} = C_p(1 - k),\ k = \frac{|\bar{y} - T|}{\frac{(USL - LSL)}{2}} \tag{3.19}$$

where $\bar{y}$ is the actual sample mean and T is the target. The value k represents the number of standard deviations the process mean is away from the target T. Typically, the target is taken to be at the center of the tolerance range, and represents the optimum control factor set point from parameter design. Thus:

$$T = \frac{USL + LSL}{2} \tag{3.20}$$

Figure 3.20 illustrates the nature of Cpk.

An equivalent form of Equation 3.20, which applies when the target is centered between the specification limits, is given by Equation 3.21, where $\bar{y}$ is the sample mean for the distribution.

$$C_{pk} = \mathrm{Min}(C_{pu}, C_{pl})$$

where,

$$C_{pu} = \frac{|USL - \bar{y}|}{3s} \text{ and } C_{pl} = \frac{|\bar{y} - LSL|}{3s} \tag{3.21}$$

Equation 3.21 states that the Cpk value is given by the smaller of the two capability indices—C_{pu}, the capability with respect to the upper limit, and C_{pl}, the capability with respect to the lower limit. Thus, the C_{pk} index measures how close to the tolerance limit a process is running.

Figure 3.20 A normal distribution that is off target

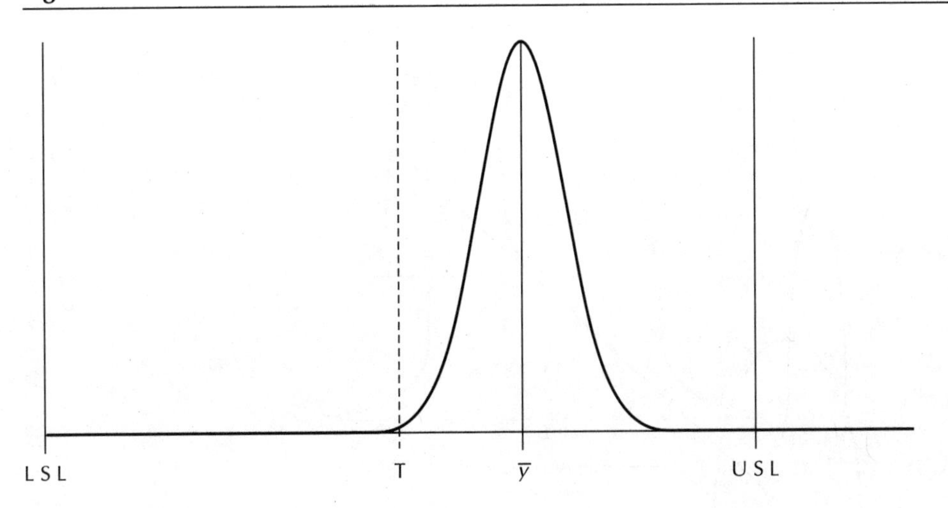

A long-term process capability index can be calculated. In this expression of process capability, the manufacturing process mean response variable, ($\bar{y}$), is found to be shifting around the target specification, (T), along with its associated standard deviation, (s). Thus, over many months of production, numerous process averages and standard deviations can be plotted (see Figure 3.22). When the process average, $\bar{y}$, is equal to the target, T, then $C_p = C_{pk}$.

The sources of variability that are time-based can be accounted for in the expression of the long-term C_{pk} capability index. Motorola University and the Six Sigma Institute have concluded that, on average, the mean of a manufacturing process shifts around by approximately 1.5 standard deviations (k) over an extended period of time. In Equation 3.22 the C_p is adjusted to account for the 1.5 standard deviation shift:

$$C_{pk} = \left(1 - \frac{1.5}{3C_p}\right) C_p \tag{3.22}$$

This enhances and broadens the estimate of the aggregate of the short-term sample standard deviations. In fact, the sample standard deviation takes on the value known as the standard error of the mean. This is developed and quantified in the central limit theorem, which, as previously mentioned, states that the mean value of a group of averages, even from nonnormal sample distributions, will produce a normal or Gaussian distribution defined by the mean of the averages, $\bar{\bar{y}}$, and the average variation about this mean, the standard error of the mean. The standard error of the mean accounts for the effect of the noises present in the manufacturing process due to external sources, unit-to-unit sources, and deterioration sources of variability. These sources of variation must be recognized and included in the manufacturing-process capability index. Figure 3.21 displays the impact of short-term versus long-term process capability.

Each month has its own short-term average and standard deviation; it is these averages that are averaged over the long term. Accounting for long-term process variability is an important task if one is to attempt to attain six-sigma component quality.

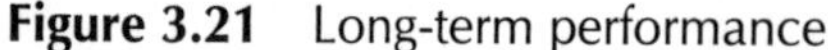

Figure 3.21 Long-term performance

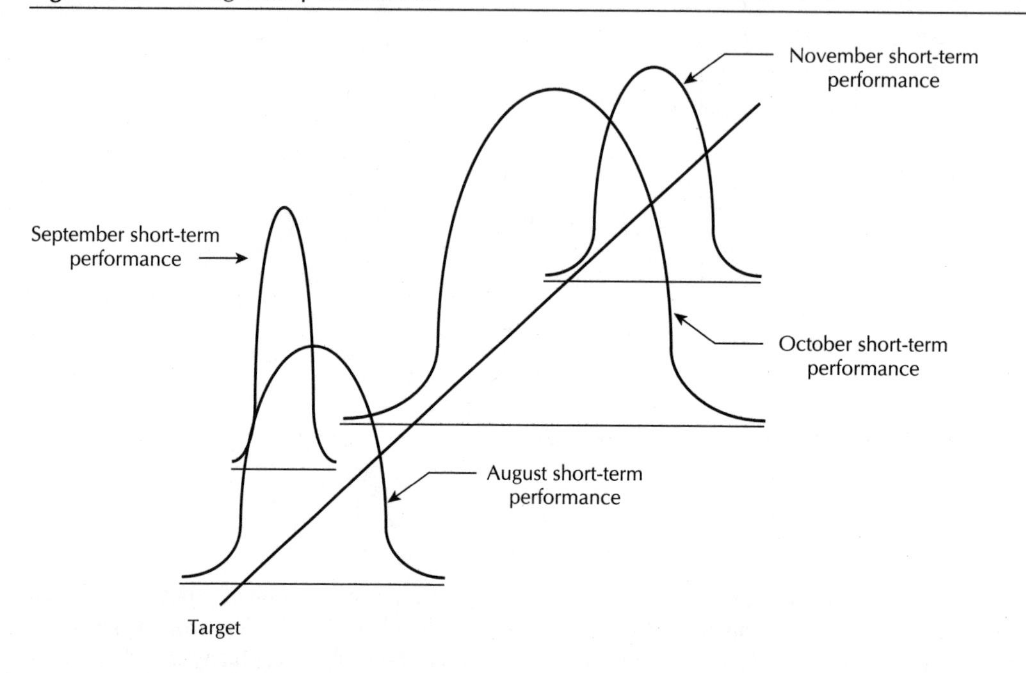

Six-Sigma Process Metrics [R3.8]

The methods of six sigma, made popular by Motorola, illustrate the practical ramifications that follow from quantifying component-part quality in terms of process capability indices. The resultant part-per-million defects can be calculated based on a well-characterized process capability.

There are several metrics in six sigma that are closely related to C_{pk}. The first one we will consider is DPMO, defects per million opportunities. This is found by calculating the probability, *p,* that a process will exceed the tolerance limits. The probability is given by the area under the normalized probability distribution curve that lies outside the tolerance limits, as shown in Figure 3.22. Recall that this probability was 0.00135 for each of the tails for a normally distributed set of data.

Assuming a normal distribution, DPMO is a simple function of the capability index. The probability, *p,* and DPMO are related by Equation 3.23.

$$p = \frac{\text{DPMO}}{10^6} \tag{3.23}$$

As previously stated, if the process is on target, then $C_p = C_{pk}$. Typically, however, the process mean will wander about somewhat over time. This is due to long-term drift, special causes, and the common causes within the control limits of statistical process control. Experience has shown that if the short-term process variability is σ, then the long-term variation of the mean about the target is typically 1.5σ. The resulting C_{pk} was defined in Equation 3.22.

$$C_{pk} = \left(1 - \frac{1.5}{3C_p}\right) C_p$$

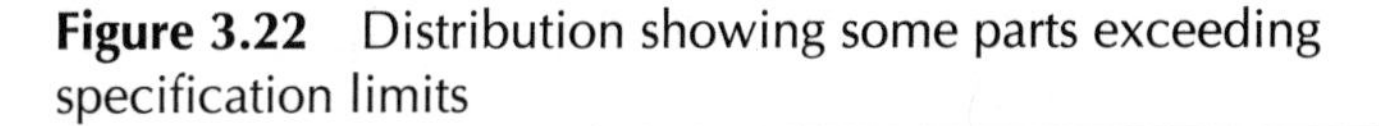

Figure 3.22 Distribution showing some parts exceeding specification limits

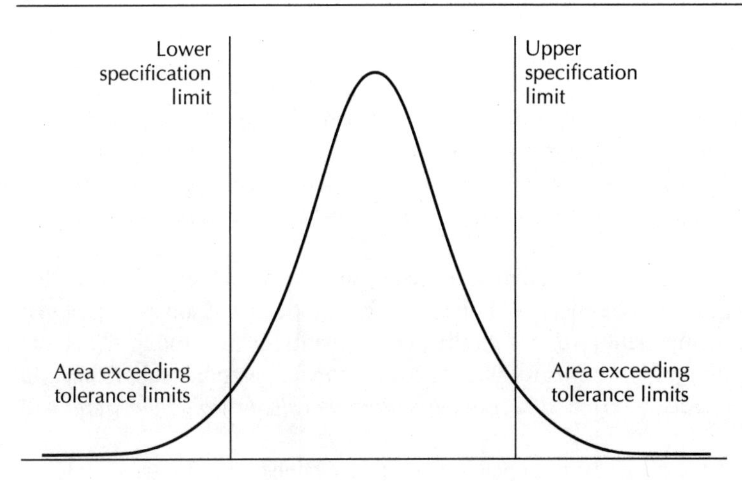

This results in a higher probability of defects than would be expected before accounting for this drift. Table 3.4 shows the expected defects per million opportunities at different quality or distribution widths (σ levels) as a function of C_p, assuming on-target performance, and C_{pk}, assuming 1.5σ drift around the target.

If we assume that there are n items in a complex system, that the probability of failure, p, is the same for each of them, and that there is no correlation between the individual component failures, then the average number of defects in the system, or defects per unit (DPU), is given by Equation 3.24.

$$\text{DPU} = \text{np} \qquad \textbf{(3.24)}$$

The probability, P{q}, that a complex system consisting of n components will have q defects is given by the Poisson distribution, shown in Equation 3.25.

$$\text{P}\{\text{q}\} = \frac{(\text{DPU})^{\text{q}}\, e^{-\text{DPU}}}{q!} \qquad \textbf{(3.25)}$$

Table 3.4 DPMO versus Cp and Cpk (with and without the mean drifting)

	Mean on target		*1.5σ drift around target*	
Quality	*Cp*	*DPMO*	*Cpk*	*DPMO*
3σ	1	2700	0.5	66,800
4σ	1.33	63	0.83	6210
5σ	1.67	0.57	1.17	233
6σ	2.0	0.002	1.5	3.4

The probability of a unit being completely free of defects as manufactured (requiring no rework) is related to the average DPU. It is found by taking the limiting case of $q = 0$ in Equation 3.25. This probability of *zero defects,* referred to as the first time yield (FTY), is given in Equation 3.26.

$$\text{FTY} = e^{-\text{DPU}} \tag{3.26}$$

The preceding equations provide a link between number of defects, yield, probability of defects, and capability indices. Now we have a set of metrics that can tell us how many out-of-specification parts can be expected when a mismatch occurs between tolerances and manufacturing variability. The effect of the number of component parts in a product (complexity) can be included in this metric, as shown in Figure 3.23.

Here we can clearly see that attaining six-sigma performance at the product level for complex designs is very difficult, if not impossible. What is really important is to notice the dramatic effect that attaining six-sigma quality at the component level has on the overall quality of the product. This text emphatically supports the application of six-sigma techniques to component parts and subsystems. In this way the aggregate beneficial effects, referred to in six-sigma terms as *rolled throughput yield,* will be maximized.

It is not difficult to estimate the cost of unacceptable parts to a business. In this era of intense focus on customer satisfaction and competition for market share, quantifying the cost of part quality with respect to manufacturing tolerances and capability is only a portion of the information needed to design correctly. These metrics are too inwardly focused on the cost structure of the product with respect to the business's profit. Today, a business cost structure is inextricably linked to the customer's cost structure. Any industry that is out of touch with its customer's cost of product ownership is already laying the foundation for failure. One cannot tolerance effectively in the current (and future) economic environment based solely on process capability indices or six-sigma goals. We need to align these metrics with a customer-focused set of metrics. A more sensitive and economically comprehensive metric is required to tell us where to accurately set our tolerances. Such a metric exists—it is called the quality-loss function.

Figure 3.23 Overall yield versus sigma level

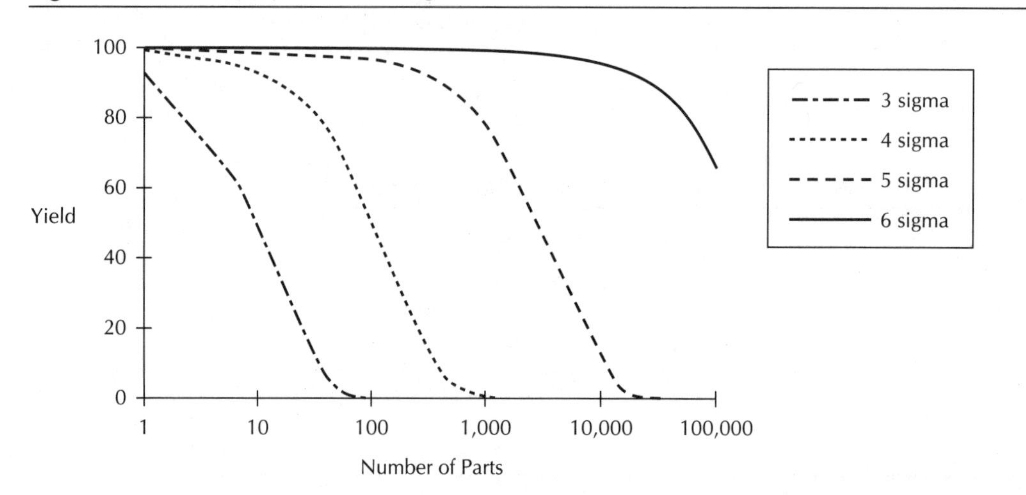

The Relationship between the Quality-Loss Function, C_p, and C_{pk}

Tolerances must be linked to more than the variability that originates in the manufacturing environment; they must have some basis in the costs that are incurred to make the product. These costs are primarily represented by the term *unit manufacturing cost* (UMC). Tolerances must further be developed in the context of two more costs.

The *life-cycle cost* (LCC) of the design will account for the broader—that is, repair and replacement—costs associated with the use of the product. This metric is particularly important in industries that must repair and service the products they sell to satisfy customer expectations. Think of it this way: You own a car dealership and you sell cars with a guarantee on parts for the first 30,000 miles. If you are paying for the repair costs, including replacement parts, your interest in part quality is going to be quite different than if the customer were bearing the costs. Who pays for off-target performance, as well as outright product failure, and how often, is a major component of modern business dynamics. LCC is a good economic metric to aid the engineering team in understanding the ramifications of tolerances on the internal costs of the business and the effect of service issues on the customer.

The cost to the customer, when a product's performance drifts off target (it does not conform to advertised features) or fails altogether, must play an equal role in the quantification of the cost of quality. Here, we add the quality-loss function as the third, balancing leg of the economic model that drives our engineering decision-making process in tolerancing (Figure 3.24).

The quality-loss function is defined as:

$$L(y) = \frac{A_0}{(\Delta_0)^2}[y_i - T]^2 \tag{3.27}$$

$L(y)$ is the loss in dollars due to off-target functional performance $(y_i - T)$. A_0 is the customer loss related to Δ_0, a physical measure of the intolerable *functional performance deviation* from the target T. The value y_i is the actual measured deviation from the target in engineering units. If y_i reaches Δ_0, then A_0 dollars are spent to react to the off-target performance. Δ_0 is typically established by assessing when approximately 50 percent of the customers will be motivated to take economic action due to the off-target performance.

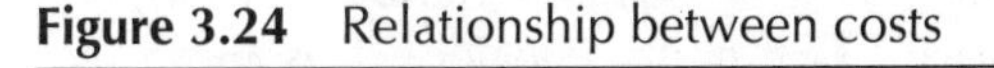

Figure 3.24 Relationship between costs

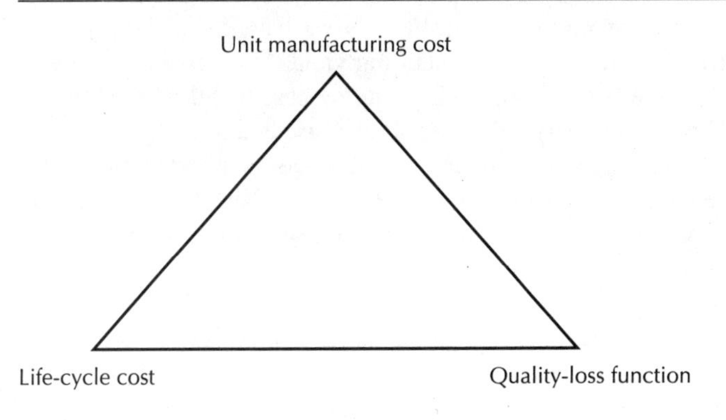

The loss function shown in Equation 3.27 is used for a single product; a more realistic expression for quality loss would include a sample of numerous products used under a variety of circumstances.

The average quality loss over a number of units is given by:

$$L(\bar{y}) = \frac{A_0}{(\Delta_0)^2}[(\bar{y} - T)^2 + (s)^2] \tag{3.28}$$

where $\bar{y}$ is the average value of the off-target performance of the samples and s is the standard deviation of the sample of products. Linking customer economic loss to the engineering metrics of average deviation from the target and the variance of performance is the key strength of the quality-loss function. The anatomy of the loss function is divided into three key values:

1. The ratio of customer loss to the customer tolerance limit for an engineering parameter, expressed by

$$\frac{A_0}{(\Delta_0)^2}$$

2. The square of the difference between the engineering parameter and the target value for that parameter, expressed by

$$(\bar{y} - T)^2$$

3. The variance of the engineering parameter relative to its measured mean, $\bar{y}$, expressed by

$$(s)^2$$

When these economic and engineering metrics are combined in the quality-loss function, the engineer can co-optimize the design for quality and cost, because all of the dynamic elements for this co-optimization are resident in the loss function.

The C_{pk} metric shares some properties with the average quadratic-loss function. Equation 3.29 shows the quadratic loss function using notation consistent with the short-term process capability indices.

$$L(\bar{y}) = \frac{A_0}{[(\text{USL} - \text{LSL})/2]^2}[(\bar{y} - T)^2 + s^2] \tag{3.29}$$

The key distinguishing characteristic of the loss function is the explicit inclusion of A_0, the customer dollar loss incurred at the functional limits (USL and LSL in this notation). This allows *customer financial loss* to be the driving force for improving quality. As we have seen, the power of six-sigma derives from a series of metrics that relate to manufacturing yield (freedom from defects). Through Equation 3.29 we can build a bridge between an engineering process based on quality loss function and one based on six sigma. This will be seen in the case study presented in Chapter 17.

If we take long-term process variability, C_{pk}, into account, then the values of $\bar{y}$ and s^2 must come from samples that are appropriately taken over reasonably long periods of time. The process capability referred to here is the design process capability. This is the ratio of the design output response tolerances to six standard deviations of the measured design output response.

Many companies today are focusing on the cost of quality. The actual accounting of the costs associated with the effects of poor quality, in a manufacturing context, are fairly easy to isolate and explain. There are few manufacturers that cannot tell you, in detail, their scrap rate, rework labor expenses, process yields, defects per unit, warranty costs, returns from the retail market, and the aggregate dollar amounts associated with such metrics. Most manufacturing firms are hard-pressed to explain the relationship between the cost of performing their engineering processes and the money saved by getting the quality designed into the product long before it goes into production. In addition, few companies can lay their production blueprints, CAD files, and specification sheets on a table and quantitatively explain how each component was toleranced to provide the maximum benefit to the *customer and the business*. If the tolerances were not developed to minimize both customer losses and internal corporate losses, they are likely contributing negatively to the *real cost of quality*.

The key to understanding the true cost of quality lies in comprehensively accounting for all the sources that contribute to it—both positively and negatively. These sources will be both internal and external to the company—in the customer's economic environment. Proper alignment of tolerances to the business's fundamental capability to make the component parts and assemble them into a robust product is the basis of many subsequent quality issues.

Summary

Chapter 3 contains a broad range of introductory material to prepare the reader for the statistical calculations required for competent tolerance analysis. It also provides an introduction to manufacturing process capability metrics, six-sigma metrics, and the Taguchi loss function. The following outline describes the material contained in Chapter 3.

Chapter 3: Introductory Statistics and Data Analysis for Tolerancing and Tolerance Design

- The Role of Data in Tolerance Analysis
- Graphical Methods of Data Analysis
  - The Frequency Distribution Table
  - The Histogram
- Quantitative Methods of Data Analysis
- Introduction to the Fundamentals of Descriptive Statistics
  - The First Moment of the Data about the Mean: The Arithmetic Average
  - The Second Moment about the Mean: The Variance
- The Standard Deviation
  - The Third Moment about the Mean: The Skew
  - The Fourth Moment about the Mean: The Kurtosis
  - The Coefficient of Variation
- The Use of Distributions
  - The Use of Distributions in Taguchi's Parameter and Tolerance Design
  - The Use of Distributions in Traditional Tolerance Analysis
- Introduction to the Fundamentals of Inferential Statistics
  - The Z Transformation Process
  - The Student-t Transformation Process
  - The Adjusted Z-Transformation Process for Working with Nonnormal Distributions

- Manufacturing Process Capability Metrics
- Six-Sigma Process Metrics
- The Relationship between the Quality-Loss Function, Cp, and Cpk

Each of the topics covered in this chapter is important for the process of analytically or experimentally developing tolerances. Chapter 3 is designed to serve as a reference until the reader becomes completely proficient in these topics.

Robust design and tolerance design procedures use experimentation to produce data that define the distribution width of key performance characteristics while noise is active during the experiment. The engineer's job is to discover ways to narrow these distributions and put them on target for the design or process being optimized. Thus, the statistical techniques for describing data distributions are equally important concepts for robust design and tolerance design. These include graphical methods such as the histogram, but much more important are the quantitative descriptions—that is, the mean, variance, standard deviation, skew, kurtosis, and coefficient of variation. These are summarized in Table 3.5. Quantitative statistics, which are calculated from sample data, are often obtained from the application of designed experiments to the engineering optimization process. The responses that will be analyzed, especially the loss function and signal-to-noise ratio, are based upon these fundamental statistical quantities. The measurement of samples of data coming from manufacturing processes will also require these statistics to be calculated to help in the definition of the nature of the manufacturing process capability.

This chapter introduces the concept of statistical distributions. Chapter 8 contains a wealth of information on the application of many of the major distributions that are available to be used in statistical computer-aided tolerance analysis.

In concluding this chapter on data and how one can begin to process it, a few words are in order on the general strategy engineering practitioners must employ in producing useful data. When con-

Table 3.5 Formulas for sample summary statistics

Name	*Symbol*	*Formula*
Mean	$\bar{y}$	$= \frac{1}{n}\sum_{i=1}^{n} y_i$
Variance	$\mu_2 = S^2$	$= \frac{1}{n-1}\sum_{i=1}^{n} (y_i - \bar{y})^2$
Standard Deviation	S	$= \sqrt{S^2}$
Skew	$\frac{\mu_3}{S^3}$	$= \frac{1}{n}\sum_{i=1}^{n} \left(\frac{y_i - \bar{y}}{s}\right)^3$
Kurtosis	$\frac{\mu_4}{S^4}$	$= \frac{1}{n}\sum_{i=1}^{n} \left(\frac{y_i - \bar{y}}{s}\right)^4$
Coefficient of Variation	COV	$= \frac{S}{\bar{y}}(100)$

ducting experiments to gather data, how variability is allowed or forced to occur is crucial to effective tolerance optimization. In order to optimize a design or process the same sources of variation that occur during component manufacturing, product assembly and customer use must be present in the tolerance experiments. Taking random samples of data without attempting to stimulate the various types of noises that can occur in real-use conditions is an inadequate practice. In this text, collecting random samples, in the absence of noise, is not a priority when performing product or process tolerance optimization. The focus is on producing and analyzing strategically biased samples of data where the engineer is intentionally forcing variability to occur by controlling and stimulating many kinds of variation during the experimentation. Tolerances that are developed in this context will themselves become supportive of robustness and will be capable of supporting the long-term reliability goals for the product.

References

R3.1: *Modern Elementary Statistics;* J. Freund; Prentice Hall, 1988, pg. 104.

R3.2: *Guide to Quality Control;* K. Ishikawa; Asian Productivity Organization, 1982, pg. 104.

R3.3: *Mathematical Statistics 5 Ed.;* J. E. Freund; Prentice Hall, 1992, pg. 108.

R3.4: *Statistical Distributions 2nd Ed.;* M. Evans, N. Hastings, B. Peacock; J. Wiley & Sons, 1993, pg. 123.

R3.5: *Statistics for Experimenters* (pages 89 – 91 Appendix 3B. Robustness of Some Statistical Procedures); G. E. P. Box, W. G. Hunter & J. S. Hunter; J. Wiley & Sons, 1978, pg. 124.

R3.6: *Modern Elementary Statistics 7th Ed.* (pg. 262); J. E. Freund; Prentice Hall, 1988, pg. 125.

R3.7: *Statistics and Principles of Tolerance Analysis;* M. Hackert-Kroot and J. Nelson; Eastman Kodak Technical Training Manual, 1995, pg. 134.

R3.8: *Six Sigma Producibility Analysis and Process Characterization;* M. J. Harry, J. R. Lawson; Addison-Wesley, 1992, pg. 151.

SECTION II

Traditional Tolerance Analysis

The word traditional, as it is used to describe the type of tolerance analysis methods discussed in Section II, refers to the practices that have become commonly associated with how tolerances are developed in most Western industries. In comparison with the experimental methods of tolerance development, these analytical methods are indeed traditional and much more commonly practiced. In general, experimental tolerance analysis has been practiced more in Japan than in the West.

Traditional does not mean antiquated or dated; the analytical methods of tolerance development have many modern techniques and applications. With the ever increasing capabilities resident in CAD and CAE work stations, tolerance analysis is becoming more and more automated. Performing highly complex mechanical stack analyses has become much easier for more engineering and design personnel to perform. The same is true for electrical and electronics design applications.

Referring back to the product evaluation processes introduced in Chapter 2, we will further demonstrate how traditional tolerancing techniques relate to and contribute to the goals in the overall product development process.

Product Evaluation Processes:

- Monitoring Functional Change
    - **<u>Evaluating Performance</u>**
        - **<u>Analytical Model Development</u>**
        - Physical Model Development
        - **<u>Graphical Model Development</u>**
            - **<u>Evaluating Costs</u>**
                - **<u>Designing for Assembly</u>**
                    - Designing for the "ilities"
                        - Reliability and Failure Analysis
                        - Testability
                        - **<u>Maintainability</u>**

Traditional tolerancing methods are quite valuable in the overall tolerance development process. They aid in the establishment and evaluation of performance in relation to the analytical models of the design, the graphical models of the design, cost analysis, design for assembly, and design for maintainability. This is not to say that they do not play a role in the other areas of product evaluation—they do. It is the areas where they play a dominant role that are worthy of our attention at this point.

This section seeks to provide a good basis in the fundamental topics of analytical tolerance development. It explores the following topics:

- Chapter 4: Using Standard Tolerance Publications and Manufacturer's Process Capability Recommendations
- Chapter 5: Linear and Nonlinear Worst-Case Tolerance Analysis
- Chapter 6: Linear and Nonlinear Statistical Tolerance Analysis
- Chapter 7: Sensitivity Analysis and Related Topics
- Chapter 8: Computer-Aided Tolerancing Techniques
- Chapter 9: Introduction to Cost-Based Optimal Tolerancing Analysis
- Chapter 10: The Strengths and Weaknesses of the Traditional Tolerance Approaches

An understanding of these topics is essential for entering the final phase of product development discussed in Chapter 2. The intent of Section II is to instill a desire and capability to apply these techniques *before* the design is tooled and put into production.

Section II starts with the basic approaches to tolerance and sensitivity analysis. It then progresses to a discussion on computer-aided tolerance topics, briefly reviews some advanced topics, and concludes with a chapter on the strengths and weaknesses of the traditional tolerancing methods. Section III will answer many of the questions raised in Section II—particularly when the analytical methods appear to offer no clear path forward to adequately evaluating a particularly complex tolerance-definition problem between technologies.

CHAPTER 4

Using Standard Tolerance Publications and Manufacturer's Process Capability Recommendations

Starting the Tolerance Design Process

This chapter discusses the processes for the initial stage of tolerance design. The material will help you to define optimum tolerances for the components of a complete product. Undertaking this process assumes that the optimum control-factor set points (nominal design specification targets) have been determined through parameter design or some other suitable optimization method. This initial set of tolerancing activities involves looking up the recommended or standard tolerances that are typically assigned to a part or component when using an established manufacturing process with appropriate material. The costs associated with standard tolerances are also usually easy to obtain.

There are a great many resources to help the engineer to identify initial tolerances. The accrued knowledge of engineers from universities and industries over the years has produced many tables and charts containing information on standard tolerances. Machine-design textbooks often contain a number of tolerance tables for common design elements and processes. Even more tolerance data can be found in standard handbooks of engineering design (in all disciplines). For very specific applications of standard design elements, it is common for the component suppliers to publish up-to-date, detailed tolerance recommendations. When employing such recommended tolerances from a supplier it is always wise to ask how the tolerances were determined. Hopefully, they were derived from statistically valid samples of data. Just how should a manufacturer develop standard tolerances? How can one tell what they mean relative to the design requirements? These questions will be answered in this chapter.

There is also a good deal of discussion concerning how to work with component suppliers on the issues related to establishing, measuring, and improving manufacturing-process capability.

One common technique for defining standard tolerances is to quantify the various levels of standard deviations of the measurable characteristics of a component as a given manufacturing process is applied to a particular raw material. This technique also includes quantifying increasing cost to make the components as the manufacturing process is forced to produce increasingly smaller standard deviations around a given target value. Eventually the manufacturer runs out of things that can be done to shrink variability.[1] The minimum variance is defined along with its expensive price tag. Manufacturing process texts often refer to "typical" and "feasible" tolerances [R4.1]. Typical tolerances are defined as routinely economical for the manufacturer to hold. Feasible tolerances are defined as the extreme (state-of-the-art) of what can physically be obtained when cost is of little concern.

The process of shrinking variance should include *manufacturing-process parameter-robustness optimization.* In this way the supplier can offer a range of variance-control capabilities over a range of costs. Applying dynamic robustness techniques to manufacturing processes can develop robustness and tunability. When queried, the manufacturer can readily inform a design engineer about the realistic limits (in standard deviation units) that can be expected for a given component specification.

The Three Sigma Paradigm

Most standard tolerances are assumed to be derived by taking the standard deviation about the mean of a given response from a manufactured component and multiplying it by a factor of 3. Thus, the production tolerance (T) equals ± 3 standard deviations. This convention is the fundamental basis for the way statistical tolerancing processes have been applied over a number of years. Harry and Stewart [R4.2] sum it up in *Six Sigma Mechanical Design Tolerancing:*

> *Over the course of many years, use of the $\pm 3\sigma$ limits as related to the normal distribution, has become a baseline convention in tolerance analysis work, as well as in Statistical Process Control (SPC) applications. Because of this we see T/3 substituted for σ when it comes to using the RSS (Root Sum of Squares) analysis method. . . . From a mechanical point of view, this would be to say that the tolerance zone is consumed by 99.73% of the process capability because the natural $\pm 3\sigma$ limits of the manufacturing process coincide with the design tolerance limits. . . . Thus, without any a priori (previous) knowledge of the true process standard deviation, the design engineer could rationally postulate a probability model and process capability for the purpose of conducting a statistical tolerance study—always assuming a Cp = 1.0, regardless of how the design tolerance might be exercised.*

Recalling the definition of the process capability index given in Chapter 3, we are now beginning to see just how important it is in the determination of tolerances. The people at Motorola were quite right to step back and question the status quo in manufacturing process capability. Every manufacturing process is different but can be improved to provide a smaller sigma. With respect to what an engi-

1. Some suppliers resort to sorting parts. This is common in the electrical component industry. The highest quality parts are sold at a premium, while the remaining parts form a bimodally distributed lot that are sold at a lower price.

neer typically knows about the variability inherent in common part manufacturing processes, Harry and Stewart [R4.2] go on to say:

> *. . . the assumption that $\sigma = T/3$ does not always hold true. To provide the engineer greater design flexibility when conducting RSS (Root Sum of Squares) analysis, we may adjust σ by . . .*
>
> $$\sigma_{adj} = T/3Cp$$
>
> *where T is some design tolerance and Cp is the capability of an actual or postulated manufacturing process. Of course, if σ is known, there would be no need for such an adjustment. In this event, $T = 3\sigma \times Cp$.*

It is imperative that the design team carry out the research required to know the sigmas of the manufacturing processes being employed in the selection, optimization (robustness), and balancing (tolerancing) of a design. Lots of manufacturing processes have many *possible* standard deviations. It depends on myriad variables within and around the manufacturing process. Many of these variables can be optimized for robustness and later placed under statistical process control to help keep the process on target in light of special causes of variation (noises) during the manufacturing process. It is incumbent upon the design engineer to research specific manufacturing-process capabilities and determine just how large the baseline sigma is and what value Cp should actually be assigned. Then he or she can begin to work on a plan to shrink the value of sigma and increase the design's tolerance latitude while trying to keep the cost of the component at a reasonable level. But before this can be done, it is wise to see just where the current variability of the process resides. Harry and Stewart [R4.2] see it this way:

> *If a manufacturing organization embodies SPC, there is a strong possibility that such data already exists, thereby allowing the design engineer to substitute the known process standard deviation(s) in the full tolerance analysis model. As a consequence, the engineer could provide more realistic analyses. . . . It is far easier (not to mention less costly) to create robust designs based on known process capabilities than it is to track down and subsequently reduce sources of variation during the manufacturing phase. This often happens when designs are "created in the dark" or when grossly erroneous assumptions are made about process capabilities. In other words, you only get out of a design what you put into it—garbage in, garbage out.*

The engineering team must not let a design or a Cp just happen by being passive in the quantification of variability. The team must take the initiative to engineer within the context of the real manufacturing-process capability. Just as practitioners of robust design should not develop designs that cannot be easily measured for their fundamental engineering input to the functional response of the product, engineers should thoroughly understand the current measure of standard deviations of component dimensions coming from production samples that are representative of the type under design. This includes the effect of the material being selected for use. It is not uncommon for a design prototype to be made from nonproduction materials. This must be taken into account as a source of noise not only in parameter design but in tolerance design as well. When it is possible, work with the production materials so that there are no surprises later when the real material is used in the manufacturing process.

What are we to make of the fact that much of Western industry is rooted in the three-sigma paradigm? We need to look at a successful company that can compete with the masters of process capability—the Japanese. Probably the best example of a company coming to grips with the three-sigma paradigm is Motorola. It has completely rebuilt its quality, engineering, and business processes to deal with the issues latent in the three-sigma paradigm. It is no accident that the whole company has a culture built around the vision of six-sigma processes. The methods developed by Motorola have greatly enhanced the practices of tolerancing. The discussions throughout this text will include the methods of six sigma. Much will be said about how to break out of the three-sigma paradigm later in this chapter and in the rest of this text. Chapters 11 through 15, on Taguchi's tolerance design, will show a specific process for upgrading the sigma level of a design (toward, to, or past six sigma), while simultaneously balancing cost.

Returning to the topic at hand, the material in this chapter presents the *initial* steps in the overall tolerancing process, because the specific application of the tolerance may require further engineering analysis to optimize the exact tolerance that helps define the component's final specifications. Remember that published or recommended tolerances are industry standards that must be *translated* into your specific design application.

On the other side of the three-sigma issue,

> *. . . studies have shown that less than 20% of the tolerances on a typical component actually affect its function. The remainder of the dimensions on a typical product could be outside the range set by their tolerances and it would still operate satisfactorily. Thus, when specifying tolerances for non-critical dimensions, always use those that are nominal for the manufacturing process specified to (economically) make the component [R4.3].*

As this quote implies, further steps may or may not be required to arrive at the right tolerance for each dimension for the component or assembly. These steps are some or all of the five tolerancing processes, discussed in Chapters 5 through 9, that take into account the fact that a component is often employed in an assembly and ultimately may end up part of an entire system. Four of these five (Chapters 6 through 9) tolerancing processes also take into account the nature of probability in the manufacture of the components. The standard $\pm 3\sigma$ tolerances at the component level are acceptable *starting points* to begin the tolerancing process and may, in fact, be the optimum tolerance. This must be determined by mathematical, probabilistic, empirical engineering analysis—which is why the product evaluation must include analytical, physical, and graphical models in the tolerancing process. These five processes draw upon the analytical and graphical models of the product to calculate the appropriate tolerances. The final process required to optimize tolerances draws upon the physical model of the product. It involves the Taguchi tolerancing and tolerance design processes, which will consume the five chapters (Section III) after the traditional tolerancing examples featured in this section (Section II) of the text.

Processes for Establishing Initial Tolerances

As we begin to define the initial phase of traditional tolerancing it is important that a distinction be made between designs that have been developed using quality engineering from the beginning of Phase 1, as defined in Chapter 2, and those that have not. Designs that have been developed using quality engineering will have gone through concept design and parameter design processes that employ concurrent engineering. This means that the manufacturing processes, their measures of variability (Cp and sigma) and initial tolerances (USL and LSL) will already have been *initially* explored. In fact, designs that are unrealistic or too expensive, or that make it difficult to get an early understanding of tolerance implications will need to be weeded out of consideration by the concept design process. A superior design is one that can be analyzed at the concept stage for manufacturing and tolerance capability. If a design cannot be intelligently reviewed for tolerance estimates and general manufacturing feasibility, the concept is not mature enough for proper inclusion in a commercialization effort.

After concept design, the initial tolerances—based on known, standard manufacturing-process capabilities—will have been defined and used in parameter design to help yield optimal control-factor nominal set points. This is done while the tolerances remain at preliminary (and least expensive) levels, until Phase 3. The initial tolerances will then be reevaluated in light of Phase 2 system robustness optimization data, where the nominal set points and the initial component tolerances may need to be altered to facilitate system performance requirements due to interactions and sensitivities from the integration of the subsystems into the complete product configuration.

For designs that have not gone through concept design and parameter design, or designs that may have to be purchased from external sources that may not use quality engineering processes, the initial tolerances may not be considered until Phase 3 in the design cycle. This means that the design concept may not be truly superior to other designs, and the nominal control-factor set points may or may not be optimized for insensitivity to noise. In this situation one may be working with a time bomb. Traditional tolerance analysis can become a difficult task because one may have a hard time telling whether the tolerance latitude is appropriately constructed for cases in which stressful noises will cause the design's response to veer off target indiscriminately. It is likely that a nonrobust design is going to force the tightening of many tolerances to compensate for the design's sensitivity to various noise factors. Had a superior design concept been selected and optimized for robustness, using preliminary tolerances as one source of noise, the risk of going forward with the initial tolerances would be much more manageable. This is so because the team would have a good idea of how various manufacturing noises would affect the design's performance. Parameter design defines the control-factor set points that would leave the design minimally sensitive to noise and thus stable for the ensuing tolerancing process. This is one of the many reasons that tolerance design should always follow parameter design. Figure 4.1 illustrates

Figure 4.1 Comparison of engineering processes

Designs based on quality engineering

Phase 1: Concept design and subsystem robustness optimization

- Manufacturing process selected, Cp estimated, initial sigma defined and initial tolerances defined (used as noise in parameter design).

Phase 2: System robustness optimization

- System interactions and sensitivities are discovered. Nominal set points adjusted and system tolerances defined.

Phase 3: Tolerance design

- Initial tolerances reevaluated for the application. Tolerance design used to optimize the tolerances.

Designs not based on quality engineering

Traditional "build-text-fix" engineering processes

- May or may not begin with a superior design concept, and may or may not optimize the design against the effects of noise.
- Typically progresses from getting the designs to work as a system as early as possible with nominal set points, with little or no stressing with appropriate noise factors, to tolerance analysis.

Tolerance analysis

- Initial tolerances are defined for the application. Tolerance design is used to compensate for components that are sensitive to noise.

when initial tolerances are defined for the two options we have for design processes (robust designs versus nonrobust designs).

The previous comparison is used to show a general contrast between a disciplined engineering process that is designed to use concurrent engineering, with a focus on quality engineering, versus the traditional build-test-fix engineering process. Every engineering team is different and will display varying levels of structure and discipline in its application of the engineering processes. In the long run, the company that runs a loose ship with respect to engineering-process focus and sequential engineering will not be able to compete with an organization that has internalized quality engineering and concurrent engineering into its own culture. Quality engineering is quite adaptable and has been included into many companies' business processes. It is important to step back from our focus on tolerancing from time to time and look at how it fits within the entire engineering process (see Chapter 2). Let's continue our discussion concerning initial tolerance selection now that the general stage has been set for this process.

Virtually every book that discusses modern tolerance analysis recommends beginning the process by understanding the current state of the process capabilities that exist for the manufacture of specific component parts. This presumes the process has been proved to be under statistical control. Statistically valid process capability data will state, for a specific tolerance latitude (USL and LSL), what the relative process standard deviation will be for the given application of the process. Some manufacturing-process variances are independent of the target response values, while many others possess dependencies between the magnitude of the output response and the process variance (see the discussion of coefficient of variation in Chapter 3). We are primarily interested in the tolerance latitude values (numerator of the Cp: USL − LSL) that are recommended for the given process. These typically come from two sources: reference books and standard publications available commercially, or from a manufacturing organization that is making components similar to the one being developed. These tolerance values are obtained through the steps shown in Figure 4.2.

The search for published tolerances is going to be effective for determining initial tolerances if the part geometries and materials are relatively standard in nature. The more unique the component is, the less information is going to be available in standard tolerance tables. This is one reason it is easier and more efficient to design with standardized components that will still facilitate the design's functional requirements. In an ideal case, it is again recommended that the standard tolerance limits be defined prior to parameter design, where they can be included as noise in the parameter optimization experiments. Then further refinements can be made to these tolerances during the tolerance design process. In this way the noise around the nominal set points can be evaluated, in a preliminary sense, during parameter optimization, and one learns how sensitive the design is to drifts in the nominal set points. Figure 4.3 [R4.4] illustrates a typical example of a published set of dimensional tolerances.[2]

The following sources [R4.5] [R4.6] are highly recommended for aiding in initial tolerance selection:

- *Handbook of Product Design for Manufacturing: A Practical Guide to Low-Cost Production* James C. Bralla, Editor in Chief; McGraw-Hill, 1986
- *Manufacturing Processes Reference Guide* R. H. Todd, D. K. Allen & L. Alting; Industrial Press Inc.; 1994

2. Figure 4.3 is used by permission of McGraw-Hill. From Douglas C. Greenwood, *Engineering Data for Product Design,* McGraw-Hill, New York, 1961.

Figure 4.2 The process of searching for published tolerances

1. Define the ***nominal*** (target) set points that promote the optimal function of the product. These set points are the outcome of parameter design, which includes analytical and empirical engineering analysis.

2. Reaffirm the manufacturing processes best suited to making the component, based on the specific part geometry and material requirements. Hopefully the team did not wait until this point to give the selection of manufacturing processes careful thought.

3. Search the current published literature for the applicable manufacturing process you intend to use. (A number of sources are cited at the end of this section.)

4. Look up the recommended tolerances associated with the economical application of the selected process, including the current standard deviation to be expected from the dimensional distribution associated with applying the process to components similar to the type being designed.

5. If you cannot obtain a current standard deviation for the existing process, a statistically valid process capability analysis will need to be completed. (Material at the end of this chapter explains how this is done.)

6. Use the published tolerance as the baseline, or initial tolerance, that will be applied to the nominal set points. This tolerance band will likely require revision for final application to the design.

7. Move on to assembly tolerance analysis.

The following are additional resources [R4.7] [R4.8] [R4.9] [R4.10] listed for use in looking up standard tolerances for many different types of manufacturing processes:

- *Machinery's Handbook;* Industrial Press Inc.
- *Standard Handbook of Machine Design;* McGraw-Hill
- *Standard Handbook for Mechanical Engineers;* McGraw-Hill
- *Design of Machine Elements;* Spotts, Prentice Hall

For nonstandard and unique components, the likelihood of published tolerances meeting the specification requirements is probably going to be low. The engineering team will have to spend time reviewing the component design with the manufacturing engineer responsible for the processes that will be used to manufacture the part. Through these discussions and any ensuing analysis and experimentation, the engineering team should be able to establish a baseline tolerance band that can be used to develop the final tolerances. This process works as shown in Figure 4.4.

As discussed in earlier chapters, much of the published tolerance literature is based on three-sigma standards. This is one of the reasons the design team will want to be careful about applying a

Figure 4.3 Standardized tolerance table

Drilled holes

Drill size, in	*Tolerance, in*	*Drill size, in*	*Tolerance, in*
0.0135–0.0420	+0.003 −0.002	0.2660–0.4219	+0.007 −0.002
0.0430–0.0930	+0.004 −0.002	0.4375–0.6094	+0.008 −0.002
0.0935–0.1560	+0.005 −0.002	0.6250–0.8437	+0.009 −0.003
0.1562–0.2656	+0.006 −0.002	0.8594–2.0000	+0.010 −0.003

Diameter or stock size, in	*To 0.250*	*0.251 to 0.500*	*0.501 to 0.750*	*0.751 to 1.000*	*1.001 to 2.000*	*2.001 to 4.000*
Reaming						
Hand	±0.0005	±0.0005	±0.0010	±0.0010	±0.0020	±0.0030
Machine	±0.0010	±0.0010	+0.0010 −0.0015	±0.0010 −0.0020	±0.0020	±0.0030
Turning		±0.0010	±0.0010	±0.0010	±0.0020	±0.0030
Boring		±0.0010	±0.0010	±0.0015	±0.0020	±0.0030
Automatic screw machine						
Internal	Same as in drilling, reaming, or boring					
External forming	±0.0015	±0.0020	±0.0020	±0.0025	±0.0025	±0.0030
External shaving	±0.0010	±0.0010	±0.0010	±0.0010	±0.0015	±0.0020
Shoulder location (turning)	±0.0050	±0.0050	±0.0050	±0.0050	±0.0050	±0.0050
Shoulder location (forming)	±0.0015	±0.0015	±0.0015	±0.0015	±0.0015	±0.0015
Milling (single cut)						
Straddle	±0.0020	±0.0020	±0.0020	±0.0020	±0.0020	±0.0020
Slotting (width)	±0.0015	±0.0015	±0.0020	±0.0020	±0.0020	±0.0025
Face	±0.0020	±0.0020	±0.0020	±0.0020	±0.0020	±0.0020
End (slot widths)	±0.0020	±0.0025	±0.0025	±0.0025		
Hollow		±0.0060	±0.0080	±0.0100		
Broaching						
Internal	±0.0005	±0.0005	±0.0005	±0.0005	±0.0010	±0.0015
Surface (thickness)		±0.0010	±0.0010	±0.0010	±0.0015	±0.0015
Precision boring						
Diameter	+0.0005 −0.0000	+0.0005 −0.0000	+0.0005 −0.0000	+0.0005 −0.0000	+0.0005 −0.0000	+0.0010 −0.0000
Shoulder depth	+0.0010	+0.0010	+0.0010	+0.0010	+0.0010	+0.0010
Hobbing	±0.0005	±0.0010	±0.0010	±0.0010	±0.0015	±0.0020
Honing	+0.0005 −0.0000	+0.0005 −0.0000	−0.0005 −0.0000	−0.0005 −0.0000	−0.0008 −0.0000	+0.0010 −0.0000
Shaping (gear)	±0.0005	±0.0010	±0.0010	±0.0010	±0.0015	±0.0020
Burnishing	±0.0005	±0.0005	±0.0005	±0.0005	±0.0008	±0.0010
Grinding						
Cylindrical (external)	+0.0000 −0.0005	+0.0000 −0.0005	+0.0000 −0.0005	+0.0000 −0.0005	+0.0000 −0.0005	+0.0000 −0.0005
Cylindrical (internal)		+0.0005 −0.0000	+0.0005 −0.0000	+0.0005 −0.0000	+0.0005 −0.0000	+0.0005 −0.0000
Centerless	+0.0000 −0.0005	+0.0000 −0.0005	+0.0000 −0.0005	+0.0000 −0.0005	+0.0000 −0.0005	+0.0000 −0.0005
Surface (thickness)	+0.0000 −0.0020	+0.0000 −0.0020	+0.0000 +0.0030	+0.0000 −0.0030	+0.0000 −0.0040	+0.0000 −0.0050

Figure 4.4 The process of defining initial tolerances through discussions with the manufacturer

1. Contact the manufacturing engineer responsible for the process used to make the component. Hopefully you have met previously to establish the proper manufacturing process for the design and are refining the discussions to include the Phase 2 system integration and interaction information that may require adjustment of the component nominals to support systemwide performance goals.

2. Review the functional requirements of the component so that the manufacturing engineer understands how it is to be used in the product, including the nominal set points previously determined through parameter design and system-integration analysis, material requirements, part life and reliability goals, cost targets and any assembly constraints that exist.

3. Request the process capability data that currently exists for components that are similar in nature to the new design. This will define the standard deviations historically associated with parts coming from the manufacturing process. Statistical process control is a common method used to develop this data on process variation. Process control charts will define the stability of the process. The manufacturing engineer will be able to describe the history of special-cause analysis carried out to keep the process on target and under control from a statistical perspective.

4. If the component is so unique that the manufacturing process engineer is not comfortable interpolating or extrapolating an estimate of variability, the team will have to take steps to gain credible estimates of the variance. This may include rapid prototyping, soft tooling, and other simulations that will provide reasonable insight.

5. Select the recommended tolerance as the baseline that will be used for application to the nominal set points. This tolerance band will likely require revision for final application to the design.

6. Move on to assembly tolerance analysis.

published tolerance without scrutinizing its implications relative to the Cp and Cpk required for the design. But there are reasons for concern over direct application of published or recommended tolerances—mainly regarding the stability of the design's functional performance in the face of noise in the customer's use environment over the product's intended life. Obviously, if every component came off the production line right on target, with no deviation into the tolerance range, and went into service with no physical degradation, then tolerancing would be of no concern. But in the real world, the sigmas from manufacturing processes (the source of measure for unit-to-unit noise) are very important, as are the design's sensitivities to external and deterioration noises. Because of these issues we need to define tolerance processes that take manufacturing sigmas into account in both the short and long term, as well as the effect of external and deterioration noise on tolerance selection. The chapters on traditional tolerance processes begin to help us deal with these issues.

Establishing Process Capability and Process Control for Identifying Initial Tolerances

We will now review the study of process capability so the reader can properly determine just how capable any manufacturing process really is. Richard Barrett Clements has written one of the most practical books on the market for applying the tools of statistics in the manufacturing environment. His text [R4.11], *Handbook of Statistical Methods in Manufacturing,* will serve as a primary source for the reader's understanding of how to properly conduct a process capability analysis. According to Clements:

> *A process capability study is really a prediction of process adequacy. This prediction is not formed by directly studying the variation in the process (itself), but by examining the variation the process creates in the product.*
>
> *What a capability study does is to randomly sample production parts from a specific process and statistically examine the variation between, and sometimes within, parts. This data is your estimate of process variation. This amount of inherent process variation is compared to the corresponding specifications.*
>
> *There are two basic types of capability studies. The first is called a* process performance study. *It measures capability while many sources of variation are present, such as different operators, different lots of materials being used, settings on the machine adjusted as needed, and so on. This type of capability study measures the full range of inherent variation the process will probably experience during normal production runs.*
>
> *The second type of capability study is the* machine capability study. *It measures the variation in the process after all assignable causes are eliminated from the process. An assignable cause is a non-random, identifiable source of variation, such as the use of an incorrect tool, quality differences in raw materials, and unnecessary changes in machine settings.*

Clements recommends the following seven steps to properly define manufacturing process capability:

1. Select the process to be studied.
2. Define and document the existing conditions during the study.
3. Select the operator for the machine. Use a single operator unless you can assure that multiple operators make no difference in process variation.
4. Use one batch of raw materials for the study.
5. Test the measurement gauges for repeatability and reproducibility, and calibration.
6. Record sample data and the time it was sampled.
7. Calculate the capability of the process (see Chapter 3 for calculating Cp).

These seven steps help assure the evaluation will obtain statistically valid samples of data that do not possess any bias due to assignable cause variation. In this sense the study will reveal the inherent, random variation resident in the process output. This is considered "clean" data—representative of the best the process can do without undergoing robustness optimization. If you plan to put the process parameters through robustness optimization, Clements recommends performing the previous seven steps to document the "normal" behavior. This data can be compared to the new data obtained by repeating the same steps after robustness optimization of the manufacturing process parameters.

Figure 4.5 SPC diagram

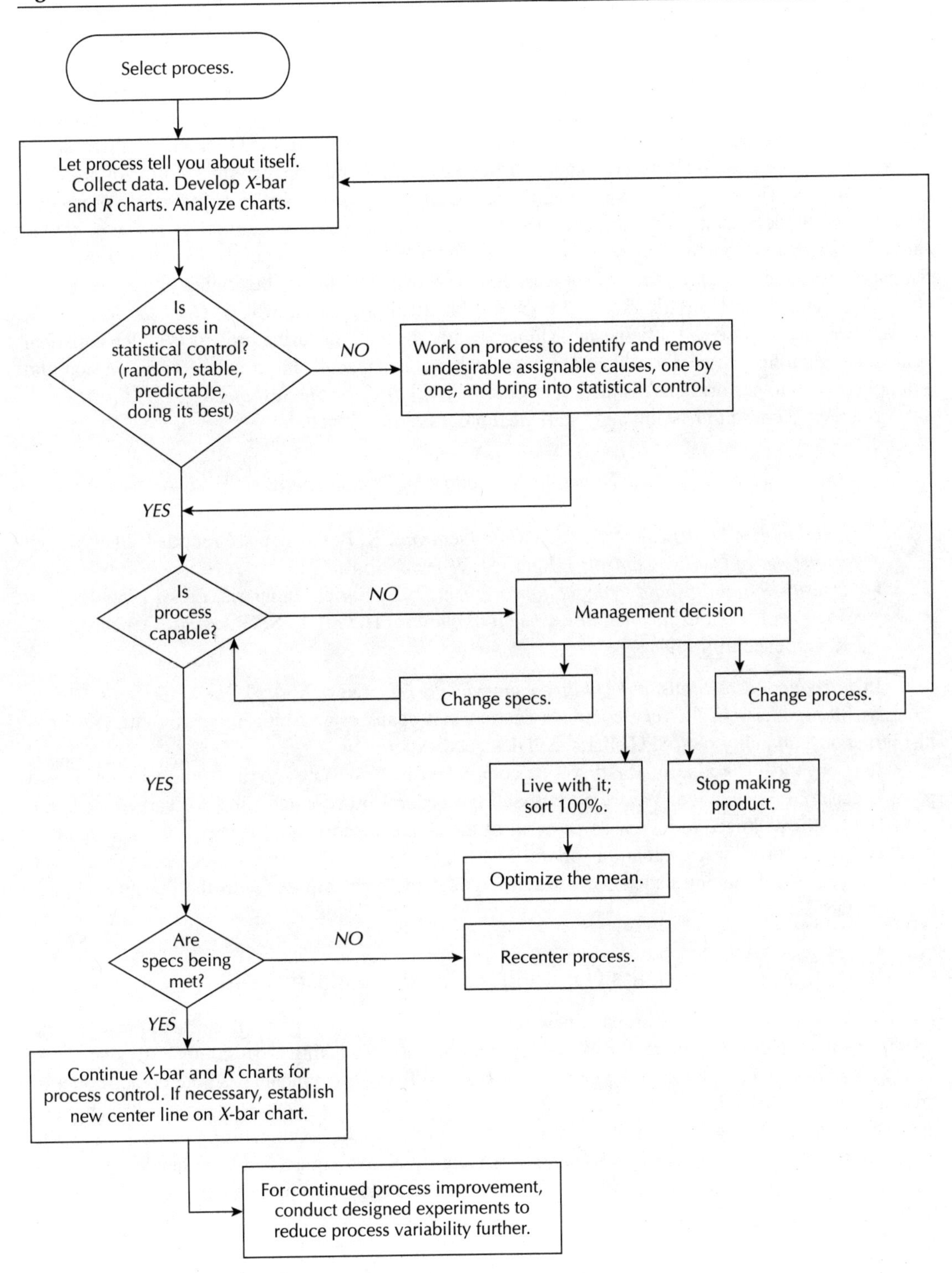

The recommended number of samples to be gathered during a process capability study is at least 30, though almost all professional sources recommend more than that. Numbers ranging from 50 to 100 are not unusual recommendations. Remember that the test for normality of a sample distribution of data is more precise as the number of samples approaches the population. The student-t distribution converges with the standard Z distribution at 30 samples of data; thus the recommendation of no fewer than 30 samples. The underlying requirement for conducting a process capability study is to prove that the process is under statistical control and that no assignable sources of variations are active in it. One can tell if assignable sources are present by looking at the pattern of data points over time from a statistical process control chart built from at least 30 days worth of data (see beginning of Chapter 6).

Figure 4.5, from Hy Pitt's excellent book on statistical process control, *SPC for the Rest of Us,* shows a very useful process for first establishing whether the manufacturing process is under statistical control and then how to go about determining whether it is capable.[3] For a complete discussion and example of how to perform this evaluation refer to [R4.12] *SPC for the Rest of Us.*

Sources to consult in establishing statistical process control include:

- *SPC for the Rest of Us;* H. Pitt; Addison-Wesley, 1994
- *The Power of Statistical Thinking;* M. Leitnaker, R. Sanders & C. Hild; Addison-Wesley, 1995
- *Handbook of Statistical Methods in Manufacturing;* R. B. Clements; Prentice Hall, 1991
- *Principles of Quality Control;* J. Banks; J. Wiley & Sons, 1989
- *Quality, Reliability, and Process Improvement;* N. L. Enrick; Industrial Press, 1985
- *Tools and Methods for the Improvement of Quality;* H. Gitlow, S. Gitlow, A. Oppenheim, & R. Oppenheim, 1989

In *Introduction to Statistical Quality Control 2nd Ed.,* (New York: J. Wiley & Sons, 1991), Douglas Montgomery has a very complete section that spans establishing process control using a histogram or probability plot, SPC charts, and designed experiments.

It is beyond the scope of this book to comprehensively cover statistical process control and process capability evaluation. We have discussed the general process and rules for carrying out the task. The reader will need to review some or all of the above texts to gain the depth of understanding necessary to perform these evaluations or work with others who will perform them. Being well read will help you avoid making bad choices when aligning the right process with the design tolerance requirements.

Creating a Library of Process Capabilities as a Data Base

As one builds up a data base with each new process capability study, one will need to preserve and update this information for future use. Wise corporations are establishing process capability data bases as key resources for competing in the global marketplace. Texas Instruments is a noted practitioner of this strategic initiative. It saves a lot of time and confusion when the development and design engineering teams can simply use their PCs or workstations to call up the manufacturing process capability library and find out the Cp, Cpk, representative probability distribution, and recommended processes for specific applications. Defining the specific shape of the distribution of data coming from a manufacturing process, once it is proved to be under statistical control, is very helpful. This will be seen

3. H. Pitt, *SPC for the Rest of Us* (Figure 23.4 from page 332, © 1988 Hy Pitt, reprinted by permission).

in Chapter 8, where we put these distributions to work in Monte Carlo analysis. Process capability libraries are excellent places to store this information.

Summary

This chapter is designed to introduce the reader to the initial tolerance selection processes. Numerous suggestions are presented to help the design team to begin the process of aligning standard tolerances to the nominal set points for each design element. It emphasizes working within a disciplined, quality engineering-focused product-development process, and reinforces the concepts of design superiority, parameter robustness optimization, and design balancing through tolerance analysis.

Two processes are suggested for defining initial tolerances from either standard tolerance literature or manufacturer's recommendations. Numerous sources of standard tolerances for various manufacturing processes and components are cited. A review of the processes for developing manufacturing process capability (Cp and Cpk) data is illustrated. The method of statistical process control (SPC) is also reviewed as a means of quantifying process capability. Once process capability data is developed it is recommended that a library of such information be saved and updated for future reference.

The following is an outline of the chapter.

Chapter 4: Using Standard Tolerance Publications and Manufacturer's Process Capability Recommendations

- Starting the Tolerance Design Process
- The 3 Sigma Paradigm
- Processes for Establishing Initial Tolerances
- The Process of Searching for Published Tolerances
- The Process of Defining Initial Tolerances through Discussions with the Manufacturer
    - Establishing Process Capability and Process Control for Identifying Initial Tolerances
    - Creating a Library of Process Capabilities as a Data Base

The team is going to perform additional analysis on many of the initial tolerances to define the final, optimized tolerances. We have reviewed how to get the tolerance process started from the standpoint of standard components and manufacturing processes. Chapter 18 contains a real case study from Kodak that demonstrates the methods discussed in this chapter. The following chapters are designed to help aid the team to refine and ultimately optimize each of the design tolerances.

References

R4.1: *Manufacturing Process Reference Guide;* Industrial Press, 1994, pg. 170.

R4.2: *Six Sigma Mechanical Design Tolerancing,* M. Harry and R. Stewart; Motorola, Inc., 1988, pg. 171.

R4.3: *The Mechanical Design Process;* (pg. 240) D. Ullman, McGraw-Hill, 1992, pg. 175.

R4.4: *Handbook of Product Design for Manufacturing, A Practical Guide to Low-Cost Production;* James C. Bralla, Editor in Chief; McGraw-Hill, 1986, pg. 182.

R4.5: Ibid.

R4.6: *Manufacturing Processes Reference Guide;* R. H. Todd, D. K. Allen and L. Alting; Industrial Press Inc., 1994, pg. 183.

R4.7: *Machinery's Handbook;* E. Oberg, F. D. Jones & H. L. Horton; Industrial Press Inc., 1980, pg. 183.

R4.8: *Standard Handbook of Machine Design;* J. E. Shigley and C. R. Mischke; McGraw-Hill, 1986, pg. 183.

R4.9: *Standard Handbook for Mechanical Engineers;* T. Baumeister, E. A. Avallone and T. Baumeister III; McGraw-Hill, 1979, pg. 183.

R4.10: *Design of Machine Elements;* M. F. Spotts, Prentice Hall, 1985, pg. 183.

R4.11: *Handbook of Statistical Methods in Manufacturing;* R. B. Clements; Prentice Hall, 1991, pg. 185.

R4.12: *SPC for the Rest of Us;* H. Pitt; Addison-Wesley, 1994, pg. 188.

See page 98 for a list of recommended texts for Statistical Process Control and Process Capability Analysis.

CHAPTER 5

Linear and Nonlinear Worst-Case Tolerance Analysis

Standard Worst-Case Methods

This approach is most often employed when several components are placed together to form an assembly. It is called worst-case analysis because it is used to add or subtract all the maximum or minimum tolerances associated with the nominal set point for each component. The worst-case tolerance buildup leaves the assembly at either its largest or smallest allowable dimension. It does not take into account the laws of probability—at least not in a realistic sense.

This process should be used sparingly because it is not really representitive of the way tolerances actually build up in the probablistic environment in which parts are made and assembled. It is very rare that all the components in an assembly will actually be at their maximum or minimum tolerance levels *at the same time.* The only time a worst-case analysis is really necessary is when the assembly is made up of very few parts that have a critical interface with some other product feature that cannot be allowed to interfere or be spaced too far apart. By critical we mean either a safety issue or a customer-preference issue for which a specific payment is being made.

The general model for worst-case analysis is simply the sum of all the component dimensions at their worst-case maximum or minimum values:

$$d_{\text{total}} = d_1(+ \text{ or } -)d_2(+ \text{ or } -)d_3(+ \text{ or } -)\ldots(+ \text{ or } -)d_n \tag{5.1}$$

Be careful to note that some assembly components geometrically double back on themselves such that you may have to subtract a dimension as you calculate d_{total}. This is why we include the (+ or −) sign in Equation 5.1. The assembly will likely have to fit within a space defined by another feature on another component or assembly. This space is often called the mating dimension, d_{mating}. To define the gap between the mating dimension, the following adjustment must be made to the general worst-case tolerance model:

$$d_{gap} = d_{mating} - [d_1(+ \text{ or } -)d_2(+ \text{ or } -)d_3(+ \text{ or } -)\ldots(+ \text{ or } -)d_n] \quad (5.2)$$

The six-sigma methods express all of this in a slightly different format, explained in the following section.

Six-Sigma Worst-Case Tolerance Analysis

The following equations are the basis for six-sigma worst-case analysis [R5.1]. We will define each of the terms at this point and will work through an example with these equations later in this chapter.

Equation 5.3 expresses the worst-case for all the component dimensions at their *maximum* size. N_{pi} is the nominal or target value of the component dimension of any part we shall generically refer to as part *i*. T_{pi} is the initial tolerance assigned to N_{pi}. This tolerance value is typically defined by the first step (tables and recommendations) discussed in Chapter 4. The *m* represents the total number of parts in the assembly.

$$WC_{max} = \sum_{i=1}^{m} (N_{pi} + T_{pi}) \quad (5.3)$$

Equation 5.4 defines the worst-case *minimum* dimension that can occur for the entire assembly due to each part being at its *minimum* allowable tolerance.

$$WC_{min} = \sum_{i=1}^{m} (N_{pi} - T_{pi}) \quad (5.4)$$

We now introduce three new terms to account for the geometry of the mating part within which our assembly is going to fit. Motorola refers to this geometry as the *envelope.* To remain consistent with the Six Sigma Institute's terms we will define the nominal envelope dimension as Ne and the envelope tolerance as Te. The subscript e refers to the envelope. The third new term, *Q,* refers to the *minimum* gap constraint, or smallest allowable gap. In some applications we will use $Q = 0$, which means we are technically at the point of a line-to-line fit. As *Q* gets bigger we will see a finite growth in the assembly gap.

$$Q \leq N_e - T_e - \sum_{i=1}^{m} (N_{pi} + T_{pi}) \quad (5.5)$$

Equation 5.6 quantifies an upper boundary on the allowable assembly gap. In many cases the assembly gap could get too big for the proper functionality of the assembly within the envelope. *R* is the *upper bound* on the assembly gap. *Q* and *R* are technically related to the assembly process and are thus critical boundary parameters for it.

$$N_e + T_e - \sum_{i=1}^{m} (N_{pi} - T_{pi}) \leq R \quad (5.6)$$

We have been discussing the assembly gap without calling out a specific term for it. G will be used to represent the assembly gap in the following equations. Here we define the maximum assembly gap as G_{max}, which is also equal to R.

$$G_{max} = N_e + T_e - \sum_{i=1}^{m} (N_{pi} - T_{pi}) \quad \textbf{(5.7)}$$

The minimum assembly gap is defined by G_{min}.

$$G_{min} = N_e - T_e - \sum_{i=1}^{m} (N_{pi} + T_{pi}) \quad \textbf{(5.8)}$$

The difference between G_{max} and G_{min} defines the acceptable range of assembly gap, G_{range}.

$$G_{range} = G_{max} - G_{min} \quad \textbf{(5.9)}$$

Employing the nominal terms from the previous equations we can now define the nominal assembly gap that is the desired target for the assembly.

$$G_{nominal} = N_e - \sum_{i=1}^{m} (N_{pi}) \quad \textbf{(5.10)}$$

We can do some rearranging to make it easy to calculate the *nominal* envelope dimension:

$$N_e = \sum_{i=1}^{m} (N_{pi} + T_{pi}) + Q + T_e \quad \textbf{(5.11)}$$

The final gap dimension we are left to define is the *tolerance* on the gap, $G_{tolerance}$.

$$G_{tolerance} = \frac{G_{max} - G_{min}}{2} \quad \textbf{(5.12)}$$

You will find, in the literature from Motorola, a unique convention to account for both the magnitude and directionality of a dimension. This is done to speed up and simplify calculations involving a complex assembly. The term V_i is used to define a *unit vector* that quantifies the *algebraic sign* of the component as it is stacked in the assembly. Typically, dimensions stacking to the right are positive and those stacking to the left are negative. Equation 5.13 will yield the same result as Equation 5.10 for the nominal gap.

$$G_{nominal} = \sum_{i=1}^{m} N_i V_i \quad \textbf{(5.13)}$$

The six-sigma literature also accounts for a feature that is located by true position with respect to a datum as is employed in GD&T. The tolerance zone must be converted from a diameter to a radius when used in the vectored summation process. A correction factor, B_i, is added to equation 5.13 to account for such an occurrence. Harry and Stewart [R5.1] recommend such a correction factor when working with "fixed or floating fasteners."

$$G_{nominal} = \sum_{i=1}^{m} N_i V_i B_i \quad \textbf{(5.14)}$$

Analysis for Nonlinear Tolerance Stacks Using the Worst-Case Method

There are going to be cases where the relationship between the component parts in an assembly are not going to stack up in a *linear functional relationship* with respect to the assembly gap or other geometric-dependent variables. When this is the case the following general expressions can be used to define the relationship between the dependent assembly variable and the independent component variables:

$$y = f(x_1, x_2, x_3, \ldots x_n) \tag{5.15}$$

The engineering team must define the functional relationship between the component dimensions (assumed to be independent variables) and the dependent assembly variable, y. In linear problems this is simple to do; in nonlinear problems it can be quite challenging. In *Worst Case Tolerance Analysis with Nonlinear Problems,* Greenwood and Chase [R5.2] state that

> *Small changes in the assembly or dependent dimension may be expressed by a Taylor's series expansion as given by . . .*
>
> $$\Delta y = \sum \frac{\partial f}{\partial x_i} \Delta x_i + \frac{1}{2} \sum_{i=1}^{n} \sum_{j=1}^{n} \frac{\partial^2 f}{\partial x_i \partial x_j} \Delta x_i \Delta x_j + \ldots \tag{5.16}$$
>
> *The common practice for tolerance analysis is to substitute tolerances for the delta quantities. For a worst-case tolerance analysis, only the first order terms are used and absolute values are placed on these terms (also see Fortini, E. I.,* Dimensioning for Interchangeable Manufacturing, *Industrial Press, New York, 1967, p. 48).*[1]

Greenwood and Chase go on to define a general worst-case tolerance equation as follows:

$$\text{tol}_y = \left|\frac{\partial f}{\partial x_1}\right| \text{tol}_1 + \left|\frac{\partial f}{\partial x_2}\right| \text{tol}_2 + \left|\frac{\partial f}{\partial x_3}\right| \text{tol}_3 + \ldots + \left|\frac{\partial f}{\partial x_n}\right| \text{tol}_n \tag{5.17}$$

$$\text{Nom}_y \approx \left|\frac{\partial f}{\partial x_1}\right| x_1 + \left|\frac{\partial f}{\partial x_2}\right| x_2 + \left|\frac{\partial f}{\partial x_3}\right| x_3 + \ldots + \left|\frac{\partial f}{\partial x_n}\right| x_n \tag{5.18}$$

In a nonlinear problem we must define specific values for the partial derivatives because they represent the particular sensitivities that each component dimension and tolerance will induce on the assembly dimension and tolerance. It is these sensitivities that often cause the need for asymmetric tolerances on assemblies, as we shall see in a later example.

If we look closely at Equation 5.17 we see that for a linear problem the partial derivative terms are simply plus or minus *one* depending upon the direction of the stack-up. The six-sigma technique uses the vector term v_i to call the sign to our attention. The second-order terms and above are all equal to zero in a linear case. When a tolerance stack is indeed a simple linear function of the components there is no risk of error showing up in the mathematical sum of the component dimensions or tolerances. When the stack is nonlinear and the worst-case analysis is applied as if it were linear, meaningful errors will occur that could jeopardize the functional capability of the assembly. This is so because the individual component sensitivities are not represented in the linear math model being used to sum up the stack that defines the assembly dimension and tolerances. These problems require a lot more engineering analysis and more extensive math modeling techniques to properly analyze the tolerances.

Techniques involving simultaneous solutions of partial derivatives will require either closed-form or numerical analysis. This may require the designer to seek assistance from other engineering-team members who have specific analytical skills required to solve such equations. Rather than bypassing the process of nonlinear analysis by approximating the problem as linear, it is recommended that the engineering team take the time to work out the problem together with analysts, designers, and drafters to get the right tolerances defined for the problem at hand. Engineers must not hand off the tolerancing process to the designers and drafters. They must blend skills to define and communicate the proper tolerances to the manufacturing community.

Process Diagrams for Worst-Case Analysis for the Stackup of Tolerances in Assemblies

As mentioned previously, worst-case tolerance analysis should be used sparingly in most modern tolerance cases. Its value lies in its ability to quantify the harshest limits that an assembly can attain if every component arrives at either the largest or smallest dimensional extreme. As we have mentioned, this is an extremely rare occurrence. When one is working where a worst-case stack-up is possible and it has the potential for causing some form of human disaster or harm then the worst-case approach can be quite helpful. It is also useful when the economic implications of a nonfunctioning assembly due to a worst-case occurrence is very large. Worst-case analysis is defined by the steps shown in Figure 5.1.

Figure 5.2 illustrates the worst-case tolerance analysis process for a one-dimensional problem.

Step 1: Parts A, B, and C are the components that define an assembly. When they are joined together they must then be placed inside the boundary established by part D. This boundary is often called a design envelope or an *assembly constraint.* The nominal dimensions typically come from some form of engineering analysis. In a quality engineering context they would come from Parameter Design optimization.

Step 2: We are interested in having the assembly of parts A, B, and C fit into the boundary of Part D without any interference. Thus, the problem is to define the proper specifications for Part D so that when the worst-case stackup of the assembly of parts A, B, and C occurs there is no interference.

Step 3: The initial tolerances have been assigned to the nominal dimensions after seeking the recommendation of the manufacturing engineer responsible for the fabrication of the component parts (see Figure 5.1). They represent ± 3 standard deviations from the measured samples of actual production parts.

Step 4: We can now calculate the worst-case stack-up dimension for the assembly of parts A, B, and C:

The rules for a tolerance stack analysis require that a sign convention be established. Typically, displacements to the left are negative and displacements to the right are positive. The displacements are called *vectors.* A series of dispacements is called a *vector chain.* A vector chain analysis, for positive direction being to the right, begins at the left side of the gap. When all the vectors in the chain are summed, a net positive value indicates clearance, and a net negative value indicates an interference.

For the case at hand we see d_{total} equals the following value:

$$d_{\text{total}} = d_1 \pm d_2 \pm d_3 \pm \ldots \pm d_n$$

$$d_{\text{total}} = -(1.001 + 2.001 + 1.501)$$

$$d_{\text{total}} = -(4.503) \text{ in.}$$

Figure 5.1 Worst-case tolerance analysis

1. Analyze the geometric and spatial relationship between components to make the assembly, then assess the assembly's geometric and spatial relationship to the product's functional capability. What is the functional purpose of the stacked up components (in very specific engineering terms)?

2. Define the specific functional requirement of the assembly. This may be done by quantifying the overall assembly specifications (Gnom, Gmax, and Gmin).

3. Either *estimate* (from look-up tables, manufacturing literature, or recommendations) or *allocate* (from a previously determined overall assembly specification) an initial tolerance for each component in the assembly using the aforementioned tolerance identification process. Each component must be assessed on an individual basis.

4. Each tolerance is to be added to and subtracted from the associated nominal set point so that each component dimension can be added together to form either a maximum possible assembly value (Gmax) or a minimum assembly value (Gmin). All the largest possible components are added together to arrive at the maximum assembly size. All the smallest components are added together to yield the smallest possible assembly size. Select a point on one side or the other of the unknown gap being evaluated as a starting point.

5. The space, or assembly gap (G), can now be assessed because it is now clear what the extremes of the component assembly can be. The nominal gap dimension (Gnom) can be calculated and its tolerances assigned. Sometimes Gnom and its tolerances will be specified as a requirement, and the component parts have to be developed to meet Gnom and its tolerance values. In this process the component tolerances will have to be tuned to support the gap specifications. This is referred to as *tolerance allocation.*

Figure 5.2 Complete assembly

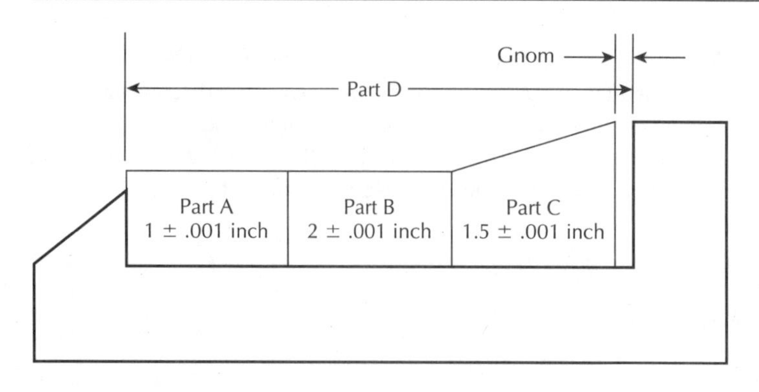

In some cases certain assembly components can geometrically double back on themselves such that one may have to subtract a dimension as one calculates d_{total}, as shown in Figure 5.3.

Again, this is why we include the plus or minus sign in the equation. The assembly will likely have to fit within a space defined by another feature on another component or assembly. This space is often called the *assembly gap, mating dimension,* or *boundary dimension,* and is abbreviated d_{mating}. To define the mating dimension, the following adjustment must be made to the general worst-case tolerance model:

$$d_{gap} = d_{mating} - (d_1 \pm d_2 \pm d_3 \pm \ldots \pm d_n)$$

We are now left with an engineering decision. If no specification has been determined for the mating dimension, one will have to be determined at this point. The issue now regards the manufacturing standard deviation for the fabrication of the mating dimension (d_{mating}). After consulting the manufacturing engineer we find that the mating dimension can be made with tolerances of $\pm$.001 inch. If we want the assembly to always fit into the boundary, the gap dimension must be greater than zero. If zero gap is unacceptable, the gap will have to be assigned a realistic value greater than zero. In this case we will select the minimum gap value (d_{gap}) of .001 inch. The nominal mating dimension is defined by the following calculations:

$$.001 = d_{mating} - (4.503)$$

$$d_{mating} = 4.503 + .001$$

Recognizing that the mating dimension has a tolerance range of $\pm$0.001 inch, the *nominal mating dimension* is finally calculated to be:

$$d_{nom.mating} = 4.503 + .001 + .001$$

$$d_{nom.mating} = 4.505$$

The literature from Motorola University and the Six-Sigma Institute [R5.1] expresses all of this in a slightly different format, illustrated next.

Equation 5.3 expresses the worst case for all the component dimensions at their maximum size.

$$WC_{max} = \sum_{i=1}^{m} (N_{pi} + T_{pi}) = 4.503 \text{ in.}$$

This expression defines the worst-case minimum dimension that can occur for the entire assembly due to each part being at its minimum allowable tolerance.

$$WC_{min} = \sum_{i=1}^{m} (N_{pi} - T_{pi}) = 4.497 \text{ in.}$$

Figure 5.3 Assembly that doubles back

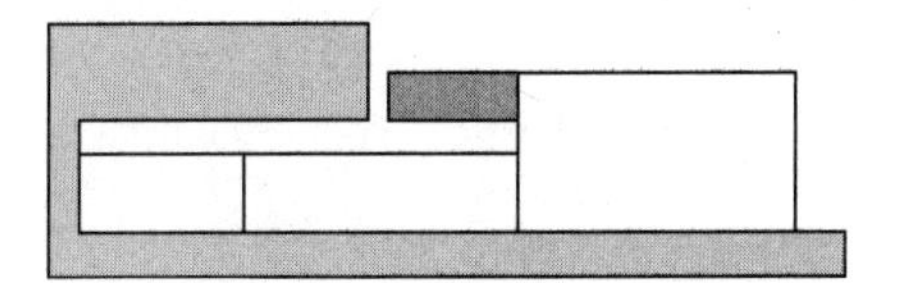

Q refers to the minimum gap constraint or smallest allowable gap. As Q gets bigger then we will see a growth in the assembly gap.

$$Q \leq N_e - T_e - \sum_{i=1}^{m} (N_{pi} + T_{pi})$$

As previously mentioned we will assign Q the value of 0.001 inch.

The next equation quantifies an upper boundary on the allowable assembly gap:

$$N_e + T_e - \sum_{i=1}^{m} (N_{pi} - T_{pi}) \leq R$$

In this problem we are not really concerned about the upper bound on the assembly gap, since it has been determined that it is of no functional consequence.

Here we define the maximum assembly gap as $G_{\max}$, which is also equal to R.

$$G_{\max} = N_e + T_e - \sum_{i=1}^{m} (N_{pi} - T_{pi})$$

In this example $G_{\max}$ cannot be calculated until $G_{\min}$ is used to define N_e, the nominal dimension for Part D. $G_{\max}$ is then calculated to be $4.505 + 0.001 - 4.503 = 0.003$ inch. The minimum allowable assembly gap is defined by $G_{\min}$, which is also equal to Q.

$$G_{\min} = N_e - T_e - \sum_{i=1}^{m} (N_{pi} + T_{pi}) = 0.001 \text{ in.}$$

$$N_e = G_{\min} + T_e + \sum_{i=1}^{m} (N_{pi} + T_{pi}) = 4.505 \text{ in.}$$

$$N_e = 0.001 + 0.001 + 4.503 = 4.505 \text{ in.}$$

The difference between $G_{\max}$ and $G_{\min}$ defines the acceptable range of assembly gap, G_{range}.

$$G_{\text{range}} = 0.003 - 0.001$$

$$G_{\text{range}} = 0.002$$

The final gap dimension we are left to define is the tolerance on the gap, $G_{\text{tolerance}}$.

$$G_{\text{tolerance}} = \frac{G_{\max} - G_{\min}}{2}$$

$$G_{\text{tolerance}} = \frac{.002}{2} = 0.001$$

This matches what we were told the manufacturing process could hold for Part D. This concludes a simplified example for worst-case analysis.

A Case Study for a Linear Worst-Case Tolerance Stackup Analysis [R5.3]

Let's look at a more realistic example of linear worst-case tolerance analysis. The case involves two mating parts that must fit together, as shown in Figure 5.4.

Figure 5.5 shows two worst-case scenarios for the mating parts. Figure 5.6 illustrates one possible set of specifications for parts A and B. We shall explore these tolerances to see if they work for a worst-case scenario.

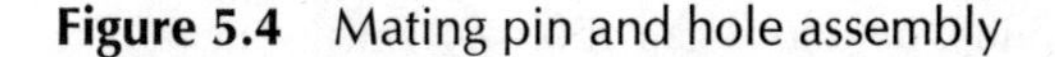

Figure 5.4 Mating pin and hole assembly

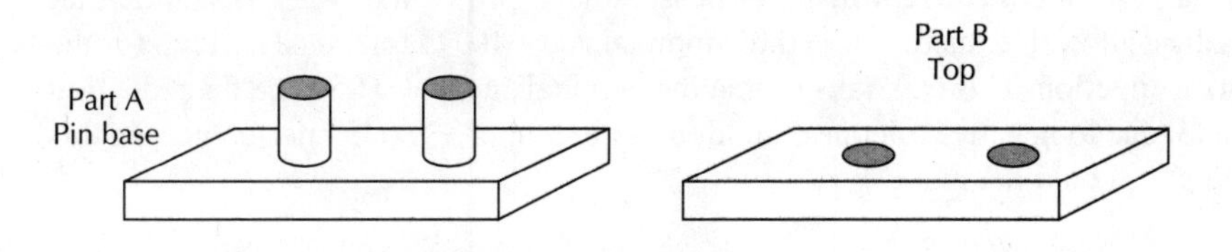

Figure 5.5 Two worst-case scenarios

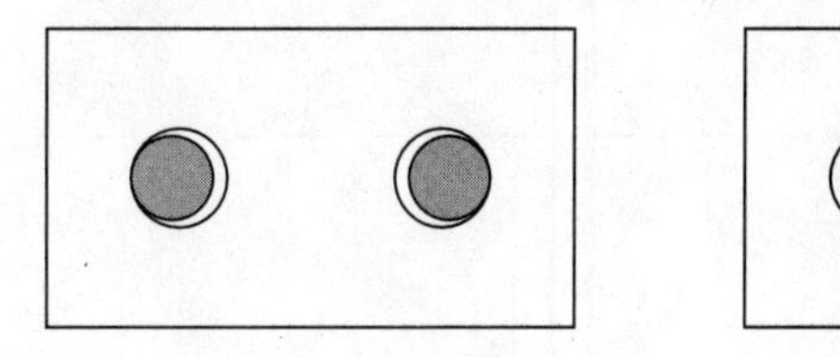

Figure 5.6 Specifications for part A and part B

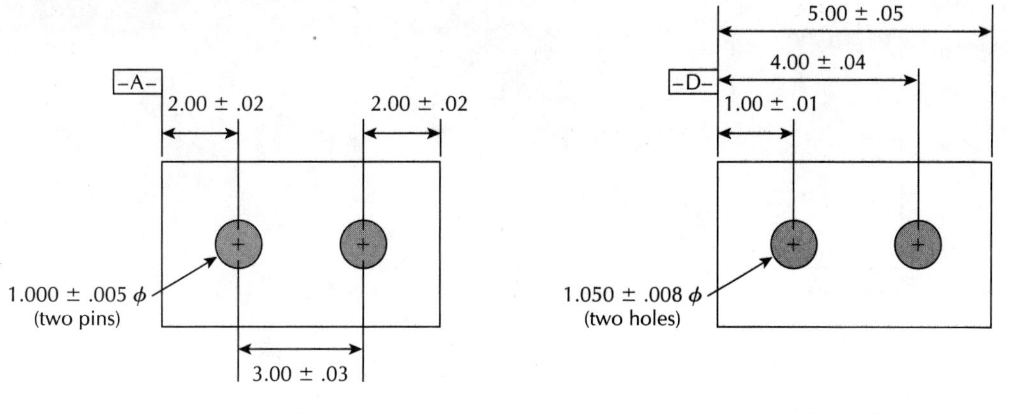

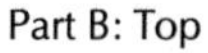

In Figure 5.7, the analysis begins at the right edge of the right pin. One must always select a logical starting point for a stack analysis. Typically, it will be at one side of an unknown critical gap dimension (the distance from the start of a to the end of g), as shown in Figure 5.7. We will step across the assembly from point a to point g to sum up all the stackup dimensions according to their sign. Each arrowed line in Figure 5.7 is called a *displacement vector.* The sign for each displacement vector is illustrated in the figure.

The displacement vectors and their associated tolerances are quantified as shown in Figure 5.8. Notice that each tolerance for a diameter has been divided by 2 to convert it to a radial tolerance. Analyzing the results we find that there is a +0.05 nominal gap and +0.093 tolerance buildup for the worst case in the positive direction. This case tells us that the nominal gap is +0.050, and the additional growth in the gap will be +0.093, due to tolerance buildup, for a total gap of 0.143. Thus, when all the tolerances are at their largest values, there will be a measurable gap. For the worst case when all dimensions are the smallest allowable, there is a + 0.05 nominal and −0.093 tolerance buildup for the worst case in the negative direction. This case tells us that the nominal gap is +0.050 and the reduction in the gap equals −0.093, due to negative tolerance buildup, for a total of −0.043 interference. In this worst-case scenario the parts could not mate.

Figure 5.7 Worst-case tolerance stackup example

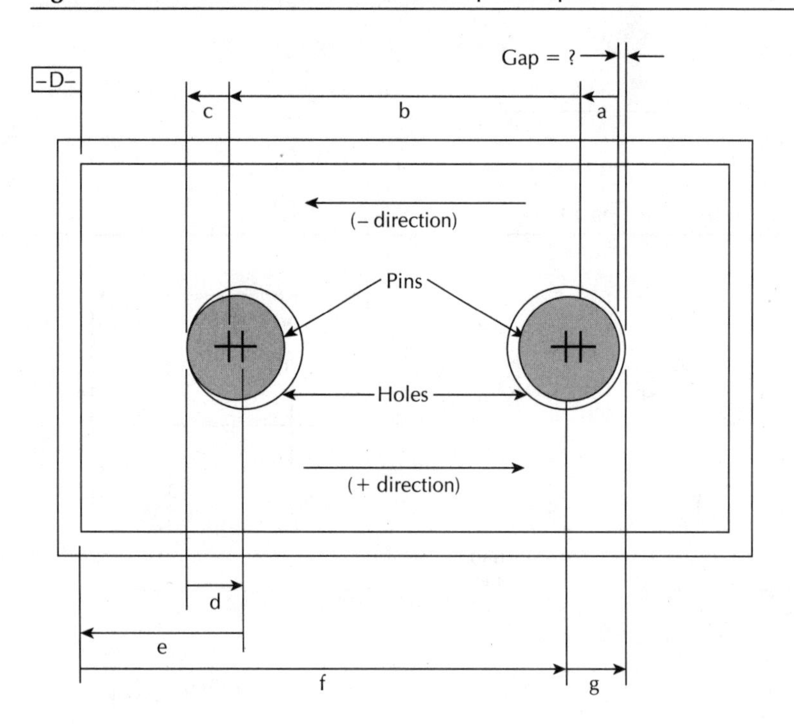

Figure 5.8 Tolerance calculation

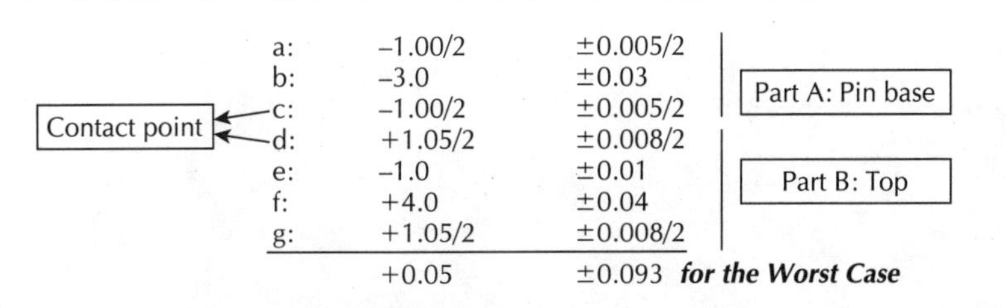

a:	–1.00/2	±0.005/2	Part A: Pin base
b:	–3.0	±0.03	
c:	–1.00/2	±0.005/2	
d:	+1.05/2	±0.008/2	Part B: Top
e:	–1.0	±0.01	
f:	+4.0	±0.04	
g:	+1.05/2	±0.008/2	
	+0.05	±0.093	***for the Worst Case***

A Process for Modeling Worst-Case Nonlinear Tolerance Stackup Problems

As previously mentioned, there are going to be cases in which the relationship between the component parts in an assembly are not going to stack up with a linear functional relationship with respect to the assembly gap or to other geometrically dependent variables.

We will now outline a standard process that any engineering team can follow to generate the requisite mathematical model for the assembly geometry. The team will sequentially construct a geometric model that will then be used to construct a math model to optimize the tolerances. After we review the model-generating process we will provide an example of a nonlinear problem along with the appropriate math model.

Figure 5.9 Nonlinear tolerance process

1. Construct a geometric layout of each component with its nominal set point and the initial recommended (bilateral) tolerances. Define the critical *functional* features, sizes, and surfaces. Some dimensions on the component will not contribute to the assembly stackup. A common error in tolerance analysis is the inclusion of tolerance parameters that are not functionally involved in the assembly tolerance.

2. Construct an assembly drawing of the components; this will graphically illustrate the geometric relationships that exist between them. Again define the critical functional features, sizes, and surfaces. At this point the engineering team should define the *required* assembly tolerance range or the nonlinear tolerance adjustment process to hit.

3. Employ plane geometry, trigonometry, algebra, and calculus as necessary to define the arithmetic (functional) relationship between the component dimensions and the assembly dimension(s).

$$y = f(x_1, x_2, x_3, \ldots x_n)$$

The math model must directly express the critical path of the functional interfaces.

4. Use the math model to solve for the nominal assembly dimension by using the nominal component values. Reaffirm the initial assembly tolerance required for proper functional product performance. This will typically involve some form of engineering analysis, including experimental evaluations.

(*Continued*)

Figure 5.9 (*Continued*)

5. Employ differential calculus to solve for the partial derivatives that represent the component-tolerance sensitivities with respect to the assembly tolerance:

$$\text{Nom}_y \approx \left|\frac{\partial f}{\partial x_1}\right| x_1 + \left|\frac{\partial f}{\partial x_2}\right| x_2 + \left|\frac{\partial f}{\partial x_3}\right| x_3 + \ldots + \left|\frac{\partial f}{\partial x_n}\right| x_n$$

The partial derivatives are solved with all the components set at their nominal values (typically at the tolerance-range midpoint). Greenwood and Chase state that the tolerances can be unilateral or equal bilateral. The differentiation is typically done using manual analytical techniques or computer-aided numerical-analysis techniques. Some problems can be difficult to differentiate. Using the definition of the derivative can simplify the process of calculating a sensitivity. Also available for help are computer-aided analytical differentiation programs like Mathematica, Macsyma, and Maple.

6. With the values for each component sensitivity quantified, the worst-case tolerance equation can be used to determine how a calculated assembly tolerance range is functionally related to the required assembly tolerance range. We do this by inserting the initial, bilateral component tolerances and their sensitivities into Equation 5.17. Note that in this process we use the absolute value of the sensitivities.

$$\text{tol}_y = \left|\frac{\partial f}{\partial x_1}\right| \text{tol}_1 + \left|\frac{\partial f}{\partial x_2}\right| \text{tol}_2 + \left|\frac{\partial f}{\partial x_3}\right| \text{tol}_3 + \ldots + \left|\frac{\partial f}{\partial x_n}\right| \text{tol}_n$$

The evaluation of the recommended initial component tolerances with respect to the required assembly tolerance, as constrained by the calculated sensitivities, will be an *iterative* process. The goal is to adjust the component tolerances to get the assembly tolerance range as close to that required as possible. The ultimate outcome of the iterations will be a revised set of component tolerances, probably a little different from the original recommended tolerances, that provide an approximate match to the required assembly tolerance. The tolerance range for the assembly, as calculated by the adjusted component tolerances, will likely be located a little above or below the required tolerance range. It is not unusual that the component nominal values have to be shifted, or *centered,* to put the actual assembly tolerance range on top of the required assembly tolerance range (or as close as is reasonably possible).

7. After shifting the component tolerances to get the assembly tolerance range near to the required range, the remaining offset between the required range and calculated range can be shifted by carefully shifting the nominal component set point values. The guiding principle in component nominal shifting is to do it in proportion to each component tolerance. Greenwood and Chase recommend that each component be shifted by an equal fraction (f) of each respective tolerance.
The needed shift in the assembly tolerance range must be calculated.

$$\Delta y = \left[\frac{(y_{\text{high}} + y_{\text{low}})}{2}\right]_{Required} - \left[\frac{(y_{\text{high}} + y_{\text{low}})}{2}\right]_{Actual}$$

The following equation is then used to determine the amount the assembly will shift for given component nominal shifts:

$$\Delta y = \frac{\partial(f)}{\partial x_1}\Delta x_1 + \frac{\partial(f)}{\partial x_2}\Delta x_2 + \ldots \frac{\partial(f)}{\partial x_n}\Delta x_n$$

When we know how far the assembly tolerance range is off from the required range (Δy), we can then go back and solve for how much to shift each component nominal value (Δx). We establish the roper direction (sign) and amount of nominal shift with the following equation:

Figure 5.9 *(Continued)*

$$\Delta x_i = \text{sign}\,(\Delta y)\ \text{sign}\,\frac{\partial(f)}{\partial x_1}\,(f)\ \text{tol}_i$$

This expression simply states that the change needed in the nominal value of parameter (i) is going to depend on the sign of the shift change required in the assembly variable (y), and the sign of the sensitivity times the fraction of the tolerance. The equation that will be used to solve for (f) will be:

$$\Delta y = \frac{\partial(f)}{\partial x_1}(\text{sign})(\text{sign})(f)(\text{tol}_1) + \frac{\partial(f)}{\partial x_2}(\text{sign})(\text{sign})(f)(\text{tol}_2) + \ldots$$

$$(f) = \frac{\Delta y}{\frac{\partial(f)}{\partial x_1}(\text{sign})(\text{sign})(f)(\text{tol}_1) + \frac{\partial(f)}{\partial x_2}(\text{sign})(\text{sign})(f)(\text{tol}_2) + \ldots}$$

8. Once the common shift fraction (f) is determined, the component nominals can be recalculated to define the new midpoint (nominal) values:

$$x_{(i\ shifted)} = x_{(i\ nominal)} \pm (f)(\text{tol}_i)$$

9. The new shifted nominals and their tolerances can now be used to calculate the new assembly nominal value and the resulting tolerance range about the assembly nominal. Referring back to equations 5.17 and 5.18:

$$\text{Nom}_y \approx \left|\frac{\partial f}{\partial x_1}\right| x_1 + \left|\frac{\partial f}{\partial x_2}\right| x_2 + \left|\frac{\partial f}{\partial x_3}\right| x_3 + \ldots + \left|\frac{\partial f}{\partial x_n}\right| x_n$$

and

$$\text{tol}_y = \left|\frac{\partial f}{\partial x_1}\right| \text{tol}_1 + \left|\frac{\partial f}{\partial x_2}\right| \text{tol}_2 + \left|\frac{\partial f}{\partial x_3}\right| \text{tol}_3 + \ldots + \left|\frac{\partial f}{\partial x_n}\right| \text{tol}_n$$

This new assembly tolerance range can then be compared to the required assembly tolerance range for suitability. The match should be reasonably close at this point. Another iteration can be performed if necessary to get the actual assembly tolerance range even closer to what is required. This can easily be done by repeating steps 7 to 9 after recalculating new sensitivity values based on the new shifted component nominal values (step 5).

Greenwood and Chase have used an overrunning clutch example (see Figure 5.10) from a case originally defined by Fortini in *Dimensioning for Interchangeable Manufacturing* to demonstrate the approach to tolerancing assemblies and components in nonlinear worst-case problems. We will feature a sample of Greenwood and Chase's work in this example [R5.2].

Step 1: Construct a layout for each component

As shown in Figure 5.11, the first component is the clutch hub. X_1 is the critical distance between the roller bearing surfaces. The initial specifications for the hub bearing surfaces are 55.29 ± 0.156 mm.

There are four roller bearings in the clutch, as shown in Figure 5.12. Because of symmetry, only *two* opposing rollers matter in the assembly tolerance stackup calculation ($X_{2\text{ and }3}$). Remember that not

Figure 5.10 An overrunning clutch assembly

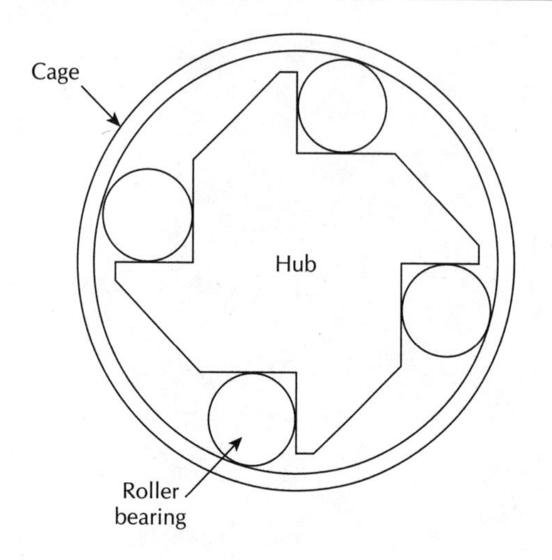

Figure 5.11 Clutch Hub

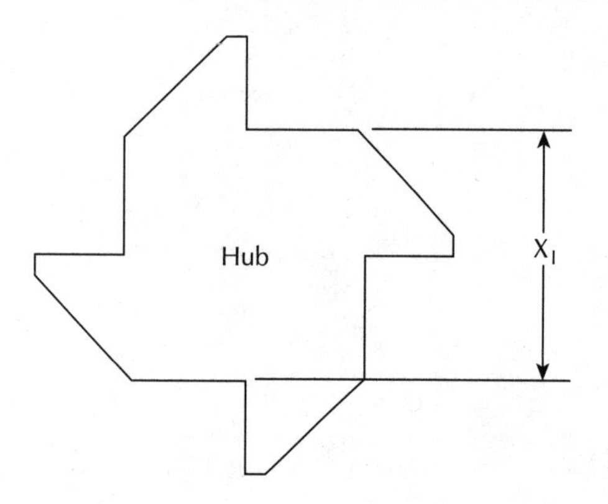

Figure 5.12 Clutch roller bearing

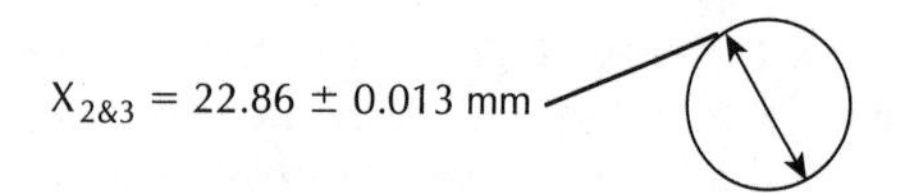

all the dimensions or components in an assembly matter in the assembly tolerance stackup model. The roller bearings are a purchased item and as such have unadjustable, and nonnegotiable specifications. For all purposes these dimensions are *fixed* and cannot be included in any component-tolerance readjustment procedures that may be required to meet assembly specification requirements.

The final component in the assembly is the cage, as shown in Figure 5.13. The initial cage specifications that are of consequence in the assembly stackup are associated with the inner diameter of the race: 101.60 ± 0.156 mm.

Figure 5.13 Clutch cage

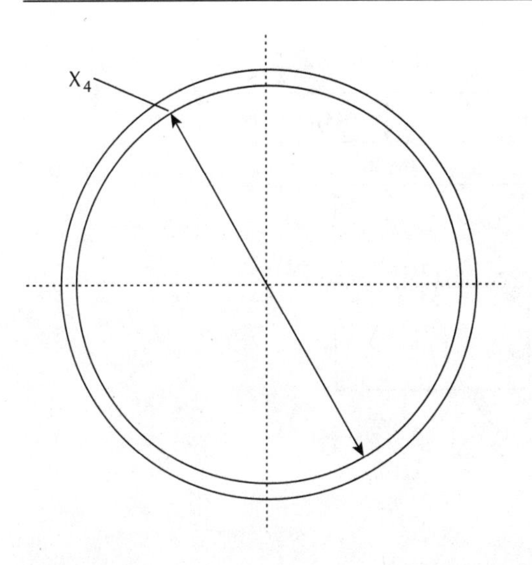

Step 2: Construct an assembly layout

Figure 5.14 shows the critical assembly dimensions and the key interfaces that determine the path of tolerance stackup within the assembly.

According to Greenwood and Chase,

> *Y is the contact angle between the roller and the cage interface. It is the dependent assembly variable. It defines the point of contact where the clutch function is initiated. We seek to define the nominal value of Y and its required (functional) tolerance range. (Greenwood & Chase [R5.2])*

The dark inclined line in Figure 5.14 illustrates the approximate critical path of component stackup. The geometric relationship between all the stackup tangency points will serve to guide us in the next step's math modeling process.

Step 3: Construct a math model that defines the assembly variable in terms of the component variables

We are now ready to construct the mathematical function between the contact angle Y (the dependent assembly variable) and the independent component parameters x_1, x_2, x_3, and x_4:

$$Y = f(x_1, x_2, x_3, x_4)$$

It is important that the designer or engineer focus on the functional geometric relationships that govern how the components directly relate to the critical assembly variable. Inspecting the assembly reveals a trigonometric relationship that allows us to relate all four component variables to the contact angle Y, as shown in Figure 5.15.

Figure 5.14 Clutch assembly layout

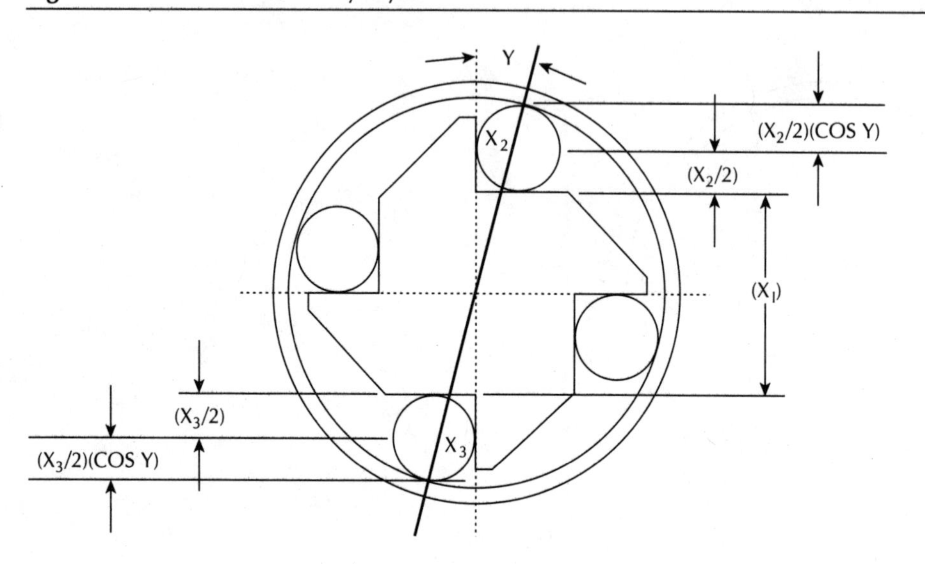

Figure 5.15

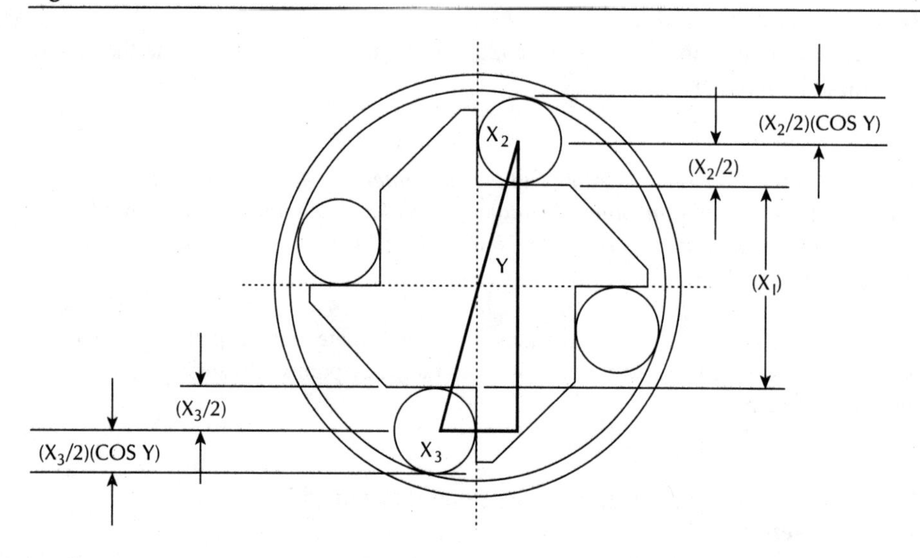

Defining the specific relationships between the components that stack up to create the assembly trigonometry is shown in Figure 5.16. Thus, we see how a nonlinear function unfolds from the trigonometry of this assembly.

$$Y = \cos^{-1}\left[\frac{x_1 + \dfrac{x_2 + x_3}{2}}{x_4 - \dfrac{x_2 + x_3}{2}}\right]$$

Figure 5.16 Trigonometric relationship

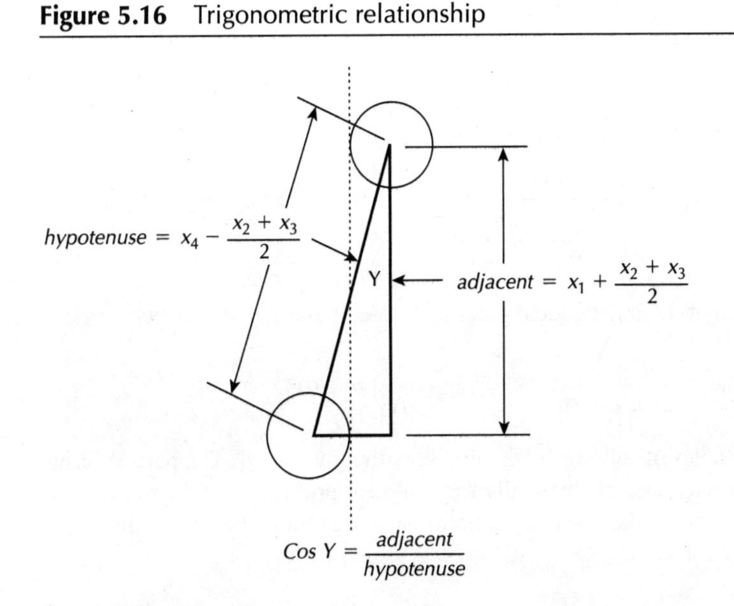

Step 4: Solve the math model for Y using the nominal component values

We had previously determined from engineering analysis (prior to or during Step 2), that the required tolerance range for the overrunning clutch needed to be within ± 0.035 radians. We now have an assembly specification, $Y = 0.122 \pm 0.035$ radians, to work with as an initial set of values:

$$x_1 = 55.29 \text{ mm}, x_2 = 22.86 \text{ mm}, x_3 = 22.86 \text{ mm}, x_4 = 101.60 \text{ mm}$$

$$Y = \cos^{-1}\left[\frac{x_1 + \dfrac{x_2 + x_3}{2}}{x_4 - \dfrac{x_2 + x_3}{2}}\right]$$

$$Y = \cos^{-1}\left[\frac{55.29 + \dfrac{22.86 + 22.86}{2}}{101.60 - \dfrac{22.86 + 22.86}{2}}\right]$$

$$Y = 0.1225 \text{ radians}$$

Step 5: Solve for partial derivatives

Step 5 is to use differential calculus to solve for the partial derivatives that represent the component sensitivities that numerically contribute to the nonlinearity of the stackup of the assembly:

$$\frac{\partial f}{\partial x_i}$$

For example, to take the partial derivative of the function with respect to x_1 we set up the problem as follows:

$$\frac{\partial_{\cos^{-1}}\left[\dfrac{x_1 + \dfrac{x_2 + x_3}{2}}{x_4 - \dfrac{x_2 + x_3}{2}}\right]}{\partial x_1}$$

The partial derivatives for each of the independent parameters are quantified as follows:

$$\frac{\partial f}{\partial x_1} = -0.1039, \quad \frac{\partial f}{\partial x_{2\&3}} = -0.1035, \text{ and } \frac{\partial f}{\partial x_4} = 0.1032$$

We will demonstrate the mechanics of solving for the first sensitivity value, x_1. The partial derivative process for working with sensitivity analysis manually for nonlinear problems will be shown step by step. The *sensitivity,* or rate of change, of the assembly function with respect to the rate of change of x_1 is defined by the partial derivative for x_1:

$$\frac{\partial f}{\partial x_1}$$

We start by defining the partial derivative for x_1 as

$$\frac{\partial_{\cos^{-1}}\left[\dfrac{x_1 + \dfrac{x_2 + x_3}{2}}{x_4 - \dfrac{x_2 + x_3}{2}}\right]}{\partial x_1}$$

We can simplify the function by noting that the roller diameters for x_2 and x_3 are the same. This allows us to modify the function for the rollers as follows:

$$\frac{x_2 + x_3}{2} = x_2$$

Since the expression for the roller geometries can be simplified to equal x_2, we can restate the equation as

$$\frac{\partial_{\cos^{-1}}\left[\dfrac{x_1 + x_2}{x_4 - x_2}\right]}{\partial x_1}$$

It is now a matter of recognizing that this function needs to be broken down into two manageable parts that can then be differentiated according to the rules of partial differentiation and general differential calculus. Using the derivative rule for ARCCOS we find that the derivative for

$$\left[\frac{x_1 + x_2}{x_4 - x_2}\right]$$

is

$$\frac{\partial_{\cos^{-1}}\left(\left[\frac{x_1 + x_2}{x_4 - x_2}\right]\right)}{\partial x_1} = \frac{-1}{\sqrt{1 - \left(\left[\frac{x_1 + x_2}{x_4 - x_2}\right]\right)^2}} \left[\frac{\partial\left(\left[\frac{x_1 + x_2}{x_4 - x_2}\right]\right)}{\partial x_1}\right]$$

For the second portion of the function we take the partial derivative of the expression:

$$\frac{\partial\left[\frac{x_1 + x_2}{x_4 - x_2}\right]}{\partial x_1}$$

This form is representative of the type of expression that can be differentiated using the quotient rule:

$$\frac{\partial\left[\frac{u}{v}\right]}{\partial x_1} = \frac{v\left(\frac{\partial u}{\partial x_1}\right) - u\left(\frac{\partial v}{\partial x_1}\right)}{v^2}$$

$$u = x_1 + x_2 \text{ and } v = x_4 - x_2$$

$$\frac{\partial u}{\partial x_1} = \frac{\partial(x_1 + x_2)}{\partial x_1} = \frac{\partial(x_1)}{\partial x_1} + \frac{\partial(x_2)}{\partial x_1} = \frac{\partial(x_1)}{\partial x_1} + x_2\frac{\partial(1)}{\partial x_1} = 1 + 0 = 1$$

$$\frac{\partial v}{\partial x_1} = \frac{\partial(x_4 - x_2)}{\partial x_1} = \frac{\partial(x_4)}{\partial x_1} - \frac{\partial(x_2)}{\partial x_1} = x_4\frac{\partial(1)}{\partial x_1} - x_2\frac{\partial(1)}{\partial x_1} = 0 + 0 = 0$$

$$\frac{\partial\left[\frac{x_1 + x_2}{x_4 - x_2}\right]}{\partial x_1} = \frac{(x_4 - x_2)(1) - (x_1 + x_2)(0)}{(x_4 - x_2)^2} = \frac{1}{x_4 - x_2}$$

The final sensitivity for the hub is then expressed as:

$$\frac{\partial f}{\partial x_1} = \frac{-1}{\sqrt{1 - \left(\left[\frac{x_1 + x_2}{x_4 - x_2}\right]\right)^2}}\left(\frac{1}{x_4 - x_2}\right)$$

Solving this expression by plugging in the nominal values for all four of the components gives the following sensitivity value:

$$\frac{\partial f}{\partial x_1} = -0.1039$$

The other sensitivities are solved in a similar manner. The sensitivities for the rollers and the cage are repeated below.

$$\text{Roller sensitivity } \frac{\partial f}{\partial x_{2\&3}} = -0.1035 \quad \text{Cage sensitivity } \frac{\partial f}{\partial x_4} = 0.1032$$

As you can see, nonlinear problems can require a good grasp of mathematics. It is recommended that the engineering team work together to develop these models so that each member can help add to the proper derivation and everyone has an understanding of how sensitivities are defined.

Step 6: Solve for the actual assembly tolerance

In Step 6, we use the tolerance Equation 5.17 to solve for the *actual* assembly tolerance as defined by the initial component tolerances and their individual assemblies.

$$\text{tol}_y = \left|\frac{\partial f}{\partial x_1}\right| \text{tol}_1 + \left|\frac{\partial f}{\partial x_2}\right| \text{tol}_2 + \left|\frac{\partial f}{\partial x_3}\right| \text{tol}_3 + \ldots + \left|\frac{\partial f}{\partial x_n}\right| \text{tol}_n$$

$$\text{tol}_y = |-0.1039|0.156 + |-0.1035|0.013 + |-0.1035|0.013 + |0.1032|0.156 = 0.035$$

The extreme limits may be calculated by substituting maximum or minimum values for x_1, x_2, x_3, and x_4 in Equation 5.18 as appropriate:

$$y_{\text{hi}} = .154 \text{ radians for the high limit of } Y \text{, and}$$

$$y_{\text{low}} = .080 \text{ radians for the low limit of } Y$$

The required tolerance range for the assembly is 0.157 to 0.087. Thus, we see that the actual assembly tolerance due to the component sensitivities and tolerances is shifted to the low side for both the high and low limits of the range. This is not a problem for the high limit but is a problem for the low limit, since we are 0.007 radians below the allowable limit from the low assembly tolerance based on engineering analysis. We will need to adjust the component nominal set points slightly to get the actual assembly tolerance range shifted up to coincide with the required range.

The first strategy is to attempt to adjust the component tolerances by negotiating with the component manufacturer. This will create a need for asymmetric tolerances on the components. It is simplest if the supplier can economically accommodate this request; if this is a problem, the component nominals can be shifted provided the design has the proper robustness latitude to provide adequate functional performance with small changes in the nominal set points.

Step 7: Shift the nominal set points

In Step 7 we shift the nominal set points to get the assembly tolerance range to adequately coincide with the required assembly tolerance range. The needed shift in the assembly tolerance range must be calculated.

$$\Delta y = \left[\frac{(y_{\text{high}} + y_{\text{low}})}{2}\right]_{\text{Required}} - \left[\frac{(y_{\text{high}} + y_{\text{low}})}{2}\right]_{\text{Actual}}$$

Entering the required and actual tolerance ranges reveals that there is shift difference equal to

$$\Delta y = \left[\frac{(.157 + .087)}{2}\right]_{\text{Required}} - \left[\frac{(.154 + .080)}{2}\right]_{\text{Actual}} = .005 \text{ radians}$$

We proceed by determining the common fraction (f) of the component tolerances that we will use to shift each component mean value.

$$(f) = \frac{\Delta y}{\frac{\partial(f)}{\partial x_1}(\text{sign})(\text{sign})(\text{tol}_1) + \frac{\partial(f)}{\partial x_2}(\text{sign})(\text{sign})(\text{tol}_2) + \ldots}$$

We will omit the rollers because they are a standard purchased part and have fixed tolerances that cannot be changed for this application; we will be basing (f) only on the values associated with the hub and cage:

$$(f) = \frac{0.005}{(-0.1039)(+)(-)(.156) + (0.1032)(+)(+)(.156)}$$

The value we will use to adjust the hub and cage nominal set points is

$$(f) = .1548$$

Step 8: Recalculate the new nominal set points for each of the components

The shifted nominal value for the hub and cage are now calculated:

$$x_{(i \text{ shifted})} = x_{(i \text{ nominal})} - (f)(\text{tol}_i)$$

For the hub (represented here by x) we get a new nominal to be set at

$$x = 55.29 - (.1548)(.156) = 55.2659$$

For the cage, represented here by y, we get a new nominal to be set at

$$y = 101.6 + (.1548)(.156) = 101.6241$$

Step 9: Recalculate the new assembly nominal value and tolerance range

$$\text{Nom}_y \approx \left|\frac{\partial f}{\partial x_1}\right| x_1 + \left|\frac{\partial f}{\partial x_2}\right| x_2 + \left|\frac{\partial f}{\partial x_3}\right| x_3 + \ldots + \left|\frac{\partial f}{\partial x_n}\right| x_n$$

$$\text{Nom}_y \approx |-0.1039|55.2659 + |-0.1035|22.86 + |-0.1035|22.86 + |0.1032|101.6241$$

The new nominal assembly value for the contact angle is:

$$\text{Nom}_y \approx 0.1274 \text{ radians}$$

$$\text{tol}_y = \left|\frac{\partial f}{\partial x_1}\right| \text{tol}_1 + \left|\frac{\partial f}{\partial x_2}\right| \text{tol}_2 + \left|\frac{\partial f}{\partial x_3}\right| \text{tol}_3 + \ldots + \left|\frac{\partial f}{\partial x_n}\right| \text{tol}_n$$

$$\text{tol}_y \approx |-0.1039|.156 + |-0.1035|.013 + |-0.1035|.013 + |0.1032|.156$$

The new assembly tolerance range is:

.1574 to .0874

The required assembly tolerance range is:

.1570 to .0870

As you can see, we have come very close to matching the worst-case tolerance range for this application. We will stop at this point in our discussion of nonlinear tolerance analysis. Greenwood and Chase go on [R5.2] to compute a third iteration and bring the calculated range even closer to matching the required range by recalculating the hub and cage component sensitivities (since the nominal values for two of the four components have been shifted).

This brings us to the end of the worst-case tolerance analysis material. While many may say this technique is not representative of statistical reality, it can and does play a valuable (though limited) role in modern tolerance analysis.

Summary

This chapter has served as an introduction to the various topics and techniques that are available for worst-case tolerance problems. We have focused on the fundamental aspects of worst-case analysis by working exclusively with linear and nonlinear one-dimensional cases. The Motorola notation for worst-case analysis has been reviewed, since many companies are using the methods available from Motorola University and the Six Sigma Institute.

Chapter 5: Linear and Nonlinear Worst-Case Tolerance Analysis

- Standard Worst-Case Methods
  - Six-Sigma Worst-Case Tolerance Analysis
  - Analysis for Nonlinear Tolerance Stacks Using the Worst-Case Method
  - Process Diagrams for Worst-Case Analysis for the Stackup of Tolerances in Assemblies
  - A Case Study for a Linear Worst-Case Tolerance Stackup Analysis
  - A Process for Modeling Worst-Case Nonlinear Tolerance Stack-Up Problems

Some of the key skills that one must build to adequately work with worst-case methods are:

- generating clear and precise graphical models of the assembly being toleranced,
- conducting a vector-chain analysis to quantify the assembly gap, and
- generating a math model that represents the geometry of the component interfaces.

We have used several examples and case studies to illustrate the basic methods of worst-case tolerance analysis for one-dimensional problems. Chapter 18 contains a real case study from Kodak that demonstrates the methods discussed in this chapter. The topics of two- and three-dimensional tolerance analysis are discussed in Chapters 6 and 8. See Appendix D for additional recommended reading on the topic of analytical tolerance analysis.

References

R5.1: *Six Sigma Mechanical Design Tolerancing;* M. J. Harry & R. Stewart, Publication No. 6σ-2-10/88, Motorola Inc., 1988, pg. 195.

R5.2: *Worst-Case Tolerance Analysis with Nonlinear Problems;* W. H. Greenwood and K. W. Chase, pg. 200, 217.

R5.3: *Statistics and Principles of Tolerance Analysis;* M. Hackert-Kroot & J. Nelson; Eastman Kodak Technical Training Manual, 1995, pg. 210.

CHAPTER 6

Linear and Nonlinear Statistical Tolerance Analysis

The real world of component manufacturing is dominated by the laws of probability, random chance, and special causes. This means that it is not possible to make a component exactly on target every time. All component manufacturing processes produce a distribution of output that is spread around the targeted output specification. When a manufacturing process is under statistical control it will produce a random, naturally occurring variability in the output values. That is why the responses are often called output *variables.* So it should come as no surprise that every process contains an inherent variability, as shown in Figure 6.1. [R6.1] [1]

Manufacturing processes and product designs can have another source of variability associated with their output, traditionally called *special-cause* variability. Quality-control engineers define this as any external or deteriorative source that moves the process from a state of random variation (statistical control) to a new, nonrandom state of variability that is beyond what is occurring due to natural (random) events. These special causes are represented by batch-to-batch variability, damaged or worn tools, contaminated raw materials, and numerous other noises. In the Taguchi paradigm, noises are always active in manufacturing processes and therefore all SPC charts are viewed as measuring external, unit-to-unit, and deterioration noise effects all the time. That is not to say all noise sources are active all the time. If one noise source is intensified or is somehow initiated within the manufacturing process or product design, then a special cause can and should be detected. Special causes are usually sources of variability that can be corrected during the manufacturing process. Typically, production engineers are always on the alert for sources of special-cause variability. The SPC charts shown in Figure 6.2 demonstrate what a process with and without special-cause variation can look like.[2]

1. H. Pitt. *SPC for the Rest of Us,* © 1994 Addison-Wesley Publishing. Reprinted by permission of Addison-Wesley Publishing Co., Inc.
2. Ibid.

Figure 6.1 Statistical process control charts [R6.1]

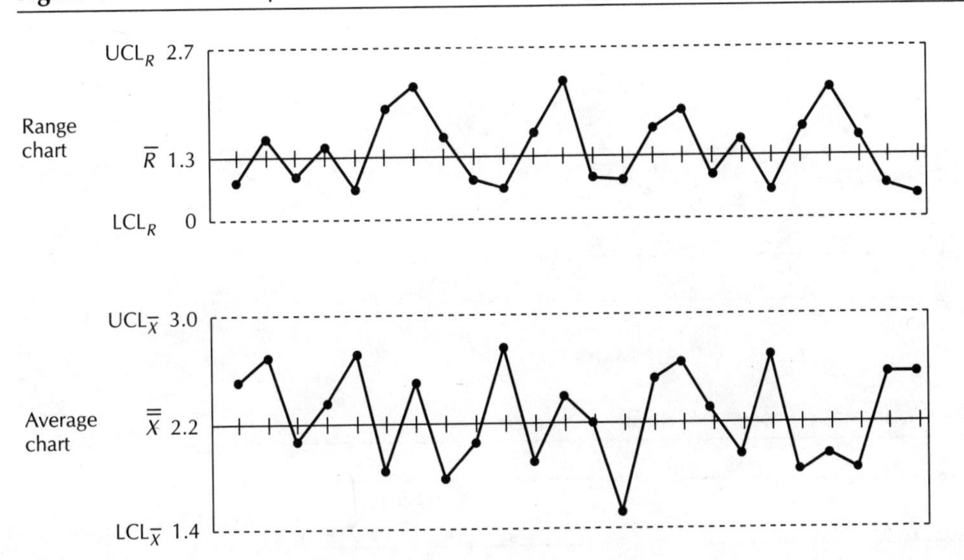

Figure 6.2 Process control charts [R6.1]

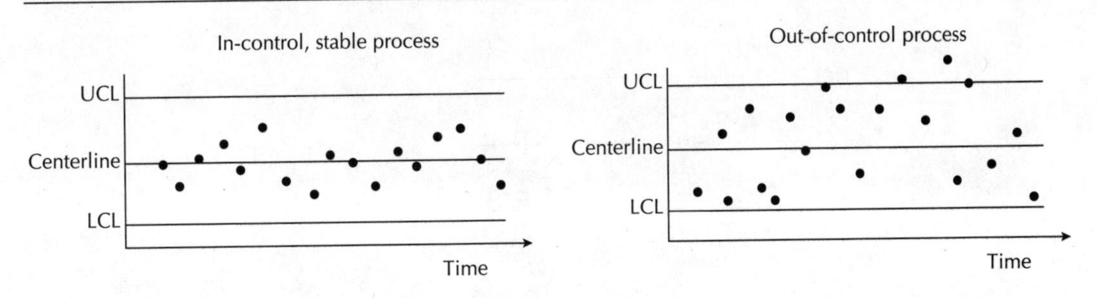

In the SPC chart shown in Figure 6.3 [R6.2] we see the famous "rule of seven" illustrated.[3] The rule of seven states that when seven or more measures fall above or below the mean on the chart, it is likely that a special cause (noise) has arisen in the design or manufacturing process.

Figure 6.4 shows other unique forms of performance to watch for in processes that are being monitored statistically.[4]

SPC is used to bring a process under control—that is, we mean to give it stable, natural variation about the mean, which is bounded by natural control limits. Statistical tolerance analysis assumes that all processes are in control, so all estimates of tolerance accumulation are due to natural process capabilities.

3. Ibid.
4. Ibid.

Figure 6.3 Rule of Seven [R6.2]

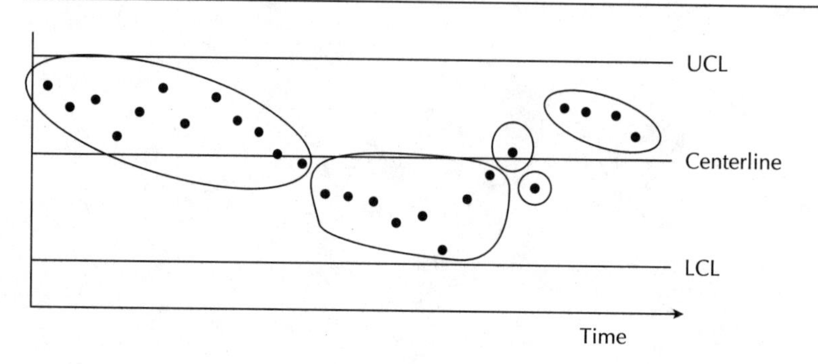

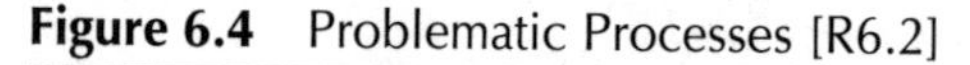

Figure 6.4 Problematic Processes [R6.2]

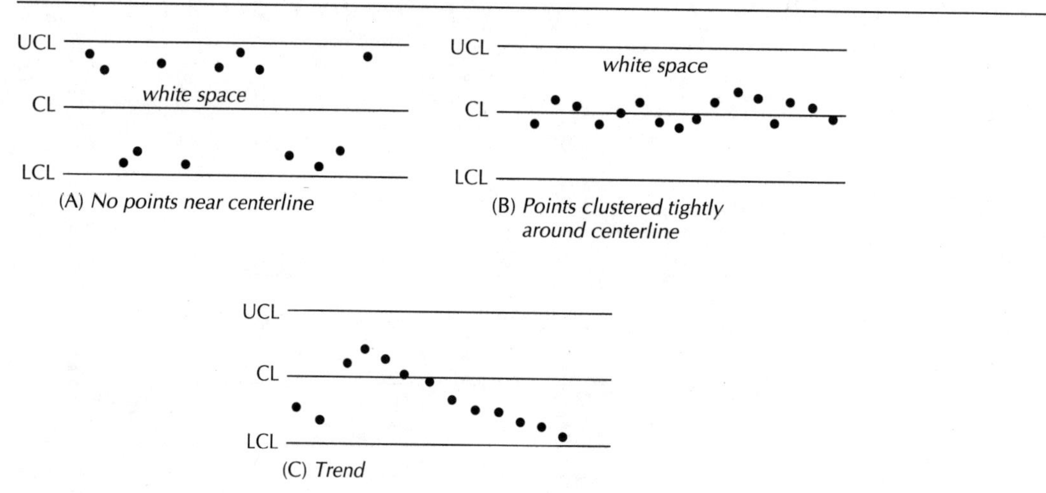

Over the years, the design engineering community in industry and academia has taken into account the probabilistic nature of manufacturing variability (unit-to-unit noise) and the outcome is our next topic—statistical tolerancing. The most widely used form of statistical tolerancing is called the root sum of squares (RSS) approach.

The Root Sum of Squares (RSS) Approach

The RSS approach is employed to account for the low likelihood of all dimensions occurring at their extreme limits simultaneously. The sum of squares is a mathematical treatment of the data to facilitate the legitimate addition of measures of variability. The RSS method is used to add up tolerance stacks when more than two components are assembled together. Often these assemblies must then fit into another assembly.

The RSS method is used to determine if a *functional* fit is going to occur between the mating assemblies. A probability of assembly success or failure can be calculated using the standard Z transform for population data, or by using the student-t transform for sample data. Refer to Chapter 3 to familiarize yourself with the simple methods of working with data in a probabilistic format using the Z and t transforms. The normal distribution is the basis for the use of these two transforms. We will assume that the sample data we are working with comes from reasonable approximations of normal distributions. The student-t distribution is designed to be more realistic in this respect. These methods will ultimately enable us to estimate, in parts per million, the number of parts that will fail to fall within the tolerances specified. We will not use the t transform, as it is insufficient for modern levels of quality (Six Sigma).

The Z Transformation Process

In Chapter 3 we demonstrated the use of the Z transform to convert population data that has been summarized into population statistics mean and sigma. One may ask why we use the Z transform when we almost never really have a complete population data set. It is common practice to assume that when one has gathered a large sample and there is evidence that the data came from a normal distribution, the Z transform is appropriate to use. This means that for any data set in excess of 30 samples that provides a distribution with an average and standard deviation that fit within the context of a normal distribution, we can convert the mean to equal zero ($\mu = 0$) and the standard deviation to equal one ($\sigma = 1$). Performing this transformation allows one to use the probability values associated with the area under the curve for a standard normal distribution to define the likelihood of an event's occurrence. In our case, the event is whether a part is likely to be within or outside the tolerances. If the data set is smaller than 30 samples it is recommended that one use the student-t distribution and transform.

We will use Z transforms when working with the RSS techniques for analyzing tolerances. They will help us to fully understand the quality ramifications of our tolerance limits and our stance with respect to six-sigma quality levels. We will return to the six-sigma terminology and techniques, including the Z transform, later, as we progress in our discussion of traditional RSS tolerancing techniques. The general model for an RSS analysis is shown in the material *directly* following:

One-Dimensional Tolerance Stacks

A given assembly is comprised of (n) components. Each component has a measured average and standard deviation associated with it due to variation in the manufacturing process. The sum or difference of each of these average component dimensions will define the average dimension, $\bar{d}$, of the assembly:

$$\bar{d} = \bar{d}_1 \,(+ \text{ or } -)\, \bar{d}_2 \,(+ \text{ or } -)\, \bar{d}_3 \,(+ \text{ or } -) \ldots (+ \text{ or } -)\, \bar{d}_n \qquad \textbf{(6.1)}$$

The sum of the squares of each of the dimensional standard deviations defines the overall assembly *variance:*

$$s^2 = s_1^2 + s_2^2 + s_3^2 + \ldots + s_n^2 \qquad \textbf{(6.2)}$$

It is statistically invalid and arithmetically wrong to add standard deviations (s). One must always convert standard deviations to variances (s^2) to add variability in this fashion. In statistical terms, this process is called the *summing of variances*. Physically speaking, standard deviations are not additive entities. The easiest way to be convinced of this is to run an experiment and take the measurements yourself. It can be proven analytically [R6.3], but reading about it does not have the same impact on your understanding of the concept.

The basis for all statistical tolerancing is defined by the following assembly tolerance stack equation:

$$T = \sqrt{T_1^2 + T_2^2 + T_3^2 + \ldots + T_n^2} \tag{6.3}$$

T_n represents the nth component tolerance in the assembly. The component tolerances are typically set equal to three standard deviations of the output from a given manufacturing process. Again, it is okay to begin at three sigma to start tolerance development and analysis (see Chapter 4); however, this *rarely* should be where the tolerance engineering stops. There are likely to be good reasons for tightening certain tolerances from this initial setting to improve quality or loosening them from the three sigma point, where functional quality is not affected and money can be saved. This leads us into the six-sigma literature and how it handles the RSS method.

Recall that the six-sigma literature employs an adjusted standard deviation if the actual process standard deviation is not known.

$$\sigma_{\text{adjusted}} = \frac{Tol}{3Cp}$$

This more realistic form of variation can now be used in the pooling of variances. The variance is merely the square of the standard deviation. Each component variance in an assembly is assumed to be independent of every other component variance. In this way each component will possess its own Cp value, which can be taken into account in the pooled variance calculation.

$$\sigma_{\text{Gap}} = \sqrt{\left(\frac{T_e}{3Cp}\right)^2 + \sum_{i=1}^{m}\left(\frac{T_{pi}}{3Cp_i}\right)^2} \tag{6.4}$$

The standard deviation of the gap of an assembly is expressed by the *square root* of the pooled variances from the envelope the assembly is to fit within and the sum of the component variances. Again, (i) represents each component from the complete group of (m) total components. Harry and Stewart [R6.4] draw an important conclusion about calculating the pooled variance in this fashion:

> *Because σ_{Gap} is positively correlated to Cp, the probability of exceeding a given gap constraint decreases as Cp increases, assuming a constant nominal condition for each of the components. Thus, for any given design, $\pm 6\sigma$ process performance gives us a higher probability of assembly than $\pm 3\sigma$ performance. [R6.4]*

We can now employ the Z transform to calculate the actual *probability of interference,* the probability of falling below the minimum allowable gap, and the probability of surpassing the maximum allowable gap. This analysis is equally useful for press fits where interferences are desired, but must be carefully controlled.

Harry and Stewart [R6.4] relate the Z transform to an assembly gap analysis through the following equation:

$$Z_Q = \frac{Q - \left(N_e - \sum_{i=1}^{m} N_{pi}\right)}{\sqrt{\left(\frac{Te}{3Cp}\right)^2 + \sum_{i=1}^{m}\left(\frac{Tpi}{3Cpi}\right)^2}} \tag{6.5}$$

where Z_Q is the number of standard deviations away from the nominal that relates to a condition of interference. Q represents the assembly gap between the envelope and the mating assembly. In this case $Q = 0$, since we are working with an interference represented by line-to-line contact. N_e is the nominal envelope dimension. N_{pi} represents the nominal dimension of any one (i) of (m) component parts. T_e is the tolerance of the envelope. Cp is the process capability index associated just with the envelope. T_{pi} represents the tolerance of any one (i) of (m) component parts. Cp_i is the individual process capability of any one (i) component part out of (m) total component parts.

This equation can be simplified into the following form:

$$Z_Q = \frac{Q - G_{\text{nom}}}{\sigma_G} \tag{6.6}$$

The table value (see Appendix A) associated with the calculated value of Z_Q is the probability of having a gap value equal to or less than $Q = 0$ (line-to-line contact).

Just as we can calculate a Z value that allows us to look up the probability of interference (see Chapter 3), we can calculate a Z value that permits us to look up the probability of exceeding an allowable maximum assembly gap. We use the term Z_R to represent the number of standard deviations away from the nominal gap that relates to the allowable *maximum* assembly gap $G_{\text{max}} = R$.

$$Z_R = \frac{R - \left(N_e - \sum_{i=1}^{m} N_{pi}\right)}{\sqrt{\left(\frac{T_e}{3Cp}\right)^2 + \sum_{i=1}^{m}\left(\frac{T_{pi}}{3Cp_i}\right)^2}} \tag{6.7}$$

The probability of being under a minimum gap can be found through Equation 6.8.

$$Z_{G\,\text{min}} = \frac{G_{\text{min}} - G_{\text{nom}}}{\sqrt{\left(\frac{T_e}{3Cp}\right)^2 + \sum_{i=1}^{m}\left(\frac{T_{pi}}{3Cp_i}\right)^2}} \tag{6.8}$$

The probability of exceeding a maximum gap can be found through Equation 6.9.

$$Z_{G\,\text{max}} = \frac{G_{\text{max}} - G_{\text{nom}}}{\sqrt{\left(\frac{T_e}{3Cp}\right)^2 + \sum_{i=1}^{m}\left(\frac{T_{pi}}{3Cp_i}\right)^2}} \tag{6.9}$$

All of the above equations relating probabilities to gap conditions, tolerances, and process capabilities are built on the assumption that the nominal conditions are, in fact, stable over time, and the means are not shifting about as a consequence of noise. This, we know, is simply not realistic over any rational length of time. These previous equations are built upon the short-term process capability index, Cp. Even though these equations all allow individual Cp's to be associated with individual component parts, which is a significant improvement over the basic "one-size-fits-all" approach embodied in the three-sigma paradigm, it still leaves the analysis at risk of being inaccurate in the context of long-term process capability. Recall that long-term process capability is accounted for by the factor Cpk. This is the next topic in our RSS discussion.

Motorola's Dynamic Root Sum Square Approach [R6.4]

The dynamic RSS (DRSS) approach is included in this discussion because it is rare, over time, for the manufacturing process mean to remain centered between the bilateral tolerance limits at the nominal design target. Process means will shift around during the weeks, months, and years of component-part production. External and deterioration noises active in the process are what cause the mean shifts to occur. How does this approach relate to the RSS approach? Harry and Stewart [R6.4] state that

> *. . . it [RSS] does not adequately account for shifts and drifts in the process mean. It assumes the process mean is always centered on the nominal design specification. This is why $\pm 3\sigma$ capability initially appears more than adequate. Unfortunately this is seldom the case, particularly in situations where the manufacturing operation is subject to tool wear. [R6.4]*

We will demonstrate that $\pm 3\sigma$ may be an inadequate tolerance base upon which to build process capability. A Cp of 1 will be problematic when the process mean starts to shift. The best way to deal with this problem is to engineer concepts that can be developed to a level that permits the tolerance to be set at $\pm 6\sigma$. This will produce a Cp of 2—by design. Then, when known sources of manufacturing noise become active, the shifts in the mean will not be anywhere near as troublesome as they would have been for a Cp of 1.

At Motorola, this process is called "designing for manufacturability." It is also another way of designing for robustness. In this case, the design is engineered to be robust against the mean shifts associated with long-term manufacturing of the component parts. In this way the development engineering team designs the quality in so that it is not an overwhelming issue when manufacturing begins. The six-sigma approach defines an additional set of tolerancing equations that take the mean shift into account. The long-term process capability index, Cpk, is the basis for the dynamic RSS approach (see Chapter 3). The following equations are simple modifications of the RSS equations 6.4 through 6.9.

$$\sigma_{\text{adjusted}} = \frac{\text{Tolerance}}{3Cpk}$$

The pooled variance model is now compensated by including Cpk. The standard deviation of the gap is defined in Equation 6.10:

$$\sigma_{\text{Gap}} = \sqrt{\left(\frac{T_e}{3Cpk}\right)^2 + \sum_{i=1}^{m}\left(\frac{T_{pi}}{3Cpk_i}\right)^2} \qquad \textbf{(6.10)}$$

The Z value is also altered to account for the long-term shifting of the mean. The Z transform has been made quite flexible by inclusion of the subscript F, which is used to represent any gap constraint such as Q, R, G_{max}, or G_{min}:

$$Z_F = \frac{F - \left(N_e - \sum_{i=1}^{m} N_{pie}\right)}{\sqrt{\left(\frac{T_e}{3Cpk}\right)^2 + \sum_{i=1}^{m}\left(\frac{T_{pi}}{3Cpk_i}\right)^2}} \quad \textbf{(6.11)}$$

Equation 6.11 can be further generalized:

$$Z_F = \frac{F - \sum_{i=1}^{m} N_i V_i B_i}{\sqrt{\sum_{i=1}^{m}\left(\frac{T_i B_i}{3Cpk_i}\right)^2}} \quad \textbf{(6.12)}$$

where N_i is the nominal component part target, V_i is the direction vector (± 1), and B_i is the correction factor for true position applications (see Chapter 5 for more on these terms).

The result of using Cpk instead of Cp is that all the standard deviations will increase, including the standard deviation of the gap. The Z value will drop in magnitude and the probability associated with it will indicate a lower likelihood of proper assembly between the mating parts. This is reality in the world of long-term production. As stated in Chapter 3, Cpk is typically understood to account for the general fact that long-term process mean shifts are on the order of ± 1.5 standard deviations. This is the approximate amount of degradation one can typically expect to encounter. In this chapter we will illustrate the difference in assembly probability between the RSS and DRSS approaches.

Realistic tolerances are very important in the communication of design specifications to the manufacturing community. It should be quite clear to the reader that *what* tolerances are communicated is as important as *how* they are communicated.

Motorola's Static Root Sum Square Approach [R6.4]

The static root sum of squares approach is another way of accounting for the fact that an individual component's manufacturing process mean can shift. This too will have an effect on the overall assembly. In the static case, the term employed in the six-sigma literature is *sustained process mean shift.* This means that the manufacturing process undergoes a transition so that the nominal set point is no longer where the targeted output response resides—in fact, the process mean moves (k) standard deviations away from the nominal target and then assumes a *static* condition. A dynamic case is defined as a continuous shifting of the process mean where, over the long run, the average of all the short-term mean shifts tends to fall near the nominal target set point. This is in contrast with a static mean shift, where a component mean moves off target and stays off target.

The following Z transform is used as a general model to account for the static mean shift. The difference is that we have replaced Cpk (k) in the denominator with Cp and placed (k) in the numerator as it pertains to each component part. The denominator is now based on the short-term process capability Cp. W_i is a new term that is called the *shift vector* that is associated with each (i) nominal dimension. This vector has the informational content pertaining to the *directionality* of the sustained mean shift. The term k_i defines the magnitude of the sustained mean shift and is used to express this magnitude in terms of some proportion of the tolerance T_i; Z_F, V_i, and B_i have already been defined.

$$Z_F = \frac{F - \sum_{i=1}^{m} (N_i + W_i k_i T_i) V_i B_i}{\sqrt{\sum_{i=1}^{m} \left(\frac{T_i B_i}{3Cp_i}\right)^2}} \tag{6.13}$$

Harry and Stewart [R6.4] are careful to state that using k provides us with a plausible and flexible mechanism for both dynamic and static analyses, while simultaneously ensuring independence between the process mean and variance.

The probability of assembly associated with the Z transform adjusted to account for sustained shifts in the mean will be lower than the RSS Z transform that assumes the mean stays centered on the nominal. Again, this is a sound dose of reality for the good of all who know enough to account for the debilitating effects of noise within the tolerancing process. We will have more to say on this unique process later in this chapter.

The Nonlinear RSS Case Method

We will close the RSS section of this chapter by extending our discussion around the solution of nonlinear tolerance cases to include the RSS approach. We can simply modify the general tolerance equations defined by Greenwood and Chase [R6.5] to include the statistical measures required to carry out an RSS analysis. The variances of the component dimensions are substituted in Equation 6.14:

$$\text{var}_y = \left(\frac{\partial f}{\partial x_1}\right)^2 \text{var}_1 + \left(\frac{\partial f}{\partial x_2}\right)^2 \text{var}_2 + \left(\frac{\partial f}{\partial x_3}\right)^2 \text{var}_3 + \ldots + \left(\frac{\partial f}{\partial x_n}\right)^2 \text{var}_n \tag{6.14}$$

This is done by bringing the variance into the first order terms of a Taylor series expansion as described by Hahn and Shapiro [R6.6]. Greenwood and Chase [R6.5] use this expression to define the tolerance equation below:

$$\text{tol}_y = \sqrt{\left(\frac{\partial f}{\partial x_1}\right)^2 \text{tol}_1^2 + \left(\frac{\partial f}{\partial x_2}\right)^2 \text{tol}_2^2 + \left(\frac{\partial f}{\partial x_3}\right)^2 \text{tol}_3^2 + \ldots + \left(\frac{\partial f}{\partial x_n}\right)^2 \text{tol}_n^2} \tag{6.15}$$

They state:

> *In the special case when the assembly equation is linear, the expression for the assembly mean and variance are exact. It is common practice to model the component variability as normal distributions with the mean at the tolerance zone midpoint, and the tolerance limits at plus and minus three standard deviations. The variance equation for an assembly,*

shown [in Equation 6.14], can then be written in terms of the tolerances as shown [in Equation 6.15]. The symbol (tol) can refer either to the full tolerance, corresponding to six standard deviations, or to equal bilateral tolerances, which corresponds to three standard deviations.

The sensitivity factors represented in the partial derivatives are exactly like those discussed in the worst-case analysis (see Chapter 5). They will be determined in the same way. The tolerances can be expressed as necessary by the six-sigma approach to account for long-term mean shifts where Cpk is employed to modify the expression of the variance. Thus, the tolerance terms can be replaced by using a modification of the adjusted standard deviation from each component as necessary:

$$\sigma_{\text{adjusted}} = \frac{\text{Tol}_i}{3C_{pk}}$$

$$\text{Tol}_i = 3C_{pk} \times \sigma_{\text{adjusted}}$$

We now have a fairly diverse set of tools to employ for statistical tolerancing of components and assemblies. We can account for both linear and nonlinear stackups in a statistical tolerancing process. In addition, we can see the effect of including either short-term variation, which doesn't include mean shifts, or long-term variation, which does include the harsh reality of dynamic mean shifts. We can also account for a sustained (static) process mean shift during the normal variation of the component during manufacture.

Process Diagrams for the Statistical Methods of Tolerance Stackup Analysis

The methods we have employed thus far have allowed us to account for worst-case situations in tolerance analysis. We will alter the tolerance process diagrams to account for the fact that variability occurs in a distributed fashion. We will add the tools of statistics to the process of developing tolerances.

The Linear RSS Method

The application of the root sum of squares method to tolerance analysis is common in industry today. When the components have a direct or linear stackup, we can simply alter the worst-case approach to account for the probabilistic or random nature of how dimensions can vary in the real world. The process shown in Figure 6.5 is representative of how a general linear RSS tolerance analysis should be performed.

Figure 6.6 illustrates the root sum of squares tolerance analysis process, including the six-sigma approach, for a one-dimensional problem:

The steps for the non-linear process are illustrated in the following example.

Step 1: Parts A, B, and C are the components that define an assembly. When they are joined, they must then be placed inside the boundary established by Part D, often called a design envelope or an assembly constraint. The nominal dimensions have come from parameter design

Figure 6.5 Nonlinear tolerance process

1. Analyze the geometric and spatial relationship between the way the components fit together to make up the assembly. Then assess the assembly's geometric and spatial relationship relative to the product's functional capability. Define the functional purpose of the stacked components (relate geometry to functionality).

2. Define the specific geometric requirement of the assembly. This may be done by quantifying the overall assembly specifications (G_{nom}, G_{max}, and G_{min} from Chapter 5).

3. Either *estimate* (from lookup tables, manufacturing literature, or recommendations) or *allocate* (from a previously determined overall assembly specification) an initial tolerance for each component in the assembly using the aforementioned initial tolerance identification process. Each component must be assessed on an individual basis.

4. Each tolerance is initially assumed to be set at plus or minus three standard deviations from the nominal set point. Thus, the standard deviation of the production process output is *assumed* to be approximately tol/3. It is highly recommended that one obtain manufacturing data to support this assumption and make adjustments.

5. The assembly tolerance can now be assessed because it is now clear that the component-part tolerances are based upon three sigma levels from normal distributions. The assessment is made by employing the RSS equation:

$$T_{assy} = \sqrt{T_1^2 + T_2^2 + T_3^2 + \ldots + T_n^2}$$

The nominal gap dimension (G_{nom}) can be calculated and its tolerances assigned. Sometimes G_{nom} and its tolerances will be specified as a requirement and the component parts have to be developed to equal G_{nom} and its tolerance values. When an allocation process must be used, the component tolerances will have to be tuned to support the gap specifications. This is referred to as *tolerance allocation*. The disciplined summation process defined in Chapter 5, including using the sign convention of displacements to the right being positive, should be used.

Figure 6.6

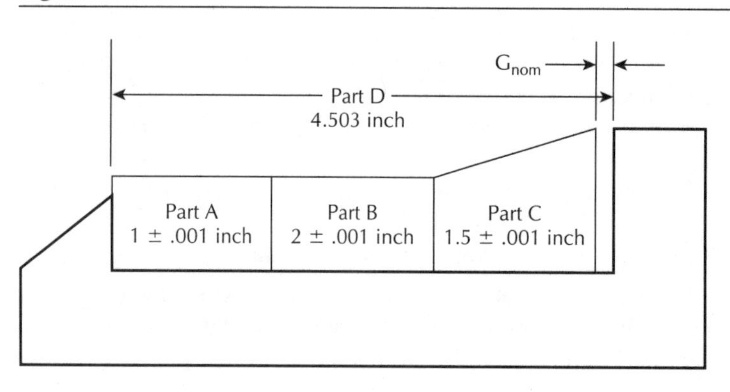

6. Six-sigma literature allows us the option of using the relationship:

$$\sigma_{\text{adjusted}} = \frac{Tol}{3Cp}$$

to state that an adjustable format for assigning tolerances is available based upon the short-term process capability (Cp). This is done when we have process data to prove that the three sigma paradigm is *not* in effect ($Cp \neq 1$). The known Cp can be used to adjust each component tolerance so that the affect is aggregated up to the assembly tolerance. Then we can calculate an estimate for the gap standard deviation:

$$\sigma_{\text{gap}} = \sqrt{\left(\frac{T_e}{3C_p}\right)^2 + \sum_{i=1}^{m}\left(\frac{T_{pi}}{3C_{pi}}\right)^2}$$

From this expression we can then calculate the probability of being above or below a specified gap using the Z transform from chapters 3 and 5:

$$Z_{G\min} = \frac{G_{\min} - G_{\text{nominal}}}{\sqrt{\left(\frac{T_e}{3C_p}\right)^2 + \sum_{i=1}^{m}\left(\frac{T_{pi}}{3C_{pi}}\right)^2}}$$

and

$$Z_{G\max} = \frac{G_{\max} - G_{\text{nominal}}}{\sqrt{\left(\frac{T_e}{3C_p}\right)^2 + \sum_{i=1}^{m}\left(\frac{T_{pi}}{3C_{pi}}\right)^2}}$$

In general form,

$$Z_{\text{gap}} = \frac{G_{\text{max or min}} - G_{\text{nominal}}}{\sigma_{\text{gap}}}$$

optimization. The relationship is a simple linear tolerance stackup. This means the sensitivities (partial derivatives) are all equal to 1.

Step 2: We are interested in having the assembly of parts A, B, and C fit into the boundary of Part D without any interference. Thus, the problem is to define the proper specifications for Part D so that when the RSS stackup of the assembly of parts A, B, and C occurs, there is no interference. Often, a finite but small assembly gap is required to facilitate physical assembly during the product-manufacturing process.

Step 3: Assign the initial tolerances to the nominal dimensions according to the recommendation of the manufacturing engineer responsible for the fabrication of the component parts. The initial tolerances represent ±3 standard deviations from the measured samples of actual production parts.

Step 4: We can now calculate the RSS stackup dimension for the assembly of parts A, B, and C.

This is what can be reasonably expected from random selection of components during assembly.

$$T_{\text{assy}} = \sqrt{T_1^2 + T_2^2 + T_3^2}$$

$$T_{\text{assy}} = \sqrt{0.001^2 + 0.001^2 + 0.001^2}$$

$$T_{\text{assy}} = 0.001732 \text{ in.}$$

Step 5: Assess the RSS problem when the Cp is not equal to 1. This will require the use of the six-sigma approach to RSS tolerance analysis. Calculate the number of parts that will fall outside of the tolerance limits using the Z transformation process.

We can decompose the three component tolerances A = ± 0.001 in., B = ± 0.001 in., and C = ±0.001 in. into their adjusted standard deviations using the following expression:

$$\sigma_{\text{adjusted}} = \frac{\text{Tol}}{3Cp}$$

When the Cp = 1—three-sigma-level quality (Z = −3 and +3, with the sum of the two tail probabilities for exceeding tolerance = 0.00135 + 0.00135 = .0027) where 2,700 parts/million will exceed the tolerance limits—the three components will have the following standard deviations:

$$\sigma_{\text{A}} = 0.000333 \text{ in.}$$

$$\sigma_{\text{B}} = 0.000333 \text{ in.}$$

$$\sigma_{\text{C}} = 0.000333 \text{ in.}$$

When the Cp = 2 (six-sigma level quality, where about 0.002 parts/million will exceed the tolerance limits), the three components will have the following standard deviations:

$$\sigma_{\text{A}} = 0.0001667 \text{ in.}$$

$$\sigma_{\text{B}} = 0.0001667 \text{ in.}$$

$$\sigma_{\text{C}} = 0.0001667 \text{ in.}$$

What does this mean in terms of our capability to calculate Z values for the probability of the assembly gap exceeding some tolerance limit?

Let's look at the Z values for a gap scenario in which Cp = 1, and then in which Cp = 2 for all the components and the assembly envelope. We will assess the tolerances and the number of assemblies falling outside the gap tolerance based on the gap statistic σ_{gap}.

We must first calculate the σ_{gap} based on the square root of the square of the tolerance of the assembly envelope and the *sum* of the squares of the component tolerances.

$$\sigma_{\text{gap}} = \sqrt{\left(\frac{T_e}{3Cp}\right)^2 + \sum_{i=1}^{m}\left(\frac{T_{pi}}{3Cp_i}\right)^2}$$

For Cp = 1:

$$\sigma_{\text{gap}} = \sqrt{\left(\frac{T_e}{3}\right)^2 + \sum_{i=1}^{m}\left(\frac{T_{pi}}{3}\right)^2}$$

For Cp = 2:

$$\sigma_{\text{gap}} = \sqrt{\left(\frac{T_e}{6}\right)^2 + \sum_{i=1}^{m}\left(\frac{T_{pi}}{6}\right)^2}$$

Calculating the Z value for the minimum gap we want to tolerate (in this case let's use 0.001 in. for $G_{\min}$ and 0.003 in. for G_{nom}) when the Cp = 1:

$$Z_{G\min} = \frac{G_{\min} - G_{\text{nom}}}{\sqrt{\left(\frac{T_e}{3Cp}\right)^2 + \sum_{i=1}^{m}\left(\frac{T_{pi}}{3C_{pi}}\right)^2}}$$

$$Z_{G\min} = \frac{0.001 - 0.003}{\sqrt{\left(\frac{0.001}{3}\right)^2 + \left(\frac{0.001}{3}\right)^2 + \left(\frac{0.001}{3}\right)^2 + \left(\frac{0.001}{3}\right)^2}}$$

$$Z_{G\min} = \frac{-0.002}{0.000667}$$

$$Z_{G\min} = -2.9985$$

Consulting a standard Z table (see Appendix A) for the tail probability associated with a Z value of −3 standard deviations away from the nominal gap, normalized to equal zero, we find a value of approximately 0.00135. This is the probability for only the left, or lower, side of the normal distribution we are using to represent the variability associated with the possible gap widths in our assembly and the part within which it is intended to mate.

Considering both sides of the tails of the normal distribution we see that the sum of the upper and lower tail probabilities states that:

$$0.00135 + 0.00135 = 0.0027$$

This means that when we consider one million assemblies, 1,350 will fall out of the lower tolerance boundary. This is equivalent to three-sigma assembly gap quality. Now let's see how things work out with a manufacturer that can supply six-sigma part quality.

Z for Cp = 2:

$$Z_{G\min} = \frac{G_{\min} - G_{\text{nom}}}{\sqrt{\left(\frac{T_e}{3Cp}\right)^2 + \sum_{i=1}^{m}\left(\frac{T_{pi}}{3C_{pi}}\right)^2}}$$

$$Z_{G\min} = \frac{0.001 - 0.003}{\sqrt{\left(\frac{0.001}{6}\right)^2 + \left(\frac{0.001}{6}\right)^2 + \left(\frac{0.001}{6}\right)^2 + \left(\frac{0.001}{6}\right)^2}}$$

$$Z_{G\min} = \frac{-0.002}{0.000333}$$

$$Z_{G\min} = -6.0$$

Again, consulting a standard Z table for the tail probability associated with a Z value of −6 standard deviations away from the nominal gap, normalized to equal zero, we find that the table value is 1.248 E – 09. This means that there is going to be little problem with the mating process.

Considering both sides of the tails of the normal distribution we see that the sum of the upper and lower tail probabilities states that:

$$1.248\ \text{E}-09 + 1.248\ \text{E}-09 = 2.496\ \text{E}-09$$

This means that when we consider one million assemblies, 0.001248 will fall out of the lower tolerance boundary. This is equivalent to six-sigma assembly gap quality. This illustrates the dramatic effect that six-sigma component quality has on the number of tolerable assemblies.

Concept design and parameter design can pay great dividends in broadening the tolerance limits (numerator of Cp). Concurrently optimizing both the tolerance limits for broad latitude and the variation in component-part quality will provide the fewest assembly defects at the lowest cost. So far, in this case, we have assessed the number of assemblies that will fall below the assembly gap tolerance with the short-term process capability (Cp). This is helpful but is short-sighted. We must consider the long-term manufacturing process capability, Cpk, to see the true ramifications of ongoing component part production on assembly quality.

The Dynamic RSS Method [R6.4]

The dynamic method of RSS tolerance analysis is the most realistic and demanding of the traditional methods we will discuss. It accounts for long-term process variability through the quantification of the effect of the dynamic shifting of the mean over long periods of time. It is demanding because it requires discipline to gather and process the data required to determine the Cpk for the components in question. In addition to the short-term adjusted standard deviation, we must establish a long-term adjusted standard deviation:

$$\sigma_{\text{adjusted}} = \frac{\text{Tol}}{3C_{pk}}$$

With this in hand we can work with a Cpk-based version of Equation 6.4 to determine the gap standard deviation:

$$\sigma_{\text{gap}} = \sqrt{\left(\frac{T_e}{3C_{pk}}\right)^2 + \sum_{i=1}^{m}\left(\frac{T_{pi}}{3Cp_i}\right)^2}$$

For this case we will define a Z value based on the Cpk = 1.5 (equivalent to a Cp = 2 but adjusted for the anticipated long-term shift of 1.5 sigma around the mean or target value):

$$Z_{G\,\min} = \frac{G_{\min} - G_{\text{nom}}}{\sqrt{\left(\frac{T_e}{3C_{pk}}\right)^2 + \sum_{i=1}^{m}\left(\frac{T_{pi}}{3C_{pk_i}}\right)^2}}$$

$$Z_{G\,\min} = \frac{0.001 - 0.003}{\sqrt{\left(\frac{0.001}{4.5}\right)^2 + \left(\frac{0.001}{4.5}\right)^2 + \left(\frac{0.001}{4.5}\right)^2 + \left(\frac{0.001}{4.5}\right)^2}}$$

$$Z_{G\,\min} = \frac{-0.002}{0.00044}$$

$$Z_{G\,\min} = -4.545$$

The standard Z table for the tail probability associated with a Z value of −4.545 standard deviations away from the nominal gap, normalized to equal zero, gives a value of approximately 2.475 E−06.

This means that when we consider one million assemblies, about 2.5 are estimated to fall out of the lower tolerance boundary. This is roughly equivalent to 4.5-sigma quality at the assembly gap level. This illustrates the effect that long-term shifts in the mean will have on the number of tolerable assemblies. Unacceptable assemblies are going to occur; they are the evidence of the fact that variability is going to be active in the ongoing production of the design. It is the team's responsibility to account for this phenomenon in the up-front design processes during product commercialization. Hopefully it is becoming clear why it is important to consider manufacturing process capabilities early, when specific components are still at the concept development stage. Long-term process variation can potentially be frozen into the design during concept design. The only true hope for substantial improvement without significant cost is to perform robustness optimization on the manufacturing process parameters and raw materials concurrent with design parameter optimization. The tolerancing process will only allow the team to balance costs as the design ends its maturation process just prior to production.

When a process mean shifts at the individual component level but does not do so in a manner that is considered dynamic at the assembly level then we must use the six-sigma methods to look at the problem a little differently.

The Static RSS Method [R6.4]

In this case the individual component means may shift, and their shift may be sustained over an extended period. This is handled by employing the general static RSS Equation 6.13:

$$Z_F = \frac{F - \sum_{i=1}^{m} (N_i + W_i k_i T_i)\, V_i B_i}{\sqrt{\sum_{i=1}^{m} \left(\frac{T_i B_i}{3C_{pi}}\right)^2}}$$

F represents some quantifiable assembly geometry (in our case it is the minimum assembly gap $G_{\min}$), N_i is the nominal component dimension, W_i is the sign (vector) associated with the tolerance (in this case we use the signs that provoke the worst-case assembly scenario), k_i is the sustained mean shift for any individual component (in this case we will use 1.5 standard deviations to represent the mean shift), and T_i is the tolerance for any individual component. We will not use V_i or B_i since we are not working with true positioning in this case.

For the assembly problem at hand we see that the Z value depends on a mix of both a component-based mean shift factor k_i that is applied as a proportion of the tolerance in the numerator (along with a worst-case directionality vector ± 1 represented by W_i) and the short-term process capability (Cp) in the denominator. We will assume the Cp = 1 and k = 1.5σ for all three components. It is typical to employ the static RSS method in such a manner as to provoke the worst-case scenario. We do this by selecting the sign of the tolerance (W_i) that forces the stack to sum up to a small value (Part D is made small and parts A, B, and C are made large), thus driving Z_F to be low in value:

$$Z_F = \frac{0.001 - \sum_{i=1}^{m} (N_i + W_i (1.5) T_i)}{\sqrt{\sum_{i=1}^{m} \left(\frac{T_i}{3}\right)^2}}$$

Summing for each worst-case component mean shift in the stack circuit in the numerator according to Equation 6.13:

$$\text{Part A: } 1.000 + (+)(1.5)(.001) = 1.0015 \text{ in.}$$

$$\text{Part B: } 2.000 + (+)(1.5)(.001) = 2.0015 \text{ in.}$$

$$\text{Part C: } 1.500 + (+)(1.5)(.001) = 1.5015 \text{ in.}$$

$$\text{Part D: } 4.503 + (-)(1.5)(.001) = 4.5015 \text{ in.}$$

$$4.5015 - (1.0015 + 2.0015 + 1.5015) = -0.003 \text{ in.}$$

Remember that the assembly gap circuit requires us to *subtract* the assembly from Part D.

Summing to pool the variances for each component in the denominator:

$$\sqrt{\sum_{i=1}^{m} \left(\frac{T_i}{3C_{pi}}\right)^2} = \sqrt{4\left(\frac{.001}{3}\right)^2} = .000667$$

$$Z_F = \frac{0.001 - (-0.003)}{.000667}$$

$$Z_F = 5.997$$

Thus, we see that when all three components undergo process mean shifts equivalent to 1.5 standard deviations, the probability of the assembly gap being within tolerance is affected. We see six-sigma-level quality at the gap assembly. This means there will be approximately 0.001 assemblies per million that will not fall within the lower tolerance limit. The static RSS approach is yet another way to alert our engineering team to the effects of variability in the tolerancing of components and assemblies. It is helpful to have some idea of how many discrepant parts or assemblies can be anticipated as a consequence of the tolerancing and manufacturing processes we choose to employ.

A case study for a linear RSS case tolerance stackup analysis [R6.7]

Let's look at a realistic case of linear RSS case tolerance analysis.[5] The case involves two mating parts that must fit together, as shown in Figure 6.7.

Figure 6.8 shows two scenarios for the mating parts. Figure 6.9 illustrates one possible set of specifications for parts A and B. We shall explore these tolerances to see if they work for an RSS scenario.

---

5. This case was developed at Eastman Kodak Company by Margaret Hackert-Kroot and Juliann Nelson.

Figure 6.7 Mating pin-and-hole assembly

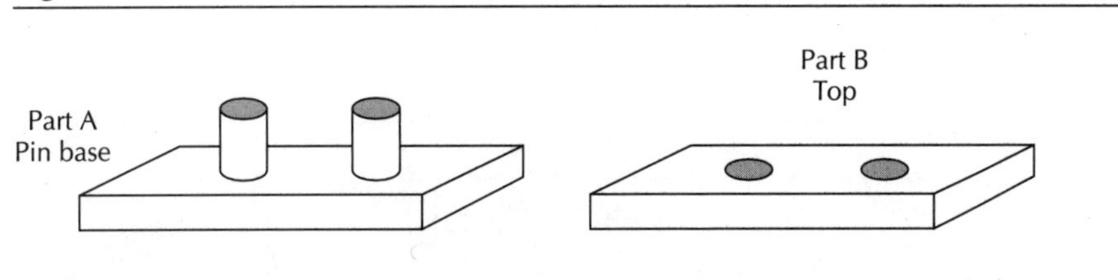

Figure 6.8 Two RSS case scenarios

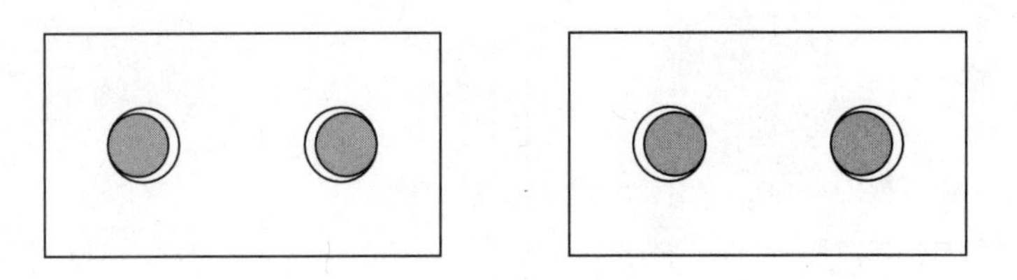

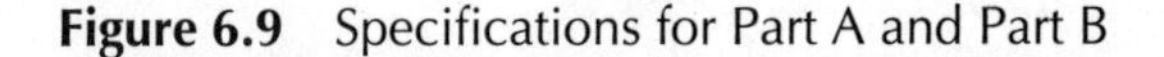
Figure 6.9 Specifications for Part A and Part B

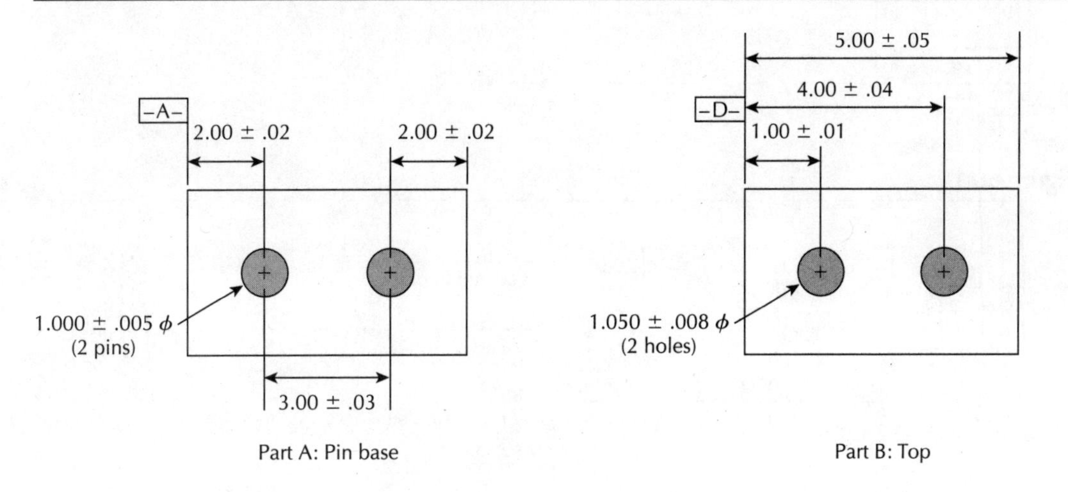

In Figure 6.10, the analysis begins at the right edge of the right pin. One must always select a logical starting point for a stack analysis. Typically, it will be at one side of an unknown critical gap dimension (the distance from the start of **a** to the end of **g**), as shown in Figure 6.10. We will step across the assembly from point a to point g to sum up all the stackup dimensions according to their sign. Recall that the arrowed lines in Figure 6.10 represent displacement vectors, and that the sign for each displacement vector is illustrated in the figure. The displacement vectors and their associated tolerances are quantified as follows. These tolerance values are squared by the rule presented in Equation 6.3, presented on pg. 241. Notice that each tolerance for a diameter has been divided by 2 to convert it to a radial tolerance. Analyzing the results we find that there is a +0.05 nominal gap and +0.051 tolerance buildup for the RSS case in the positive direction.

Figure 6.10 RSS case tolerance stackup example

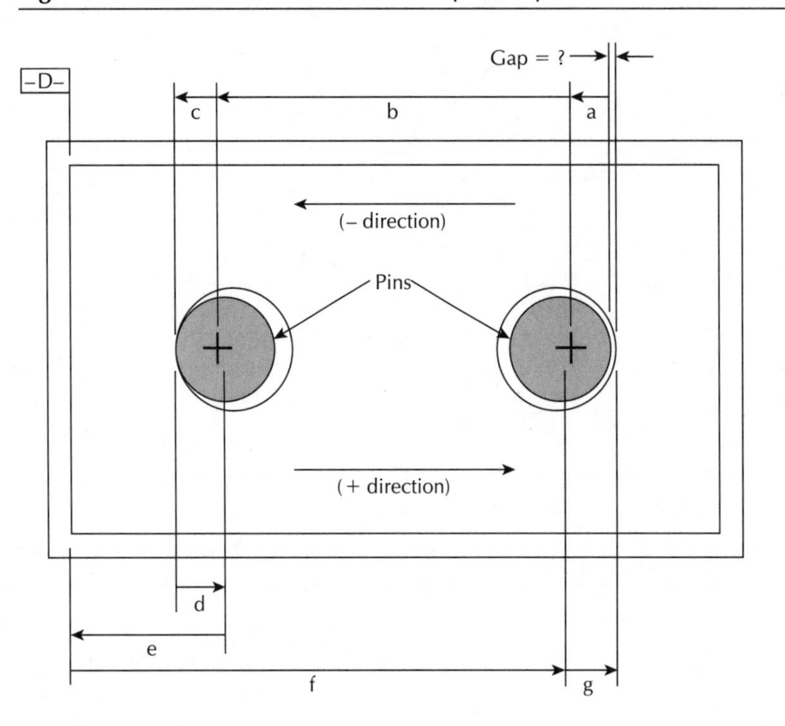

Figure 6.11

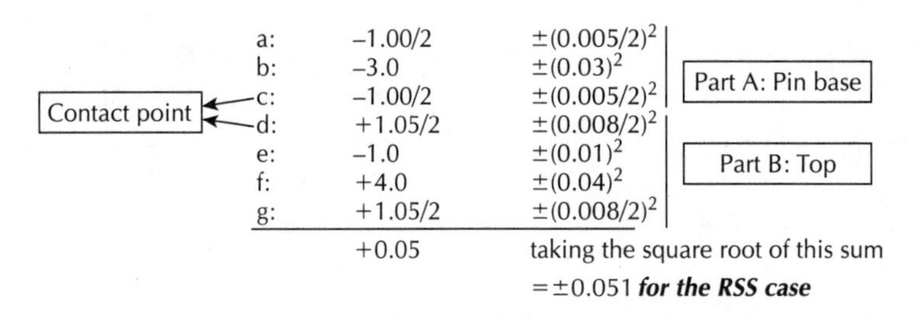

a:	−1.00/2	$\pm(0.005/2)^2$	Part A: Pin base
b:	−3.0	$\pm(0.03)^2$	
Contact point ← c:	−1.00/2	$\pm(0.005/2)^2$	
Contact point ← d:	+1.05/2	$\pm(0.008/2)^2$	Part B: Top
e:	−1.0	$\pm(0.01)^2$	
f:	+4.0	$\pm(0.04)^2$	
g:	+1.05/2	$\pm(0.008/2)^2$	
	+0.05	taking the square root of this sum $= \pm 0.051$ ***for the RSS case***	

This tells us that the nominal gap is +0.050 and the additional growth in the gap will be +0.051, due to tolerance buildup, for a total gap of 0.101. Thus, when all the tolerances are at their largest values, there will be a measurable gap.

For the RSS case when all dimensions are the smallest allowable, there is a + 0.05 nominal and −0.051 tolerance buildup for the RSS case in the negative direction. This tells us that the nominal gap is +0.050 and the reduction in the gap will be −0.051, due to negative tolerance buildup, for a total of −0.001 interference. In this scenario the parts could not mate. If the worst-case conditions are such that the pins are biases toward contacting the inner surfaces of the holes (the second scenario shown in Figure 6.8), the gap results are the same as the scenario we just worked out, because of the symmetry of the tolerancing scheme shown in Figure 6.9 (this need not be true of other pin-and-hole situations).

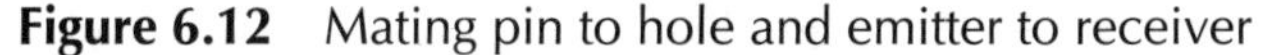
Figure 6.12 Mating pin to hole and emitter to receiver

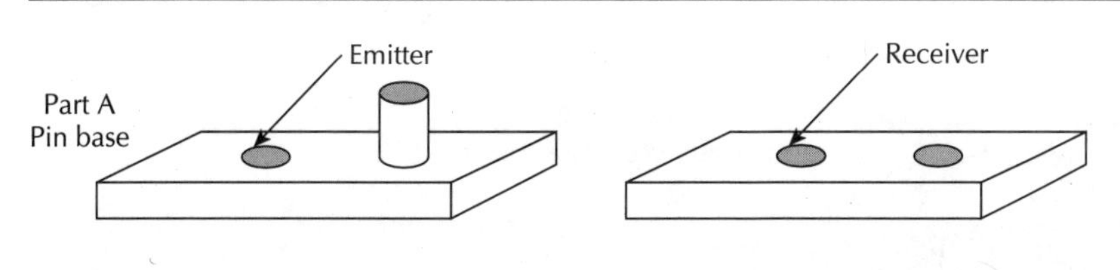

Additional considerations for RSS tolerance stack analysis [R6.7]

Another useful form of the mating of pins and holes can be demonstrated by evaluating a light emitter and receiver assembly.

The desired output of the assembly is to have the emitted light shine directly into the receiver. This means we would like to have the center line of the emitter fall within 0.08 inch of the centerline of the receiver. Let's begin to set up the problem by looking at two worst-case conditions for the pin and hole. We need to look at this worst-case scenario to set the stage for how we need to look at the RSS application correctly.

Variation between centers of a pin and hole (one-dimensional case)

What is the nature of variation between the pin centerline and the hole centerline when we consider the tolerances and slop? Slop is defined using the following illustration.

We must set up the pin and hole at their least material conditions and nominal conditions. This will define the *maximum distance* between the pin and hole centerlines.

A similar figure can be drawn for the case where the hole is biased to the left. The equation for D_{max} in that case is:

$$D_{max} = -Tol_{pin} + R_{pin} - R_{hole} - Tol_{hole}$$

With this worst-case information in hand we can now state the difference between the left- and right-hand biased examples:

$$\text{(worst-case right)} - \text{(worst-case left)} = \text{total interplay or throw between worst cases}$$

$$\text{Total throw} = (+Tol_{pin} - R_{pin} + R_{hole} + Tol_{hole}) - (-Tol_{pin} + R_{pin} - R_{hole} - Tol_{hole})$$

$$\text{Total throw} = 2Tol_{pin} - 2(R_{pin} + R_{hole}) + 2Tol_{hole}$$

The total throw can be expressed as a bilateral worst-case interplay:

$$\pm(Tol_{pin} - R_{pin} + R_{hole} + Tol_{hole})$$

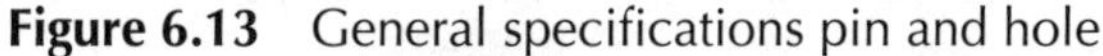
Figure 6.13 General specifications pin and hole

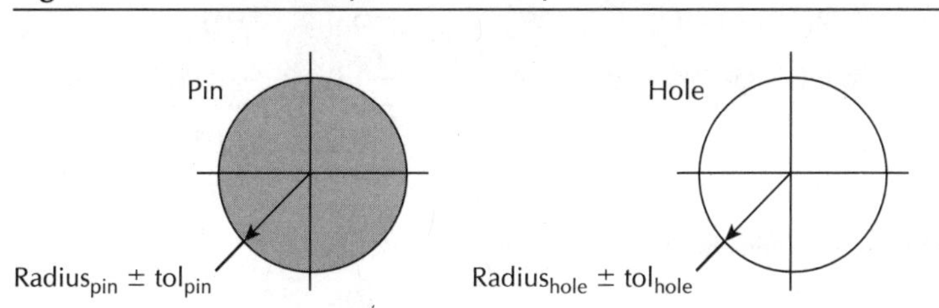

Figure 6.14 Maximum distance between pin and hole centerlines

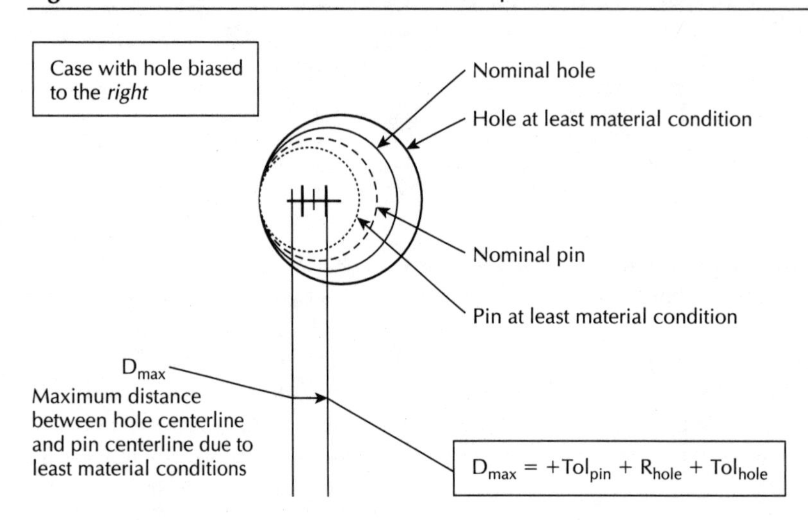

This is simply the worst case from the *right-biased* hole case (not the case if the tolerances were asymmetrical). From this total throw tolerance we must recognize that, for RSS analysis, three items must be included in the stackup calculations:

1. Radial pin tolerance
2. Radial hole tolerance
3. Nominal radial slop, or ($Radius_{hole} - Radius_{pin}$)

Let's return to the emitter and reflector example and apply what we have learned.

Using the same specifications as shown in Figure 6.14, the displacement vectors and their associated tolerances are quantified for the RSS case as follows (note that we are working with diameters that have been *divided by two* to express them as radial distances):[6]

a: Left edge of left hole to centerline of the left hole in the top	$+1.05/2 \pm (0.008/2)^2$
b: Centerline of left hole in top to datum -D-	$-1.0 \pm (0.01)^2$
c: -D- (datum) to centerline of right hole in top	$+4.0 \pm (0.04)^2$
d: Centerline of right hole on top to centerline of pin on base	
Radial hole tolerance	$+0.0 \pm (0.008/2)^2$
Radial slop	$+0.0 \pm ((1.05-1)/2)^2$
Radial pin tolerance	$+0.0 \pm (0.005/2)^2$
e: Centerline of pin in base to centerline of emitter in base	$-3.0 \pm (0.03)^2$
f: Centerline of emitter in base to left edge of emitter in top	$-1.00/2 \pm (0.005/2)^2$
RSS gap total:	$+0.025 \pm 0.057$

6. Recall that this is the square root of the sums of squared terms from the stack.

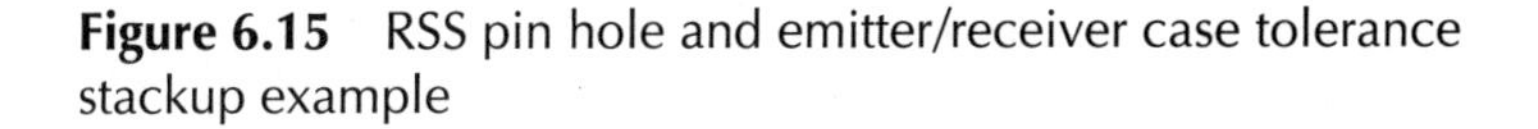

Figure 6.15 RSS pin hole and emitter/receiver case tolerance stackup example

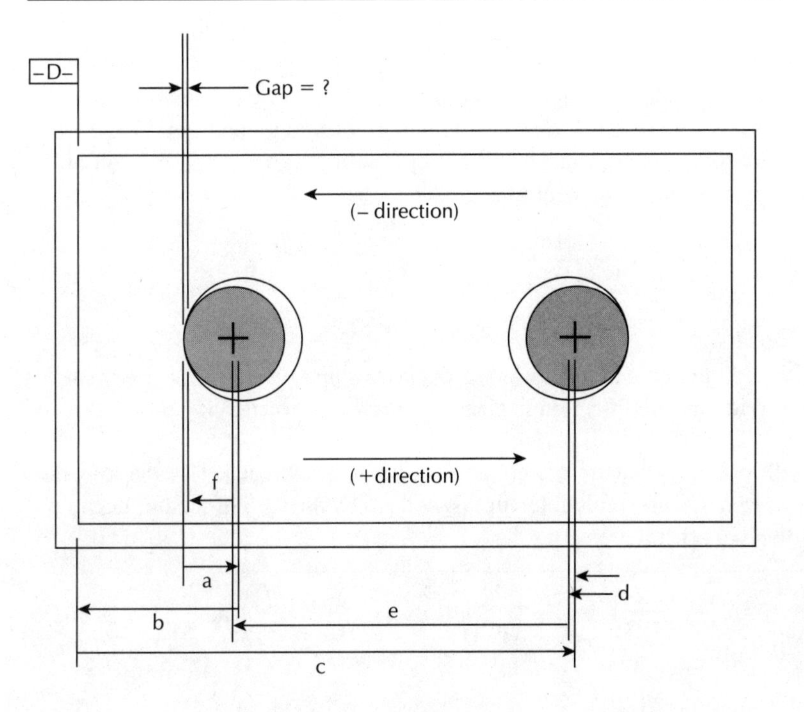

We have introduced the reader to a variety of RSS analyses. The examples have all been one-dimensional and linear in nature. We will proceed with nonlinear cases and will discuss two- and three-dimensional cases in Chapter 8.

The Nonlinear RSS Method

With a few changes the nonlinear approach to tolerance analysis can be used in the root sum of squares format. We will return to the over running clutch problem (Chapter 5) to illustrate the nonlinear RSS approach.

We have already stated that the variances associated with manufactured component dimensions can be summed in a process called *pooling of variances.* If we further assume that the tolerances are initially approximated by the expression $T_i = 3$ standard deviations, then we can define a RSS approach to tolerancing based on the Equation 6.14:

$$\text{var}_y = \left(\frac{\partial f}{\partial x_1}\right)^2 \text{var}_1 + \left(\frac{\partial f}{\partial x_2}\right)^2 \text{var}_2 + \left(\frac{\partial f}{\partial x_3}\right)^2 \text{var}_3 + \ldots + \left(\frac{\partial f}{\partial x_n}\right)^2 \text{var}_n$$

Substituting the tolerances into the expression yields Equation 6.15:

$$\text{tol}_y = \sqrt{\left(\frac{\partial f}{\partial x_1}\right)^2 \text{tol}_1^2 + \left(\frac{\partial f}{\partial x_2}\right)^2 \text{tol}_2^2 + \left(\frac{\partial f}{\partial x_3}\right)^2 \text{tol}_3^2 + \ldots + \left(\frac{\partial f}{\partial x_n}\right)^2 \text{tol}_n^2}$$

As previously mentioned, the tolerances can be modified as necessary as defined by the six-sigma approach to account for long-term mean shifts where Cpk is employed to modify the expression of the variance. Thus the three-sigma tolerance terms can be replaced by using a modification of the Cpk adjusted standard deviation for each component as necessary:

$$\sigma_{\text{adjusted}} = \frac{\text{Tol}}{3C_{pk}}$$

$$\text{Tol}_i = 3C_{pk} \times \sigma_{\text{adjusted}}$$

With this knowledge we can define a modification to the *sixth* step in the nonlinear worst-case process steps from Chapter 5. With this modification in place, the rest of the tolerancing process can be followed as described in Figure 5.

Here is how Step 6 will look now that we are employing the RSS approach: Use the tolerance equation 6.15 to solve for the *predicted* assembly tolerance as defined by the RSS of the initial component tolerances and their individual sensitivities:

$$\text{tol}_y = \sqrt{\left(\frac{\partial f}{\partial x_1}\right)^2 \text{tol}_1^2 + \left(\frac{\partial f}{\partial x_2}\right)^2 \text{tol}_2^2 + \left(\frac{\partial f}{\partial x_3}\right)^2 \text{tol}_3^2 + \ldots + \left(\frac{\partial f}{\partial x_n}\right)^2 \text{tol}_n^2}$$

Figure 6.16

6. (Modified for the RSS case). With the values for each component sensitivity quantified, the RSS tolerance equation can be used to determine how a calculated assembly tolerance range functionally and statistically relates to the required assembly tolerance range. We do this by inserting the square of the initial (bilateral) component tolerances and the square of their sensitivities into Equation 6.15. We then take the square root of the sum of squares to define the assembly or dependent tolerance.

$$\text{tol}_y = \sqrt{\left(\frac{\partial f}{\partial x_1}\right)^2 \text{tol}_1^2 + \left(\frac{\partial f}{\partial x_2}\right)^2 \text{tol}_2^2 + \left(\frac{\partial f}{\partial x_3}\right)^2 \text{tol}_3^2 + \ldots + \left(\frac{\partial f}{\partial x_n}\right)^2 \text{tol}_n^2}$$

The evaluation of the recommended initial component tolerances with respect to the required assembly tolerance, as constrained by the calculated sensitivities, will be an *iterative* process. In this step the goal is to adjust the component tolerances to get the assembly tolerance range as close to the required range as possible. The ultimate outcome of the iterations will be a revised set of component tolerances, probably a little different from the original recommended tolerances, that provide an approximate match to the required assembly tolerance. The tolerance range for the assembly, as calculated by the adjusted component tolerances, will likely be located a little above or below the required tolerance range. It is not unusual that the component nominal values have to be shifted to put the assembly tolerance range on top of the required assembly tolerance range (or as close as reasonably possible). Remember the additional option of including Cp or Cpk values in defining the tolerance values in this process:

$$Tol_i = 3C_{pk} \times \sigma_{adjusted}$$

Returning to the one-way clutch problem from Chapter 5:

$$\text{tol}_y = \sqrt{(0.1039)^2(0.156)^2 + 2[(0.1032)^2(0.013)^2] + (0.1035)^2(0.156)^2}$$

$$y_{\text{high}} = .1448 \text{ radians for the high limit of } Y\text{, and}$$

$$y_{\text{low}} = .0991 \text{ radians for the low limit of } Y$$

The RSS tolerance range for the assembly is now 0.1448 – 0.0991 radians, which is well within the required tolerance limits of 0.157 – 0.087 (the worst-case process calculated tolerance limits of 0.154 – 0.080, and thus required further iterations to adjust the nominal set points). The nominal value for y remains .1225 radians. Thus, we see that the actual assembly tolerance due to the RSS of the component sensitivities and tolerances is not a problem. With the worst-case approach we needed to adjust the component nominal set points slightly to shift the actual assembly tolerance range to coincide with the required range. This work may not need to be done depending on the criticality of the case at hand; the RSS approach is often good enough.

Harry and Stewart go on to further refine and develop optimizing tolerances. Their paper, *Six Sigma Mechanical Design Tolerancing,* is listed in the references for this chapter and is highly recommended for additional study [R6.4].

Summary

Chapter 6 reviewed the fundamental concepts of statistical tolerance analysis. Comparisons have been made between the worst-case methods of Chapter 5 and the probabilistic methods expressed in the root sum square approach. The six-sigma approach to Statistical Tolerance Analysis has also been reviewed. Several examples and case studies illustrated the basic approach to working with the various methods for RSS tolerancing. We concluded the chapter by discussing the nonlinear case as modified to handle the RSS approach.

Chapter 6 is outlined as follows:

Chapter 6: Linear and Nonlinear Statistical Tolerance Analysis (Root Sum of Squares Analysis)

- The Root Sum of Squares (RSS) Approach
    - The Z Transformation Process
    - One-Dimensional Tolerance Stacks
    - Motorola's Dynamic Root Sum of Squares Approach
- Motorola's Static Root Sum of Squares Approach
- The Nonlinear RSS Case Method
- Process Diagrams for the Statistical Methods of Tolerance Stackup Analysis
    - The Linear RSS Example
    - The Dynamic RSS Example
    - The Static RSS Example
        - A case study for a linear RSS case tolerance stackup analysis
        - Additional considerations for RSS tolerance stack analysis

- Variation between centers of a pin and hole
- The Nonlinear RSS Method

The use of statistical analysis helps to make the development of tolerances more realistic. Probability plays a very real role in how manufactured components are going to vary. The RSS methods we have discussed provide the reader with a wide variety of options to deploy to account for the nature of statistical variation in assembly tolerance analysis. Chapter 18 contains a real case study from Kodak that demonstrates the methods discussed in this chapter.

References

R6.1: *SPC for the Rest of Us;* H. Pitt; Addison-Wesley, 1994, pg. 235

R6.2: *The Power of Statistical Thinking;* M. Leitnaker, R. Sanders and C. Hild; Addison-Wesley, 1995, pgs. 236, 237

R6.3: *Mathematical Statistics;* J. E. Freund; Prentice Hall, 1992, pg. 241

R6.4: *Six Sigma Mechanical Design Tolerancing;* M. J. Harry and R. Stewart, Publication No. 6σ-2-10/88, Motorola Inc., 1988, pg. 242

R6.5: *Root Sum Squares Tolerance Analysis with Non-Linear Problems;* W. H. Greenwood and K. W. Chase; 1988, pg. 251

R6.6: *Statistical Models in Engineering,* Hahn, G. J. and Shapiro, S. S.; J. Wiley and Sons, 1967, pg. 251

R6.7: *Statistics and Principles of Tolerance Analysis;* M. Hackert-Kroot and J. Nelson; Eastman Kodak Technical Training Manual, 1995, pg. 268

Appendix D contains additional references for analytical tolerancing methods.

CHAPTER 7

Sensitivity Analysis and Related Topics

Sensitivities are a frequent topic in this book. A sensitivity quantifies the change in the output variable relative to a change in a *single* input variable. Sensitivity analysis is an extremely important topic in both the analytical and empirical methods of tolerance analysis.

The Various Approaches to Performing Sensitivity Analysis

There are two kinds of analysis that can be used to identify the sensitive dimensions or parameters in a design or process: mathematical sensitivity analysis, and empirical sensitivity analysis.

Mathematical Sensitivity Analysis

Until the advent of personal computers, engineering workstations, and CAD workstations, mathematical or analytical sensitivity analysis had to be done manually. One form of manual mathematical sensitivity analysis was introduced during the discussions on worst-case (Chapter 5) and RSS (Chapter 6) nonlinear tolerance analysis. Manual mathematical analysis is typically used when working with mechanical nonlinear stackup problems that are amenable to math modeling procedures. The power of the math model is in the accuracy of the partial derivatives that calculate the specific component sensitivities; if these are properly defined from the geometric or physical assembly model, much time and effort can be saved by establishing, manipulating, and optimizing the critical tolerances analytically.

The three characteristic forms of mathematical sensitivity analysis

Mathematical sensitivity analysis has three defining characteristics, as shown in Figure 7.1:

1. First is the simplest form of sensitivity, which can be measured in both linear and nonlinear cases. This form is based on the nature of individual component *tolerance range magnitudes.* These component-to-component tolerance range differences can be thought of as sensitivity factors in tolerance stackups for an overall assembly.
2. Second is the scenario where the probabilistic nature of each component's manufacturing output distribution can influence the component sensitivities within the assembly stackup.
3. Third is the sensitivity that exists due to nonlinear functional relationships that are inherent in how the components are physically related in an assembly (as discussed in Chapter 5).

Tolerance-Range Sensitivity Analysis

Sensitivity analysis is often used as an extension of the statistical tolerancing process. It commonly builds upon the equations used in the traditional RSS analysis. This form is associated with the first characteristic of sensitivity—the magnitude of the individual component tolerance ranges.

$$s = \sqrt{s_1^2 + s_2^2 + s_3^2 + \ldots + s_n^2} \tag{7.1}$$

Equation 7.2 is used to define a statistically based expression that allows the engineer to relate each of these component variances, which can also be used to express the tolerances by virtue of the assumption that $s = T/3$, to the overall tolerance.

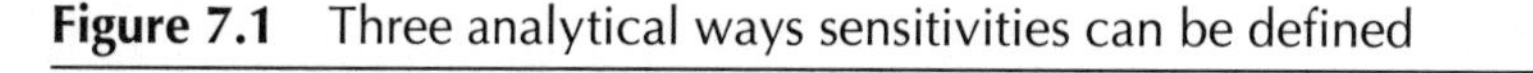

Figure 7.1 Three analytical ways sensitivities can be defined

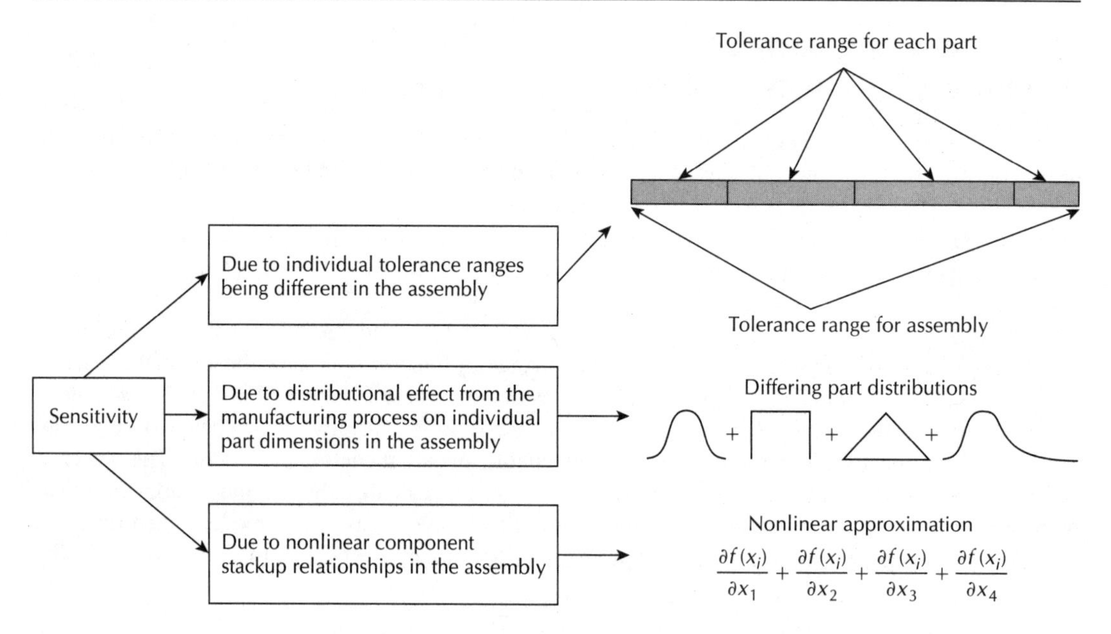

$$t_{total} = \sqrt{t_1^2 + t_2^2 + t_3^2 + \ldots + t_n^2} \tag{7.2}$$

The overall proportion of all the components contributing to the overall tolerance is 100 percent, or, as it is expressed in Figure 7.3, in decimal form, as 1.

$$1 = P_1 + P_2 + P_3 + \ldots + P_n \tag{7.3}$$

The proportion value for each of the individual components is called P_i, and is mathematically defined as the ratio of the square of the individual component tolerance (t_i^2) to the square of the overall assembly tolerance (t_{total}^2).

$$P_i = \frac{t_i^2}{t_{total}^2} \tag{7.4}$$

Summing up the contribution to the overall assembly tolerance coming from each component will always yield a value of 1 or 100 percent, depending on whether you use decimals or percentages. Equation 7.4 will produce a decimal less than 1; if you want to represent the value as a percentage simply multiply by 100.

Once the largest contributors to the magnitude of the overall assembly tolerance (and thus the allowable variability) are determined, they can be individually targeted for reduction to adjust the overall tolerance to any required functional level. In this way the overall assembly can be balanced for functional requirements and cost goals simultaneously. The tolerances can be adjusted in relation to one another for cost and manufacturing feasibility.

Sensitivity analysis can be performed on two-dimensional and three-dimensional problems as well. Obviously the complexity of these types of problems is greater than that of a one-dimensional case. This is where computer-aided tolerancing is quite useful, if not essential. We will discuss the CAT process in Chapter 8.

Component Manufacturing Process Output Distribution Sensitivity Analysis

The second characteristic of sensitivity analysis is defined by the nature of the manufacturing process output distribution for each component. Here, one must gather a statistically significant sample of output response data (>>30 samples) and construct a histogram for each component. The statistical parameters for the histogram, along with its visible shape, will enable the engineering team to estimate the type of distribution the data suggests for each component. When a stackup is made from components that have different underlying distributions, each component will have a particular sensitivity within the assembly due to the effect of these distributional characteristics. Chapter 8 contains an example of this effect in the section on sensitivity analysis.

Nonlinear Tolerance Sensitivity Analysis

The family of nonlinear tolerance cases (1-D, 2-D, and 3-D worst-case and statistical cases) cannot be *quantitatively* understood without employing sensitivity analysis. A nonlinear sensitivity is commonly used to evaluate the *first* term from a Taylor series expansion (see Equation 5.16). A sensitivity is manually calculated by taking the partial derivative of the functional relationship between the dependent (assembly) dimension and the independent (component) dimensions:

$$S_i = \frac{\partial y}{\partial x_i} \tag{7.5}$$

where $y = f(x_1, x_2, x_3 \ldots x_n)$

Thus, each sensitivity coefficient can be calculated to express the unique relationship each component change has with respect to a change in the assembly dimension.

The calculation of sensitivities by the method of partial derivatives is required only when working with nonlinear cases. In linear cases, the partial derivatives are all equal to 1. Ullman [R7.1] and Greenwood and Chase [R7.2], among others, describe a general process for handling these types of problems manually that can deal with any nonlinear functional relationship. The challenging task is often defining the proper math model to account for all the components in the nonlinear assembly. A paper [R7.3] was published by Chase, Gao, and Magleby that presents a simplifying technique, based on a method called the direct linearization method (DLM), for calculating sensitivities for two-dimensional problems. This approach employs matrix algebra to bypass the need to calculate cumbersome partial derivatives. While numerous CAD-based software packages exist to facilitate this process automatically on a graphical model, Chase et al. is recommended for those interested in the mechanics of calculating 2-D tolerances and their sensitivities for a *mathematical* model.

The Steps for Basic Nonlinear Sensitivity Analysis

Step 1: Identify a general function where (y) is any *dependent* functional variable. Typically these are dimensions, but they can be any engineering unit (either fundamental or derived units such as stress, energy, or volume, to name a few). The independent factors are represented by the x_i terms:

$$y = f(x_1, x_2, x_3 \ldots x_n) \tag{7.6}$$

Step 2: Define the mean and standard deviation for each x_i term. This can be translated directly into the mean of the dependent variable:

$$\bar{y} = f(\bar{x}_1, \bar{x}_2, \bar{x}_3, \ldots \bar{x}_n) \tag{7.7}$$

Ullman [R7.1] states that:

> *. . . the standard deviation is more complex:*
>
> $$s = \left[\left(\frac{dy}{dx_1}\right)^2 \times s_1^2 + \ldots + \left(\frac{dy}{dx_n}\right)^2 \times s_n^2\right]^{\frac{1}{2}} \tag{7.8}$$
>
> *Note that if $dy/dx_i = 1$, as it must in a linear equation, then [Equation 7.8] reduces to [Equation 7.1]. [Equation 7.8] is only an estimate based on the first terms of a Taylor series approximation of the standard deviation. It is generally sufficient for most design problems.*

This is the same math that Greenwood and Chase employed in their work with nonlinear worst-case and RSS stackups. Chapter 5 contains an example of how to calculate sensitivities using this approach (the overrunning clutch).

As mentioned at the beginning of the chapter, the advent of computer-aided engineering (CAE) and computer aided design (CAD) workstations has provided strong capabilities for simplifying the process of analytically determining sensitivities. This is done through the use of graphical, as opposed to mathematical, models. A mathematical modeling case can be resolved by using PC-based applications such as *Crystal Ball,* the Monte Carlo simulation program highlighted in Chapter 8. This computer-aided approach to sensitivity analysis is very helpful to those who have difficulty with manual or numerical analysis approaches to solving the partial derivatives. The team will still need to generate the nonlinear math model for *Crystal Ball.* There are also several engineering workstation-based tolerance software packages that readily perform sensitivity analysis on graphical models; these too are covered in Chapter 8.

Empirical Sensitivity Analysis

If the design involves mixtures of technology such as an assembly made up of optics, mechanics, electronics, thermodynamics, and magnetics, for example, the problem often can become exceptionally difficult to model analytically. Subsystems of this nature are best analyzed through designed experiments. These types of assemblies are frequently nonlinear with respect to tolerancing issues, and will have very important sensitivities to identify and quantify.

ANOVA Sensitivity Analysis

Empirical sensitivity analysis is a proven method of employing designed experiments and the ANOVA (Analysis of Variance) data-processing method for the identification of the relative magnitude of parameter sensitivities. *The method we will use does not define the actual sensitivity, as a partial derivative would—just the relative sensitivity of one component relative to another in terms of variance contribution to the measured response variable.* We will show how this method works in Chapter 16, which covers ANOVA. ANOVA is an extremely flexible method for analyzing highly complex assemblies that contain components and process set points.

One-Factor-at-a-Time Sensitivity Analysis

If one is not using a designed experiment to generate empirical data, there is a simple manual approach. The following technique will allow the team to empirically determine sensitivities by changing one factor at a time. In *Probabilistic Systems Analysis,* A. M. Breipohl states:

> *In many practical problems an analytical model either is unknown or the model is only a gross approximation of the actual hardware, that is,* $y = g(x_1, x_2, x_3, \ldots x_n)$ *is not known. In these instances experimental methods may be used to find the partial derivatives.*
>
> *The mean output can be measured with all the variables at their mean values. The partial derivatives may be approximated by holding all but the Kth variable at its mean value and then varying the selected variable* X_k *about one standard deviation and measuring the variation of the output. That is*
>
> $$\frac{\delta y}{\delta X_k} \cong \frac{\Delta y}{\Delta X_k} \tag{7.9}$$

From Breipohl's explanation we can see a practical way to obtain reasonable approximations for the sensitivities of components in an assembly. It is up to the engineering team to find out what "about one standard deviation" is for the variation about each component nominal. This means the team must obtain variance data from the manufacturing process that is representative of the component they are working with (see Chapter 4 for guidance on obtaining this information). One must also recognize that the delta (Δ) notation simply means the physical measure of some change in a variable. The team's craftsmanship in changing the component nominals and measuring the corresponding change in the dependent variable (y) will predominately determine how closely the data will be able to approximate the sensitivity. Remember to take careful measurements with instruments of documented precision and accuracy.

It is sufficient at this point to recognize that there are two approaches to sensitivity analysis and that both are very helpful provided they are linked to the appropriate type of problem. Problems that are amenable to analytical modeling should, for the sake of economy and time, be pursued by the mathematical sensitivity analysis approach. Problems that are too difficult to model should be approached by the empirical sensitivity analysis method. The empirical approach will require physical models and time-consuming experiments in a laboratory or industrial setting—but it will pay big dividends in providing the necessary data to define the relative critical sensitivities that will lead to enlightened decision-making (see Chapter 17) as the design is optimized for economic production.

Before we leave this topic there is one more use that can be made of sensitivity analysis.

Using Sensitivity Analysis in Concept Design

Ullman [R7.5] states . . .

> *It must be noted that [Taguchi's] philosophy is different from that traditionally used by designers, where parameter values are determined without regard for tolerances or other noises and the tolerances added on afterward. These tack-on tolerances are usually based on company standards. This philosophy does not lead to robust design and may require tighter tolerances to achieve quality performance.*

Ullman has a unique vision of how one should approach a tolerancing problem from the very onset of the design process. His approach is forward thinking and as such is useful long before one traditionally begins to consider tolerance activities. Ullman and Zhang [R7.5] propose a method of minimizing a critical parameter's variability (such as an overall assembly dimension) by considering the noncritical nominal set points associated with the components that make up the critical assembly. One would normally view these as fixed values that will be assigned tolerances based on functional requirements and constraints within the manufacturing processes used to make the components. Ullman and Zhang took a fresh look at the functional relationship that exists between the dependent assembly variable (Y) and the independent component variables (x_i) through the already familiar expression

$$y = f(x_1, x_2, x_3 \ldots x_n)$$

They also appropriately recognized that the independent component variables would manifest themselves in the real world not as single nominal values but as distributions around the nominal dimension based on random and special-cause variations present in the manufacturing process (unit-to-unit noise) and the customer use environment (external and deterioration noise). Ullman and Zhang state that . . .

> *If the non-critical [component] dimensions are distributions then the critical [assembly] dimension (Y), as a function of the x_i values, must also be a distribution and have a deviation. Each [component] dimension may effect the deviation of the critical [assembly] dimension through its own variation from the nominal value.*

This situation is expressed arithmetically by the following functional deviation relationship:

$$\delta Y = f(S_1\,\delta x_1, S_2\,\delta x_2, S_3\,\delta x_3, \ldots S_n\,\delta x_n) \tag{7.10}$$

where (δY) is the deviation of the critical dimension (Y), (δx_i) is the deviation of the dimension (x_i) and (S_n) is the *sensitivity*. We have explained that the sensitivity is a coefficient representing the nature of the change in the critical dimension relative to a unit change in the noncritical component dimension x_i. The sensitivity is calculated by taking the partial derivative of the functional relationship between the critical (assembly) dimension and the noncritical (component) dimensions:

$$S_i = \frac{\partial Y}{\partial x_i}$$

Thus, each sensitivity coefficient can be calculated to express the unique relationship each component change has with respect to its effect on a change in the assembly dimension. From the previous equations, Ullman and Zhang go on to define another very common expression that represents the *variance* in the critical dimension as a sum of the component variances as attenuated by their respective sensitivity coefficients:

$$dY^2 = \sum_{i=1}^{n} (S_i \times dx_i)^2 \tag{7.11}$$

From this expression Ullman and Zhang isolate a unique design opportunity:

> *From this equation it is obvious that there are two ways to control the deviation of critical dimensions. The first is to keep them small by tightening manufacturing tolerances and controlling aging and environmental effects (tighten dx_i). This is a traditional approach and is generally expensive as it requires tighter manufacturing tolerances and protection from aging and the environment. The second way to control critical dimension deviation is to make these dimensions insensitive to the non-critical dimensions (reduce S_i). . . . reducing sensitivity can often be accomplished without tighter manufacturing tolerances simply by altering the value of $S_i \times dx_i$.*

This approach is very helpful in avoiding the cost associated with quality improvement based on buying down variance by tolerance tightening. One can analyze any number of conceptual designs very early in the concept development stage of a project using Ullman and Zhang's approach. This can be done by using computer-aided numerical optimization techniques to minimize the critical dimension variance through noncritical component nominal set point adjustments. Specific noncritical component nominals can be assessed for their ability to produce *low* sensitivity in the assembly function. This process essentially depends on sensitivity minimization as a means of prescreening a given set of nominals for tolerance latitude. Arguably, this is a good way to design for economic tolerances long before tolerance analysis is traditionally carried out. In this way the engineering team can proactively avoid a number of instances that could otherwise require tight tolerances as the design is readied for production.

Is this approach right for every assembly case? It depends. Searching for nominals that minimize sensitivity using computer-aided analytical methods is an approximation at best, because the math model used in the optimization may not contain terms that adequately account for variation that is going to affect the assembly due to environmental and deteriorative effects. The math models used in assembly cases are generally geometry-based and do not contain terms that allow many kinds of variation to be induced in the calculations.

An Example of the Use of Sensitivity Analysis in Concept Development

The example is from Ullman and Zhang* [R7.5].

> *Consider the assembly shown in Figure [7.2]. The height of the center of the cylinder, Y, the only critical dimension in this problem, is dependent on the value of dimensions x_1, x_2, y_1, y_2, h and r. The smaller the deviation of Y, the higher the quality in the product.*
>
> *Assuming the values for non-critical dimensions [Figure 7.2] and their deviations as shown in Table [7.1], the sensitivities to each dimension, as calculated using $S_i = \delta y/\delta x_i$, are all shown in the fourth column. The contribution of each dimension to the deviation of the critical dimension Y, $|S_i \times dx_i|$, and the percentage contribution are shown in the fifth and sixth columns.*
>
> *The results in Table [7.1] show how to adjust dimensions to reduce the deviation of the critical dimension. Dimensions y_1 and r have the largest contribution to the deviation of Y. Thus, reducing the values of $\delta y/\delta y_1$ and $\delta y/\delta r$, if possible or tightening their deviations [tolerances] will have the largest effect on decreasing the deviation of the critical dimension Y.*
>
> *Calculating the second partial derivatives shows that reducing the value of x_1, y_1 or h will cause $\delta y/\delta y_1$ and $\delta y/\delta r$ to drop. Also increasing the values of x_2, y_2 or r will have the same effect. Thus, there may be other sets of dimensions which, without changing the deviations [tolerances] on the non-critical dimensions will give a smaller deviation for the critical dimension.*
>
> *The next step in this example is to minimize the deviation of the critical dimension by changing the non-critical dimensions. Keeping the target value of the critical dimension at 166.67 and keeping the deviations [tolerances] of the non-critical dimensions the same, we can find the set of values for the non-critical dimensions that will give the minimum deviation of the critical dimension. Thus, this is an optimization problem with the objective function to minimize the deviation of the critical dimension. The target value of the critical dimension is an equality constraint and, in order to give an allowable range to each non-critical dimension, bounds on these values serve as inequality constraints. Thus, the optimization problem is as follows:*
>
> $$\text{minimize } dY = F(x_1, x_2, y_1, y_2, h, r)$$
>
> $$\text{subject to: } Y = Y_{target} = 166.67$$

* Reprinted by permission of Ullman, D. G. and Zhang, Y.-K., Oregon State University.

where: $x_L = x_i$ *lower bound*

$x_U = x_i$ *upper bound*

and: $x_2 - x_1 > 0$

$y_2 - y_1 > 0$

The last two inequality constraints are necessary to keep the stairs stepping upward as in the figure. Optimizing, the results shown in Table [7.2] were obtained. By optimizing in this way the deviation in Y has been reduced from 0.496 (in Table [7.1]) to 0.396 (in Table [7.2]).

This was accomplished by reducing the dimensions on x_1, y_1 and h and increasing dimensions x_2, y_2 and r. The deviations (tolerances) on the manufactured parts have not *been altered. Thus there has been little or no additional cost to improve the design quality by 20%.*

Figure 7.2

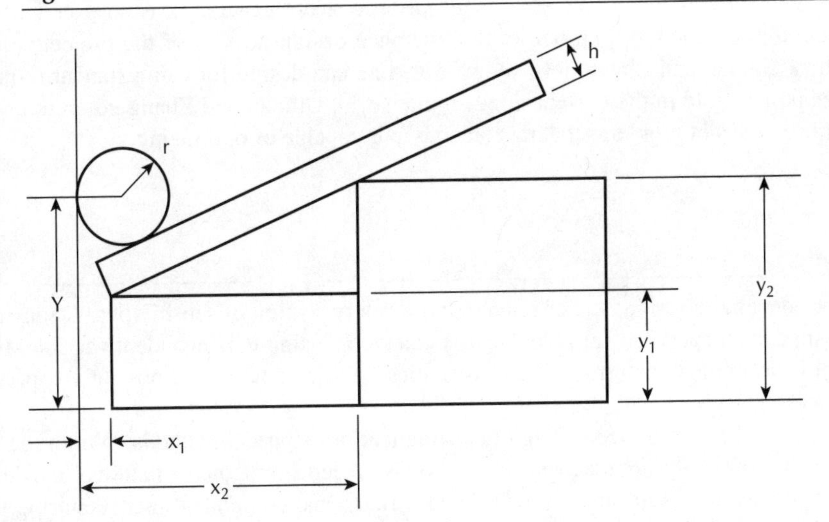

Table 7.1

Name	*Value*	*Deviation (Tolerance)*	*Sensitivity S_i*	*Deviation $\|S_i \times dx_i\|$*	*Percent*
x_1	100	0.200	−0.6132	0.122	6.11%
x_2	260	0.400	−0.0117	0.004	0.01%
y_1	80	0.300	0.9812	0.294	35.20%
y_2	180	0.500	0.0384	0.019	0.15%
h	50	0.100	1.1792	0.118	5.65%
r	50	0.200	1.8042	0.360	52.89%

Critical dimension: Y = 166.67
Critical dimension deviation: dY = 0.496

Table 7.2

Name	*Value*	*Deviation (Tolerance)*	*Sensitivity* S_i	*Deviation* $\lvert S_i \times dx_i \rvert$	*Percent*
x_1	70	0.200	−0.2121	0.042	1.15%
x_2	338	0.400	0.1445	0.058	2.07%
y_1	65	0.300	0.9326	0.279	49.24%
y_2	128	0.500	0.1080	0.054	1.82%
h	35	0.100	1.0253	0.102	6.29%
r	65	0.200	1.2519	0.250	39.43%

Critical dimension: Y = 166.67
Critical dimension deviation: dY = 0.396

This completes the example of how to improve an assembly variable without tightening tolerances. Shifting nominal component values around during concept generation is an excellent way to prepare for economic production long before the tolerance design portion of the project is actually undertaken. Thus, Ullman and Zhang have shown how one can design for using standard tolerances that are the cheapest to hold in the current three-sigma world. Ullman and Zhang go on to develop a number of unique classes of problems that this technique is capable of optimizing.

Summary

This chapter introduced the various topics related to the determination of sensitivities. A sensitivity is defined as a change in a dependent variable due to a change in a single independent variable. In tolerance analysis, it is desirable to determine the sensitivities associated with a number of components as they relate to an assembly variable.

Sensitivities can be determined through mathematical or empirical methods. Within the mathematical methods there are three approaches that can be taken, depending on the nature of the design. If the assembly of components is linear then the tolerance range and component manufacturing process output distribution approaches will be useful in accounting for the sensitivities of the components. If the assembly is nonlinear, the steps within the nonlinear sensitivity approach can be used to quantify the sensitivities.

Computers are very useful in simplifying the determination of component sensitivities. Chapter 8 (the Monte Carlo method) and Chapter 16 (the ANOVA method) are useful in defining computer-aided sensitivity analysis processes, which can be done on a math or geometric model.

When the design is difficult to model mathematically or geometrically we can resort to experimentally based methods to determine the sensitivities. One can use a designed experiment or a one-factor-at-a-time approach to develop data that can be used to calculate the sensitivities for the respective components in an assembly or a subsystem. For a designed experiment, the ANOVA data decomposition process is used to isolate the parameter sensitivities (see Chapter 16).

The chapter ends with a computer-aided application of an optimization method for sensitivity analysis. This approach is designed for concept optimization early in the development process and was developed by Ullman and Zhang. It aids the engineering team in defining design concepts that are pre-engineered for low sensitivities that will promote the use of standard or loose tolerances.

The contents of Chapter 7 are outlined as follows:

Chapter 7: Sensitivity Analysis

- The Various Approaches to Performing Sensitivity Analysis
  - Mathematical Sensitivity Analysis
    - The three characteristic forms of mathematical sensitivity analysis
    - Tolerance-Range Sensitivity Analysis
  - Component Manufacturing Process Output Distribution Sensitivity Analysis
  - Nonlinear Sensitivity Analysis
  - The Steps for Basic Nonlinear Sensitivity Analysis
  - Empirical Sensitivity Analysis
  - ANOVA Sensitivity Analysis
  - One-Factor-at-a-Time Sensitivity Analysis
- Using Sensitivity Analysis in Concept Design
- An Example of the Use of Sensitivity Analysis in Concept Development

With the capability to perform sensitivity analysis comes the capacity to adjust and optimize component tolerances in support of the economic and functional integrity of an assembly, subsystem, or product.

References

R7.1: *The Mechanical Design Process;* David Ullman; McGraw-Hill, 1992, pg. 287

R7.2: *Design Issues in Mechanical Tolerance Analysis;* K. W. Chase and W. H. Greenwood; ASME, 1988, pg. 287

R7.3: *General 2-D Tolerance Analysis of Mechanical Assemblies with Small Kinematic Adjustments;* K. W. Chase, J. Gao and S. P. Magleby; Journal of Design and Manufacturing, 1995, pg. 287

R7.4: *Probabilistic Systems Analysis;* A. M. Breipohl; J. Wiley and Sons, Inc., 1970, pg. 291

R7.5: *A Computer-Aided Design Technique Based on Taguchi's Philosophy;* D. Ullman and Y. Zhang (unpublished), pg. 293

CHAPTER 8

Computer-Aided Tolerancing Techniques

The use of computer-aided design workstations, computer-aided engineering workstations, and personal computers (microcomputers) in the analysis and determination of tolerances has become a powerful tool in modern design practice. Everything we have discussed thus far in traditional tolerance analysis can be automated by adapting the statistical and mathematical relationships defined by component geometry, cost, and various sources of manufacturing variability for use on a computer. Figure 8.1 illustrates the various outputs that a computer can produce in tolerance analysis and communication.

Each of these processes or activities is of value to the overall development and communication of tolerances. It is often necessary to employ most if not all of them in the process of developing a complete product.

It is beyond the scope of this book to provide a comprehensive tutorial for all the various types of computer-aided tolerance (CAT) software programs. There are not a lot of CAT products on the commercial market. It is up to the engineering team to assess the features, complexity, capabilities, and cost of the few available software packages. For the sake of being thorough in explaining how computers can aid in the tolerancing process, we will explore one commercially available PC-based software package that uses the methods of Monte Carlo analysis: *Crystal Ball* [R8.1]. This program is designed to work with the popular spreadsheet programs Excel and Lotus 1-2-3. It is available for either Macintosh or PC (Windows) platforms. It can aid in the tolerance analysis of any system of elements that can be related by a single mathematical function or related set of mathematical functions. We will examine how the Monte Carlo method works after a few more comments related to general CAT issues.

Figure 8.1 Computer-aided tolerancing techniques

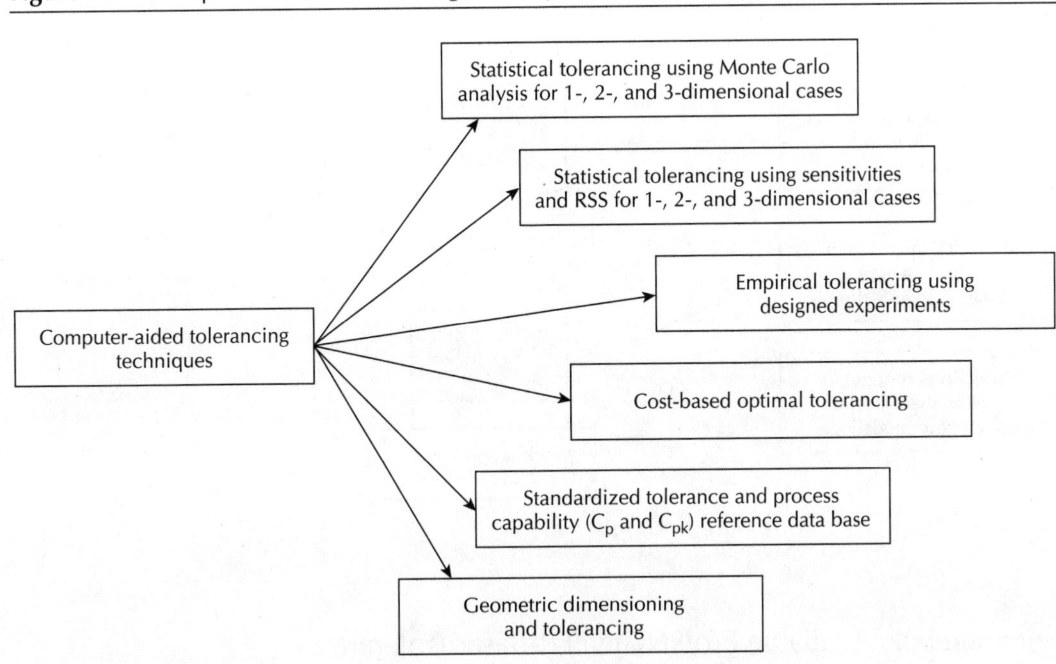

Various Software and Platform Options to Support CAT Analysis

Computer algorithms are quite adaptable to the various calculations involved in tolerance analysis. They are excellent means for handling the complex mathematical relationships between diverse geometries in one, two, or three dimensions. There are three levels of sophistication associated with software and computer platforms, as shown in Figure 8.2.

Commercial software is available to provide tolerance analysis capability either through add-ons to existing spreadsheet applications or through Pro/ENGINEER-based parametric design analysis applications. The spreadsheet add-on software packages typically are designed to work on PC or Macintosh computer platforms. The Pro/ENGINEER packages are more diverse; they are typically available in formats that run on various workstation computer platforms.

Custom, proprietary software is sometimes developed within an industry and used in conjunction with the more advanced CAD systems such as Unigraphics 3-D. Proprietary software for design analysis is typically developed for workstation-based computer platforms. Simpler, PC-based applications and the more complex workstation-based applications provide a good contrast between basic and advanced CAT platforms. Before we move on to specific examples, let's review some of the commercially available stand-alone tolerance analysis software.

Figure 8.2 Software and platforms for computer-aided tolerance analysis

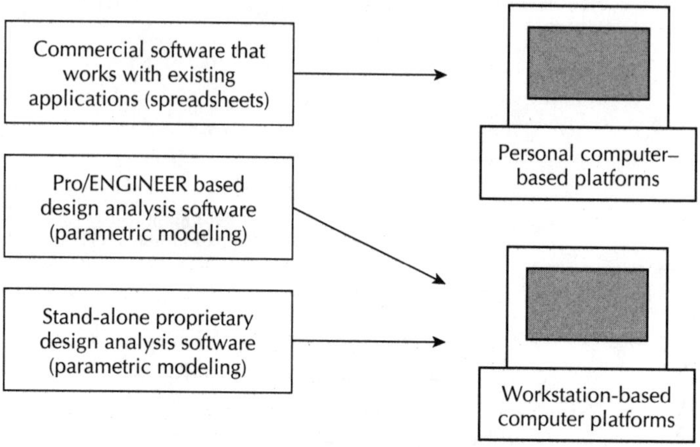

Commercial software that works with existing applications (spreadsheets)

Pro/ENGINEER based design analysis software (parametric modeling)

Stand-alone proprietary design analysis software (parametric modeling)

Personal computer–based platforms

Workstation-based computer platforms

Commercially Available Pro/ENGINEER-Based Tolerance Analysis Software

There are two well-known tolerance analysis software packages that are readily available in the commercial market. The packages are run within the well-known Pro/ENGINEER[1] design analysis application from Parametric Technology Corporation. The two packages are called VSA 3-D Tolerance Analysis Tool Kit[2] from Variation Systems Analysis and TI/TOL 3D+[3] from Texas Instruments. The distinguishing features of these two packages are as follows:

- Both directly integrate with Pro/ENGINEER solid model design analysis software.
- Both packages can model one-, two-, and three-dimensional tolerance problems.
- Both have optimization capabilities to reduce the complex process of tolerance analysis.
- Both can reduce design-analysis cycle time.
- Both can provide a percentage contribution to assembly variation for the individual design elements.
- VSA has some unique benefits in its capability with GD&T symbology.
- TI/TOL+ has some unique benefits relative to six-sigma analysis and capability.

These two packages are very capable of aiding engineers and designers in the complex task of optimizing tolerances, particularly in a 3-D scenario. It is up to the engineering team to evaluate which package best serves the design goals for a particular application. The packages work with the geometric model created by the designer. They are *not* dependent on a user-generated math model.

1. Pro/ENGINEER is a registered trademark of the Parametric Technology Corporation.
2. VSA is a registered trademark of Variation Systems Analysis Inc.
3. TI/TOL+ is a trademark of Texas Instruments Inc.

Monte Carlo Simulations in Tolerance Analysis

The Monte Carlo Method, as it pertains to engineering applications, has been aptly represented by S. S. Rao [R8.2]:

> *. . . as an experiment performed on a computer rather than performed in an engineering laboratory. If a system parameter is known to follow (a) certain probability distribution, the performance of the system is studied by considering several possible values of the parameter (or tolerance), each following (a) specific probability distribution. . . . Since physical experimentation on real systems is a time consuming and costly affair, this experimentation on trial systems generated with random numbers has proven to be very convenient. The Monte Carlo approach, thus, requires considerable calculations and is useful only for complex interrelated systems or for verification of other analyses.*

Rao's text contains a chapter on the theory behind Monte Carlo methods. His work delves into how random number generators are actually derived, constructed, and deployed within a Monte Carlo simulator algorithm. This material is very helpful if one desires a detailed, inside view of the mathematical mechanics of how to build a functional Monte Carlo simulator for any given application. For those who want to use an existing Monte Carlo simulator, we will review the Monte Carlo method through the step-by-step description of the *Crystal Ball* software package and the process one needs to follow to successfully complete a basic linear tolerance analysis. *Crystal Ball* is the recommended software to be used with the examples in this text. Another PC-based Monte Carlo package that is commercially available is called @ *Risk*.

Crystal Ball is not a dedicated CAT program, but rather a spreadsheet add-on that can serve as a multiple-application Monte Carlo and Latin Hypercube simulator that is very useful in tolerance analysis. It is technically a forecasting tool for risk analysis and decision-making. *Crystal Ball* is used, in this text, to statistically assess the effect tolerances have on inputs to an assembly of stacked components or process parameters. This assessment is aided by creating a distribution of randomly generated possible values, which occur as a result of assembling the components as they are manufactured under the constraints imposed by the proposed tolerances. The trial values that are placed within the simulation are generated from an appropriate *statistical distribution* that is representative of the component values actually occurring from the variation inherent in production processes. Technically, this process is facilitated by carefully selecting representative probability distributions to generate input values to exercise the math model representing the component tolerance stack being analyzed. It works with any system that can be expressed with a math model based on explicit expressions. As Chase says, "the technique is limited to cases where all variables on the right side of a stack equation are of known variation. Many problems in 2-D and 3-D geometry cannot readily be posed in this way. One ends up with a set of implicit equations ($w = f(x_i, y_i, z_i)$) *which must be solved simultaneously by a nonlinear equation solver.*"[1]

Monte Carlo simulations are typically used to produce many thousands of sample values, referred to as *trials,* for the response variable. From this large set of response data one can get an idea of how the tolerances can affect the long-term output response due to the natural variation resident in the production processes. A word of caution is in order. The response data is largely void of many real (special-cause) sources of variability due to environmental and deterioration noise factors that are not

4. From personal communications with Dr. K. Chase, Department of Mechanical Engineering, Brigham Young University.

going to be easy to represent in the math model or in the distribution used to randomly perturb the response values. This is why Taguchi's empirical tolerance analysis approach is so helpful; it alone has the capacity to account for these noise factors. Nonetheless, the Monte Carlo technique is highly useful in developing valuable insight into the performance of an assembly or system with regard to tolerance sensitivity. As previously mentioned, statistically based tolerance analysis is an excellent way to study unit-to-unit noise effects on a design or an assembly. It can be done quickly and cheaply using these simulation techniques.

In general terms, *Crystal Ball* requires the engineer or designer to input nominal setpoints, tolerance ranges, probability distributions that are representative of manufactured component variability, and the number of simulations that are to be run. It will output the individual simulated responses in a frequency distribution, minimum and maximum responses, a sensitivity analysis, histograms of the responses along with histogram descriptors, and statistical data that can provide an estimate of the process capability index (Cp). The simulated Cp will be based solely on the effects of the selected distributions that randomly exercise the math model. Real Cp's have the effects of deterioration and environmental noise included because they come from data gathered from real production processes. Again, be careful to note this important difference.

The *Crystal Ball* manual [R8.1][5] states that Monte Carlo simulation

> *is a simple technique that requires only a random number table or a random number generator on a computer. Instead of calculating all possible combinations of input values, Monte Carlo simulation randomly chooses a relatively small number of values as inputs (assumption values) to the (math) problem to generate a very good approximation of the answer (to an equation). During a simulation,* Crystal Ball *uses Monte Carlo simulation to randomly create inputs to your problem that look like real-life possibilities, calculate the results, and plot them on a graph. Random numbers are distributed according to range estimates you define for the (input) assumption cells. . . . The spreadsheet (math model) is recalculated to produce results for the forecast cells (output).* Crystal Ball *charts the forecast results in an easy-to understand graphical format (forecast chart). As the numbers change in the assumption cells (inputs), the values in the forecast cells (outputs) change, and the forecast chart displays these (updated) values.*
>
> . . . Crystal Ball *implements Monte Carlo simulation in a repetitive three step process. 1.) For every assumption cell,* Crystal Ball *generates a number according to the probability distribution you defined and placed into the spreadsheet. 2.)* Crystal Ball *commands the spreadsheet to recalculate itself. 3.)* Crystal Ball *then retrieves a value from every forecast cell and adds it to the graph in the forecast windows.*

The iterative nature of the Monte Carlo process is shown in Figure 8.3.[6]

It would not be unusual to specify that the Monte Carlo routine run through 1,000 or more iterations. This will provide an ample set of data to help estimate a useful portion of the real world effects that can occur from random unit-to-unit variation based on some identifiable set of probability distributions.

---

5. The *Crystal Ball* User Manual [R8.1] is quoted extensively throughout Chapter 8 with permission from Decisioneering, Boulder, Colorado.
6. Used by permission from Decisioneering, Boulder, Colorado.

Figure 8.3

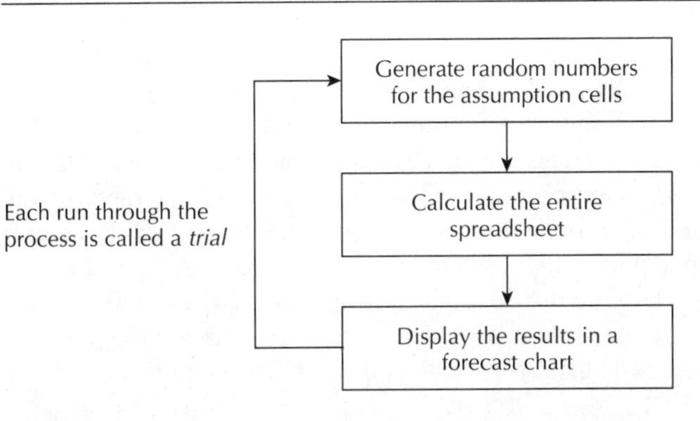

A *probability distribution* can be determined either from sample data or from the output of a mathematical equation that expresses the output as a value between zero and one or a percentage between 0 percent and 100 percent. The probability is based on independent parameters we typically refer to as *sample* or *population statistics*. Typical input parameters for probability distributions are based on whether the distributions are discrete (binary) events or continuous (analog) events. For discrete distributions we often see input parameters stated as the probability of an event (P) and the total number of trials (n) available for the event to occur or not occur. There are other parameters for discrete distributions. You will need to consult the *Crystal Ball* manual or a text on probability and statistics to obtain additional information concerning the discrete distribution statistics. The continuous distributions typically use the well-known parameters μ (the population mean) and σ (the population standard deviation). You will need to consult the *Crystal Ball* manual or a text on probability and statistics to obtain additional information.

The shape of each distribution provides a graphic illustration of how likely a particular value or range of values is to occur. Probability distributions show us the pattern that nature has assumed for certain phenomena as they disperse themselves when allowed to occur in a random or unbiased condition.

The following types of probability distributions are available in *Crystal Ball:*

- Uniform
- Normal
- Triangular
- Binomial
- Poisson
- Hypergeometric
- Lognormal
- Exponential
- Weibull (Rayleigh)
- Beta
- Gamma
- Logistic
- Pareto
- Extreme value

Crystal Ball can also permit custom (user-specified) distributions. This unique portfolio of user adjustable distributions allows one to provoke a very diverse set of variations in the model trial runs through the program's random number generator.

Each of the probability distributions has a specific nature and set of parameters that give it its distinctive form and that represent the entire range of probabilities for an event or value. As previously mentioned, there are two types of probability distributions: the *discrete* form, which accounts for integer values such as any one value appearing on a die being thrown, and the *continuous* probability distribution, which can be described by any rational value that might occur such as in reading an analog thermometer, which outputs a continuum of measured responses. Discrete distributions do not have decimals associated with a response value. The events they measure are individual occurrences, such as five flights leaving an airport every hour or three deaths a year. The continuous distributions will typically use decimals since they work with continuous phenomena that can have a very fine range of values that can be infinitely spread from one value to another. For example, we might find that a car's velocity may be 72.454 miles an hour at one point and 66.375 miles an hour at another.

We will briefly discuss the main probability distributions used in *Crystal Ball*'s Monte Carlo simulator to introduce the general shape of the distributions and the reason they are used to represent actual phenomena. Figure 8.4 is the first of two "galleries" displaying 12 of the 16 probability distributions resident in *Crystal Ball.*

Although *Crystal Ball* has 16 distributions available for use, we will focus on those displayed in Figure 8.4. There are certain distributions that are more commonly used in Monte Carlo analysis. The first three distributions we will discuss are referred to in *Crystal Ball* as the *common* probability distributions, because it is not always easy to assign a distribution to a particular situation without first knowing something about the nature of variability resident in the production process of the component being toleranced. Thus, the common distributions are the default selections to employ until one has a better alternative.

Figure 8.4 *Crystal Ball*'s main distribution gallery

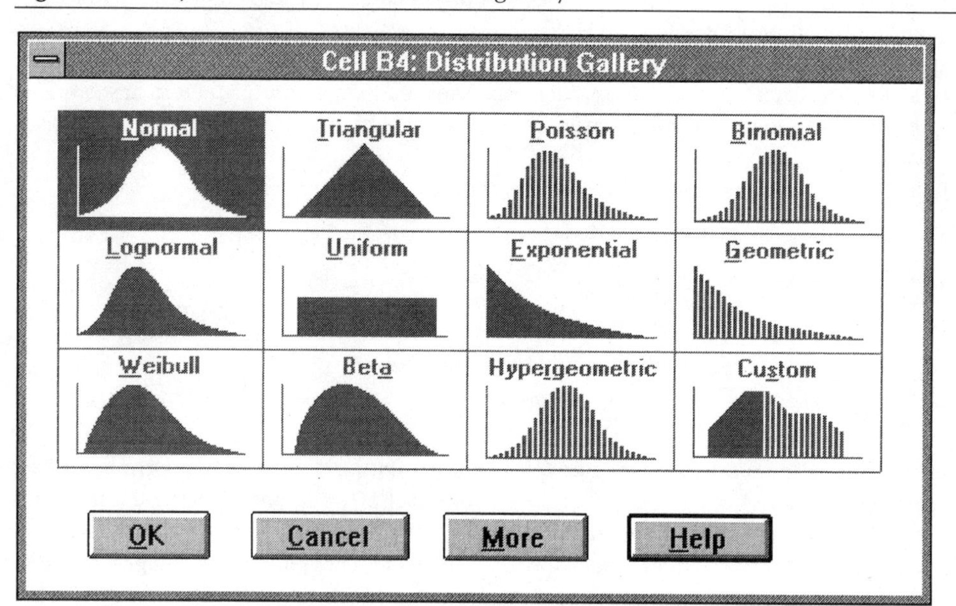

Characterizing Probability Distributions for Tolerance Analysis Applications

The usual method of identifying the distribution type that matches or at least approximates the behavior of an input variable (typically a component dimension) being used in a Monte Carlo simulation is to gather a large, statistically valid sample of data (30 to 100 samples) from a similar component that is currently being produced. This data is used to construct a histogram, which can then be used in comparison with one or more of the distributions displayed in the distribution gallery shown in Figure 8.4. The histogram, along with engineering judgment, will generally suffice to make a reasonable selection for the tolerance analysis input assumptions. A professional statistician with manufacturing experience can be of significant help in the process of identifying the appropriate probability distribution to use in many situations.

The Common Probability Distributions Available in *Crystal Ball*

The normal distribution

The normal, or Gaussian, distribution is a very common method of expressing the variation and central tendency of the output from many types of manufacturing processes. The normal distribution has three distinctive characteristics about how it expresses uncertainty:

1. Some value of the uncertain (input) variable is the most likely (the mean of the distribution).
2. The uncertain (input) variable could as likely be above the mean as below (symmetrical about the mean).
3. The uncertain variable is more likely to be in the vicinity of the mean than far away, which accounts for the bell shape of the distribution.

The uniform distribution

The uniform distribution is helpful in cases where it is just as likely that the input variable will assume one value in the range of possible values as any other. The uniform distribution has three distinctive characteristics about how it expresses uncertainty:

1. The minimum value in the range is fixed.
2. The maximum value in the range is fixed.
3. All values between the minimum and the maximum values occur with equal likelihood.

The triangular distribution

The triangular distribution accounts for the case where there is clear knowledge of a range between which likely values will occur with a single *most likely* value located somewhere within the range of possible values. The triangular distribution has three distinctive characteristics about how it expresses uncertainty:

1. The minimum value in the range is fixed.
2. The maximum value in the range is fixed.
3. The most likely value falls between the minimum and maximum values, forming a triangular-shaped distribution, which shows that values near the minimum or maximum are less likely to occur than those near the most likely value.

The Less Common Distributions Available in *Crystal Ball*

The binomial distribution

The binomial distribution describes the number of times a particular event occurs in a fixed number of trials, such as the number of heads in 10 flips of a coin or the number of defective items in a lot containing 50 items. The binomial distribution has three distinctive characteristics about how it expresses uncertainty:

1. For each trial, only two conditions are possible.
2. The trials are independent; what happens in the first trial does not effect the second trial, and so on.
3. The probability of an event occurring remains the same from trial to trial.

The Poisson distribution

The Poisson distribution describes the number of times an event occurs in a given interval, such as the number of telephone calls per minute or the number of errors per page in a document. The Poisson distribution has three distinctive characteristics about how it expresses uncertainty:

1. The number of possible occurrences in any unit of measurement is not limited to a fixed number.
2. The occurrences are independent; the number of occurrences in one unit of measurement does not affect the number of occurrences in other units.
3. The average number of occurrences must remain the same from unit to unit.

The geometric distribution

The geometric distribution describes the number of trials until the first successful occurrence, such as the number of times you need to spin a roulette wheel before you win. The geometric distribution has three distinctive characteristics about how it expresses uncertainty:

1. The number of trials is not fixed.
2. The trials continue until the first success.
3. The probability of success is the same from trial to trial.

The hypergeometric distribution

The hypergeometric distribution is similar to the binomial distribution; both describe the number of times a particular event occurs in a fixed number of trials. The difference is that the binomial distribution trials are independent, while hypergeometric distribution trials change the probability for each subsequent trial and are called *trials without replacement.* For example, suppose a box of manufactured parts is known to contain some defective pieces. You choose a part from the box, find it defective, and remove it from the box. If you choose another part from the box, the probability that it will be defective is somewhat lower than for the first part because you have already removed a defective part. If you had replaced the defective part, the probabilities would have remained the same, and the process would have satisfied the conditions for a binomial distribution. The hypergeometric distribution has three distinctive characteristics about how it expresses uncertainty:

1. The total number of items or elements (the population size) is a fixed number—a finite population.
2. The sample size (the number of trials) represents a portion of the population.
3. The known initial probability of success in the population changes slightly after each trial.

The lognormal distribution

The lognormal distribution is widely used in situations where values are positively skewed (most of the values occur near the minimum value). Some manufacturing processes produce results that are positively skewed, rather than normally (symmetrically) distributed. Parts or design elements exhibit this trend because they cannot fall below a lower limit of zero but may increase to higher values at or beyond the upper limit. The lognormal distribution has three distinctive characteristics about how it expresses uncertainty:

1. The uncertain variable can increase without limits but can not fall below zero.
2. The uncertain variable is positively skewed, with most of the values near the lower limit.
3. The natural logarithm of the u

The exponential distribution

The exponential distribution is widely u ... ch as the time between failures of electronic equ ... a design element undergoing fluctuating stress. ... ibes the number of times an event occurs in a g ... ed to get approximate solutions to difficult distri ... concerning how it expresses uncertainty:

1. The exponential distribution
2. The distribution is not affecte

The Weibull distribution

The Weibull distribution describes da ... y used to describe failure time in reliability studi ... and quality control tests. Weibull distributions ... s, such as wind speed.

The Weibull distribution is a fa... ... of several other distributions. For example, depending on the shape of the parameter you define, the weibull distribution can be used to approximate the exponential and Rayleigh distributions among others.

The Weibull distribution is very flexible. When the Weibull shape parameter is equal to 1.0, the Weibull distribution is identical to the exponential distribution. The Weibull location parameter lets you set up an exponential distribution to start at a location other than 0.0. When the shape parameter is less than 1.0, the Weibull distribution becomes a steeply declining curve. A manufacturer might find this effect useful in describing part failures during a burn-in period. When the shape parameter is equal to 2.0, a special form of the Weibull distribution, called the Rayleigh distribution, results. A researcher might find the Rayleigh distribution useful for analyzing noise problems in communication systems or radial error in positional tolerances, or for use in reliability studies. When the shape parameter is set to 3.25, the Weibull distribution approximates the shape of a normal distribution; however, for applications when the normal distribution is appropriate, using the Weibull distribution is not recommended.

Due to the relatively complex nature of the Weibull distribution, it is recommended that the reader review the details of its definition and application in a formal statistics text, which will provide guidance in defining and determining shape parameters.

The Beta distribution

The Beta distribution is a very flexible distribution commonly used to represent variability over a fixed range. One of the more important applications of the beta distribution is its use as a conjugate for the parameter of a Bernoulli distribution. In this application, the beta distribution is used to represent the uncertainty in the probability of occurrence of an event. It is also used to describe empirical data and predict the random behavior of percentages and fractions.

In *Crystal Ball*, the beta distribution supports noninteger parameters and has a scale parameter. These features offer more flexibility in defining a beta distribution. The value of the beta distribution lies in the wide variety of shapes it can assume when one varies alpha and beta parameters. If the parameters are equal, the distribution is symmetrical. If either parameter is 1 and the other parameter is greater than 1, the distribution is j-shaped. If alpha is less than beta, the distribution is said to be positively skewed (most of the values are near the minimum); if alpha is greater than beta, the distribution is negatively skewed (most of the values are near the maximum value). Again, due to the relatively complex nature of the beta distribution, it is recommended that the reader review the details of its definition and application in a formal statistics text, which will provide guidance in defining and determining the two shape parameters.

There are two conditions that underlie the beta distribution:

1. The uncertain variable is a random value between zero and a positive value.
2. The shape of the distribution can be specified using two positive values.

Now that you have a feel for what distributions are available to account for how manufacturing or production processes can cause input variables to behave, we will move on to define the general process to follow in performing a Monte Carlo tolerance analysis.

Step-by-Step Process Diagrams for a *Crystal Ball* Monte Carlo Analysis

Step 1: Create a spreadsheet

Open/create a spreadsheet in Excel or Lotus 1-2-3 under the *Crystal Ball* Function window.

Step 2: Set up a problem

Set up the problem by defining the input assumptions.

1. Define your assumptions. An assumption in *Crystal Ball* is an estimated value that will come from a specified probability distribution that you select as an input to the spreadsheet math model. *Crystal Ball* has 16 distributions.
2. Select and enter the distribution for your assumption. In this application the selected probability distribution will represent the nature of the production process output distribution for the dimension or characteristic to which you are applying the tolerance. You can select the distribution by using the Cell menu and the Define Assumption command.
3. Repeat Step 2 for each component or element in your tolerance design case.

Step 3: Define the output forecasts

Set up the problem by defining the output forecasts.

1. Establish the functional relationships (math models) that exist between the various input assumption terms to produce the output response. A *forecast* is simply the value taken on by the dependent variable, y, as each independent variable, x_i (input assumption value), changes as the Monte Carlo simulation runs through each trial. The forecast formula is represented in general form as:

$$y = f_1(x_1, x_2, x_3, \ldots x_n)$$

More than one forecast can be generated by defining several forecast formulas within the same spreadsheet:

$$z = f_2(x_1, x_2, x_3, \ldots x_n)$$

2. Select Define Forecast from the Cell menu. Specify the engineering units being calculated in the forecast formula.
3. Repeat step 2 for each forecast formula in your tolerance analysis.

Step 4: Run the simulation

Running a Monte Carlo simulation means that you will exercise the spreadsheet math model(s) at least 200 times by using the software's random number generator in conjunction with the distributions that represent your input assumptions. Thus, at least 200 trials will be calculated to represent the uncertainty or likelihood of what can happen in your tolerance case.

1. Choose Run Preferences from the *Crystal Ball* Run menu.
2. Type in the number of trials you would like to run (200 minimum).
3. Enter a seed value to initialize the random number generator. A *seed value* is an arbitrary integer that is the first number in a sequence of random numbers. A given seed value will produce the same sequence of random numbers every time you run a simulation.

Step 5: Interpret the results.

Determine the level of certainty in the forecasted results desired.

Crystal Ball reports the forecasted outputs from the math models in a graphic display that contains both the probability and frequency distributions of the results. The *probability axis* (left side vertical axis) of the chart states the likelihood of each output value or group of values in the range of zero probability to 100 percent probability (0.000 to 1.000). The *frequency axis* (right side vertical axis) of the chart states the number of times a particular value or group of values occurs relative to the total number of runs in the simulation.

This process involves changing the display range of the forecast to see specific portions of the output response distribution and identify the probability frequency of certain responses of interest occurring. In tolerance analysis these responses happen to be the values that are occurring within the tolerance range. *Crystal Ball* will show you 100 percent of the distribution of trials, then it is up to you to modify the range of values *Crystal Ball* includes in the Forecast Certainty window.

Crystal Ball allows the user to alter the minimum and maximum values (tolerance range) of the range of output to be selected and evaluated. The Certainty Box responds by displaying the percentage certainty of occurrence of the values within your specified range. In this way we can quantify the percentage of values falling within the required range of tolerance for an assembly or system.

This concludes the basic capability of *Crystal Ball* to analyze a tolerance problem. The program has additional capabilities to take the user deeper into tolerance analysis. The user can, for instance, receive the following statistical information concerning the output response data and distribution:

- mean
- median
- mode
- standard deviation
- variance
- skew
- kurtosis
- coefficient of variability
- range minimum
- range maximum
- range width
- standard error of the mean

All of these quantitative statistical measures are defined in Chapter 3. Crystal Ball can also construct a table of the distributed results by increments of 10 percent (or by other increments set by the user).

Crystal Ball allows you to compare multiple tolerance analyses within a large system as you alter the assumptions or tolerance ranges. Crystal Ball uses *trend charts* to facilitate this graphical analysis process; we will discuss these later in this chapter.

Sensitivity Analysis Using Crystal Ball

We have spoken at length about sensitivity analysis in Chapters 5, 6, and 7. Crystal Ball can help with this process by allowing the user to quantify the effect of each of the input assumptions on the overall output response. That is, the software can calculate the sensitivity coefficients, including the proper sign associated with the directionality of their effect. In Crystal Ball terms, this means we can assess the sensitivity of the forecast to each assumption. Some practitioners of tolerance sensitivity analysis may be uncomfortable with the process of the derivation and solution of partial derivatives. Crystal Ball is a friendly addition to the user's tolerance analysis toolkit because it does all the sensitivity analysis within the software algorithms. It ranks each assumption cell (tolerance distribution input) according to its importance to each forecast (output response distribution) cell. The program's sensitivity chart displays these rankings in bar form, indicating which assumptions are the most and least important in affecting the output response.

According to the Crystal Ball manual, the sensitivity analysis chart feature provides three key benefits:

1. You can find out which assumptions are influencing your forecasts the most, reducing the amount of time needed to refine estimates.
2. You can find out which assumptions are influencing your forecast the least, so that they may be ignored (or at least strategically identified for alternative action) or discarded altogether.
3. As a result, you can construct more realistic spreadsheet models and greatly increase the accuracy of your results because you will know how all of your assumptions affect your model.

Crystal Ball uses a technique that calculates *Spearman rank correlation coefficients* between every assumption and every forecast cell while the simulation is running. In this way the sensitivities can be quantified within Crystal Ball for any case being considered. The Crystal Ball manual discusses how sensitivities are calculated:

> *Correlation coefficients provide a meaningful measure of the degree to which assumptions and forecasts change together. If an assumption and a forecast have a high correlation coefficient, it means that the assumption has a significant impact on the forecast (both through its uncertainty and its model sensitivity [from the physical and statistical parameters embodied in the system being analyzed]).*

The physical parameters are embedded in the math model placed in the forecast cell; the statistical parameters are embedded in the distributions selected as assumptions:

> *Positive coefficients indicate that an increase in the assumption is associated with an increase in the forecast. Negative coefficients imply the reverse situation. The larger the absolute value of the correlation coefficient, the stronger the relationship.*

In Chapter 5, on worst-case tolerancing, we discussed defining signs for the directionality of the sensitivity coefficients in nonlinear evaluations. This can be a little confusing—but with *Crystal Ball* the sign is automatically calculated right along with the sensitivity value. Again, the concern for the tolerance analyst is in properly setting up the mathematical relationship (forecast) between the assumptions so that the program correctly calculates the magnitude and direction of the sensitivity coefficient.

Crystal Ball can be of help even if you are assessing tolerance issues between components, subassemblies, and major assemblies as you integrate an entire product or system:

> Crystal Ball *also computes the correlation coefficients for all pairs of forecasts while the simulation is running. You may find this sensitivity information useful if your model contains several intermediate forecasts (assemblies or subsystems) that feed into a final forecast (system).*

The program has an option that helps clarify the output for sensitivity analysis:

> *The Sensitivity Preference dialog box lets you display the sensitivities as a percentage of the contribution to the variance of the target forecast. This option, called Contribution to Variance, doesn't change the order of the items listed in the Sensitivity Chart and makes it easier to answer questions such as "what percentage of the variance or uncertainty in the target forecast is due to assumption x?" However, it is important to note that this method is only an approximation and is not precisely a variance decomposition [such as a formal ANOVA would provide].* Crystal Ball *calculates Contribution to Variance by squaring the rank correlation coefficients and normalizing them to 100 percent.*

There are two scenarios that can occur in the application of *Crystal Ball*'s sensitivity analysis option that are important to recognize.

In the first scenario, the sensitivity calculation may be inaccurate for correlated assumptions. For example, if an important assumption were highly correlated with an unimportant one, the unimportant assumption would likely have a high sensitivity with respect to the target forecast (output). Assumptions that are correlated are flagged on the Sensitivity Chart. In some circumstances, turning off correlations in the Run Preference dialog box may help you to gain more accurate sensitivity information.

For tolerance applications this warning will be valid when a design assembly is made up from components that are antisynergistically interactive. This means that one component is dependent on another for its effect on the response (Figure 8.5). In practice, this type of problematic interaction should be identified and corrected long before tolerancing is undertaken. Some designs are inherently interactive; in such cases one must be vigilant as to how these interdependencies will effect the response variable. Sometimes the interactions are necessary and helpful. A word to the wise is sufficient: Know where the design interactions exist and at what strength they act to influence the output response. Fortunately, *Crystal Ball* has a mechanism to decouple the assumptions from their correlations so that you can assess each one in relative independence. In general, it is considered good design practice to create and optimize components, subsystems, and systems that are free from elements that have strong antisynergistic interactions. When design elements act with interdependence that is synergistic and monotonic, performance generally becomes more predictable, especially when strong noises are present. (See [R8.3] for help in developing products and processes that do not possess strong interactions.)

With super-additivity or synergistic interactivity the correlation is still present, but Figure 8.5 shows how the elements work together to drive the response in a useful direction. Antisynergistic inter-

Figure 8.5 Assumptions with correlations

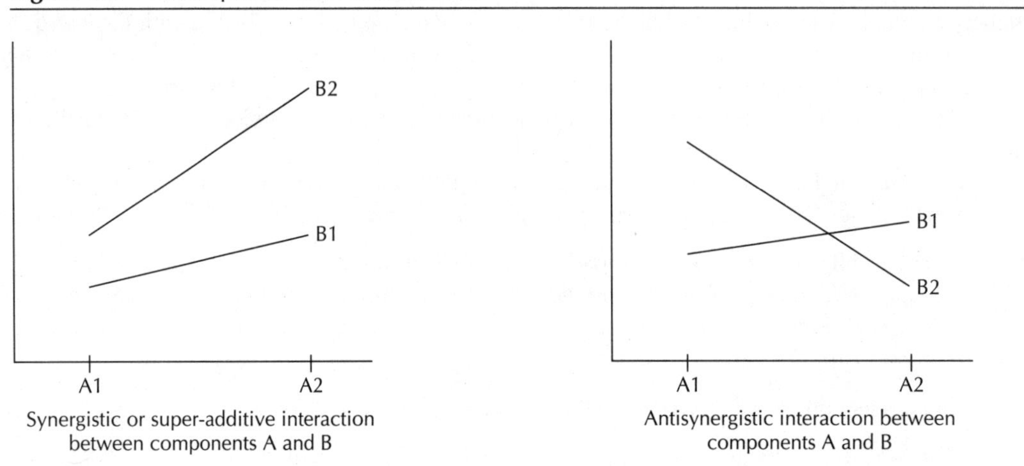

Synergistic or super-additive interaction between components A and B

Antisynergistic interaction between components A and B

activity between two component contributions to the response is counteractive. The directionality (monotonicity is demonstrated in Figure 8.6) of the response is reversed as one factor changes. This situation often forces the design team to apply very tight tolerances as a countermeasure to this problematic form of sensitivity. Also, many designers end up tightening tolerances on unimportant parameters because they are correlated to important ones. This leads to unnecessary expense during manufacturing.

In the second scenario, the sensitivity calculation may be inaccurate for assumptions whose relationships with the target forecast are not monotonic. A monotonic relationship means that an increase in the assumption (input) tends to be accompanied by a strict increase or decrease in the forecast (output).

If you're worried about being dependent on your calculus skills and knowledge of parameter interactivity in tolerance analysis, you can relax and focus your mental energy on the real issue of properly setting up the assumption cells with the appropriate probability distributions and defining the right mathematical relationship between each assumption cell (component tolerance) and the forecast cell (assembly or system tolerance); this is where the tolerance analysis is going to be set up either right or

Figure 8.6 Monotonicity

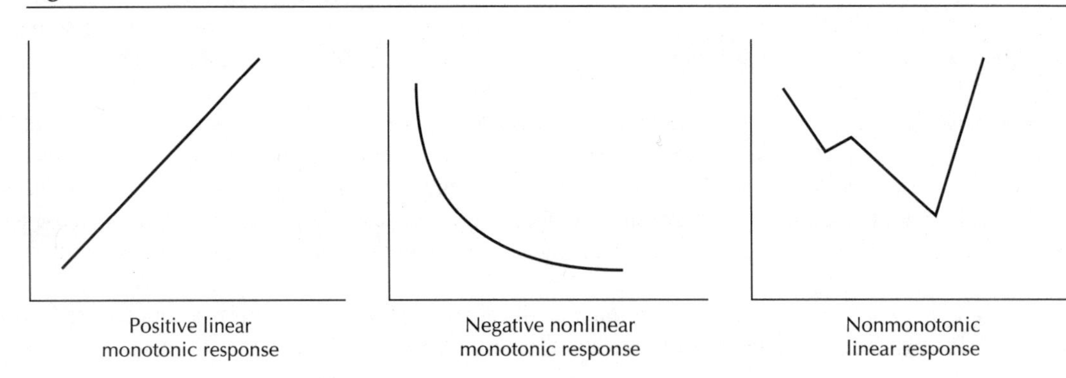

Positive linear monotonic response

Negative nonlinear monotonic response

Nonmonotonic linear response

wrong. You need look no further than these two critical areas for the basic skills necessary for competent tolerance analysis.

How to Use *Crystal Ball*

We are going to run a Monte Carlo simulation on the linear example used in Chapters 5 and 6. The simple linear stack problem is shown in Figure 8.7.

We begin by opening *Crystal Ball* by double clicking on its icon in the applications window. The program will open on an Excel spreadsheet, as shown in Figure 8.8.

We will save this file by opening the File menu and selecting Save As. We will name this file Tolex1.xls.

The next step is to set up the *Crystal Ball* spreadsheet to work with a tolerance problem.

1. In cell A1 type *Tolerance Study for a Linear Case.*
2. In cell A3 type *Part A:.*
3. In cell A4 type *Part B:.*
4. In cell A5 type *Part C:.*
5. In cell A6 type *Part D:.*

A, B, C, and D are the names of the four components for which we will enter six pieces of data that *Crystal Ball* will need to do the tolerance analysis. These are as follows:

1. Nominal dimension for each part (nom).
2. Tolerance on each nominal dimension (tol).
3. The type of distribution that describes the uncertainty in the nominal dimension (distr).
4. The tolerance divided by 3 standard deviations (Tol /3).
5. The nominal dimension plus the tolerance (nom + tol).
6. The nominal dimension minus the tolerance (nom − tol).

To enter and align these six pieces of data with the four components:

1. In cell B2 type *Nom.*
2. In cell C2 type *Tol.*
3. In cell D2 type *Distr.*
4. In cell E2 type *Tol/3.*

Figure 8.7 Complete assembly

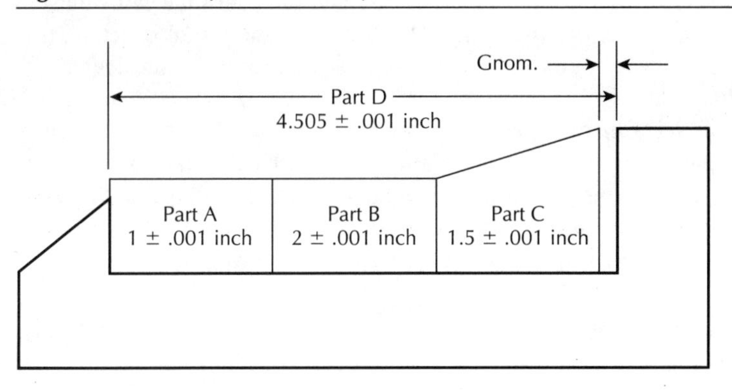

Figure 8.8 The opening screen in *Crystal Ball*

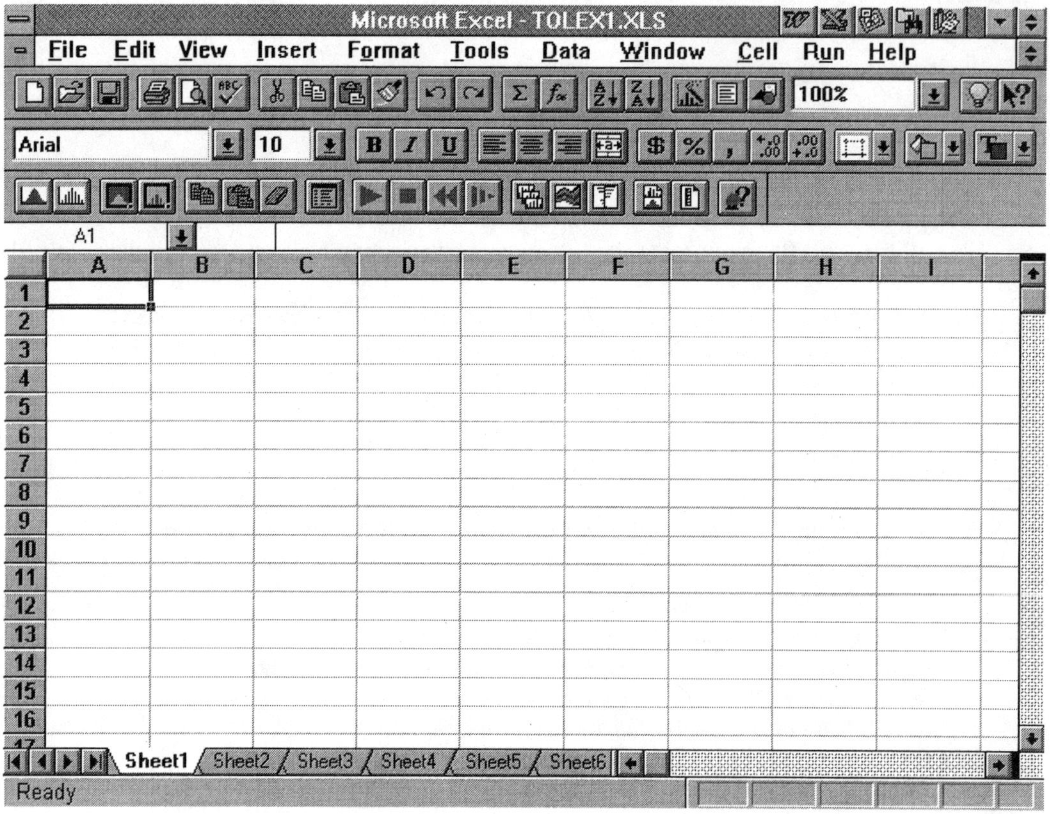

5. In cell F2 type *Nom* + *Tol.*
6. In cell G2 type *Nom* − *Tol.*

Cells B3 through B6 hold the *assumption values.* In this case the assumption values are the nominal values 1.0000, 2.0000, 1.5000, and 4.5050; enter them in cells B3 through B6, then enter the appropriate tolerances into the cells C3 through C6.

It is assumed that you understand the basics of spreadsheet use and can adjust the cell widths to accommodate the entries so that the *Crystal Ball* spreadsheet has an orderly and useful structure. It is also assumed that you have a basic knowledge of how to program a spreadsheet to do calculations; however, we will illustrate how this is done.

Next we will enter the simple linear math model that sums the individual component contributions to form a residual assembly gap dimension. This is the output response called the forecast. It is the uncertainty associated with this gap dimension that we would like to study. Recall that in the original problem defined in Chapter 5, the gap cannot be allowed to fall below 0.001 inch. To enter the math model we perform the following process, entering the model information in the appropriate cell:

1. Select cell B8 and type *=B6 – B3 – B4 – B5* (this is the assembly gap math model).
2. Click on the checkmark in the box to the left of the bar used to enter the equation (note that it is made up from the assumption value cells).

3. Cell B8 now holds the tolerance equation that employs probabilistic values that will be determined by the random number generator in conjunction with the specific distribution you assign to parts A, B, C, and D. This value is the assembly gap dimension in units of inches, and is called the *forecast value.*

Our final task is to enter the appropriate distribution for each component and program the rest of the spreadsheet cells to hold their proper respective values based on the nominals and tolerances assigned to each component part.

Parts A and C are known to be represented by uniform distributions; parts B and D are represented by normal distributions. This is entered into cells D3 through D6. Later, we will ask *Crystal Ball* to assign these distributions to actively control the values generated in the assumption cells. Cells in columns E, F, and G are calculated from cells in columns C and D.

Column E holds the calculated standard deviations based on the three-sigma tolerance assumption. Each cell for column E is calculated by dividing the respective tolerance from column C by 3. Thus, column E will have the following entries:

- E3 will hold =*B3/3*
- E4 will hold =*B4/3*
- E5 will hold =*B5/3*
- E6 will hold =*B6/3*

Columns F3 through F6 and G3 through G6 hold the tolerance limits as they are either added to or subtracted from the nominal dimensions for each component:

- F3 will hold =*B3*+*C3*
- F4 will hold =*B4*+*C4*
- F5 will hold =*B5*+*C5*
- F6 will hold =*B6*+*C6*
- G3 will hold =*B3*−*C3*
- G4 will hold =*B4*−*C4*
- G5 will hold =*B5*−*C5*
- G6 will hold =*B6*−*C6*

The values in these cells will be used to modify the output from various forecasts. The modifications will help in the analysis of the output distribution from the simulation runs.

This concludes the basic spreadsheet construction process for preparing to perform a Monte Carlo–based tolerance analysis. It matters little whether the tolerance analysis is performed on linear or nonlinear cases as far as the spreadsheet setup is concerned.

The key issues revolve around the model the user derives and enters into the Forecast cell and the distributions assigned to the assumption cells. If the functional relationship correctly expresses the nature of the component stackup or interface geometry and the appropriate distributions are properly assigned as assumptions, the rest of the analytical procedure is handled by asking *Crystal Ball* for the appropriate output charts and graphs.

The spreadsheet shown in Figure 8.9 represents how most tolerance cases will be set up.

We are now ready to assign the specific distributions to each component Assumption cell. To do this, we select the Assumption cell by clicking on it and then select the Cell menu. Under the Cell menu select Define Assumption command. A window will appear displaying the Distribution Gallery. Click on the distribution you want. This window is displayed in Figure 8.10 with the uniform distribution selected (darkened).

Click OK and the Distribution Gallery will disappear. Since we have selected the Uniform distribution, a new window will appear, as shown in Figure 8.11. This is the window for entering specific parameter information for the uniform distribution.

Figure 8.9 Typical spreadsheet

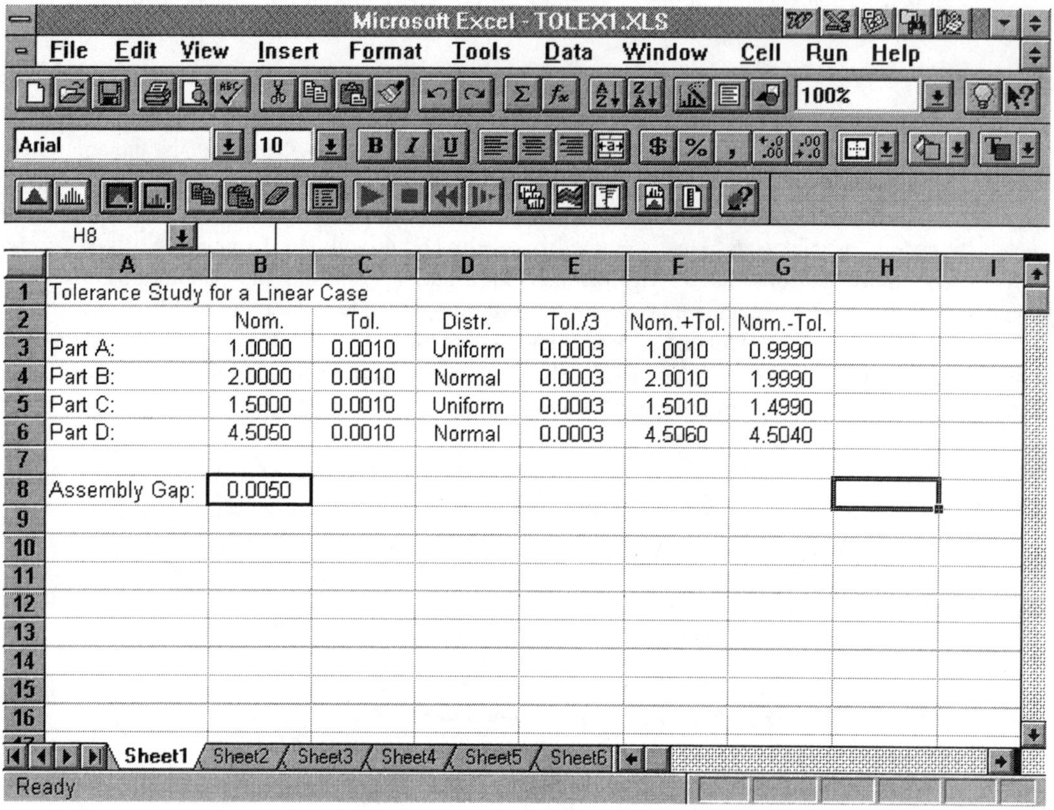

Microsoft Excel - TOLEX1.XLS

File Edit View Insert Format Tools Data Window Cell Run Help

	A	B	C	D	E	F	G	H	I
1	Tolerance Study for a Linear Case								
2		Nom.	Tol.	Distr.	Tol./3	Nom.+Tol.	Nom.-Tol.		
3	Part A:	1.0000	0.0010	Uniform	0.0003	1.0010	0.9990		
4	Part B:	2.0000	0.0010	Normal	0.0003	2.0010	1.9990		
5	Part C:	1.5000	0.0010	Uniform	0.0003	1.5010	1.4990		
6	Part D:	4.5050	0.0010	Normal	0.0003	4.5060	4.5040		
7									
8	Assembly Gap:	0.0050							

Sheet1 Sheet2 Sheet3 Sheet4 Sheet5 Sheet6

Ready

Enter the values for the upper and lower tolerances into the boxes labeled Min and Max. Click Enter and then click OK.

We have entered the uniform distribution for the Assumption cell for Part A. We'll now be returned to the *Crystal Ball* Spreadsheet window, and we can proceed with defining the Assumption cell for Part B. Simply repeat the process of opening the Cell menu and selecting Define Assumption, the window shown in Figure 8.12 will open. Click on the normal distribution since it represents how Part B will express its uncertainty in the assembly.

Click OK to tell *Crystal Ball* to select the normal distribution. Figure 8.13 displays the Normal distribution window.

In this window enter the Mean = 2.0000 and Std. Dev = 0.0003 values for Part B. Click Enter and then OK; this returns you to the main *Crystal Ball* spreadsheet, where you can repeat this process for Parts C and D, respectively.

We are now ready to use the Define Forecast command located in the Cell menu just below the Define Assumption command. First, click on cell B8 to activate the Forecast cell that we are defining. Then click on the Cell menu and select Define Forecast. The window shown in Figure 8.14 will appear.

The Unit box will need to be filled out by typing in inches. Click OK. This returns you to the main *Crystal Ball* spreadsheet and signals the successful definition of the Forecast cell.

Figure 8.10

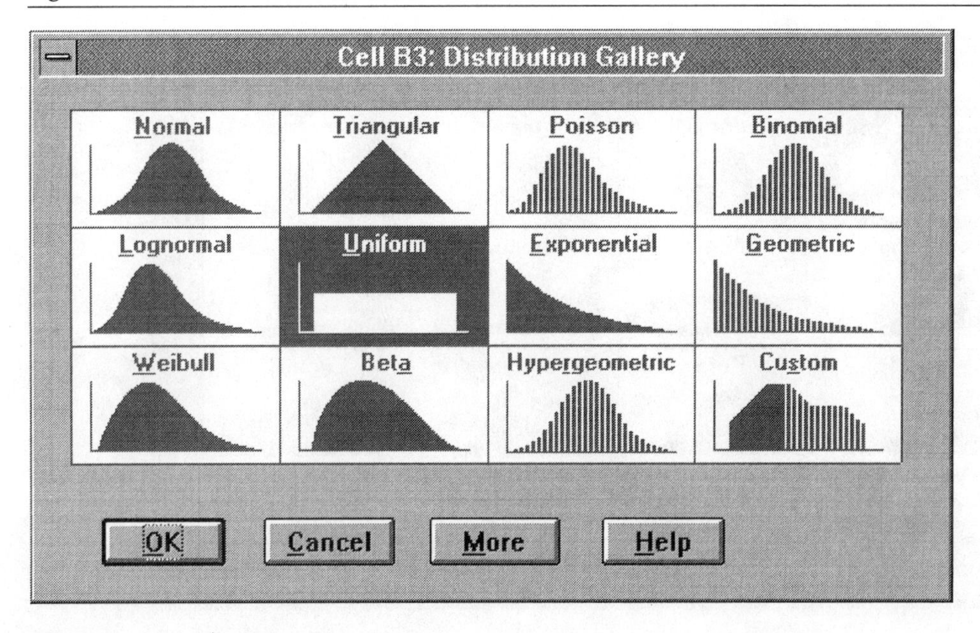

Figure 8.11

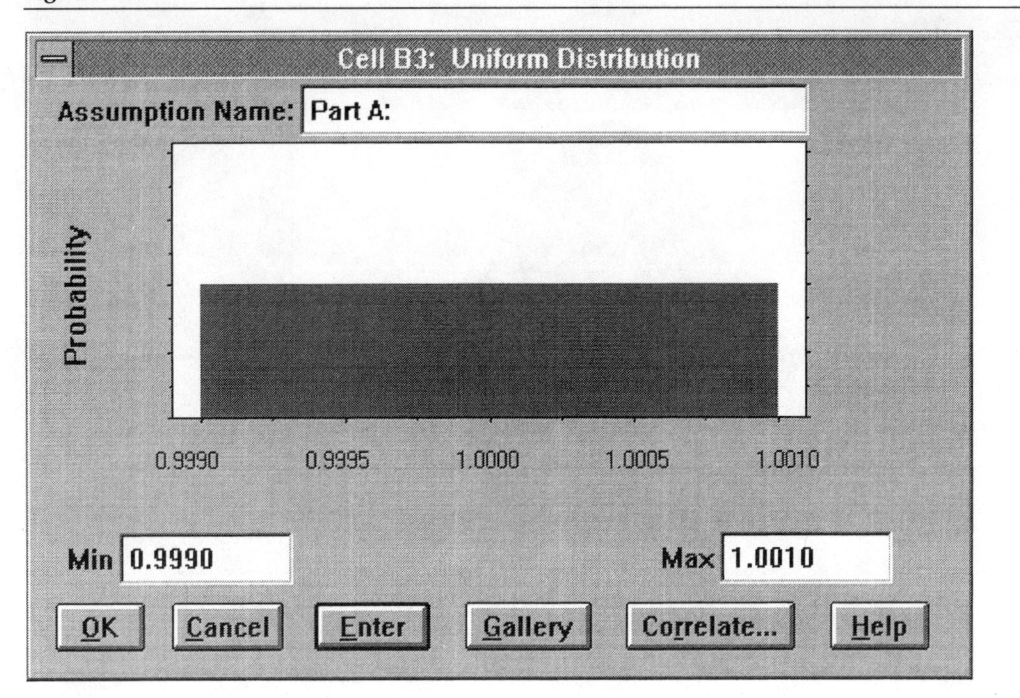

Figure 8.12

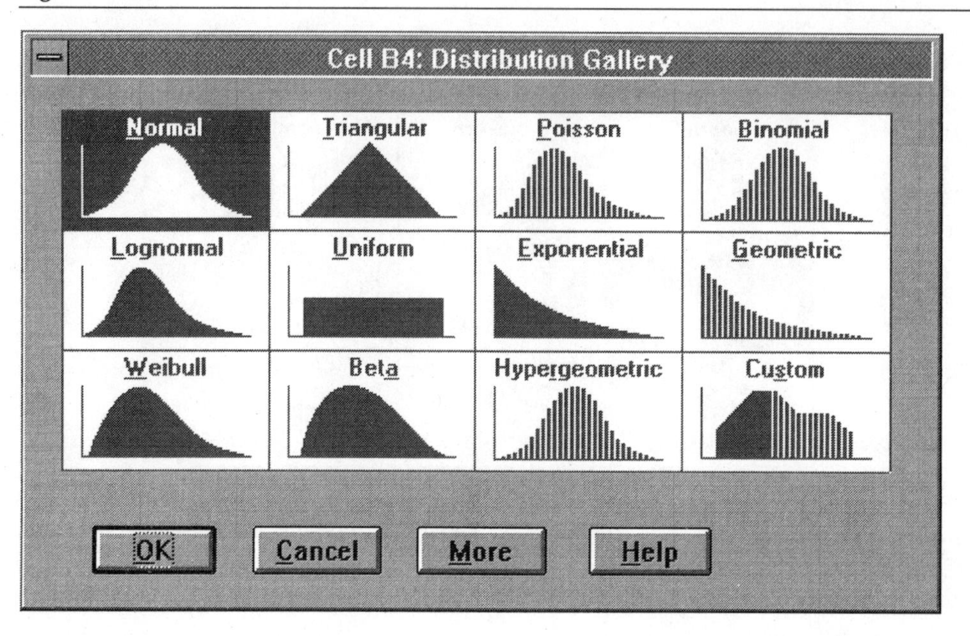

Figure 8.13

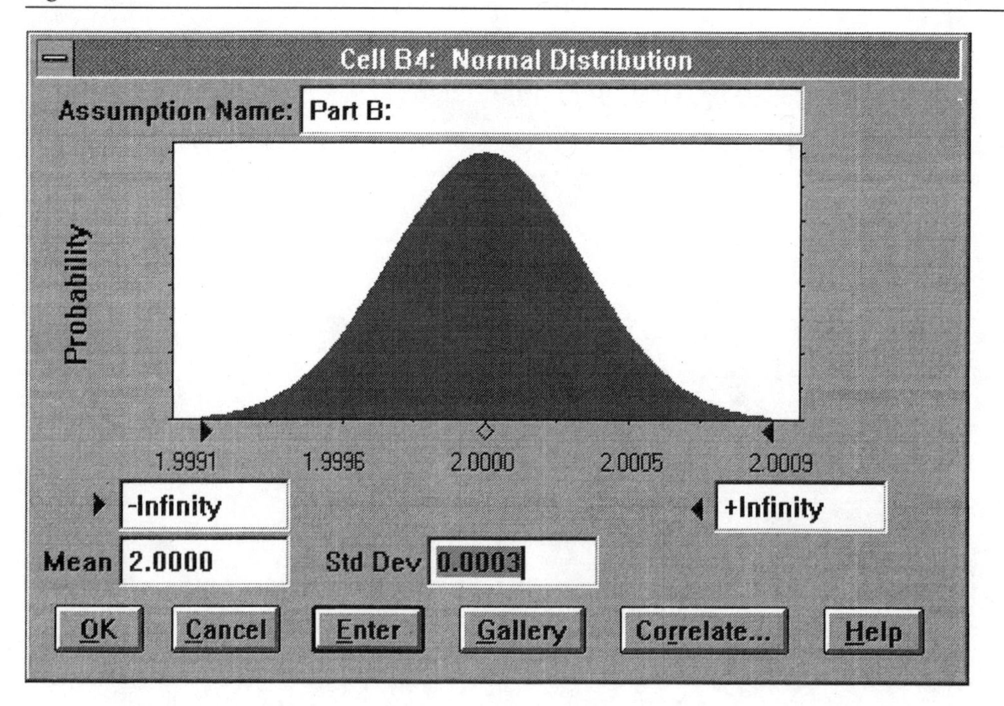

Figure 8.14

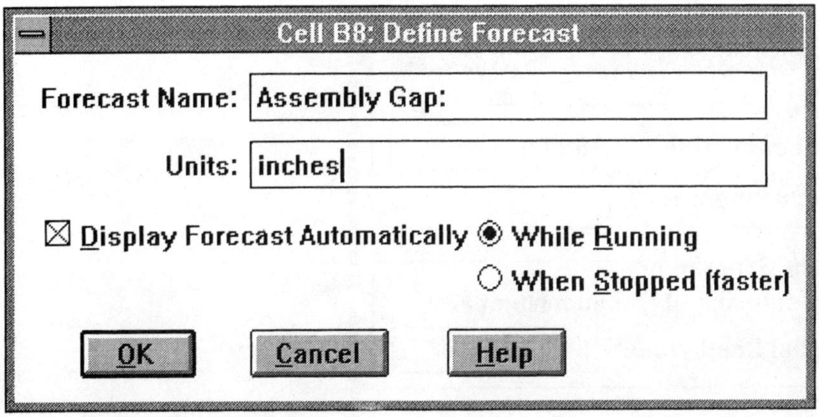

Cell B8: Define Forecast

Forecast Name: Assembly Gap:

Units: inches

☒ Display Forecast Automatically ◉ While Running

○ When Stopped (faster)

OK Cancel Help

Running the Monte Carlo Simulation

The first thing you need to do before running the simulation is to specify a few *preferences*. These are user-defined items that are necessary before *Crystal Ball* can properly iterate through the calculations.

First, select the Run menu and then the Run Preferences options. Figure 8.15 shows what the Run Preferences box looks like. Note that there are a number of items that can be changed in this window.

For the case at hand, the only preferences we need to set are the:

- Maximum number of trials: 5,000
- Sampling method: Monte Carlo
- Check box to turn on sensitivity analysis

Now that these settings are enabled, click OK. To run the simulation:

- Select the Run menu.
- Select the Run command to start the simulation.
- The Run menu is altered once the simulation is running. The Run command is replaced by a Stop command, which allows you to temporarily interrupt the simulation.
- Once the simulation is stopped the Run menu is again altered. The Stop command is replaced by the Continue command. This reactivates the simulation where it left off. Once the simulation is running again, the Stop command is back in place. You can stop and start the simulation at will until the specified number of trials is completed.

You can view one of two windows to see how the simulation is progressing:

- Under the View menu select Statistics to see how the Forecast statistics are evolving.
- Under the View menu select Frequency Chart to see how the Forecast frequency and probability distribution are evolving.

The completed run of the simulation for 5,000 trials is shown in both the Frequency Chart (Figure 8.16) and Statistics Table (Figure 8.17) formats.

Figure 8.15

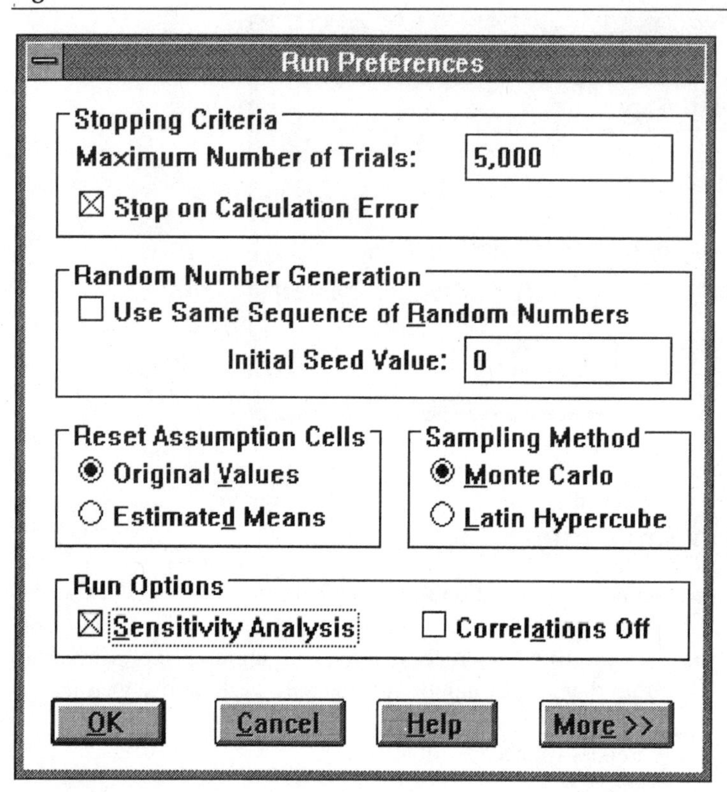

Figure 8.16

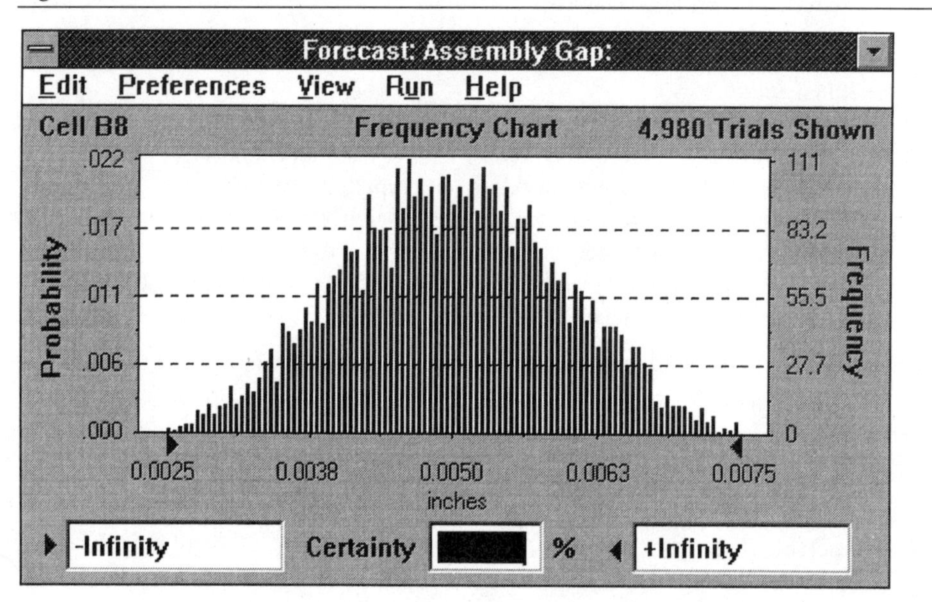

Figure 8.16 shows the graphical form of the distribution of response values generated by the simulation both in frequency (right axis) and probability units (left axis). Notice that there are two arrows located on the horizontal axis of the chart; these are adjustable and are linked to the Certainty box, in which you can enter the upper and lower tolerance limits being evaluated for the assembly gap. In this case, we wanted all values to remain > 0.001 inch. The Certainty box will display the percent certainty of occurrence for the range of values contained within the arrows. In this example the values displayed between 0.0025 and 0.0075 contain 100 percent of the results from the simulation. Thus, we are very certain that we will not have any significant occurrence of assembly interference. This is how the software helps you to determine the number of assemblies that will be outside the tolerance limits.

The statistical table shown in Figure 8.17 displays the results of the simulation in various quantitative measures that provide the descriptive statistics to help you understand the nature of variability associated with the assembly.

The sensitivity analysis is displayed in the Sensitivity Chart shown in Figure 8.18.

The Sensitivity Chart displays, graphically and quantitatively, just how much each component in the assembly affects the output response (the assembly gap). Notice that *Crystal Ball* automatically applies the correct sign to each component sensitivity coefficient. Note the affect each assumed distribution had on the sensitivities (see Chapter 7 for a discussion on this distributional form of sensitivity). The uniform distribution has twice the sensitivity effect of the normal distribution for equal tolerance ranges of 0.0006 inch.

As you can see, *Crystal Ball* is versatile in its ability to display the data from the Monte Carlo simulation. This versatility allows you to explore many different forecast scenarios. In tolerance analysis, complex assemblies can exist where there are multiple assemblies, each made up of subgroups of components. This scenario means that for each assembly you will have a forecast equation and response. If you are designing a system made up of two or more assemblies that are going to interface with one another, you can use *Crystal Ball*'s trend analysis to evaluate the effects of multiple assembly

Figure 8.17

Forecast: Assembly Gap:

Edit Preferences View Run Help

Cell B8 Statistics

Statistic	Value
Trials	5,000
Mean	0.0050
Median (approx.)	0.0050
Mode (approx.)	0.0046
Standard Deviation	0.0009
Variance	0.0000
Skewness	0.04
Kurtosis	2.65
Coeff. of Variability	0.19
Range Minimum	0.0022
Range Maximum	0.0079
Range Width	0.0057
Mean Std. Error	0.0000

Figure 8.18

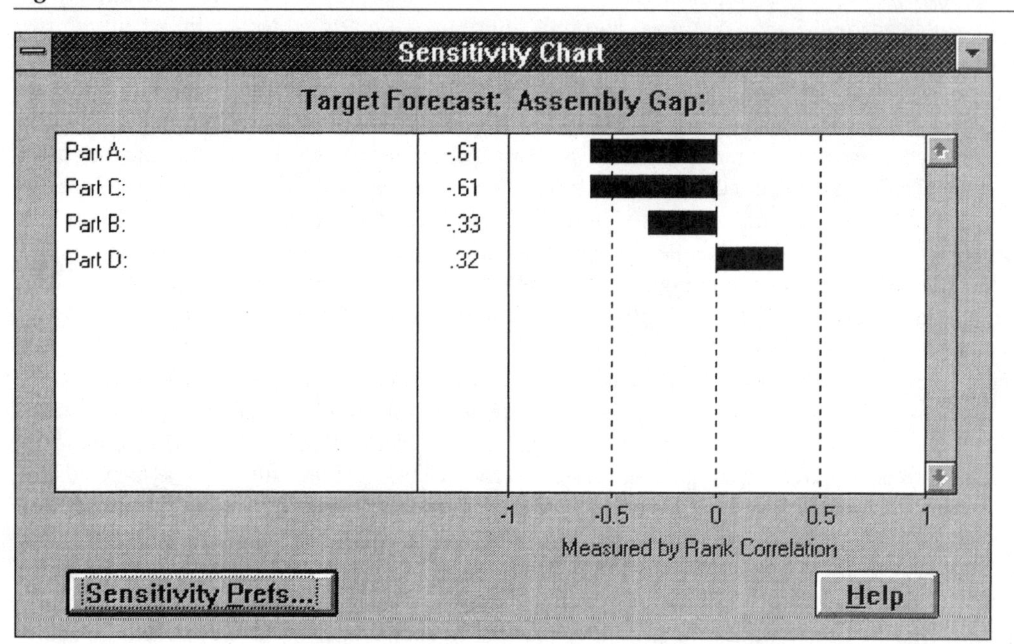

forecasts. This is one important way in which *Crystal Ball* can support system tolerance activities. According to the program's manual, once you have defined the multiple forecasts and their individual assumptions and complete the simulation,

> *you can create a trend chart to view certainty ranges of all the forecasts on a single chart. A trend chart summarizes and displays information from multiple forecasts, making it easy to discover and analyze trends that may exist between related forecasts. You can customize your trend chart to display the probability that given forecasts will fall in a particular part of a value range.*

Preparing Engineering Analysis Reports

Once you've completed the simulations and are ready to summarize the results, *Crystal Ball* has a Report feature that will automatically create a comprehensive set of information:

- Trend charts
- Sensitivity charts
- Forecast charts
- Frequency counts
- Assumption charts
- Forecast summaries
- Statistics
- Percentiles
- Assumption parameters

You can customize the report to contain just what must be included in an engineering analysis report, a team presentation, or a management review of the team's progress. An important part of any engineering analysis is the documentation of the team's work. There may be many others within the corporation or business who need to refer to this work to complete their tasks, either immediately or some time in the future. It is the team's responsibility to leave high-quality documentation of the results as they move on to other projects. This will be of value both to the team and those who must work with its designs.

This concludes the introduction to Monte Carlo simulation for tolerance analysis. If *Crystal Ball* is a tool that you believe will add value to the tolerancing process, you can obtain a copy of this software by referring to Appendix E.

Another Computer-Aided Tolerance Approach

Another excellent source of information concerning computer-aided tolerance analysis comes from Oyvind Bjorke, a professor of computer-aided manufacturing in Norway.

His book, *Computer-Aided Tolerancing* [R8.4], is an advanced "*presentation of a unified method for tolerance calculations developed by [Bjorke] and an interactive computer system by which real-life tolerance calculations could be performed.*" Bjorke develops tolerance chains as a fundamental concept upon which his approach is based. His text explains the PC-based software system called TOLTECH, which uses a non-Monte Carlo approach. It is designed to work with tolerance chains, links, and their interrelationships. Though this text uses a unique vocabulary and notation, it is an excellent example of a European approach to tolerance engineering. I encourage you to obtain this material as yet another resource to employ in the traditional analytical approach to stackup tolerance analysis; it is one of the very few modern sources dedicated to traditional tolerance analysis with computer-aided applications. Bjorke can be contacted for further information at: University of Trondheim, The Norwegian Institute of Technology, Division of Product Engineering, Rich. Birkelandsv, 2B, N-7034 Trondheim, Norway.

Summary

This chapter has introduced the reader to a variety of the computer-aided tolerance options currently available. It introduced the software packages associated with the engineering workstation computer platforms; these powerful packages are capable of greatly facilitating the complexities associated with two- and three-dimensional tolerance stackup problems.

It focused on developing a detailed, step-by-step approach to Monte Carlo simulation-based tolerance analysis, using the *Crystal Ball* software package to show exactly how to set up, run, and interpret a Monte Carlo simulation. A very detailed description of the variety of possible distributions are reviewed. We have shown a process to follow for setting up any type of stackup problem, and followed up with a specific example application. The topic of sensitivity analysis, as it relates to the effect of components having various underlying distributions constraining their input to the assembly stackup, has been discussed in detail. This chapter has presented a very comprehensive application-based tutorial on modern Monte Carlo simulation as it relates to tolerance analysis.

The chapter concludes by reviewing a traditional computer-aided tolerance approach by Oyvind Bjorke from the University of Trondheim in Norway. His work is the only modern compilation of traditional computer-aided tolerance stack analysis in print.

Chapter 8 is outlined as follows:

Chapter 8: Computer-Aided Tolerancing Techniques

- Various Software and Platform Options to Support CAT Analysis
  - Commercially Available Pro/ENGINEER-Based Tolerance Analysis Software
- Monte Carlo Simulations in Tolerance Analysis
- How *Crystal Ball* Uses Monte Carlo Simulation
- Characterizing Probability Distributions for Tolerance Analysis Applications
  - The Common Probability Distributions Available in *Crystal Ball*
  - The Less Common Distributions Available in *Crystal Ball*
- Step-by-Step Process Diagrams for a *Crystal Ball* Monte Carlo Analysis
- Sensitivity Analysis Using *Crystal Ball*
  - How to Use *Crystal Ball*
- Running the Monte Carlo Simulation
- Preparing Engineering Analysis Reports
- Another Computer-Aided Tolerance Approach

Computer-aided tolerance analysis is a powerful tool that simply cannot be emphasized enough for use in the rapid and accurate development of tolerances that are associated with any technology that can be evaluated analytically or geometrically on a PC or engineering workstation. Chapter 18 contains a real case study (case #4) from Kodak that demonstrates the methods discussed in this chapter.

References for Chapter 8:

R8.1: Crystal Ball *User Manual V4.0;* R. Sargent and Eric Wainwright; Decisioneering, 1996, pg. 306
R8.2: *Reliability Based Design;* S. S. Rao; McGraw Hill, 1992, pg. 309
R8.3: *Engineering Methods for Robust Product Design;* W. Y. Fowlkes and C. M. Creveling; Addison-Wesley, 1995, pg. 335
R8.4: *Computer-Aided Tolerancing* 2nd Ed.; O. Bjorke; ASME Press, 1992, pg. 354

CHAPTER 9

Introduction to Cost-Based Optimal Tolerancing Analysis

The economic implications associated with tolerancing components and the effect on assembly quality and cost has been a topic of study for some time now. Taguchi has become well known for his pioneering work in linking customer-loss costs to functional variability in modern product development processes. His approach to tolerance analysis is primarily focused on the application of *experimental* methods. Because of his ability to make balancing cost and quality during tolerance analysis simple, his techniques are generally well received by practicing engineers. But just because a method is relatively easy does not always mean it alone is the best approach for all cases.

In the U.S. academic community, a few professors and graduate students have linked with industrial practitioners of analytical tolerance analysis to define a rational approach to cost-based optimal tolerance allocation, which is typically used when discussing cost-based optimal tolerancing methods. Greenwood, Chase, Loosli, and Hauglund [R9.1] define tolerance allocation as "the distribution of the specified assembly tolerance among the components of the assembly." This top-down allocation is necessary because the driving force behind the assembly optimization process is a critical assembly function. The components must be defined in light of the balance between cost and quality that must exist to identify local minimums in an assembly-cost model. The following material is a brief introduction to the Western analytical approach to balancing cost and quality during traditional tolerance analysis. It is beyond the scope of this text to delve into the details of the latest research in cost-optimizing tolerance methods. These methods are under development and are not widely known, accepted, or practiced in U.S. industry. This is unfortunate. I would like to challenge readers to obtain the papers that explain these methods and put the cost-based optimal tolerancing methods to the test of industrial application.

Skills Required for Cost-Based Optimal Tolerance Analysis

There is a unique requirement for practicing the methods of cost-based optimal tolerance allocation. It has to do with the set of skills required to apply the methods. This skill set is not trivial. Cost-based optimal tolerance allocation requires some if not all of the following engineering skills:

- the ability to generate, derive, and manipulate complex mathematical models,
- the ability to solve closed-form nonlinear geometric equations,
- the ability to write computer algorithms to perform numerical analysis to solve assembly equations that require iterative analysis,
- the ability to perform classical optimal design analysis, and
- the ability to relate various cost-versus-tolerance functions to a specific problem.

This list illustrates that this particular form of tolerance analysis clearly resides in the domain of engineering and is probably only in the comfort zone, for the time being, of those who have done post-graduate work. Nonetheless, these methods are important and worthy of review for use as each tolerance case requires. Part of being a competent engineer is to recognize the need for advanced analysis and seeking assistance when it is clear that it must be performed. So, if the following processes seem a bit daunting, do not be intimidated; seek aid from your colleagues when in doubt. In chapters 11 through 17 we will thoroughly review an empirical approach that will provide a way to balance cost and quality outside of the math modeling context in which these optimal tolerance allocation methods reside. See Chapter 10 to find which approach suits your skill set best.

The Various Approaches to Cost-Based Optimal Tolerance Analysis

Greenwood, Chase, Loosli, and Hauglund [R9.1] develop a rigorous approach to cost-based optimal tolerancing by first defining an assembly tolerance that is acceptable from a functional standpoint. They then define the following allocation scheme:

> *The component tolerances could be distributed equally among all of the parts in an assembly. However, each component tolerance may have a different manufacturing cost associated with it due to part complexity or process differences. By defining a cost-versus-tolerance function for each component dimension, the component tolerances may be allocated to minimize the cost of production.*

Greenwood, Chase, Loosli, and Hauglund provide a useful overview of a number of proposed cost-versus-tolerance models that can be employed in the optimization process.

Table 9.1 shows a partial listing from their paper that includes the math principle the model is based upon, the model, the method of analysis, and the author of the model. The model terms are defined on a cost-per-part basis and are defined as follows: **A** quantifies the typical fixed costs (per part) in the manufacturing process, **B** quantifies the cost of making a single part or component, and **T** quantifies the tolerance.[1]

1. These models can be further studied in the article "Evaluation of Cost-Tolerance Algorithms for Design Tolerance Analysis and Synthesis," *Manufacturing Review* (ASME, vol. 1, no. 3, Oct. 1988, pp. 168–179).

Table 9.1

Math principle	*Cost model*	*Analysis method*	*Model author*
Reciprocal	$A + B/T$	Lagrange multiplier	Chase and Greenwood
Reciprocal squared	$A + B/T^2$	Lagrange multiplier	Spotts
Reciprocal power	$A + B/T^k$	Lagrange multiplier	Sutherland and Roth
Reciprocal power	B/T^{ki}	Nonlinear programming	Lee and Woo
Exponential	Be^{-mT}	Lagrange multiplier	Speckhart
Exponential/reciprocal power	$Be^{-mT/Tk}$	Nonlinear programming	Michael and Siddall
Piecewise linear	$A_i - B_i T_i$	Linear programming	Bjork and Patel
Empirical data	Discrete parts	Zero-one programming	Ostwald and Huang
Empirical data	Discrete parts	Combinatorial	Monte and Datseris
Empirical data	Discrete parts	Branch and bound	Lee and Woo

Greenwood et al. review the Lagrange multiplier method as applied to tolerance allocation; combinatorial process selection methods, including exhaustive search, zero-one discrete search, and univariate search; and nonlinear programming methods, including continuous zero-one selection coefficients. Their paper outlines the concept behind these methods to expose the reader to some of the options available.

Vasseur, Kurfess, and Cagan of Carnegie Mellon University put forth a unique cost-based tolerancing process (hereafter referred to as the VKC process) that requires a rigorous approach to accounting for costs [R9.2]. Their model is defined in the spirit of Taguchi's quality-loss function in that it relies heavily on the dependency between functional quality and cost to the customer; the fundamental metrics used in the VKC process are clearly influenced by Taguchi's work, as evidenced in the derivations of the terms they employ throughout their work. The economic terms they use include: N, the number of products in demand by the customers; P, the price of the product; F_c, the fixed cost of part production (production wages, tools, facilities, and inspection costs); C, the variable costs associated with part production (such as scrap costs); and

$$V\left(\frac{\sigma_{out}}{\Delta}\right),$$

the value a customer gives to a product based on the ratio of the process standard deviation and the tolerance. The VKC process provides an insightful mechanism to allocate assembly tolerances down to component tolerances on the basis of solving optimization equations that seek to minimize the cost based on customer value and demand. To sum up the focus of the VKC process we can do no better than to read their intent in their own words:

> *Industrial quality is presented as a direct result of manufacturing practices (e.g., tolerance allocation, process selection, inspection procedures). A quality indicator is developed that allows for the computation of a function representing the cost of an industrial product versus its quality level. This cost function is then incorporated in an economic model that estimates the consumer's demand for the product as a function of both its price and quality level. The profit maximizing values of the price and the quality level are derived, and in turn indicate the optimum production mode.*

Figure 9.1 Cost versus tolerance plot

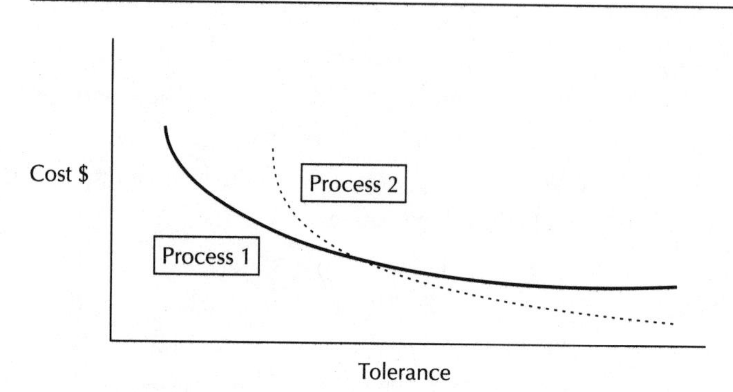

Cost versus Tolerance Plots

Both Greenwood et al. [R9.1] and Vasseur et al. [R9.2] make use of a common plot of cost-versus-tolerance functions. Such a plot is shown in Figure 9.1.

These plots typically express hyperbolic cost-versus-tolerance functions that are available from various manufacturing processes. These curves are used to aid in the solution of the optimal tolerance problems. If you are doing business with manufacturing companies that do not possess the capability to define these functions you have choices: help them do it or find someone who already possesses this knowledge.

Summary

It has been our purpose in this chapter to briefly review the equations and functional relationships discussed by the approach of Greenwood et al. and of the VKC process. With this basic understanding in hand I recommend you explore the cost-based optimal tolerance processes presented by these two approaches. I highly recommend that you obtain these papers and review the example problems in detail. The current text seeks to only provide a basic awareness of these approaches.

References

R9.1: *Least Cost Tolerance Allocation for Mechanical Assemblies with Automated Process Selection;* K. W. Chase, W. H. Greenwood, L. F. Hauglund and B. G. Loosli; 1989, pg. 357

R9.2: *Optimal Tolerance Allocation for Improved Productivity;* H. Vasseur, T. Kurfess and J. Cagan; IFAC Workshop on Automatic Control for Quality and Productivity, 1992, pg. 3

Additional references for the analytical optimization of costs and tolerances are contained in Appendix D.

CHAPTER 10

The Strengths and Weaknesses of the Traditional Tolerance Approaches

As we bring this review of traditional tolerancing methods to a close, it is appropriate to provide a sense of what strengths and weaknesses reside in each of these approaches. The first thing that should be noted, on the positive side, about all six approaches is that each is capable of helping an engineering team do a better job of developing tolerances. In the absence of these approaches, the design will be vulnerable to quality and cost problems either at the manufacturing and assembly phases or in the consumer's hands. The worst thing an engineer can do is to assign the task of tolerancing to a member of the design team who may lack the skills necessary to adequately fulfill the requirements of the tolerance engineering process. All members of the design team should be aware of these methods and work together to bring their unique skills to bear on the critical task of developing and properly communicating tolerances. It is recommended, in general, that the design engineer focus on developing a strong skill set in the area of tolerance engineering processes and tools, and that the drafter/designer, in general, focus on becoming a consummate expert in the communication of tolerances through the methods of geometric dimensioning and tolerancing. Both should cross train so that all can participate in the overall process.

On the negative side, none of these six methods is capable of doing the *complete* job of developing tolerances to balance cost and quality across an entire product or system. These are primarily mechanical stack tolerancing methods, and there are two things they cannot account for in their application.

First, none of the six traditional techniques is capable of dealing with multidisciplinary functions—that is, the use of diverse technologies in forming a subsystem or system. For example, in electrophotography, one way to form an image is by placing toner, a form of plastic dust, onto a film in a very controlled manner. This deposition process depends upon an assembly of mechanical, electrical, chemical, electrostatic, optical, and magnetic components *and* process set points. The entire assembly of complex and interactive elements must be toleranced at the component and process set-point level, as well as at the aggregate level as a subsystem that delivers toner to another subsystem, the film. This must be done in the debilitating presence of noise. The six traditional approaches can be a great deal of help in this endeavor, but they alone cannot see the job through to completion.

Second, none of the six methods directly employs *physical* models to which all three types of stressful noise (external, unit-to-unit, and deterioration) can be applied to induce the critical forms of variability that will occur beyond the variation in the product's performance due to unit-to-unit variation. The thing that the traditional approaches can't *easily* tell the team is how the tolerances constrain variability in the presence of external and deterioration noises. Most of the methods are purely analytical and are utterly dependent on the accuracy of a math model that is employed to carry out the simulated analysis. The math models are not capable of expressing the effect of the noise accurately because engineers are not at a point, as mathematicians, where such a sophisticated model can be efficiently and effectively constructed. It is possible to account for deterioration in the form of wear and dimensional changes due to external or internal thermal effects. Some of the CAT methods are purely geometric-model dependent. No noise factors outside of the geometry can be included in the analysis.

Performing designed experiments with all three types of noise active during tolerance analysis is a more thorough, reliable, and efficient approach to identifying the true sensitivity. Classically designed experiments such as central composite and D-optimal designs can be employed to empirically develop analytical expressions of very complex systems of components and process set points. Accurate models can be challenging to develop without the proper inclusion of the three types of noise in the actual experiments. The latest in the proper development of empirical math models can be found in the Meyers and Montgomery's *Response Surface Methods* [R10.1], which demonstrates the techniques needed to plan, construct, and run modeling experiments, and discusses how to analyze the data and represent the results in a 3-D plot called a *response surface.*

With these two general weaknesses identified we can now look at each of the six approaches to define some of their individual strengths and weaknesses.

Using Standard Tolerance Publications and Manufacturer's Process Capability Recommendations

The strength of this approach is that it draws on the historical experience and wisdom of hundreds if not thousands of design and manufacturing engineers, production experts, and machinists. The data that currently exists is invaluable at pointing one in the right direction for *starting* a tolerance analysis. The voice of experience is always a good starting point.

The weakness of this approach is that it is largely rooted in the three-sigma paradigm. Few of the manufacturing industries in this country have applied robustness optimization to their processes. Consequently, much of today's data on process capability are latent with suboptimal performance because processes are still quite sensitive to noise. The engineering team's task, in this case, is to work with the appropriate manufacturing colleagues to help them manufacture process parameter optimization to reduce output variance. The team needs to listen closely to how their processes will match its design needs. They will rarely steer engineers into a manufacturing process that is not appropriate for a given application—but engineers must ask *early* in the design phase and then carefully proceed to develop tolerances based on initial recommendations.

Worst-Case Tolerance

The strength of this method is t or tolerancing an assembly by accounting for the tolerance values are often employed because the rovide tolerance limit recommendations they knov t; this means the components will likely assemble ach is it is often *too* conservative. It is rare that al their worst-case limits simultaneously. Thus, it m ituations at the assembly level.

The Statistical Methods c

The Root Sum Square Appr

The statistical approach to tolerance y. It is generally considered a more rational appr bability of all the component tolerances coming in n on designing assemblies in too conservative a embly tolerance—which, of course, has direct i

The downside of RSS is that it s capability of the manufacturing process into acc quality may be significantly below the level of var required to provide the functional performance the customer requires over time. The six-sigma approach to RSS does account for the Cp or short-term process capability of a manufacturing process. This is an improvement over standard RSS that is solely focused on the three-sigma paradigm. The Cp-based RSS approach allows each component to be analyzed with respect to its specific process capability. This is a strong step toward the realistic assessment of tolerance requirements.

The Dynamic Root Sum of Squares Approach

This six-sigma approach to tolerance analysis is based on the inclusion of the long-term process capability, Cpk, of the individual component manufacturing parameters. This is a great strength for this approach; this characteristic sets it far above the rest of the RSS approaches because it includes not only the aggregate effects of the short-term variances of the process, but the additional variability that enters the total long-term picture by including the numerous shifts in the mean response.

This approach has no glaring weakness except those it shares with the other analytical mechanical stack approaches: a general inability to account for external and deterioration noises and a lack of dexterity to include other technical disciplines. If a mean shift occurs over the long term but does not possess a dynamic characteristic that encourages the mean to change numerous times, this technique is not the best way to account for the variability present in the process. To properly account for this scenario the *static* RSS method should be employed.

The Static Root Sum of Squares Approach

This approach, as just mentioned, is designed to account for a shift in the long-term mean of the process from one point to another for individual components. It can accurately account for the effect of a static mean shift. This is a strength when attempting accuracy in assessing the real driving forces behind the variability occurring in the components.

The weakness of the static RSS approach is that it may be used by an impatient process engineer when a dynamic state actually exists with respect to long-term mean shifts. The key to proper application is to adequately define what a *long-term* actually entails, and to be careful to account for component mean shifts in the assembly.

The Nonlinear Tolerance Approaches

Nonlinear tolerancing is an excellent tool to account for components that depart from linear stacking relationships. Significant errors can accrue when this approach is bypassed. This approach is valuable in defining tolerances that may need to depart from the typical bilateral expression that linear cases produce. Asymmetric tolerances may be necessary, and their need will not be detected through linear approximation methods.

The downside to this approach is that it is not always easy to define the nonlinear math model that expresses the true nature of the stackup. This approach depends on a higher level of math skill and will represent a challenge to those not fluent in the solution of partial derivatives. Fortunately, several very useful software packages are commercially available to simplify this approach. The nonlinear methods can be modified to include the Cp or Cpk in the determination of the tolerances, but this adds somewhat to the complexity of the analysis.

Sensitivity Analysis

This technique is extremely useful in isolating the real contributors to the tolerance sensitivity of the assembly. Certain components may be driving a higher percentage of the variability in the assembly tolerance, and these need to be identified. Once they are, they can be targeted for variance reduction to aid in efficiently meeting or controlling the overall assembly tolerance. This provides additional insight that facilitates a tuning ability to the system of tolerances in an assembly.

Computer-Aided Tolerancing

The strength of this approach is in its ability to handle very complex tolerancing cases in one, two, and three dimensions. When a computer application employs Monte Carlo simulation, it is also prized for its ability to assess many thousands of runs through the likely distribution of the component and assembly dimensions that are a consequence of the tolerance limits. It can be used to assess numerous "what if" scenarios to aid the engineering team in the process of defining the right tolerance level. The automation of the tolerance analysis process is yet another example of the power of personal and expert computer systems in design engineering.

Again, some software packages used in this approach suffer from the bounds of math model–based analysis. Other packages are purely geometry-driven and are extremely useful in carrying out complex 3-D tolerance analysis without any modeling requirements by the user.

CAT cannot account for environmental or deterioration noises in the way that a real assembly under experimental evaluation can. Monte Carlo simulations can suffer from improper model definition and distribution characterization. A reasonable degree of statistical literacy is required for the proper application of this approach; this is not really a weakness but a liability to those who enter the process unaware of the specifics of statistical distributions and how they impact the tolerancing process. CAT is bringing the complex task of performing nonlinear 2-D and 3-D tolerance analysis down to the level where designers can be quite comfortable using it. If one trend is clear in modern traditional tolerance analysis, it is that industry is rapidly putting the power of CAT to work in meeting the ever-increasing demand to rapidly design specifications for complex assemblies.

Cost-Based Optimal Tolerance Analysis

Cost-based optimal tolerance analysis techniques are very helpful in promoting economic design for functionality. They require a good deal of insight into developing a proper math model that relates cost and functional quality; once such a model is properly defined, the power of this optimal design process becomes quite evident. This assembly tolerance allocation process for defining component tolerances is noted for its ability to quantitatively demonstrate local optimum points for component tolerances.

The weakness of this approach is twofold: First, it has typically been built around component costs to the exclusion of direct costs to consumers due to off-target performance; second, it is the most demanding of all the traditional tolerancing processes in that it requires a high level of advanced engineering-math and numerical-analysis capabilities. In my experience this technique is intimidating to many practicing design engineers, who often doubt the accuracy of the math model they would construct and the validity of their solution. This technique remains a focus for many in academia and a conundrum for many in industry.

How the Six Processes Relate to the Overall Product Tolerancing Process

If there is to be one *manual* (as opposed to computer-aided) approach to be recommended for linear tolerancing problems, it would be the six-sigma DRSS approach. It prepares the engineering team to enter the final cost-versus-quality balancing process well and easily. This is not to say the other methods are defective; however, they do suffer from application-oriented problems, particularly with respect to mathematical complexity. The six-sigma methods stand out as techniques that have a strong intuitive appeal to engineers; this will go a long way in assuring that they are used. They are easy to learn and apply, sensible to the practicing engineer, statistically sound, and far-reaching in their ability to establish rational tolerances at the assembly and component levels. They can produce a platform of tolerances that can then be carried on, when necessary, into the Taguchi tolerance design process. This point is the point at which mechanical stacks can be related to nonmechanical components and set points that must be cotoleranced as contributors to overall product performance that minimizes loss to the customer as well as the business. As previously mentioned, the nonlinear method can be modified to accept six-sigma metrics, thus increasing their flexibility. For 1-D CAT applications with reasonably easy to define stackup math models (typically linear problems), we recommend the use of Crystal Ball. For 2-D and 3-D tolerance analyses (linear and nonlinear cases), we recommend one of the Pro/ENGINEER-based software packages.

I suggest the reader refrain from switching to experimental tolerance methods, just because the assembly math models are difficult to derive. A better approach is to apply a geometry-based CAT software package to the problem. In this way the software does all the work of accounting for the sensitivities and nonlinear relationships. The time to switch over to empirical tolerance design is when the subsystem technologies are such that CAT and manual methods cannot quantify the complex relationships and their attendant noises. Another complete tolerancing approach, based on the quality-loss function, is suggested by Taguchi, and it is fully explained in chapters 11 through 17. Traditional tolerance analysis is quite compatible with Taguchi's approach provided the reader invests the time to fully appreciate the strengths and weaknesses of *both* approaches.

Every engineering problem is different and requires the corporate wisdom of the engineering team to determine the exact process that will develop the design elements to their highest potential. Section II has developed and defined the portfolio of the most widely recognized traditional tolerancing methods available today. New research is undertaken in the field of tolerance analysis each year. It is recommended that the reader stay current with the state of the art through the many engineering conferences, professional societies, and publications available throughout the United States and the world. The sources quoted throughout this book are an excellent starting point for building one's personal tolerance engineering library.

Summary

We conclude our discussion of traditional tolerance analysis methods by reviewing the strengths and weaknesses of each of the six approaches covered in Section II. This rich portfolio of tolerance analysis techniques is summarized to draw out the most salient points concerning their applicability to specific cases.

Chapter 10 is outlined as follows:

Chapter 10: Strengths and Weaknesses of the Different Traditional Tolerance Approaches

- Using Standard Tolerance Publications and Manufacturer's Process Capability Recommendations
- Worst-Case Tolerance Analysis
- The Statistical Methods of Tolerance Analysis
    - The Root Sum Squares Approach
    - The Dynamic Root Sum of Squares Approach
    - The Static Root Sum of Squares Approach

    - Nonlinear Tolerance Approaches
- Sensitivity Analysis
- Computer-Aided Tolerancing
- Cost-Based Optimal Tolerance Analysis
- How the Six Processes Relate to the Overall Product Tolerancing Process

The chapter concludes on a cautious note. Be careful not to switch from traditional tolerance analysis to empirical tolerance analysis without good engineering reasons. Traditional and empirical methods both play a key role in tolerance development. The more one knows about each method, the easier it will be to discern which tool fits the task at hand. With this in mind, let's move on to Section III: The Taguchi Methods of Tolerancing and Tolerance Design.

Reference

R10.1: *Response Surface Methods;* R. Meyers and D. Montgomery; J. Wiley & Sons, 1995, pg. 367

SECTION III

Taguchi's Approach to Tolerancing and Tolerance Design

The methods of quality engineering have gained significant popularity in Western industry since Genichi Taguchi first introduced his approach at AT&T in 1980. Many thousands of engineers have become practitioners of the experimentally based optimization process known as robust design. Far fewer technical professionals have become proficient in the follow-on process known as tolerance design. This section is designed to provide a comprehensive introduction to the quality-loss function approach to analytical and experimental tolerance development. This material is designed to bring practitioners up to speed in tolerance design, thus enabling them to complete the off-line quality engineering activities prior to the manufacture of the product.

Section III contains the following chapters:

- Chapter 11: The Quality-Loss Function in Tolerancing and Tolerance Design
- Chapter 12: The Application of the Quadratic Loss Function to Tolerancing
- Chapter 13: General Review of Orthogonal Array Experimentation for Tolerance Design Applications
- Chapter 14: Introducing Noise into a Tolerance Experiment
- Chapter 15: Setting Up an Experiment for Variance and Tolerance Analysis
- Chapter 16: The ANOVA Method
- Chapter 17: The Tolerance Design Process: A Detailed Case Study

Taguchi's methods provide a direct link between a product's technical specifications and their ability to constrain the loss incurred by customers, who use the product in a stressful environment. This link can be used to aid in the process of balancing the cost of a product with its quality.

This section will round out your ability to match the right tolerance analysis tool with the appropriate application. After reading this section, you will have been exposed to all of the major tolerance development tools in modern engineering and design science. When the case lends itself to analytical modeling, you can select the approach that best suits the nature of the design at hand. If an experimental approach is deemed most helpful, you can follow the procedures outlined in Section III to facilitate the process.

Every attempt has been made to provide a comprehensive and practical explanation of the Taguchi approach of applying designed experiments to the development of tolerances. Several software packages can help with the computer-aided design of experiments as applied to tolerance analysis. The six-sigma metrics have been used in conjunction with the quality-loss function to help quantify the changes in quality and cost as a result of the optimization of tolerances.

CHAPTER 11

The Quality-
in Toleranci
Tolerance D

Linking Cost and Functio

The quality-loss function is the culmination of years of observation and applied research by Taguchi. It is his finest contribution to a twentieth-century society that is continuously seeking to save money by purchasing products that perform on target for their intended life. His premise is simple: A product that functionally stays on target will generate the smallest loss to the consumer. As a product or process deviates from its target response or expected output, costs, in one way or another, will begin to accrue. The company that produces products that inherently possess design attributes that resist output performance variation when environmental, manufacturing, and deterioration noises present themselves will survive and may even make a reasonable profit in the long run. Whether it makes a profit or not depends, in part, on how the company develops and deploys tolerances. Many companies use tolerances to control variability as an afterthought to a development cycle that focuses only on getting the nominal component and assembly set points to produce a prototype that works. They use tolerances to force the design to stay within accepted boundaries of a minimum level of performance, and as a countermeasure for sensitivity to interactive elements within a product or process. They often apply tolerances to nominal set points without stressing the design or its components to find out what can happen when the inevitable onslaught of variation invades the design. Thus the tolerances are assigned outside of the context of the very real effects of noise.

Other companies are more strategic in developing nominal set points that are made robust against noise. These teams develop tolerances as a final quality balancing process—tightening them strategically, based on knowledge from the customer and the real performance of the design parameters in the face of debilitating noise factors. These companies resist developing tolerances until after making the design robust. Interactive elements are desensitized so that tolerances can be freely developed to enhance customer satisfaction while keeping cost under control. Finally, the cost-balancing tolerances are built on direct knowledge of customer losses due to off-target performance; they are economically derived and applied, by design, based on insight from stressful performance data. These are some of the differences one needs to recognize between tolerancing practices designed to minimize a *company's* costs versus tolerancing practices designed to minimize a *customer's* costs. Companies that tolerance based on a conscious accounting of customer loss, as a consequence of controlled expenditure of corporate dollars to improve internal design and manufacturing process capabilities, will enter the next century with a powerful edge over companies that tolerance solely on the basis of unit manufacturing cost.

An Example of the Cost of Quality

Consider the following example. Imagine that you are playing the role of a product manufacturer whose product quality, costs, and reputation depend on the quality and costs of the subassemblies being bought from a supplier. There are three suppliers who want to do business with you—which, by the way, means they are doing business with your customers. They are producing a subassembly that must perform on target in order for the rest of the product to function properly. You are sent in as part of a team of quality auditors to measure how capable each supplier actually is. Your task is to construct representative distributions of performance based on 30 samples, as shown in Figure 11.1.

As Figure 11.1 indicates, all three suppliers are able to produce *most* of the subassemblies within the specification limits. All the suppliers have methods to assure that no out-of-specification subassemblies are delivered to your plant. In fact, the worst performer, Supplier C, delivers more than 99 percent of its subassemblies within specifications. Which supplier can best help to satisfy your customers? Most engineers would focus on the percentage of subassemblies that fall *outside* of the specification limits. It is quite common in most industries to consider only the costs associated with units that are outside the specification limits and focus on what can be done to minimize that number. Instead, one should consider *all* subassemblies—those within as well as outside of specification limits. The true and complete cost of quality crosses the boundary of the tolerance limits and focuses on performance *inside* the boundaries. This point of view is not just to get all units within the specification limits, but to get all units as close as possible to being on-target.

Measuring only the number of parts within specification is a misleading and incomplete method of quality measurement. It implies that all products that meet specifications are equally good, and that all products found to be out of the specification limits are equally bad.

A major problem with "within-specification" measures of quality is that, from a customer's perspective, a product that is just within specification is not as good as a product that is perfectly on target. Consider how our educational system uses a pass/fail criterion for examinations. Is there really much difference between a student who passes a test with a 65 percent grade and a student who fails the test with a 59 percent grade? In a similar fashion, is there much difference between a product that is barely within specification and one that is barely out of specification? On the other hand, consider the difference between a student who scores a 100 percent on a test and the student who also passes, but with a 70 percent grade. There is a substantial difference in performance between these two students. Similarly, there is a very substantial difference in quality between a product that is on target and one that is barely within specification.

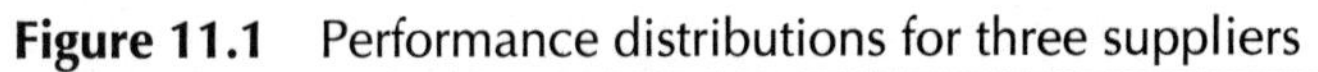
Figure 11.1 Performance distributions for three suppliers

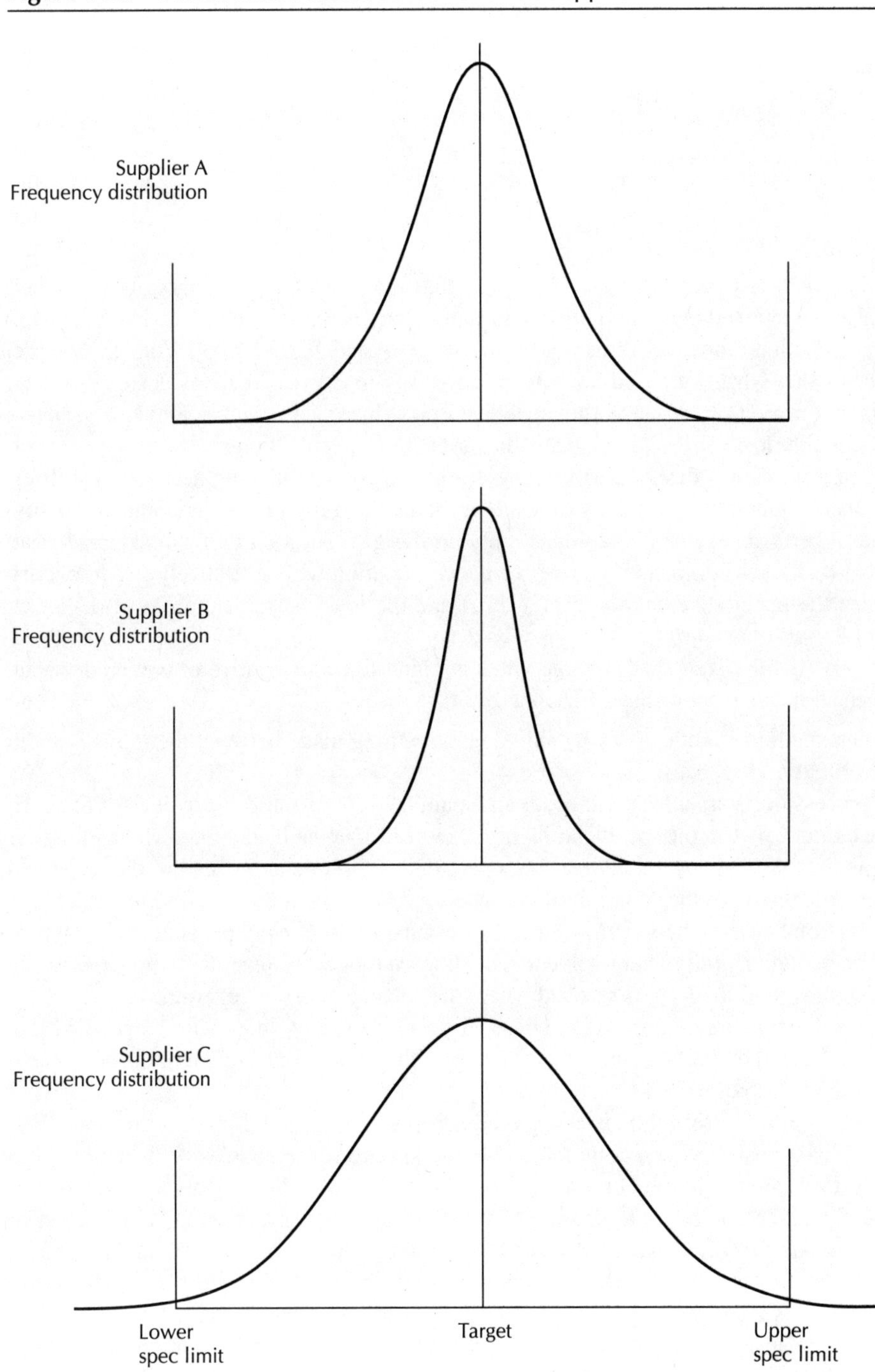

Figure 11.2 Flat distribution between specification limits

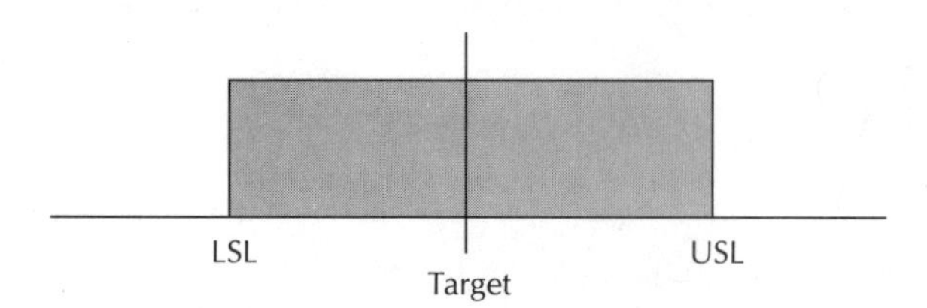

A study at Ford Motor Company [R11.1] showed that transmissions with parts close to their specified target values functioned better and operated more quietly than the same parts that were technically "within specification," but had dimensions that were spread further away from their target values. This indicates that what is needed is a quantitative way to express the loss customers incur when the products perform off target but within specifications. This requires a metric that is sensitive enough to quantify quality loss for design elements that are *within* specifications.

Tolerances are a necessary design entity. They are useful for establishing acceptability limits when tied to customer requirements. But how they are developed and used in an economical, quality-based approach has a performance *and* cost impact on your designs. All too often, companies have a quality strategy that facilitates component and performance variation with a relatively flat frequency distribution between the upper specification limit (USL) and the lower specification limit (LSL), as shown in Figure 11.2.

In this case it is *equally likely* that any one value is going to occur relative to any other value. Several types of behaviors are responsible for such a result:

1. Companies maximize productivity by shipping everything made between the limits without regard to quality differences.
2. Known process drifts are allowed to occur and manufacturing strategies are devised to maximize the output between the specification limits. For example, in machining, where tool wear causes predictable changes in a dimension, operators may set up at one end of the range and allow the process to produce parts until the other end is reached, thus achieving a relatively flat distribution between the starting and ending points. Another example is in chemical processing or photofinishing, where solvent recycling causes a uniform drift from the initial pure conditions until the limit is reached where the solvents must be renewed.
3. Out-of-specification parts are reworked after inspection to bring them within specification. Many of the reworked parts are never put on target; they are just brought into the acceptable range as quickly as possible.
4. Variation between control limits is considered acceptable. Statistical process control (SPC) is used to distinguish between common-cause and special-cause sources of deviation from the target. *Common cause* refers to random sources of variation that are always present and accepted. Special cause refers to sources of variation that are correctable by elimination or compensation *as they are detected.* A basic SPC strategy is to choose control limits based on three standard deviations of the process output about the mean. Any random variation within the control limits is judged to be due to common causes, and thus is treated as acceptable. Only variation outside of these limits or samples lacking randomness (see the SPC charts in Chapter 6) trigger corrective action. Thus, three-sigma SPC necessarily documents a distribution, whose width is related to the three-sigma control limits and whose shape depends

Figure 11.3 A distribution biased toward the target

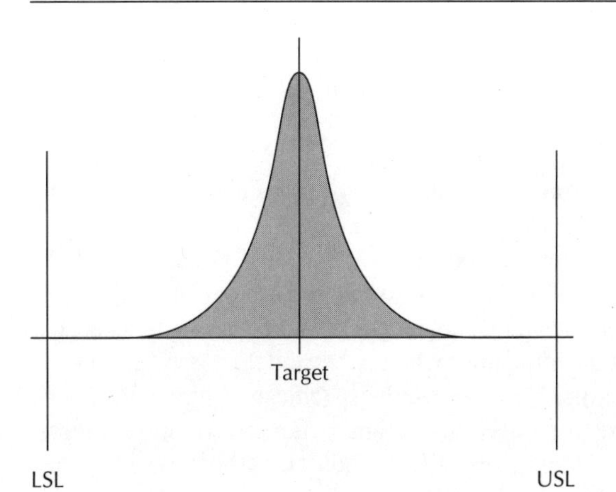

on the nature of the common causes (random noise factors). When the control limits are set according to the traditional three-sigma statistical standards, cost guidelines and quality requirements are very much constrained. As we have discussed at length in Chapters 1, 3, and 4, the three-sigma paradigm has run its course and is no longer viable in the demanding markets that have developed in the global economy.

All of these behaviors share a manufacturing philosophy that aims to maximize productivity, maximize materials utilization, and minimize defects, rejects, and waste. In principle, these are all laudable goals. But there is nothing that explicitly encourages on-target manufacturing. Companies that do not encourage on-target manufacturing do not *explicitly* reward or encourage their staffs to work beyond a "good yield" to produce on-target results.

Companies that focus on a quality strategy based on meeting target values exhibit a very different kind of frequency distribution, as shown in Figure 11.3.

Several types of behaviors are responsible for such a result:

1. A serious effort at fostering statistical literacy for everyone with product-performance responsibility. For everyone, from operators through engineers to management, the ability to acquire and analyze data using the statistical measures discussed in Chapter 3 is a critical skill requirement.
2. Measuring and displaying distributions using tables and histograms helps indicate, much more than SPC run charts alone, the specific quality of various production processes.
3. Shifting away from three-sigma to six-sigma behaviors. Achieving very high-process capabilities and using measures, such as Cpk, that penalize a process for being off target, force attention to focus on on-target performance. Six-sigma metrics are discussed in detail in Chapter 3.
4. Maintaining manufacturing-process capability data bases that assure the correct process is specified for a required level of quality and cost performance. Without such information there is a tendency to make assumptions about the nature of variability that are not realistic (see Chapter 4).

5. Using design of experiments (DOE). This is undoubtedly one of the most powerful statistical tools available for improving product quality because of its preventive nature. It is not possible to eliminate all problems purely through analysis; therefore, the empirical methods of DOE constitute critical engineering skills. Fowlkes and Creveling [R11.2] demonstrate and explain the application of DOE to robust design.
6. Comprehensively using quality engineering methods. Taguchi methods are one of the most powerful engineering tools for improving product quality early in product development by the application of on-target engineering (see Chapters 1 and 2).
7. Understanding the true cost of quality. Introduced at the end of Chapter 1 as quality loss, the true cost of quality to industry is poor profits, loss of customers, and loss of whole businesses. The companies that have suffered such a fate do not go out of business solely because of out-of-spec parts. They are displaced by companies who understand that costs go down as quality improves and that quality is measured well *within* the specification limits. Profits go up as on-target quality improves because of reduced inspection, reduced warranty and repair costs, better efficiency, and the advent of more competitive products. Ultimately, when customer losses are *rationally* minimized, the cost of quality will also be minimized.

All of these examples share a manufacturing philosophy that aims to maximize quality beyond simply measuring conformance to specifications. Problem prevention and variability reduction are greatly emphasized as being necessary to keep processes on target. The best of these companies explicitly reward and encourage their staffs to produce on-target results from concept inception all the way to product delivery.

The Step Function: An Inadequate Description of Quality

The preceding discussion reveals that a lot can be understood about a company's quality practices, reward systems, and assumptions from studying the type and width of their products' performance distributions. There is a distinct *quality culture* at companies that practice on-target engineering. The value system at companies that are satisfied by within-specification performance can be described by the *step function,* as shown in Figure 11.4.

Engineering specifications often take the form of a target value, *m,* with a bilateral tolerance, written as $m +/- \Delta_0$. Using such a specification to suggest that all values in the range of $m - \Delta_0$ to

Figure 11.4 Step function describing quality

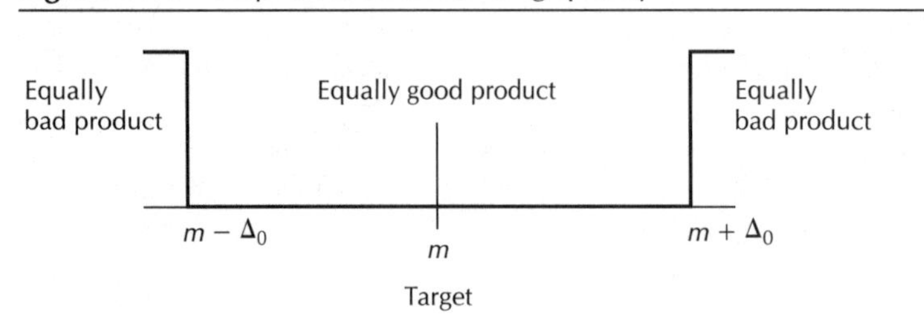

Figure 11.5 Average preference distribution

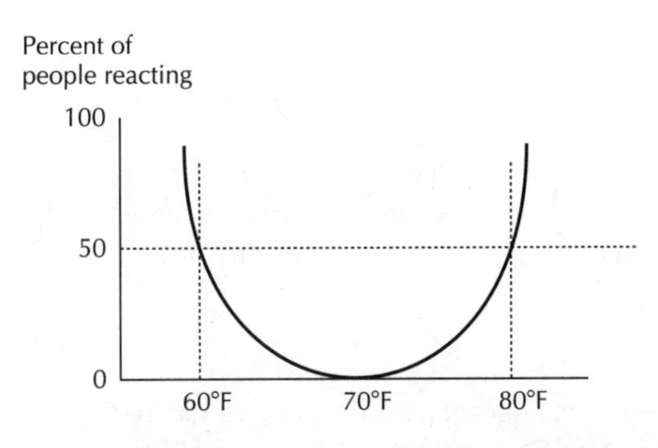

$m + \Delta_0$ are of *equally good* quality for the customer, and all values outside this range are of *equally bad* quality is an error. The step function is an inadequate description of quality.

The Customer Tolerance

Tolerances are defined as the limit at which some significant, economically measurable action is taken. On-target engineering does not eliminate the need for tolerances; it simply shifts their focus to customer costs (Chapter 12). How far from the target can the product performance get before the customer would incur intolerable monetary loss? How far from the target can a component be before manufacturing should reject the unit on behalf of the customer's best financial interest? These are important quantities to define for deciding what limiting actions need to be taken.

The customer tolerance, Δ_0, corresponds to the point at which a significant number of customers take economic action because of off-target performance. The concept of a customer tolerance limit, Δ_0, can be demonstrated with an example of human behavior. The parabolic curve is commonly applied to the frequency distribution of individual preferences for the level of a nominal target-quality characteristic. For example, consider a hypothetical poll taken of people's inclination to adjust a thermostat that is controlling the room temperature against ambient conditions. Each person states at what ambient temperature deviation from 70°F they would become uncomfortable and adjust the thermostat to compensate. Figure 11.5 represents the typical response that a large group (more than 20 people) would generate [R11.3].[1]

Notice that as the ambient temperature drifts farther away from 70°F, more and more people are inclined to adjust the thermostat. As a result, the general form of a parabola emerges. One commonly used standard for the customer functional limit is the deviation at which 50 percent of the customers *take action* to correct the deviation. This is shown by the dotted lines in Figure 11.5. It is also referred to as the LD 50 point. LD stands for *live or die,* and is taken from medical studies of the effects of certain experimental treatments. Here, LD 50 is simply when approximately 50 percent of the customers' tolerance limit is reached and they take economic action—at their loss.

1. This example was originally introduced by Peter T. Jessup, "The Value of Continuing Improvement," IEEE Conference, 1985.

The Quality-Loss Function: A Better Description of Quality

Companies that are practicing on-target engineering use an alternative approach to the limitations the step function exhibits as a measure of quality. The quadratic quality-loss function, or simply the quadratic-loss function (QLF), was developed by Taguchi to provide a better estimate of the monetary loss incurred by manufacturers and consumers as product performance deviates from its target value. The quadratic quality-loss function, shown in Equation 11.1, approximates the quality loss in a wide variety of situations.

$$L(y) = k(y - m)^2 \qquad \textbf{(11.1)}$$

where $L(y)$ is the loss in dollars due to a deviation away from targeted performance as a function of the measured response, y, of the product; m is the target value of the product's response; and k is an economic constant called the *quality-loss coefficient.*

Figure 11.6 illustrates the quality-loss function. Compare it to Figure 11.4. At $y = m$, the loss is zero, and it increases the further y deviates from m. The quality-loss curve typically represents the quality-loss for an average group of customers. The quality loss for a specific customer would vary depending on that customer's tolerance and usage environment. However, it is not necessary to derive an exact loss function for all situations. That would be too difficult and not generally applicable. The quality-loss function can be viewed on several levels:

1. As a unifying concept of quality and cost that allows one to practice the underlying philosophy driving on-target engineering.
2. As a function that allows one to relate economic and engineering terms in one model.
3. As an equation that allows one to do detailed optimization of all costs, explicit and implicit, incurred by the firm, customers, and society through the production and use of a product.

The empirical observation for the shape of the quality-loss function can be derived formally [R11.4]. The value y denotes the actual performance as measured or perceived by the customer. The value m represents the target performance the customer has paid to receive. The function $L(y)$ is given by a Taylor series expansion about $y = m$:

$$L(y) = L(m) + [L'(m)/1!][y - m] + [L''(m)/2!][y - m]^2 + \ldots \qquad \textbf{(11.2)}$$

Figure 11.6 The quadratic-loss function

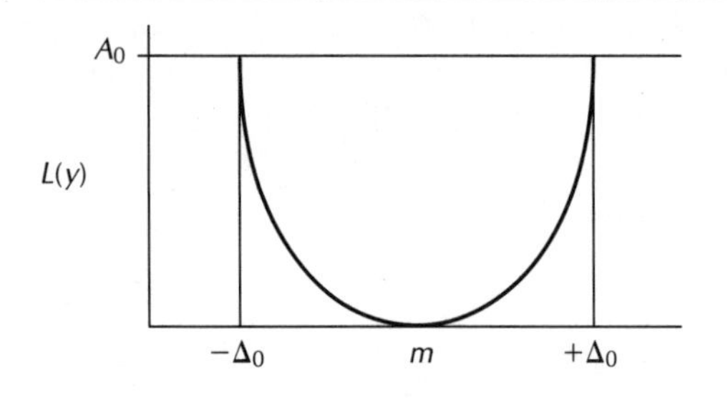

When the performance, y, is on the target, m, the quality loss should be zero (assuming the product satisfies its intended customer base when performing on target). Thus, the first term is zero: $L(m) = 0$. When the first derivative is taken at the target, the second term in the expansion is zero since the loss function is a minimum at $y = m$: $L'(m) = 0$. This leaves the third term, $[L''(m)/2!][y - m]^2$, as the leading term in the expansion. Assuming that this expansion is applied to situations where y is close to m, the defining function for customer loss as performance, y, deviates from the target, m, is given by Equation 11.3.

$$L(y) \approx [L''(m)/2!][y - m]^2 \qquad \textbf{(11.3)}$$

The higher order terms of the expansion are, in practice, inconsequential, since the loss function is generally applied *close to the target,* where higher order terms are relatively small. The term $[L''(m)/2!] = k$, the quality-loss coefficient, is an economic constant of proportionality. Thus, we are left with the quadratic quality-loss function given in Equation 11.1.

The Quality-Loss Coefficient

The quality-loss coefficient, k, is determined by first finding the functional limits or customer tolerance for y, the measured response. The *functional limits,* $m +/- \Delta_0$, are the points at which the product would fail or produce unacceptable performance in approximately half of the customer applications, as shown in Figure 11.6. These represent performance levels that are equivalent to the average customer tolerance. Tolerances have been defined as the limit at which some significant, economically measurable action takes place. In this case, the product has essentially failed, so the economically measurable consequences of failure, A_0, becomes the value of the quality-loss function at $m+/- \Delta_0$, as shown in Figure 11.6. These consequences include repair, replacement, loss of use, waste, and all the other categories discussed earlier. Let the total of all the losses at $m+/- \Delta_0$ be equal to A_0 in dollars:

$$L(y) = A_0 \text{ at } y = m +/- \Delta_0 \qquad \textbf{(11.4)}$$

Substituting the functional limits $m +/- \Delta_0$ and the total losses A_0 into equation 11.1, the quality-loss coefficient is found to be:

$$k = A_0/(\Delta_0)^2 \qquad \textbf{(11.5)}$$

Remember that A_0 is the cost to replace or repair the product, including the losses incurred by the manufacturer and customer, as a consequence of off-target performance. Some typical situations that figure into repairing or replacing a product are:

- Loss due to lack of access to the product during repair,
- The cost of parts and labor needed to make the repair, and
- The costs of transporting the product to a repair center.

Regardless of who pays for the losses—whether it be the customer, the manufacturer, or a third party—all losses should be included in A_0. Substituting Equation 11.5 into equation 11.1 defines the quality-loss function as:

$$L(y) = \frac{A_0}{\Delta_0^2}(y - m)^2 \qquad \textbf{(11.6)}$$

An Example of the Quality-Loss Function

A spring—which stores and releases energy—is used in the operation of a camera shutter. The process used to wind and form the dimensions necessary to constrain the spring wire has a certain level of variability.

Variability in the forming process results in variability of the measurable parameter called the *spring rate* (measured in ounces of force per inch of deflection). Assume that the target spring constant is $m = 0.5$ ounces per inch. The shutter speed varies due to several factors, one of which is the spring rate. Assume that the functional limits for the spring constant are m +/− 0.3 oz./in. This means that half the customers who purchase a camera with a spring constant of m +/− 0.3 oz./in. would consider the camera defective because of improperly exposed pictures. Let's also assume that the average cost for repairing or replacing a camera with unacceptable shutter speed due to an off-target spring constant is \$20. Thus the customer loss is $A_0 = \$20$. By substituting these values into the k factor ($k = 222$ \$/oz./in.) quality-loss function:

$$L(y) = \frac{\$20}{0.3^2}(y - 0.5)^2 = 222\left(\frac{\$}{\text{oz./in.}}\right)(y - 0.5)^2 \tag{11.7}$$

Equation 11.7 can be used to calculate the expected quality loss, in dollars, for various springs. For example, the quality loss to customers who purchased a camera with a spring constant of 0.25 is:

$$L(0.25) = \$222(0.25 - 0.5)^2$$

$$L(0.25) = \$222(-0.25)^2$$

$$L(0.25) = \$13.88$$

For a camera with a spring constant of 0.435, the quality loss is:

$$L(0.435) = \$222(0.435 - 0.5)^2$$

$$L(0.435) = \$222(-0.065)^2$$

$$L(0.435) = \$0.94$$

The loss function is generally used in establishing manufacturing tolerances during initial tolerancing and the formal tolerance design process. It is also used to help define signal-to-noise metrics used in parameter design (see Chapters 4 and 5 in Fowlkes and Creveling [R11.2]).

The Types of Quality-Loss Functions

There are four types of quality-loss functions. Three are derivatives of the general case called the nominal-the-best case.

The nominal-the-best case (NTB)

This is the case just used to describe the loss attributed to the off-target spring force. The loss function is shown in Figure 11.6; Equation 11.6 is the nominal-the-best function. In the NTB case the measured response, y, always has a specific target value, m. The quality loss is equally undesirable on either side of the target.

Here are some examples of the nominal-the-best type of problem:

- Control of aerosol flow from a spray can,
- Boring an engine cylinder to an aim diameter,
- Controlling the diameter of a filament for a light bulb,
- Controlling the viscosity of an automotive oil,
- Controlling the diameter of a car tire,
- Maintaining the part geometry of an injection molded part, and
- Creating a particular hue in mixing paint.

Average quality loss for nominal-the-best cases

The nominal-the-best loss function can be applied to just one part or product or to the average loss associated with more than one unit. The concept of the average loss is central to the signal-to-noise ratio concept. Therefore, it is worth the effort of deriving the average loss and seeing how it decomposes into two parts: the contribution due to the mean being off target and the contribution due to the variance.

The average quality loss is found by defining an average of the measure of off-target performance $(y - m)^2$, which is the argument of the loss function (Equation 11.1). The average off-target value is referred to as the *mean square deviation,* or *MSD.* The deviation referred to is from the target m. The average loss, $\overline{L(y)}$, is given by

$$\overline{L(y)} = k(\text{MSD}) \tag{11.8}$$

where,

$$\text{MSD} = \frac{1}{n}[(y_1 - m)^2 + (y_2 - m)^2 + (y_3 - m)^2 + \ldots + (y_n - m)^2]$$

$$= \frac{1}{n}\sum_{i=1}^{n}(y_i - m)^2$$

$$= \frac{1}{n}\sum_{i=1}^{n}(y_i^2 - 2y_i m + m^2)$$

$$= \frac{1}{n}\left(\sum_{i=1}^{n} y_i^2 - 2m\sum_{i=1}^{n} y_i^2 + \sum_{i=1}^{n} m^2\right)$$

Applying the definition of the mean to the second term in the parentheses,

$$\text{MSD} = \frac{1}{n}\sum_{i=1}^{n}(y_i^2) - 2\overline{y}m + m^2$$

where $\overline{y}$ is the average value (mean) of the individual quality characteristic values, y_i. By adding and subtracting equal terms, the MSD can be expressed as the sum of two squares.

$$\text{MSD} = \frac{1}{n}\sum_{i=1}^{n}(y_i^2) - \overline{y}^2 + \overline{y}^2 - 2\overline{y}m + m^2$$

Using an identity $\frac{1}{n}\sum_{i=1}^{n}(y_i^2) - \overline{y}^2 = \frac{1}{n}\sum_{i=1}^{n}(y_i - \overline{y})^2$, the final equation is obtained.

$$\text{MSD} = \frac{1}{n}\sum_{i=1}^{n}(y_i - \overline{y})^2 + (\overline{y} - m)^2 \quad \textbf{(11.9)}$$

Notice that the MSD is made up first of a sum that expresses the average square deviation from the mean, and second, of a term that expresses the deviation of the mean from the target. This decomposition is seen again and again as the mathematics of robust design is explored. The sum $\frac{1}{n}\sum_{i=1}^{n}(y_i - \overline{y})^2$ is the population standard deviation, σ^2. Thus:

$$\text{MSD} = \sigma^2 + (y - m)^2 \quad \textbf{(11.10)}$$

Equation 3.4, in Chapter 3, gives the expression for the variance, S^2, that is based on a sample from the total population. The variances are related by:

$$S^2 \approx \sigma^2 \text{ for } n > 30 \quad \textbf{(11.11)}$$

Therefore, the loss function for n units is

$$L(y) \approx k[S^2 + (\overline{y} - m)^2] \quad \textbf{(11.12)}$$

With this form of the loss function available, several key points can now be made that form the basis for a strategy that drives all of our engineering optimization endeavors. To reduce loss, the MSD must be reduced. This can be done in two ways:

1. Reduce the variability that is causing the dispersion of the data about $\overline{y}$, thus minimizing S^2.
2. Adjust the average response, $\overline{y}$, to fall on the target m, $(\overline{y} - m)^2 = 0$.

These concepts are used [R11.2] to derive the signal-to-noise ratio.

Equation 11.12 can be used to calculate the average quality loss for a set of products. Returning to the springs problem, let there be two machines making the springs—one new spring winder and one older model that has a good deal of wear and tear on it. Again, the target value is $m = 0.5$ oz./in. The average loss function is:

$$L(y) = \$222[S^2 + (\overline{y} - m)^2]$$

Table 11.1

Winder	*Data*	S^2	$\overline{y}$	$(\overline{y} - m)^2$	$L(y)$
New	0.37, 0.41, 0.37 0.43, 0.39, 0.35 0.40, 0.36	0.0007	0.385	0.0132	3.08
Old	0.55, 0.67, 0.70 0.54, 0.41, 0.32 0.46, 0.66	0.0184	0.539	0.0015	4.41

Figure 11.7 The smaller-the-better loss function

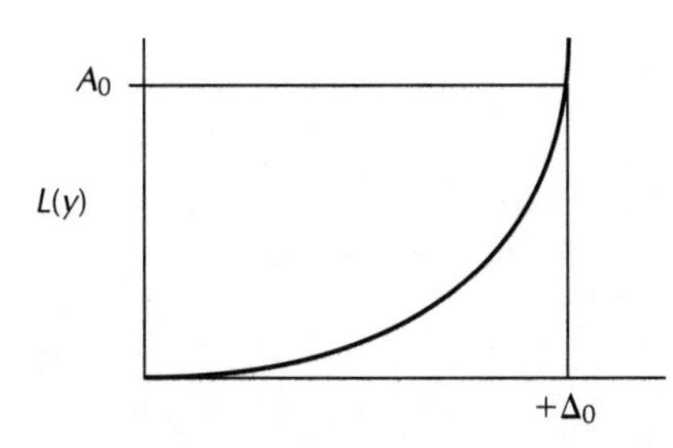

Table 11.1 shows that the new machine has the lower variance but is, on average, fairly far from the target. The old machine gives the impression that it is doing pretty well by averaging around the target—but it has a lot of variability. Notice that the older machine has the greater loss. It is easier to shift the tight variance onto the target than it is to tighten a widely variant distribution.

The smaller-the-better case (STB)

Some responses never have a negative value; their targeted response is ideally zero. These are referred to as smaller-the-better. Because $m = 0$, $L(y) = k(y - 0)^2$; thus, the STB quality loss function is given by:

$$L(y) = k(y)^2 \tag{11.13}$$

As the value of y gets farther from zero, the performance gets worse and the loss starts to increase.

Here are some examples of smaller-the-better cases:

- Microwave oven radiation leakage,
- Time it takes to get the first copy out of a copy machine,
- Paper jams in a copier,
- Defects on an image,
- Background density on an image,
- Automotive exhaust pollution,
- Steering column vibration,
- Electromagnetic interference from consumer electronics, and
- Corrosion of metals.

Fog, or background density, in an image is an example of the smaller-the-better case. In the copier industry, one measure of a copy's acceptability is the amount of background toner that adheres to the portion of the copy intended to be white. Minimizing the residual toner in white areas is a smaller-the-better objective. It has been determined that approximately half of the customers will not tolerate a background level beyond the standard measure of 1.2 background units. Beyond that level, a service call is placed at a cost of \$200 plus the cost of the down-time of the copier—typically valued at approximately \$150 per hour. If the average copier down-time is about 2.5 hours, the customer's loss is \$375. Thus, the total cost is $A_0 = \$200 + \$375 = \$575$.

$$k = \frac{A_0}{\Delta_0^2} = \frac{\$575}{1.2^2} = \$399.30\ (\$/\text{background}^2)$$

Thus the STB loss function is given by

$$L(y) = \$399.30(y)^2$$

Average quality loss for the smaller-the-better case

Just as the NTB case is modified for calculating the loss for more than one item, the same can be done for the STB case. Again, the mean square deviation is used:

$$\text{MSD}_{STB} = \frac{1}{n}\sum_{i=1}^{n} y_i^2 \approx (S^2 + \bar{y}^2) \quad \textbf{(11.14)}$$

Consequently the average loss for the STB case is:

$$L(y) = k[S^2 + \bar{y}^2] \quad \textbf{(11.15)}$$

Since the target is zero in the STB case, $\bar{y}$ is the deviation of the mean from zero. The ideal function is focused on the smallest response value possible. The losses can be added to account for total product *economic* performance as it relates to a direct measure of *physical* performance.

Equation 11.15 can be used to calculate the expected quality loss, in dollars, for the STB copier background case where $k = \$399.30$. For example, consider two machines in one print shop giving the performance shown in Table 11.2.

The table shows the first machine to be producing the smallest amount of loss.

The larger-the-better case (LTB)

Some measured responses, while never having negative values, are better as their value grows. Ideally, in LTB cases, as the performance value approaches infinity the quality loss approaches zero. This loss function is simply the reciprocal of the smaller-the-better case:

$$L(y) = k[1/(y)^2] \quad \textbf{(11.16)}$$

To determine k for this loss function, the reciprocal relationship must be used. Thus the loss A_0 is related to the functional limit Δ_0 by $A_0 = k[1/\Delta_0^2]$, and the loss function coefficient is given by

$$k = A_0\Delta_0^2 \quad \textbf{(11.17)}$$

Table 11.2

Machine	*Data*	S^2	$\bar{y}$	$S^2 + \bar{y}^2$	*L(y)*
#1	0.64, 0.56, 0.71 0.55, 0.59, 0.75 0.64, 0.76	0.0068	0.65	0.4293	\$171.41
#2	0.55, 0.67, 0.70 0.94, 0.71, 0.82 0.86, 0.96	0.0203	0.776	0.6229	\$248.70

Figure 11.8 The larger-the-better loss function

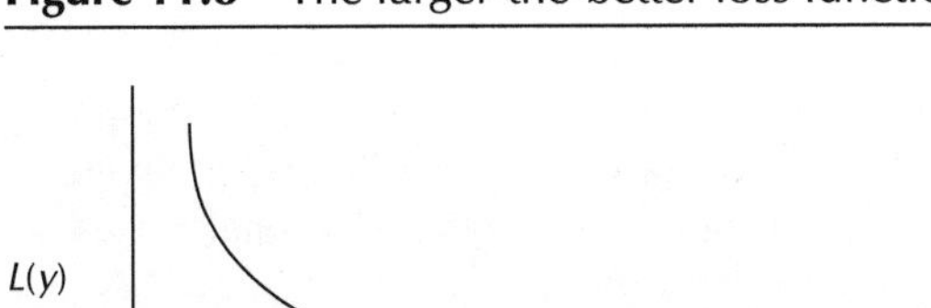

Here are some examples of larger-the-better cases:

- The strength of a permanent adhesive,
- The weld strength of a joint,
- The traction capability of a tire,
- Automobile gas mileage,
- Efficiency of a home heating furnace, and
- Corrosion resistance of an auto body.

The seal strength of a vacuum blower housing in an office copier is an example of a larger-the-better case. The better it can run under widely varying environments, the better it minimizes loss. When the blower seal fails to operate, it costs $40 to replace; $20 in part costs and another $20 for labor. While the device that uses the vacuum blower sits idle, the cost to the customer is $340 per hour. On average it takes 30 minutes to replace the blower.

$$A_0 = \$170 + \$40 = \$210$$

The seal can suffer changes in its adhesive properties as a result of vibration over time, as well as humidity and temperature. So, as the deterioration noises assail the adhesion of the seal, the vacuum loss grows to an unacceptable level that approximately 50 percent of the customers will notice. The larger the adhesive stability the better. Seal integrity is measured by testing the seal adhesion strength in pounds per square inch. The level at which the vacuum loss becomes objectionable is 20 psi. Thus, $\Delta_0 = 20$ psi.

The LTB loss function is given by:

$$k = A_0\Delta_0^2$$

$$k = (\$210)(20)^2[\$(\text{psi}^2)]$$

$$L(y) = 84{,}000[1/(y)^2]$$

In this case the seal strength of a returned machine is at 13 psi. Thus,

$$L(y) = 84{,}000[1/(13)^2]$$

$$L(y) = 84{,}000[0.0059]$$

$$L(y) = \$497$$

Therefore, when the seal strength drifts to a low of 13 psi, a total loss of about $500 is incurred. How can this be when the cost of repair is only $210? The consequences of seal failure are increased noise, loss of toner-dust containment resulting in machine and office contamination, image quality degradation, and, of course, a service call. If the machine is allowed to degrade beyond the nominal functional limit due to lack of attention or misdiagnosis of the failure mode, additional losses are incurred.

Average quality loss for larger-the-better

Just as for the nominal-the-best and smaller-the-better cases, it is useful to calculate the average LTB quality loss. The mean square deviation for the larger-the-better case is given by Equation 11.18.

$$\text{MSD} = \frac{1}{n}\sum_{i=1}^{n}(1/y_i)^2 = \frac{1}{n}[1/(y_1)^2 + 1/(y_2)^2 + 1/(y_3)^2 + \ldots + 1/(y_n)^2] \quad \textbf{(11.18)}$$

The average LTB loss function is given by Equation 11.19.

$$L(y) = k(\text{MSD}) = k\left[\frac{1}{n}\sum_{i=1}^{n}(1/y_i)^2\right] \quad \textbf{(11.19)}$$

It is possible to express the MSD in terms of the mean, $\bar{y}$, and the variance, S^2, of the quality characteristic after some algebraic manipulation [R11.5]. This is required because of the reciprocal in the LTB case. Thus, the MSD is given by:

$$\text{MSD} \cong \frac{1}{\bar{y}^2}\left(1 + \frac{3S^2}{\bar{y}^2} + \frac{4\hat{\mu}_3}{\bar{y}^3} + \frac{5\hat{\mu}_4}{\bar{y}^4}\right) \quad \textbf{(11.20)}$$

where the higher order terms are given by:

$$\hat{\mu}_3 = \frac{1}{n}\sum_{i=1}^{n}(y_i - \bar{y})^3 \text{ and } \hat{\mu}_4 = \frac{1}{n}\sum_{i=1}^{n}(y_i - \bar{y})^4 \quad \textbf{(11.21)}$$

The higher order terms can be neglected only if two conditions are met:

1. The distribution is not too skewed—that is, it is close to normal or Gaussian—so that $\hat{\mu}_3 \approx 0$.
2. $\bar{y} >> \hat{\mu}_4$—that is, the distribution width (kurtosis) is not too great.

The likelihood of these conditions being met routinely is not high. Therefore, we recommend using Equation 5.18 when calculating the larger-the-better MSD, forgoing any calculation shortcuts.

Equation 11.19 can be used to calculate the expected quality loss, in dollars, for the LTB seal adhesion problem for the vacuum blower for two separate sites. Each has eight machines operating in it. One site is in Miami, Florida, and the other is in St. Paul, Minnesota. The loss function for this case is:

$$L(y) = 84{,}000\left[\frac{1}{n}\sum_{i=1}^{n}(1/y_i)^2\right]$$

After transforming the data by taking the reciprocal of the square of the measured seal strengths:

Table 11.3

Site	*Data*	*MSD*	*L(y)*
Miami	17, 21, 30, 12		
	10, 24, 16, 27		
St. Paul	37, 28, 30, 42		
	29, 32, 36, 25		

Table 11.4

Machine	*Data*	*MSD*	*L(y)*
Miami	0.0035, 0.0023, 0.0011, 0.0069	0.0039	$323.40
	0.0100, 0.0017, 0.0039, 0.0014		
St. Paul	0.0007, 0.0013, 0.0011, 0.0006	0.0010	$84.00
	0.0012, 0.0010, 0.0008, 0.0016		

According to the data from the investigation of their seal strengths, the St. Paul site has the smaller loss. The vacuum seal is expected to deteriorate due to vibration, temperature, and humidity. The lower loss figures for St. Paul are a manifestation of the fact that the copiers there will last longer and result in fewer or later service calls, thus minimizing losses in real dollars.

The asymmetric nominal-the-best loss function

There are instances in which it is more harmful for a product's performance to be off target in one direction than in the other. There are two distinct quality-loss coefficients required for this case—one for each direction away from the target. The asymmetric loss function takes on the following form:

$$L^{+}(y) = k^{+}(y - m)^2, y > m \tag{11.22a}$$

$$L^{-}(y) = k^{-}(y - m)^2, y \leq m \tag{11.22b}$$

Some examples of the asymmetric loss function are:

- The amount of toner a copier uses to make an image. Too much toner use increases toner consumption, which is costly, and too little produces weak images that are immediately unacceptable.
- Refrigerator temperature variation. More food will spoil if the temperature is above target than will spoil if the temperature is below target.

As an example, consider the case of temperature drift in a refrigerator. The standard target for most refrigerators is approximately 40°F. Consider the consequences of being above or below this targeted temperature. When the temperature gets above 50°F, several things can happen that annoy the consumer. These include tepid food and drink that is not pleasant to the taste, and spoilage due to accelerated bacterial growth. Each of these can cause economic loss, losses due to discarding and replacement of food, and losses due to illness from ingestion of tainted food. When the temperature

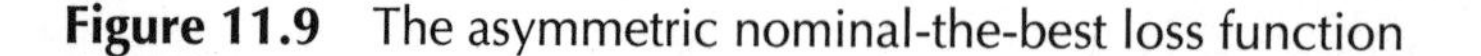

Figure 11.9 The asymmetric nominal-the-best loss function

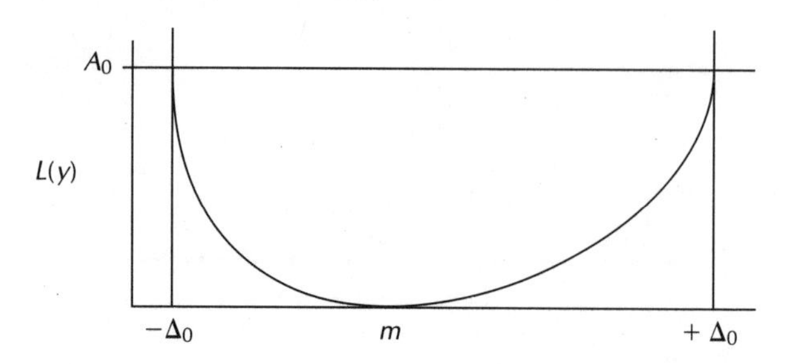

gets below 30°F there may be some damage due to ice crystals, but there should be little food lost. The loss function constant $k^- < k^+$.

The following loss function is used for the case of drifting above 40 degrees:

$$L^+(y) = k^+(y - m)^2, y > m$$

where $k^+ = A_0/\Delta_0^2$. As in the nominal-the-best case, A_0 is the monetary loss incurred by approximately 50 percent of customers when the temperature drifts to a value Δ_0 from the target. Assume an average value of \$50 for lost food replacement and \$100 for the service call and parts to correct the problem when the temperature goes to 50°F. Thus, $A_0 = \$150$ when $\Delta_0 = +10°F$.

$$k^+ = \$150/(10°F)^2 = 1.5\ (\$/°F^2)$$

The upper loss function is:

$$L^+(y) = 1.5(y - 40)^2$$

The following loss function is used for the case of drifting below 40 degrees:

$$L^-(y) = k^-(y - m)^2, y \leq m$$

where $k^- = A_0/\Delta_0^2$. Again, A_0 is the monetary loss incurred by approximately 50 percent of customers when the temperature drifts to a value Δ_0 from the target. Assume an average value of \$10 for lost food replacement and \$80 for a service call and parts to correct the problem when the temperature goes to 32°F. Thus, $A_0 = \$90$ when $\Delta_0 = -8°F$.

$$k^- = \$90/(-8°F)^2 = 1.40\ \$/°F^2$$

The lower loss function is:

$$L^-(y) = 1.40(y - 40)^2$$

This concludes the development and examples of the quality-loss function. We now need to define a process by which you can develop your own quality-loss functions; you will need them for the development of initial tolerances and for the final process of tolerance design.

Developing Quality-Loss Functions in a Customer's Environment

Developing quality-loss functions is a challenging endeavor. Very few companies in the West actively develop these valuable product-development metrics. To properly carry out the many detailed technical activities in product development one must devote a serious effort to building a data base of customer needs and tolerances. If one does not understand what is tolerable by a customer, all the engineering in the world is not going to routinely produce appealing products. To participate in on-target engineering, it is no longer enough to look at the product functionality from a step-function perspective when considering tolerances. Consumers see quality in terms of a continuum, where costs escalate as performance deteriorates. We must learn not to focus solely on metrics such as mean time to failure. We need to focus on how our products lose value in our customer's eyes as they begin to drift, sometimes slowly and sometimes rapidly, away from their intended output.

If all this concern for the customer's needs is so obvious and is becoming a major thrust in Western industry, why is there not widespread use of the quality-loss function? Part of the reason is that the current paradigm in many organizations is that marketing and sales groups have separate agendas, performance goals, vocabularies, and work processes than the product-development engineering groups. They simply are out of touch with each other's business processes, and are often at odds over what the customer really wants or needs.

The marketing people see and talk with the customers frequently. The engineering teams occasionally get out and rub elbows with the customers, but generally are isolated from detailed contact on a continuing basis. Because of this, they have different perspectives.

On the other side of the issue, the engineering teams know what technologies are available to enhance new-product performance. They are technologically in the know and are motivated by seeing their work placed into the next new product. The marketing team is usually very knowledgeable about the features and performance of the current product line. They are also very capable of producing a long list of items they believe will enhance the viability of a new product. Frequently, the marketing wish list and the engineering team's portfolio of new technologies are not in synchronization. The problem is how to get these teams to work on the right things—together, so that the customer sees a unified stream of routinely excellent products. The teams need a common metric to quantify value and quality in both economic and technical terms.

Michael Hammer and James Champy were two of the first to recognize the true nature of the gulf that was separating corporations from their customers and the reasons it existed. Hammer coauthored a very insightful book with James Champy called *Reengineering the Corporation* [R11.6]. Hammer and Champy stated that for a business to satisfy customers it had to completely change its internal physics by studying and rebuilding its processes from scratch. To Hammer and Champy, anything that is done in the total business cycle and structure is a business process—including engineering and marketing, tolerancing and tolerance design.

In order to get business processes in line with your customer's needs and expectations, things often must change internally within the business. The changes are generally going to have to be unique and radical to effect large changes in performance that will be obvious to the customers. Hammer and Champy believe that for radical change, you must nuke the current business and rebuild it around radically efficient processes that harness your corporate capabilities to delight your customers—every thing else you are doing around incremental improvement is a waste of time.

It is in a reengineering context that the quality-loss function offers tremendous opportunities for companies willing to change by focusing on developing products based primarily on customer costs and losses and secondarily on internal economic issues. The quality-loss function is radical in comparison with the usual manner in which companies quantify the cost of their design specifications. Hammer and Champy offer a way to help reengineer tolerance development. Their book is not specifically about engineering processes but about business processes in general. Business process reengineering is one way to change a company's culture to allow the quality-loss function to thrive as a unifying metric to drive customer satisfaction. To see the routine development of quality-loss function in U.S. industry will require reengineering the interface between marketing, customer support, and engineering.

That said, one way to improve the tolerance development process is to rip down the barriers between marketing and engineering teams. Remember, marketing knows customer expectations and tolerances for product performance in their own vocabularies. Engineers command the language around the laws of physics and the unique technical vocabularies that are needed to develop products to satisfy customer demands—including the constraints defined by tolerances. Often these two vocabularies are not translated back and forth very well. Voice-of-the-customer gathering techniques, along with quality function deployment translation techniques, are the best modern tools to fix this communication problem (see Chapter 2). The key issue is to relate customer tolerances and sensitivities to the tolerances and sensitivities within the product.

The common metric that tolerance development processes must focus on is a continuum, expressed as a ratio, of the *cost per change in performance* that customers observe as they use the product. When degradation or variation of a specific functional output from a product occurs, a customer will immediately think, "Well, that is going to cost me \$*X* today and \$*X* every day until it's fixed—plus what it costs me to get it fixed."

How do we begin to approach the task of constructing a quality-loss function? First we must create an environment to get the marketing and engineering people to sit down with a reasonably diverse sample of customers to quantify their losses as they relate to specific functional changes that can be translated into engineering units. Thus, the marketing and engineering groups team up internally and work with customers externally to construct the economic coefficients $(A_0/(\Delta_0)^2)$. The next section is offered as a general approach to carrying out this critical task.

Constructing the Quality-Loss Economic Coefficient $(A_0/(\Delta_0)^2)$

There are two steps involved in constructing the quality-loss economic coefficient. First, identify the customer observable critical product function that is to be evaluated. This function needs to be related to a critical engineering parameter measurable in engineering units (e.g., voltage, velocity, image density, force, volumetric flow rate, and so forth). There will be many instances where the customer observable function will have to be translated into an engineering function. The team needs to develop skills for doing this accurately.

An example is a laptop computer track ball that has a chronic slip problem. From time to time the customer can't get the pointer to move around on the screen, even though the ball rolls freely. The engineering translation: There is a time-based change in the coefficient of friction between the ball and the rollers in the tracking mechanism. The track ball and roller mechanism is sensitive to some form of noise.

This process will identify the engineering units for Δ_0, the x axis of the quality-loss function plot:

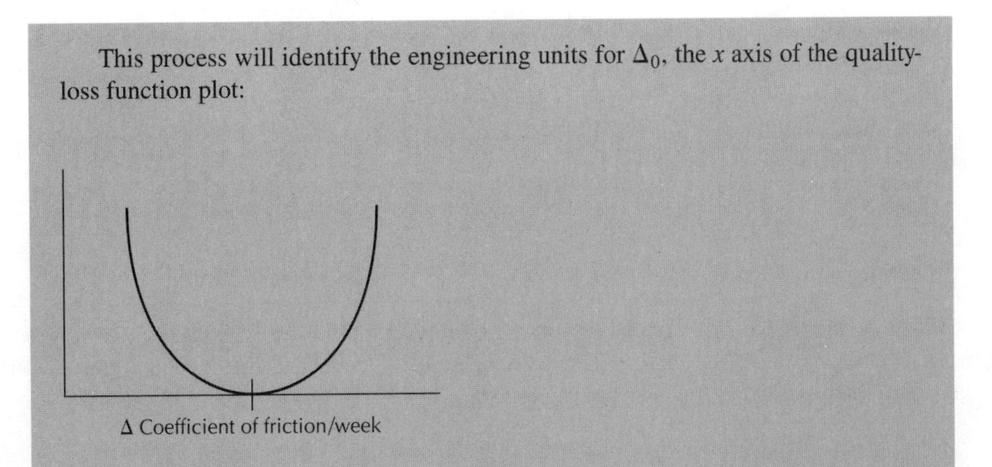

The track-ball problem is a nominal-the-best case where too low or too high a value of the coefficient of friction will result in a loss to the customer.

The next step involves polling a reasonable sample of customers as to their tolerance of off-target performance. The costs, in dollars, associated with an intolerable level of deviation of performance from the target must be defined. Each customer is going to have a different level of tolerance to off-target performance. You are seeking a reasonable average dollar value where approximately 50 percent of customers encounter some amount of loss due to the performance variation. The percentage will vary depending on the product and function under scrutiny.

Using the track-ball example, we interview 100 XYZ Brand laptop users. We find that 63 percent of the users felt that slippage occurring more frequently than every two weeks was intolerable. The customers found it necessary to return the laptop to the manufacturer when slippage occurred more often. The customers have the option to clean the track ball and tracking rollers whenever they slip. They did not mind cleaning the mechanism themselves every two weeks or more. The loss was calculated to be approximately $30 for service on the laptop at the repair center. The loss of use of the laptop for the day was a great inconvenience, and was calculated to be approximately $45 due to the manual work required in lieu of the computer's efficiency. This quantifies the loss in terms of dollars—$A_0 = \$75$. Δ_0 must be measured to determine its value at A_0.

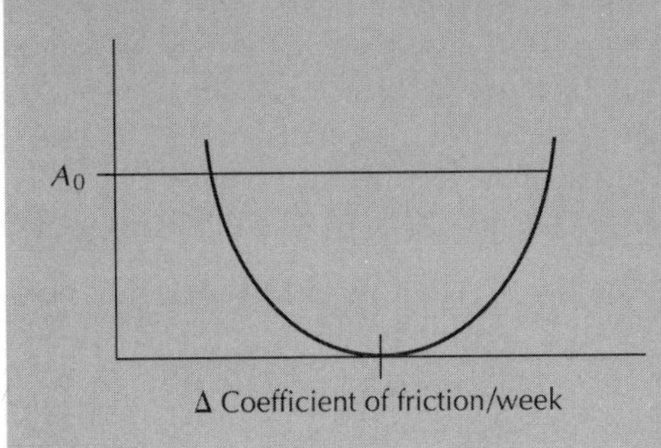

Care must be taken to build quality-loss functions for critical parameters that are high priorities for customers. Every subsystem or assembly will have critical parameters associated with its functional contribution to the product. These make excellent items for quality-loss analysis. How many QLFs one develops and how far one takes them down into the component level of a product is up to the product-development team.

Summary

Chapter 11 is a comprehensive introduction to the quality-loss function. The QLF is developed conceptually in comparison to the within-specification point of view. The mathematical and empirical derivations of the quality-loss function are presented.

Chapter 11: The Quadratic Loss Function in Tolerancing and Tolerance Design

- Linking Cost and Functional Performance
- An Example of the Cost of Quality
- The Step Function: An Inadequate Description of Quality
- The Customer Tolerance
- The Quality-Loss Function: A Better Description of Quality
    - The Quality-Loss Coefficient
    - An Example of the Quality-Loss Function
    - The Types of Quality-Loss Functions
        - The nominal-the-best case (NTB)
        - Average quality loss for nominal-the-best cases
        - The smaller-the-better case (STB)
        - Average Quality Loss for the smaller-the-better cases
        - The larger-the-better case (LTB)
        - Average quality loss for larger-the-better cases
        - The asymmetric nominal-the-best loss function
- Developing Quality-Loss Functions in a Customer's Environment
- Constructing the Quality-Loss Economic Coefficient ($A_0/(\Delta_0)^2$)

The chapter concludes with a review of how a product-development team might go about driving a quality-loss function culture into a business. One way to get a QLF culture embedded and moving quickly is to employ the methods of business process reengineering. Guidelines to carry out the construction of a loss function are provided along with an example.

The quality-loss function is very useful in the analysis of tolerances. It is fundamental to Taguchi's tolerancing and tolerance design processes. Chapters 12 and 17 will define applications based upon its fundamental principles.

References

R11.1: *Quality by Design: Taguchi Methods and U.S. Industry;* L. Ealey; ASI Press, 1988, pg. 386

R11.2: *Engineering Methods for Robust Product Design;* W. Y. Fowlkes and C. M. Creveling; Addison-Wesley, 1995, pg. 390

R11.3: *The Value of Continuing Improvement;* P. T. Jessup; IEEE Conference, 1985, pg. 393

R11.4: *Quality Engineering in Production Systems;* G. Taguchi, E. Elsayed & T. Hsiang; McGraw Hill, 1989, pg. 395

R11.5: *The Exact Relationship of Taguchi's Signal-To-Noise Ratio to His Quality Loss Function;* S. Maghsoodloo; Journal of Quality Technology 22 (1), 57 – 67, 1991, pg. 411

R11.6: *Reengineering the Corporation;* M. Hammer and J. Champy; Harper Business, 1993, pg. 417

Additional resources for the Quality-Loss Function can be found in the Experimental Tolerance Design section of Appendix D.

CHAPTER 12

The Application of the Quadratic Loss Function to Tolerancing

The Difference between Customer, Design, and Manufacturing Tolerances

We have discussed, at length, how to define initial tolerances for a component (see Chapter 4). This was done in the context of the three-sigma paradigm and working from process capability recommendations from manufacturers. There is an alternative to the three-sigma standard approach; Taguchi has proposed an approach based on the application of the quality-loss function that we will use to define initial tolerances.

When working with the quality-loss function in defining initial tolerances one must begin with the definition of two types of interrelated tolerances, high-level and low-level.

Customer Tolerances

High-level tolerances are the customer tolerances that are direct expressions of the customer's displeasure at incurring financial loss due to off-target performance. These tolerances are often referred to in terms of the variation in a critical product output response, as opposed to the technical design and manufacturing tolerances, which are "buried" in the detailed design and actually limit variability in the components, assemblies, and subsystems. Customer tolerances are typically expressed in the vocabulary of a user of the product in contrast to the vocabulary of a designer of the product. This creates the need for careful and circumspect translation of the customer's terms regarding acceptable limits in performance.

Design and Manufacturing Tolerances

A design tolerance is a functional variable within the design that is constrained during the assembly or service adjustment of the product. A manufacturing tolerance is a constraint imposed on a specification variable during the manufacture of components, assemblies, or subsystems.

Low-level tolerances are the design and/or manufacturing output limits established by the engineering team. These are the tolerances that relate to component variation control and to boundaries that are imposed at the assembly and subsystem levels to keep interacting elements from producing unacceptable performance. Sometimes these are referred to as *higher* level tolerances within the product architecture. It is recommended that the engineering team take the time to calculate or experimentally determine where product-level performance deteriorates to the point of unacceptability, by varying each critical component, assembly, or subsystem tolerance boundary. This means the team must "hunt for the cliff"—the point at which performance drops off to a level that will motivate approximately 50 percent of the customers to take action.

The high-level (product) or higher level (subsystem or assembly) tolerances should be used to drive the determination of the low-level (component) tolerances. In this way the product will be toleranced from the top down so that nothing in the detailed design is held to boundaries that don't add value to a clear customer expectation. The critical design parameters are to be toleranced based on customer loss limits that directly relate to phenomena that *customers can see.* We will continue to represent the high level customer tolerances with the symbol Δ_0. Typically these tolerances are defined by the dollar amount of loss incurred by approximately 50 percent of the customers as they become so dissatisfied with a product's output they are forced to take economic action. From time to time we will find that the customer's tolerance, Δ_0, will have to be converted into a product-level engineering tolerance called an *intermediate tolerance,* Δ_I. Sometimes this has to be done if the customer's tolerance is not expressed in engineering units. The low-level tolerances will be identified by the symbol Δ. We will demonstrate how to relate high-level to low-level tolerances quantitatively using Taguchi's tolerancing equations.

The Taguchi Tolerancing Equations

Taguchi writes about the development of a quality-loss function tolerancing process in his two most recent books, *Taguchi on Robust Technology Development* [R12.1], and *Quality Engineering in Production Systems* [R12.2], as well as in his classic works *Introduction to Quality Engineering* [R12.3]; and *System of Experimental Design* Vols. 1 & 2 [R12.4]. These are the four best sources available to read about Dr. Taguchi's tolerancing methods in his own words. This text will draw heavily from Dr. Taguchi's work. We will attempt to stay very close to Dr. Taguchi's point of view while developing a comprehensive review of his approach.

Our explanation of the Taguchi tolerancing equations will begin by defining tolerances on the basis that they serve as safety factors. One might ask, from what malady is the tolerance providing safety?

Taguchi's Economical Safety Factor

The concept of a design component tolerance being defined through the application of a factor of safety, in a quality engineering paradigm, is inextricably related to customer tolerances based on *financial* loss. A factor of safety, ϕ, is established to prevent off-target performance values, beyond a

Figure 12.1

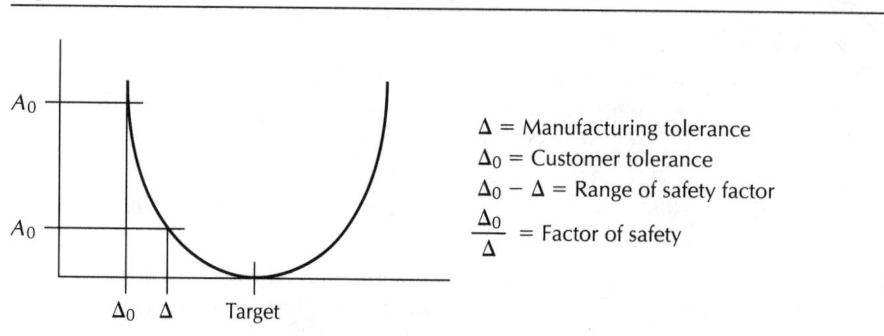

specific range, from becoming evident to customers. The economic loss associated with a range of off-target performance that begins at some manufacturing tolerance point, Δ, and extends to an intolerable point, Δ_0, where 50 percent of customers are taking action, is the amount of loss that a company defines as unacceptable for its customers to bear.

The company quantifies what expense is worth funding to remedy off-target performance at or beyond Δ. In this way the company is expending its own financial and technical resources, in a very disciplined manner, to limit the loss experienced by customers. The loss associated with off-target performance below Δ is deemed inconsequential in terms of immediate spending on the production line to correct for the small variability from the target. This type of deviation is dealt with by the long-term application of on-line quality control techniques designed to keep the output of the production process centered on the target. The loss associated with Δ_0 is to be avoided and is why a safety factor is desirable for tolerance development. Please note that all of this limiting and cost accounting is going on well within the customer's tolerance limits. Thus, the Taguchi safety factor is an economic safety factor. Initially, the tolerance for a product-level response can be established by using the loss function. From this point, the engineering team can drive the lower level tolerance limits according to what the company can rationally afford to spend on limiting variability.

The Derivation of Taguchi's Tolerance Equation and the Factor of Safety

The basis for the quality-loss function can be further defined from the following definite integral equation [R12.2].

$$L(y) = \frac{1}{n} \sum_{i=1}^{N} \int_0^T L_i(t,y)\, dt \qquad \textbf{(12.1)}$$

where $L(y)$ is the average loss in dollars due to off-target performance, n is the number of products in the market under consideration, T is the total product life, and $L_i(t,y)$ is the loss rate based on a finite length of time, t, and a measurable value, y, from the target for product i.

Equation 12.1 states that the rate of average quality loss for a product line of n total products is the sum of all the integrated losses (on a plot, as shown in Figure 12.2, of loss versus time) for each individual product over a finite design-life interval—all divided by the total number of products measured.

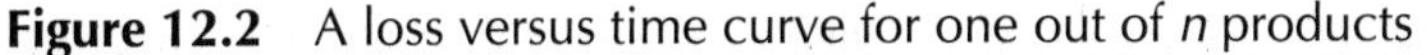

Figure 12.2 A loss versus time curve for one out of *n* products

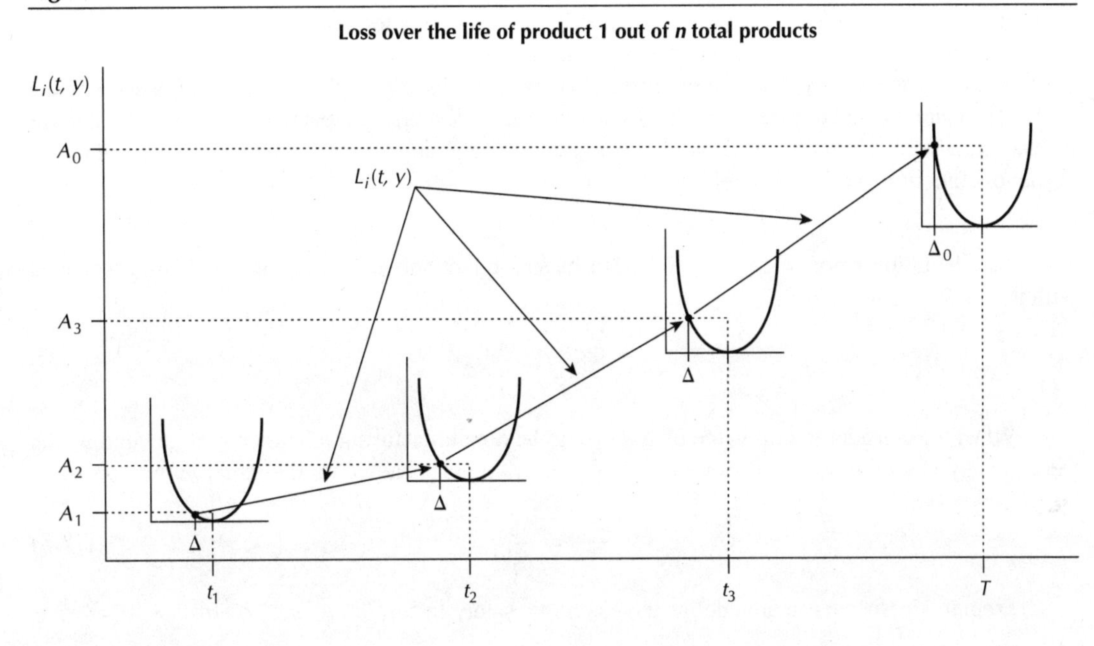

Figure 12.2 illustrates how performance and loss can be defined over time. This curve is representative of just one product out of a complete production run of n products on the market. To account for the total loss generated by a product line, one would have to sum up all the losses for the products and divide by (n) products. To account for the loss of one product over its life one must integrate, from time = 0 to time = T, the loss function $L(t, y)$ to quantify the area under the loss versus time curve. The area under the loss versus time curve will be different for every product, from 1 to n.

To solve for losses in individual cases, we can simplify Equation 12.1 through the application of a Taylor series expansion to the loss as defined in Chapter 11. The result is

$$L(y) = \frac{A_0}{\Delta_0^2}(y - m)^2 \tag{12.2}$$

We can then begin to refine the loss function to relate the deviation of y from the target with a new term Δ_i.

$$L(y_i) = \frac{A_0}{\Delta_0^2}(\Delta_i)^2 \tag{12.3}$$

Where does the term Δ_i^2 come from? If we look at the difference, Δ_i, that any value y_i has with respect to the target m, we get a quantification of off-target performance. Since the quality-loss function is a quadratic mathematical function that expresses loss and functional deviation, Δ_i, as a parabolic relationship, Δ_i^2 comes from the following equation:

$$(y_i - m)^2 = \Delta_i^2 \tag{12.4}$$

When we relate the customer's financial loss at a specific point of intolerance, or annoyance, we define a specific limit, Δ_0, that has been formally defined as the customer tolerance. It is recommended that if one does not know where this value lies for a particular design case, one needs to test the design at ever-increasing levels of Δ_i until the product fails to perform as expected—thus revealing an estimate of Δ_0. The team will have to correlate this value with the customers' perspective to generate a value for Δ_0 that is realistic. This value is squared, due to the quadratic approximation of loss, and is employed in the loss function's economic coefficient $\left(\frac{A_0}{\Delta_0^2}\right)$.

The deviation prior to Δ_0, Δ_i, will also have a quantifiable loss, A_i, that will be associated with it:

$$L(y_i) = A_i = \frac{A_0}{\Delta_0^2}\Delta_i^2$$

When we consider the ith value of Δ and A to be a manufacturing tolerance limit beyond which the company is willing to spend money to adjust the value of y back onto the target m, we can state:

$$\frac{\Delta_0^2}{\Delta_i^2} = \frac{A_0}{A_i} \tag{12.5}$$

From this basis we can now define the economic safety factor ϕ:

$$\phi = \sqrt{\frac{\Delta_0^2}{\Delta_i^2}} = \sqrt{\frac{A_0}{A_i}} \tag{12.6}$$

or, in a simplified format (also dropping the i notation):

$$\phi = \frac{\Delta_0}{\Delta} = \sqrt{\frac{A_0}{A}} \tag{12.7}$$

The product response can then be restated in economic (loss) terms to define the nature of the safety factor as shown in Equation 12.8 [R12.1]:

$$\phi = \sqrt{\frac{\textit{Average loss in dollars when a product characteristic exceeds customer tolerance limits}}{\textit{Average loss in dollars when a product characteristic exceeds the manufacturing and/or design tolerance limits}}} \tag{12.8}$$

Equation 12.9 illustrates, in short notation, how the customer loss, A_0, is related to the manufacturing loss, A. This is the safety factor, or the cost a company incurs when a value is established to prevent the customer loss from occurring. This ratio defines the safety factor:

$$\phi = \sqrt{\frac{A_0}{A}} \tag{12.9}$$

Remember, we must take the square root because this term was derived from the quadratic-loss function equation.

Relating Customer Tolerances to Engineering Tolerances

Taguchi's tolerancing approach provides a simple way to link customer limits to the development of design element limits. The design limits must be carefully and concurrently developed in harmony with the capability of the available manufacturing processes being considered to actually make the components. This is how current customer requirements are supported by the state of the art in manufacturing-process capability. When these two things get out of synchronization, and they almost always do, the manufacturing technology research community rises to the occasion to produce new and better ways of making things. One must be careful to prevent getting a product design capability requirement out ahead of current manufacturing process capabilities; thus the need for concurrent design and manufacturing process capability development, optimization, and balancing.

Equation 12.10 is used to relate a product-level customer tolerance to a specific design-level component, assembly, or subsystem. Δ is a continuous variable that expresses the allowable *manufacturing* deviation from the target performance set point.

$$\Delta = \sqrt{\frac{A}{A_0}}(\Delta_0) \tag{12.10}$$

An example of this may be a camera that produces an overexposed (washed out) image, which greatly disturbs the customer. The engineering translation could be that the shutter mechanism is being slowed by some phenomenon; in this case it may be differential spring rates coming from a purchased spring. The customer observable response of washed-out photographs has been converted into a continuously measurable engineering design variable. A dollar loss can now be tied to spring-rate variability. We can now begin to deal with the issue of how much can the company afford, A, to spend to limit the spring rate variability, Δ. In this scenario A_0 is likely to be the cost of a new camera and Δ_0 is the spring-rate differential that is creating the overexposure.

An Example of Tolerancing Using the Loss Function (Nominal-the-Best Case)

We will use a common example where the product we are making is a power supply. The output voltage is both the customer observable characteristic *and* the engineering characteristic (high-level tolerance characteristic) we are interested in evaluating. We know that the customer loss (A_0) is \$200 when the voltage is over 135. The target voltage is 115 plus or minus 20. Thus Δ_0 is 20 volts. The loss function for this case is:

$$L(y) = \frac{A_0}{\Delta_0^2}(y - m)^2$$

The economic coefficient, k, is defined by:

$$k = \frac{A_0}{\Delta_0^2} = \frac{\$200}{\$400} = 0.5\ (\$/V^2)$$

The production department has determined that it can afford to adjust the voltage back on target for a cost of \$5. This is the value of A, the company's loss, or cost, for correcting off-target performance. We now have enough information to determine the production tolerance, Δ:

$$\Delta = \sqrt{\frac{A}{A_0}}\,\Delta_0$$

Solving for Δ we get:

$$\Delta = \sqrt{\frac{\$5}{\$200}}\,20 = \pm 3.162 \text{ volts}$$

We can round off the tolerance to 3 volts. When the production line measures a power supply over 118 or under 112 volts, the power supply is put back on target at 115 volts. The safety factor that is put in place based on company and consumer economics is:

$$\phi = \sqrt{\frac{A_0}{A}} = \sqrt{\frac{\$200}{\$5}} = 6.32$$

Thus, the consumer is protected by an economic safety factor of more than 6 through strategic, economically acceptable spending within the manufacturing company. This is how we balance company spending on quality control with known customer losses. It is not hard to imagine that this practice may have a positive impact on the budget required for product advertising. When the product quality is well managed from an economic standpoint, the product will have stronger market pull, thus requiring a lower advertising cost. These savings can be passed on to shareholders or reinvested into new product development. Tolerancing based on the loss function promotes economic efficiency inside and outside the company.

The tolerance equations for smaller-the-better and larger-the-better cases are displayed below. Their application is essentially the same as NTB cases, except that STB cases seek 0 as an ideal and LTB seek the highest value possible. Note that for all cases, NTB, STB, and LTB, the safety factor is calculated by:

$$\phi = \sqrt{\frac{A_0}{A}}$$

But the lower level tolerance is calculated by:

NTB and STB cases: $\Delta = \dfrac{\Delta_0}{\phi}$ LTB cases: $\Delta = \phi\Delta_0$

Relating Customer Tolerances to Subsystem and Component Tolerances

We have focused on how the loss function can be manipulated to aid in the determination of an engineering tolerance for a single critical parameter that has a direct effect on a customer observable product response. We are now going to learn how Taguchi's loss function can be used to aid in the development of tolerances for *multiple* design elements that have a *linear* relationship with the output response. While it is important to focus on properly tolerancing critical-to-function parameters, we need to recognize that all components must be toleranced.

Taguchi, Elsayed, and Hsiang state: "*When a product consists of k components, the characteristics of the product are affected by the tolerances of these components, and tolerance design for each component is needed so that the output characteristic of the product conforms to the functional limit*" [R12.2]. Each component is going to play a role in how the product ultimately performs. In this case we are working with components that have a linear effect on the output response. We can define this linear relationship, for a single component represented by the value of x, by the general equation:

$$y = m_0 + \beta(x - m) \quad \textbf{(12.11)}$$

where y is the actual output response of the product, m_0 is the target value for y, x is a design parameter that has a linear effect on the output response, m is the target value for x, and β expresses the linear relationship between x and y.

The Linear Sensitivity Factor, β

β is the change in a high-level customer observable characteristic or a product-level engineering characteristic, y, when a unit change occurs from the target set point in a low-level engineering parameter, x. This relationship is illustrated in Figure 12.3.

The loss function for this type of problem is defined as follows:

$$L(y) = \frac{A_0}{\Delta_0^2}(y - m_0)^2 \quad \textbf{(12.12)}$$

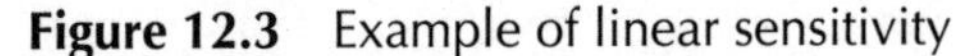

Figure 12.3 Example of linear sensitivity

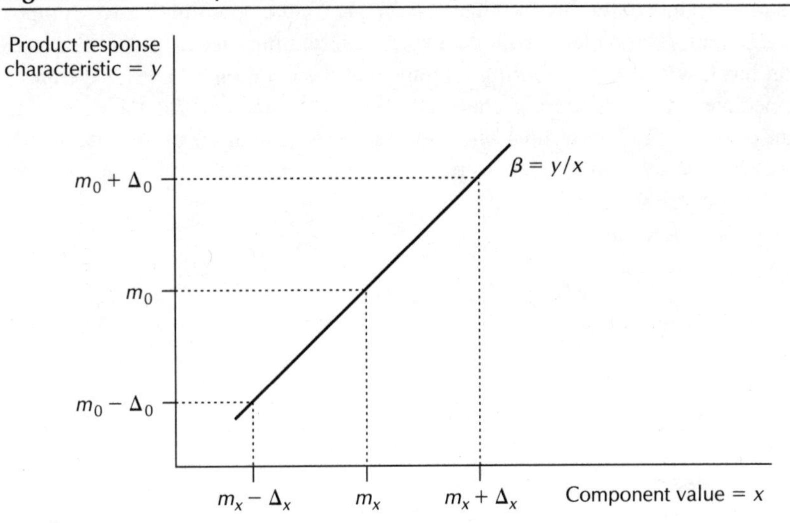

The customer loss in dollars is expressed as A_0. The customer tolerance is defined by Δ_0. The measure of off-target performance is quantified by $(y - m_0)^2$. When we substitute the effect of x from its target value m and include the linear coefficient β, we can represent how the loss due to y being off target is driven by x being off target. In this case we see how a component can be responsible for a change in a high-level response—including the economic implications.

$$L(y) = \frac{A_0}{\Delta_0^2} [\beta(x - m)]^2 \tag{12.13}$$

We can now apply the same manipulations to this case that we applied to the product-level (high-level) case and find that the manufacturing loss, A, is equal to the loss function for variability in component x.

$$A = \frac{A_0}{\Delta_0^2} [\beta(x - m)]^2 \tag{12.14}$$

Solving for x we get:

$$x = m \pm \sqrt{\frac{A}{A_0}} \left(\frac{\Delta_0}{\beta}\right) \tag{12.15}$$

When we think of this in direct tolerance terms we end up with Equation 12.16:

$$\Delta = \sqrt{\frac{A}{A_0}} \left(\frac{\Delta_0}{\beta}\right) \tag{12.16}$$

Using the Loss Function for Multiple-Component Tolerance Analysis

For k multiple components in a product assembly the changes, $\Delta_{1 \text{ to } k}$, can be quantified. Each component has a specific linear coefficient, β, associated with the expression relating the customer tolerance, at the product or subsystem level, with the engineering tolerance, at the component level. We can, if necessary, define an intermediate-level engineering characteristic, Δ_I, to account for the aggregate effect of all the component changes, Δ_k. This value will represent the translation of the customer's units into engineering units. Every case is unique; it is up to the engineering team to decide when to translate customer units into engineering units.

The low-level tolerances are shown as:

$$\Delta_1 = \sqrt{\frac{A_1}{A_0}} \left(\frac{\Delta_0}{\beta_1}\right), \Delta_2 = \sqrt{\frac{A_2}{A_0}} \left(\frac{\Delta_0}{\beta_2}\right), \Delta_3 = \sqrt{\frac{A_3}{A_0}} \left(\frac{\Delta_0}{\beta_3}\right), \ldots \Delta_k = \sqrt{\frac{A_k}{A_0}} \left(\frac{\Delta_0}{\beta_k}\right)$$

From $\Delta_k \beta_k = \sqrt{\frac{A}{A_0}} (\Delta_0)$ we get $\Delta_k^2 \beta_k^2 = \frac{A}{A_0} (\Delta_0)^2$ which is $(\beta_k \Delta_k)^2$. We can sum up all the individual component terms to define an aggregate high-level change term Δ_I.

$$\Delta_I^2 = (\beta_1 \Delta_1)^2 + (\beta_2 \Delta_2)^2 + (\beta_3 \Delta_3)^2 + \ldots + (\beta_k \Delta_k)^2 \tag{12.17}$$

Figure 12.4 Office copier cleaning subsystem

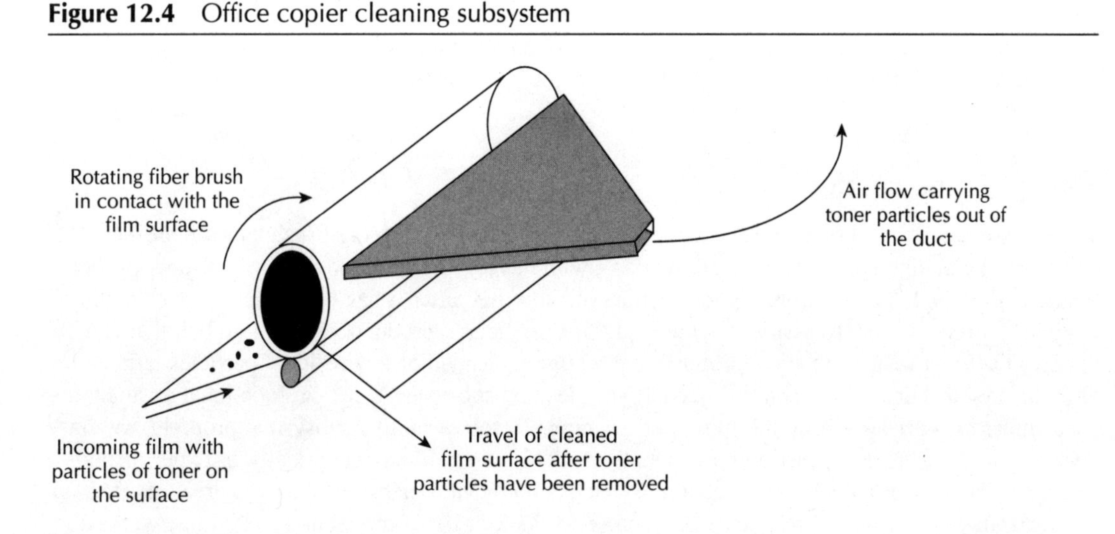

This can be restated in another form for the component manufacturing losses, customer losses, and customer tolerance limits.

$$\Delta_I^2 = \left(\sqrt{\frac{A_1}{A_0}} \times \Delta_0\right)^2 + \left(\sqrt{\frac{A_2}{A_0}} \times \Delta_0\right)^2 + \left(\sqrt{\frac{A_3}{A_0}} \times \Delta_0\right)^2 + \ldots + \left(\sqrt{\frac{A_k}{A_0}} \times \Delta_0\right)^2 \quad \textbf{(12.18)}$$

Combining terms we get:

$$\Delta_I^2 = \frac{A_1 + A_2 + A_3 + \ldots + A_k}{A_0} \Delta_0^2 \quad \textbf{(12.19)}$$

$$\Delta_I = \sqrt{\frac{\sum_{i=1}^{k} A_i}{A_0}} \times \Delta_0 \quad \textbf{(12.20)}$$

This is the intermediate higher level tolerance limit, in engineering units, that directly relates to a customer tolerance limit at the product level. We will use this in an example to illustrate how it works.

An Example of Applying the Quality-Loss Function to a Multicomponent Problem[1]

The following example is a comprehensive analysis of a subassembly that requires critical parameter definition prior to tolerance analysis. We will illustrate the process of identifying what to tolerance and what to leave out of the analysis. This process can be affected by an engineering team's judgment. This

1. Adapted with permission from Eastman Kodak Company.

approach should be done by an experienced team of engineers, designers, and technicians who understand the physics behind the technology being converted into marketable components, assemblies, and subsystems. We will define the subassembly we are going to tolerance.

Setting Up the Problem

The process we are discussing is very much like the one used in a rotating-brush vacuum cleaner. The commercial cleaning application we are working with has to do with removing plastic toner particles (sometimes called dry ink) from a photo-conductive film belt after it has been used to make a photocopy in an office copier. To assure that the copier makes clean reproductions, the film belt that is used to carry the toner image to the paper must be thoroughly cleaned of any residual particles left behind after the previous image has been transferred. The cleaning subsystem's critical functional parameter is the number of particles left on the film after cleaning. To tolerance this subsystem properly, we must define what its *critical-to-function* components are. This can be done analytically or experimentally. Sensitivity analysis is a good way to quantify which component dimensions have the greatest effect on the variability of the output response. But before we look at the component sensitivities we need to identify the key functional parameters.

Identifying Critical Parameters

In the process of removing particles from an office-copier film surface there are a few distinct forms of energy that need to be put into the removal process. The parameters that control the energy transformations and applications are going to have the biggest effect on performance. All other design parameters are passive in comparison, and serve only supporting functions to the factors that have active roles in directing energy into the subsystem function.

In this case there are three *active* applications or transformations of energy:

1. Frictional scrubbing and direct impact forces are delivered from the rotation of the fiber brush on the toner-contaminated film surface. The brush has a controlled amount of interference with the film, which is supported by a back-up roller. There is an induced normal force between the brush and film that is controlled by the interference mounting of the brush with respect to the film back-up roller. When the brush is rotated, kinetic energy called scrubbing friction is applied to the film and toner. We need to tolerance the components that control this energy input carefully.
2. An attractive pressure is exerted on the loosened toner particles' surface area from a viscous air flow controlled by the velocity of the air as it rushes by the brush and toner at the vacuum nozzle interface. The air flow is created by a pressure drop induced by a vacuum blower. This is called *capture velocity.* Again, we need to determine the key controllers of the uniformity of the air velocity, as well as the adequate tolerances for their nominal set points.
3. As the brush rotates about its axis, it imparts a tangential velocity on the toner particles accumulated on its outer surface. This velocity is defined by the angular speed of the brush and its diameter; it produces a centrifugal force that tends to fling the particles off of the brush fibers. When the brush passes the vacuum nozzle there is a place for the particles to go. The synergy between the pull of the air flow and the push of the brush fibers supplies the required energy to clean the brush.

For each of these energy applications, which are key functions in the design, we can identify components that control the actions.

1. Factors for Action 1:
   - Brush interference with the film and back-up roller (inches)
   - Brush diameter (inches)
   - Mounting bracket dimensions (inches)
   - Brush speed (rpm)
   - Brush material characteristics (density, stiffness, coefficient of friction with film)
2. Factors for Action 2:
   - Air velocity at vacuum nozzle (feet/minute)
   - Air-flow resistance due to filter loading prior to vacuum blower
   - Air-flow resistance due to brush to housing spacing
3. Factors for Action 3:
   - Brush speed (rpm)
   - Brush diameter (inches)

The measurable engineering characteristic for the cleaning subsystem is the number of particles per square inch left behind. The measurable product characteristic the customer can see is the background on the copies. Background looks like gray fog in places where there should be pure white paper. When the particles-per-inch-squared measurement is low, background will be low. Many copier customers sell the copies they make and incur financial losses when background occurs. They lose business and have high service costs when the machine is not running free of background.

Mapping the Critical Parameters and Their Sensitivities

The measured response for the tolerance analysis is as follows:

Image background is measured in units that account for the number of particles of background on the copy. To be sure that we can quantify the onset of background that a customer can see we need an engineering characteristic sensitive enough to detect impending background. The engineering characteristic we can use is a measure of the number of particles left behind after cleaning, since they will become the background that will be transferred to the customer's copies. Figure 12.5 illustrates the breakdown of the key contributors to changes in film-cleaning performance.

$$\Delta \text{ Particles/in}^2 \approx \text{f[(changes in brush interference with film: } \Delta\text{B-F}),$$
$$\text{(changes in capture velocity: } \Delta\text{V), and}$$
$$\text{(changes in brush to housing ID spacing: } \Delta\text{B-H)}$$

We can refine the issues around which parameters are critical for tolerancing and which it is reasonable to assume are not critical with respect to the engineering characteristic being measured. While the following parameters are important to the cleaning function they are not part of the tolerance problem we are developing:

- Brush speed
- Brush material characteristics
- Air-filter resistance to flow

Brush speed is not a critical factor for cleaning efficiency. It can vary quite widely and not affect the engineering characteristic. Previous experimentation provides the data to assure this is true. The brush material characteristics are important parameters to attach specifications to, but their sensitivity lies in another engineering characteristic—namely, film wear and deterioration. In this film-cleaning context they are not strong effectors of performance if the density and stiffness change within normal manufacturing variability ranges. The air flow cubic feet per minute (CFM) and velocity feet per minute (FPM) are controlled by the cleaning vacuum blower. The blower air-flow circuit does have a filter in it that can get clogged with particles and reduce air flow significantly. This is a maintenance frequency issue, not a design or manufacturing tolerance issue. Under these circumstances, the three parameters just reviewed are taken out of the analysis and held as constants since they are not deemed to be highly sensitive elements in the subsystem. This is how prior knowledge, previous experimentation, and engineering judgment come into play in setting up a tolerance-analysis problem. Figure 12.5 shows how changes (Δ) in the predetermined critical design parameters flow up to effect image background, which is a customer observable product characteristic.

Now that we have outlined the critical tolerancing parameters, we can set up the general form for a traditional tolerancing equation:

$$\Delta\text{particles/in}^2 \approx f(\Delta\text{brush diameter, }\Delta\text{housing-mount dimensions, }\Delta\text{housing diameter})$$

As you can see we have used the lowest level design parameters to express the independent variables in the approximation. The parameters in the next level up are controlled by these three basic dimensions. If they are on target then performance will be on target. The parameters we are assuming to be reasonably inconsequential for this particular engineering characteristic are brush speed, blower voltage, and filter loading. Recall that filter loading is a deterioration characteristic. Taguchi discusses

Figure 12.5

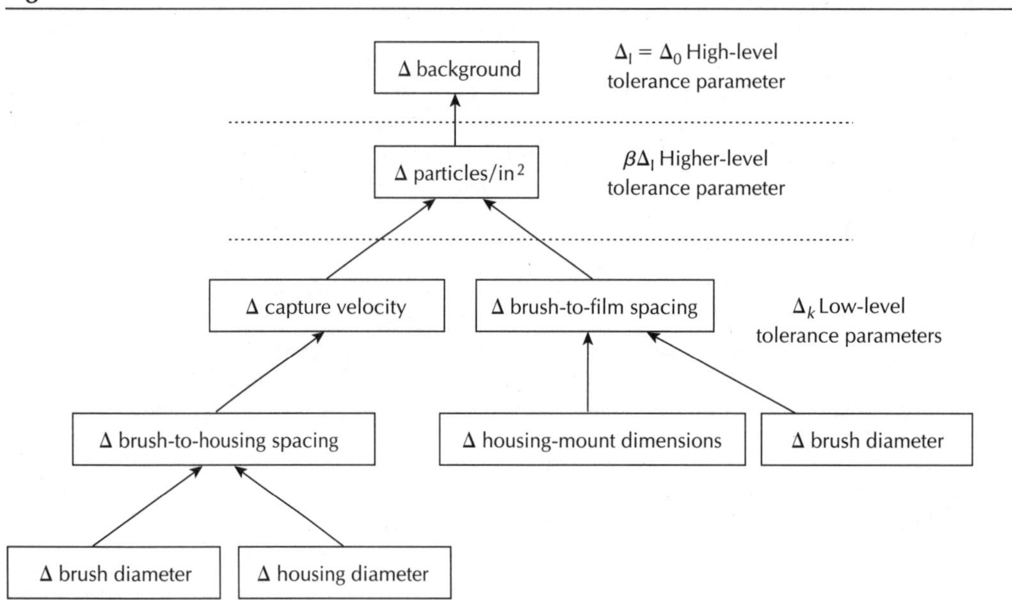

how deterioration characteristics can be included in his tolerancing approach (see [R12.2]). We will discuss tolerancing deterioration-characteristic parameters later in this chapter.

Assuming the relationship between the three tolerancing parameters to be approximately linear for small changes about the target (which *may* be a false assumption) we can write the following expression:

$$\Delta\text{Part./in}^2 = \frac{\partial\text{Part./in.}^2}{\partial\text{Brush Dia.}}(\Delta\text{Brush Dia.}) + \frac{\partial\text{Part./in.}^2}{\partial\text{Mount Hdw Dim.}}(\Delta\text{Mount Hdw Dim}) + \frac{\partial\text{Part./in.}^2}{\partial\text{Housing Dia.}}(\Delta\text{Housing Dia.})$$

This expression shows the relationship between the change in particle count after cleaning to the sum of the independent tolerance parameters, including their respective sensitivities represented by the partial derivatives. Soon we will discuss how to handle this problem in the event the relationships are not approximated properly by a linear model.

Converting the Traditional Tolerance Problem into a Quality-Loss Tolerance Problem

Continuing with our assumption of linearity between the components and the response variable, we establish the intermediate engineering characteristic Δ_I. For now, its units are the same as the customer's units, which are background on the copy.

The following equation establishes Δ_I:

$$\Delta_I^2 = \left(\sqrt{\frac{A_{brush}}{A_0}} \times \Delta_0\right)^2 + \left(\sqrt{\frac{A_{mount}}{A_0}} \times \Delta_0\right)^2 + \left(\sqrt{\frac{A_{housing}}{A_0}} \times \Delta_0\right)^2$$

where Δ_0 and A_0 are the customer tolerance and loss based on their perception of background on the copies. The A values for each of the three components is the cost of putting their values back on target. This could be their scrap cost or rework cost. Equation 12.18 can be rewritten as . . .

$$\Delta_I = \sqrt{\frac{\sum_{i=1}^{k} A_i}{A_0}} \times \Delta_0$$

The task now is to obtain the costs associated with scrapping or reworking the components. Once this is done we can assess the tolerance for the engineering characteristic Δ_I. Once the individual component losses are defined, their individual loss functions can be determined and their associated tolerances specified.

The losses that will be incurred by the company for the three components are as follows:

- A_{brush} = brush scrap cost = \$9
- A_{housing} = housing scrap cost = \$40
- A_{mount} = mount-bracket assembly scrap cost = \$5

The β_k (linearity) values were experimentally determined to be:

- $\beta_1 = 0.8$ (background units/in.)
- $\beta_2 = 0.3$ (background units/in.)
- $\beta_3 = 1.3$ (background units/in.)

The value for Δ_0 is 1.2 background units. The value of A_0 is $575.

We can now solve for each component tolerance:

$$\Delta_1 = \sqrt{\frac{A_1}{A_0}}\left(\frac{\Delta_0}{\beta_1}\right) = \sqrt{\frac{\$9}{\$575}}\left(\frac{1.2}{1.5}\right) = \pm 0.0125 \text{ in.}$$

$$\Delta_2 = \sqrt{\frac{A_2}{A_0}}\left(\frac{\Delta_0}{\beta_2}\right) = \sqrt{\frac{\$40}{\$575}}\left(\frac{1.2}{.8}\right) = \pm 0.104 \text{ in.}$$

$$\Delta_3 = \sqrt{\frac{A_3}{A_0}}\left(\frac{\Delta_0}{\beta_3}\right) = \sqrt{\frac{\$5}{\$575}}\left(\frac{1.2}{1.3}\right) = \pm 0.008 \text{ in.}$$

The specifications for the three components are found to be:

- Cleaning brush diameter = 3.5 ±0.0125 in.
- Housing diameter = 3.625 ±0.100 in. (rounding down 0.004 in. for simplicity)
- Mount-bracket assembly hardware dimension = 3.900 ±0.008 in.

Now we can solve Equation 12.1 to find the engineering characteristic tolerance for the particles per inch squared. The background tolerance will be determined as a function of the losses for the components and the customer tolerance:

$$\Delta_I^2 = \frac{A_1 + A_2 + A_3}{A_0}\Delta_0^2 = \frac{\$9 + \$40 + \$5}{\$575} 1.44$$

Using Equation 12.20 we find that the aggregated engineering characteristic tolerance is a fraction of the customer tolerance:

$$\Delta_I = \sqrt{\frac{\sum_{i=1}^{k} A_i}{A_0}} \times \Delta_0 = 0.306\, \Delta_0 \text{ background units}$$

This value indicates that if we tolerance the components based on the loss function, we will have a cleaning-process output value that will be well below the customer tolerance; in fact it is calculated to be at about one-quarter of the way to the customer's point of intolerance. To translate this value into particles per square inch we need to define the β value relating background, Δ_0, to residual particles. Then we can use this β to define the tolerance for particles per square inch.

How to Evaluate Aggregated Low-Level Tolerances

Taguchi cites three cases for what to do when the aggregated component costs relative to the cost of the high-level output characteristic are beyond their limits [R12.1]. This comparison is used in deciding when to adjust the out-of-tolerance product back on target versus when to scrap it. These tolerance limits are used to make various economic decisions. This is a fairly rigorous way to intertwine functional performance and cost.

When the sum of all the costs associated with low-level response characteristics exceeds their limits and is put in ratio with the loss associated with the higher level tolerance being surpassed, a value is produced that can help us determine what course of action to take. The low-level costs are typically going to be from manufacturing. The higher level costs will also include many customer-loss values that go well beyond just what was paid to own the product.

Case 1: Ratio of Losses Are Much Less than Unity

In this case, the ratio is found to be much less than 1. This means that when all the components exceed their limits, the aggregate cost is still much less than the cost of the higher level characteristic being beyond its limiting value.

$$\frac{A_1 + A_2 + A_3 \ldots + A_k}{A_0} << 1 \tag{12.21}$$

Here it is cheaper to make or buy a new product than to adjust the existing unit. In customer terms, it is cheaper to keep this product from getting to the market than to send it out and incur the cost to put it right. A one-time use camera is a good example of such a product.

Case 2: Ratio of Losses Are Much Greater than Unity

Just the opposite is true in this case. The aggregated costs of the components far exceeds the cost of the high-level characteristic being out of bounds. In this case one would definitely want to adjust the product to return its performance characteristic back onto the target value. There is no financial incentive to scrap the product in this scenario. The cost of purchase is high, while the cost to the owner to put it right is acceptably low in comparison. A jet aircraft, a car, and an office copier are classic examples.

$$\frac{A_1 + A_2 + A_3 \ldots + A_k}{A_0} >> 1 \tag{12.22}$$

Case 3: Ratio of Loss Approximately Equal Unity

Here we see relative parity between aggregated component costs and the loss associated with the higher level product characteristic. Since this case is straddling the fence and leaves no clear course of action, Taguchi recommends taking a second look at the tolerances to see if they are indeed set appropriately.

$$\frac{A_1 + A_2 + A_3 \ldots + A_k}{A_0} \cong 1 \tag{12.23}$$

Cases where it costs about the same to make some item as it does to adjust it are somewhat uncommon. Because of declining costs, the VCR industry has drifted into this type of situation. For example, a service technician usually suggests a limiting dollar value of diagnosis and repair before giving up and recommending the purchase of a new VCR.

These three cases serve to illustrate the stark contrast between the traditional approach to tolerance stack analysis and loss function–based tolerancing of individual components independent of other component-tolerance effects on the higher level tolerance. The true economical tolerance is built around minimizing loss in light of the customer's best interest. Traditional stack analysis simply seeks to maximize production yields. There is no direct capacity in the traditional methods to account for customer loss.

The copier-cleaning-subsystem case illustrates how to handle a multicomponent problem that has linear relationships between the components and the output variable. We will now look at how to approach a problem when this is not the case.

Using the Loss Function in Nonlinear Relationships

There are going to be many cases where the assumption of linear behavior between components and a product output characteristic is not appropriate. Taguchi, Elsayed, and Hsiang [R12.2] offer a method of identifying and working with such cases.

This approach depends on the creation of a series of graphs that relate the range of values from the output response with a range of variation from each of the components. Typically, the range of variation from the components spans not only the initial estimate of tolerance limits (either from the processes discussed in Chapter 4 or from the loss function-based process); but also the values that force the output to attain values that are considered outright failure points. That is to say, we force the component to obtain values far enough away from the target that we measure and confirm Δ_0, the customer tolerance limit for the product characteristic. Figure 12.6 shows what such a graph might look like.

Notice that the tolerance range for the component takes on asymmetric values. This is exactly what we found back in Chapter 5, where we first discussed nonlinear tolerance problems.

Taguchi, Elsayed, and Hsiang [R12.2] recommend that the engineering team use the smaller of the two tolerance values as a precautionary step. This can be evaluated later, when the tolerance-balancing experiments are conducted during tolerance design. At that point we can assess whether the

Figure 12.6 Example of nonlinear component sensitivity

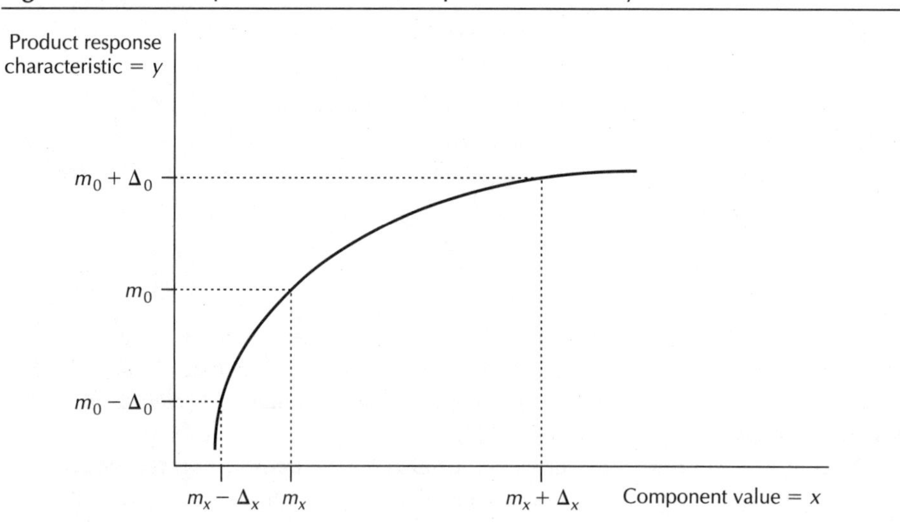

tighter tolerance that this conservative approach recommends is, in fact, appropriate. Tolerance design will be thoroughly reviewed in Chapters 13 through 17.

Just as it is important to experimentally evaluate and verify that the tolerances developed from traditional, analytical methods are correct; the team needs to do the same for tolerances developed by quality-loss function techniques. As recommended in Chapter 2, all mathematical analysis on critical, functional design elements needs to be verified by experimental analysis. When this evaluation is short-circuited by rushing to meet program schedules, the effects can be functionally and economically catastrophic. This is especially true later in the product's life cycle.

Developing Tolerances for Deterioration Characteristics in the Design

A product may function quite well as it begins its life cycle. Out-of-the-box quality may be quite high. After some time goes by and the effects of environmental and deterioration noise have had a chance to occur, the weaknesses of the initially toleranced components and assemblies will begin to show. We need to take a moment and set the stage for tolerancing in the context of the deterioration of a product's output characteristics.

We will define the product's output characteristics in two ways: 1) using the initial output characteristic and 2) using the time-based, or deteriorated, output characteristic. A robust product is one that produces on-target output not only initially, but over time, defined as *T*, the operating life of the product. This life cycle is usually defined in years but can be specified in smaller increments of time.

The initial product characteristic has already been defined as having a value *y* that can deviate from the target m_0. The customer-defined limits for the product characteristic have been defined as Δ_0 and A_0. Thus it can be said that the customer's specification for the initial product-output characteristic (at least for one of them, since there can be many) is equal to $m_0 \pm \Delta_0$.

We have also stated that there will be one or more lower level product characteristics (*x*), often referred to as engineering characteristics, that will cause a change in *y* as they themselves change. These lower level characteristics will deviate from their target values *m*. These deviations have been defined as Δ and have a loss, usually valued at a replacement, repair, or adjustment cost of *A* dollars. Thus it can be said that the specification for an initial low-level characteristic is equal to $m \pm \Delta$.

To solve tolerance-allocation problems with the loss function we needed to define a relationship between *y*, the product characteristic, and *x*, the lower level engineering characteristic. We needed to know how *y* changed for a unit change in *x*. This must be known for every critical component in the product that significantly affects *y*. The resulting term has been defined as β. It is related to the initial condition and the deterioration condition of *y*, the high-level product output characteristic.

We need to introduce a new term, $\beta_{\text{deterioration}}$. This is defined as the *time based rate of change* of *y*. In the production of the product, there must be a directly measurable value for $\beta_{\text{deterioration}}$. If none is available then a surrogate measure of deterioration rate will need to be found in order to adjust a parameter in the factory to put $\beta_{\text{deterioration}}$ on target. As time progresses, $\beta_{\text{deterioration}}$ may have a distinct, possibly changing, value that we want to decouple from the changes occurring in all but one or two of the low-level characteristics. This does not mean *y* is changing, because all the *x* values are remaining stable; they are, for the most part, going to change with time as well. We want to establish an acceptable rate of change for *y*. This rate of change becomes a nominal set point for the design and will have specific low-level characteristics that are capable of slowing or speeding the rate of

$\beta_{\text{deterioration}}$. You can think of these unique parameters as *deterioration deceleration factors,* since they are used to slow and stabilize the rate of deterioration.

One example is the specification of a bearing material that is more resistant to wear and dynamic loading effects than another, cheaper material. Rather than attempting to prolong its life by having more precision on the variability in one material, one can use a higher class of material. Using the economically determined tolerances of all the components is not the way to attenuate the rate of deterioration; the way to do it is to *isolate one or two factors that can be easily adjusted, either in the production process or in the customer's environment, and that put the rate of deterioration on its target.* Thus, we seek deterioration engineering factors (control factors) that can be adjusted, then toleranced. This is profoundly different thinking in comparison to traditional tolerance practice, where component stacks are adjusted using some if not all of the tolerances as adjustment factors to keep the product output on target. Taguchi (1993) quotes a discussion he had with a student regarding these issues [R12.1]:

> *[Taguchi:] One must consider the economic relationship between each component and its higher level characteristic individually to decide tolerance specifications for each component.*

> *[Student:] In my opinion [referring to methods based on the accumulation of variation of higher level characteristics caused by variation of lower level characteristics versus loss function-based tolerancing] these methods will make the process of specifying tolerances much easier than stacked variation methods, because one can decide the tolerances of each lower level characteristic directly from A [the cost of the lower level characteristic] and b [b = β, which is the change in a higher level characteristic resulting from a unit change in its lower level characteristic] without considering other lower level characteristics.*

Taguchi's reply to the student is very important:

> *As a matter of fact, to correctly determine the value b [β] of one lower level characteristic, one still needs to simulate the effects of several [other] lower level characteristics simultaneously [inducing noise in a tolerance evaluation]. This method might even be more reliable than the methods described in this chapter [Chapter 3] [R12.1]. This approach involves orthogonal arrays to simulate the compounding [noisy] effects of several lower level characteristics. However, the methods described in this chapter are practical and reliable enough for general application.*

Changing tolerances to balance cost and quality will be necessary at some point. In the initial determination of tolerances, finding deterioration-rate adjustment factors is much more cost effective than moving right to tolerance adjustment. The team must have faith in its loss functions and how they quantify loss and constrain each component independently. Where does this faith come from? Properly researched and constructed quality-loss functions. Otherwise it is back to the old practice of inefficient and customer-blind stackup-of-variation techniques.

Tolerancing the Deterioration Rate of a Higher Level Product Characteristic

We need to define the specification for the rate of *y*'s deterioration. The deterioration specification is a design metric that will be directly related to the long-term design process capability index ($\text{Cpk}_{\text{design}}$) due to the time-based shift in its mean. It is shown as follows:

$$[\beta_{\text{deterioration}} \pm \Delta\,\beta_{\text{deterioration}}]$$

(Taguchi refers to $\Delta\ \beta_{\text{deterioration}}$ as Δ^* in his works [R12.1 and 3].) The tolerance for $\Delta_{\text{deterioration}}$ is based on the definition of a low-level tolerance (Equation 12.24):

$$\Delta = \sqrt{\frac{A}{A_0}}\,\frac{\Delta_0}{|\beta|} \tag{12.24}$$

We can use the usual loss-function approach to define the deterioration-rate tolerance. We will now show what terms have to be added to Equation 12.24 to account for the rate of degradation.

Quoting from [R12.3], Taguchi (1986) states:

> *If the characteristic value of the product starts out at the nominal value m and changes by b [β] per year, at the end of T years it will deviate from the nominal by bT. As it gradually drifts away from the nominal value, the mean squared deviation [MSD] is given by the following integral.*

$$(\text{MSD}) = \frac{1}{T}\int (\beta\, t)^2\, dt = \frac{T^2}{3}\beta^2 \tag{12.25}$$

(Taguchi often uses the terms MSD and σ^2 synonymously. We have not followed his convention since the reader has been exposed to the term σ^2 as the sample variance from classical statistical theory. In this case we will stay with MSD terminology.)

Reworking this expression, Taguchi defines the tolerance for the rate of deterioration as:

$$\Delta^* = \sqrt{\frac{3A^*}{A_0}}\left(\frac{\Delta_0}{|\beta|T}\right) \tag{12.26}$$

The new terms we see added to Equation 12.26 are defined as:

A^* = loss or cost when y's deterioration rate exceeds its tolerance

T = the life of the product

Our intent is to illustrate Taguchi's method as clearly as possible rather than to modify it. In that spirit, we will review a classic example from his work [R12.1 and 3].

Determining Initial and Deterioration Tolerances for a Product Characteristic

This case comes from Chapter 3 of *Taguchi on Robust Technology Development.* A company in Japan is producing a lighting product. The key measure of customer tolerance is the output illuminance. This is the highest level output characteristic. The customer tolerance is found to be 50 lux $= \Delta_0$. The loss when this value is reached is \$150 $= A_0$. The next lower level characteristic associated with the output illuminance is called the *luminosity.* This is the engineering output characteristic we seek to tolerance. It is found by experimental means that the output of the illuminance changes 0.80 lux for every 1 candela of luminosity (a unit change in the lower level characteristic). Thus, β is defined as 0.80 lux/candela. The cost to adjust the lower level luminosity is $A =$ \$3. The cost of replacing the light source altogether when it surpasses the deterioration tolerance limit is $A^* = \$32$. The life of the light source is $T = 20{,}000$ hours. We can now calculate the tolerance on y, the illuminance, both as an initial high-level characteristic and on y's rate of change with respect to time as a deterioration characteristic.

$$\Delta = \sqrt{\frac{A}{A_0}}\frac{\Delta_0}{|\beta|} = \sqrt{\frac{3}{150}}\frac{50}{|0.80|} = 8.8 \text{ (lux)}$$

$$\Delta^* = \sqrt{\frac{3A^*}{A_0}}\left(\frac{\Delta_0}{|\beta|T}\right) = \sqrt{\frac{(3)(32)}{150}}\left(\frac{50}{|0.80|20{,}000}\right) = 0.0025 \text{ (lux /hr.)}$$

If, in the manufacturing plant, the illuminance deviates more than ± 8.8 lux from its target, adjustment must be made. If the rate of change in illuminance deviates more than 0.0025 (lux /hr.), then an adjustment will also have to be made to change the rate to be within its specification. The deterioration rate is usually a smaller-the-better characteristic. The illuminance performance output will follow the luminosity and will also degrade proportionally with the rate of luminosity degradation.

Taguchi's writings hold many insightful case studies and problems from which the reader can benefit. While much of his work is translated from Japanese into English, it is worth the effort to read his personal description of how the loss function applies to tolerance allocation. His work is original and profound. It is regrettable that his tolerancing techniques are largely unknown or applied in the United States. All of Japan uses the Japanese Industrial Standards as a guide for tolerancing processes; these were developed around Taguchi's quality-loss function approach. This is an important point of comparison that needs to be emphasized between economically based tolerancing practices in Japan and the U.S.

Summary

Chapter 12 has reviewed many topics associated with the application of the quality-loss function to tolerancing. A more complete derivation and explanation of the loss function is presented, particularly with respect to specific applications. An economic factor of safety is defined based on the loss-function point of view.

Chapter 12: The Application of the Quadratic Loss Function to Tolerancing

- The Difference between Customer, Design, and Manufacturing Tolerances
- Customer Tolerances
- Design and Manufacturing Tolerances
- The Taguchi Tolerancing Equations
- Taguchi's Economical Safety Factor
- The Derivation of Taguchi's Tolerance Equation and Factor of Safety
- Relating Customer Tolerances to Engineering Tolerances
- An Example of Tolerancing Using the Loss Function (Nominal-the-Best Case)
- Relating Customer Tolerances to Subsystem and Component Tolerances
- The Linear Sensitivity Factor, β

- Using the Loss Function for Multiple-Component Tolerance Analysis
- An Example of Applying the Quality-Loss Function to a Multicomponent Problem
    - Setting Up the Problem
    - Identifying Critical Parameters
    - Mapping the Critical Parameters and Their Sensitivities
    - Converting the Traditional Tolerance Problem into a Quality-Loss Tolerance Problem
- How to Evaluate Aggregated Low-Level Tolerances
- Using the Loss Function in Nonlinear Relationships
- Developing Tolerances for Deterioration Characteristics in the Design
- Tolerancing the Deterioration Rate of a Higher Level Product Characteristic
- Determining Initial and Deterioration Tolerances for a Product Characteristic

We have presented Taguchi's point of view clearly so the reader can understand the major principles of how the loss function works in the tolerancing process. Both linear and nonlinear cases have been discussed. An approach for accounting for deterioration within a product has also been reviewed.

Working with the quality-loss function has become a fairly common practice in Japan. The U.S. academic and industrial communities have a large task ahead to create the learning and application environments that will be required for current and future engineers to bring the customer's losses into the tolerancing process.

This chapter has illustrated the analytical side of Taguchi's methods for tolerance development. Chapters 13 through 17 develop the required skill set to deal with the experimental approach to tolerance design.

References

R12.1: *Taguchi on Robust Technology Development,* G. Taguchi, ASME Press, 1993, pg. 427

R12.2: *Quality Engineering in Production Systems,* G. Taguchi, E. A. Elsayed and T. C. Hsiang, McGraw-Hill, 1989, pg. 427

R12.3: *Introduction to Quality Engineering;* G. Taguchi; Asian Productivity Organization, 1986, pg. 427

R12.4: *System of Experimental Design* Vols. 1 and 2; Quality Resources and American Supplier Institute, 1987, pg. 427

CHAPTER 13

General Review of Orthogonal Array Experimentation for Tolerance Design Applications

Developing Tolerances Using a Designed Experiment

Now that we have begun to define the mechanics of how to shift from math model-based tolerance analysis to performing tolerance analysis in an experimental context, we are ready to discuss efficient means for producing experimental data for tolerance analysis. It is at this point that an engineering team plans and conducts designed experiments on actual hardware. We will see the effect of the manufacturing process standard deviation associated with each design parameter and the initially selected design tolerances on various types of output performance.

We will learn what a difficult task it can be to measure with math models and schematic representations the phenomena we have discussed in theory. Now we must experiment with what we have developed by including all three forms of noise during the tolerance analysis as they are strategically changed to balance cost and quality (see Figure 13.1).

We will introduce a special class of designed experiments called *fractional factorial orthogonal arrays* to evaluate the effect of setting nominal parameters at plus or minus one standard deviation level. We will measure the effect of these changes on a critical output response, and perform the experiments in the presence of stressful environmental, unit-to-unit, and deterioration noise factors. Once we have analyzed the data through a process called ANOVA, or analysis of variance, we will refine and optimize the parameters' standard deviations and define appropriate tolerances. We will use the quality-loss function to guide this process by concurrently balancing costs and quality. We will learn to work with component-manufacturing engineers to reduce the standard deviations coming from the manufactured-component critical parameters. The cost increases necessary to strategically enhance quality will be justified; conversely, the initial standard deviations that were thought to be too large,

Figure 13.1 The two types of tolerance experiments

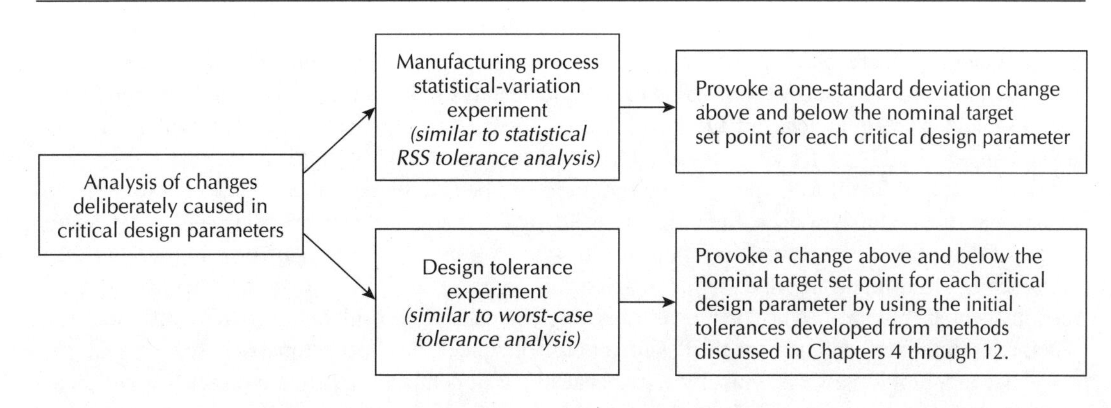

and the associated tolerances that may have been made too tight, will be identified and appropriately relaxed. The ensuing lower manufacturing cost will be used to control the overall cost of the product. Then we will return to the original designed experiment with new, upgraded components made to the new limits of their optimized specifications. Finally, we will learn how to run a verification experiment to quantify the improvements on output performance variability. The final tolerance experiment will be run within the same stressful environment of the three noise conditions as was the initial statistical variation experiment.

Use of Orthogonal Arrays in Tolerance Design

Experiments are often used by an engineering team to study the effects of parameters as they are set at various levels. The experimental boundaries are established by what parameters are included in an experiment and the set points at which they are evaluated. These boundaries are referred to as the *experimental design space.* In our case, the experimental design space is the experimental space defined by the standard deviations and associated tolerances for the design parameters. When a series of design parameters is set at one standard deviation from their nominal value or beyond (usually at their tolerance limits), the team can evaluate the parameters' ability to affect a product's overall response. There are several ways to produce the data for such an evaluation:

1. Running build-test-fix iterations (usually using one-factor-at-a-time methods) on selected parameter standard deviations or tolerances suspected of being critical.
2. Test every possible combination of design parameter first-standard-deviation set points or parameter-tolerance limits.
3. Test a strategically limited but statistically valid series of design parameter first-standard-deviation or tolerance-limit combinations.

Testing a set of parameters at their one-sigma levels is analogous to performing statistical RSS analysis (see Chapter 6); testing a set of parameters at their design tolerance limits is analogous to performing a worst-case analysis (see Chapter 5).

The Build-Test-Fix Approach

The build-test-fix approach is a common but risky method that typically leads to lengthy periods of experimentation. Its results are often difficult to reproduce because they typically are produced in an undisciplined manner that is not amenable to statistical analysis. This strategy is strongly dependent on the skill, and sometimes luck, of the individual engineer. These evaluations rarely include stressful noises, and the build-test-fix method suffers from the two additional difficulties.

First, it is difficult to know if a true optimum has been achieved since a build-test-fix cycle is considered successful as soon as functionality is attained. As soon as the performance is in the desired range of acceptability, the work is often halted while some other important task is initiated. This approach is often found in understaffed engineering organizations. The prevailing management philosophy behind this approach is "get results any way you can." The result is insufficient information on whether optimal performance, for the design and its tolerances, has been attained.

Second, build-test-fix is typically slow because it is dependent on the team finding a solution out of myriad options. Because it lacks a disciplined technique to search through the many parameter set point tolerance options, it often leads to reevaluation prior to integration of an assembly or subsystem into the product. Over time, as the design is finally integrated into the product, there is typically a need to reoptimize the parameter set points yet again. In production of the final product, fire-fighting regularly occurs due to the sensitivities and interactivity that frequently occurs in build-test-fix designs.

We need to look at the other two approaches, using as an example a *designed* experimental setup for a statistical variability or tolerance experiment. We will study seven two-level parameters that can be set to the current plus and minus one-sigma values for each parameter. The goal of the experiment is to find the individual contribution from each of the seven parameters to the output response's variability. From this evaluation the tolerances can be defined.

When an experimental analysis defines the parameters that are producing the most variability in the output response of the assembly, subsystem, or product, the task becomes one of reducing the critical parameter one-sigma values. This must be done such that the output response is acceptable if not on target. Factors of safety, customer loss, manufacturing cost control, and manufacturing process capabilities must all be considered in the determination of the design tolerances.

Introduction to Full Factorial Experiments

A full factorial experiment investigates all possible combinations of all parameter levels.

Figure 13.2 shows a few of all 128 of the combinations that are possible with seven two-level parameters. The total number of combinations that are possible can be calculated using the formula given in Equation 13.1.

$$\text{\# of combinations} = y^x \tag{13.1}$$

where x = number of parameters that have y levels. Even for the simple example of seven parameters with two levels for each parameter, there are 128 combination (2^7). In fact, full factorial experiments are practical only with small numbers of parameters and levels. For example, 13 factors with 3 levels for each factor have 3^{13} (1,594,323) combinations!

The full factorial approach investigates all possible parameter standard deviation or tolerance-limit combinations. This assumes that all other parameters, outside of the seven being studied, are held

Figure 13.2 An experimental design for a full factorial experiment

Run	*A*	*B*	*C*	*D*	*E*	*F*	*G*
1	1	1	1	1	1	1	1
2	1	1	1	1	1	1	2
3	1	1	1	1	1	2	1
.	.	.	.	.	.	.	.
.	.	.	.	.	.	.	.
126	2	2	2	2	2	1	2
127	2	2	2	2	2	2	1
128	2	2	2	2	2	2	2

at their nominal set points so that the variation in the response is due only to the *critical* parameter variational effects.

The biggest weakness of this approach is that too many experiments are used for the amount of information actually required to understand the parameter tolerance effects. It is not necessary to investigate every parameter standard deviation or tolerance combination to find the set of values that defines the contribution each parameter makes to the variability in the output response. ANOVA, a simple statistical analysis procedure, can quantify the contribution made by each parameter standard deviation or tolerance by looking at each parameter's effect on the output response variability. If we impose structure and balance on our experimentation we should have little trouble making sense out of the data that are gathered in what might appear to be a massively confusing and intertwined fashion. Designed experiments always look confusing at first, but with practice they soon become very easy to work with and apply.

Designed Experiments Based on Fractional Factorial Orthogonal Arrays

The fractional factorial orthogonal array is a method of setting up experiments that only requires a fraction of the full factorial combinations. The term *array* simply refers to a mathematically derived matrix arrangement that constrains the way all the parameter set points are set up prior to running the experiment. An orthogonal array imposes an order on the way the matrix of experiments is carried out; *orthogonal* refers to the balance between the various combinations of parameters so that no one parameter is given more or less opportunity to express its effect on the response in the experiment than any of the other parameters. Orthogonal also refers to the fact that the effect of each parameter can be mathematically assessed independent of the effects of the other parameters. Figure 13.3 shows a fractional factorial orthogonal array that would accommodate the seven parameter example.

This array describes the coding (ones and twos) for the level (standard deviation or tolerance) settings for each of the seven parameters that are to be used for the eight runs in the experiment. In this case, the treatment combinations are mathematically chosen to uniformly span the experimental space represented by the 128-run full factorial experiment from which the fractional factorial is taken. The derivation of how such an array is constructed is beyond the scope of this book; to see how this is done refer to a classical statistical design-of-experiments book. (See [R13.1 & R13.2].)

Figure 13.3 An orthogonal array design for a fractional factorial experiment

Run	*A*	*B*	*C*	*D*	*E*	*F*	*G*
1	1	1	1	1	1	1	1
2	1	1	1	2	2	2	2
3	1	2	2	1	1	2	2
4	1	2	2	2	2	1	1
5	2	1	2	1	2	1	2
6	2	1	2	2	1	2	1
7	2	2	1	1	2	2	1
8	2	2	1	2	1	1	2

Figure 13.4 Illustration of balance in an orthogonal array

Run	*A*	*B*	*C*	*D*	*E*	*F*	*G*
1	1	1	1	1	1	1	1
2	1	1	1	2	2	2	2
3	1	2	2	1	1	2	2
4	1	2	2	2	2	1	1
5	2	1	2	1	2	1	2
6	2	1	2	2	1	2	1
7	2	2	1	1	2	2	1
8	2	2	1	2	1	1	2

The array shown in Figure 13.4 can be used to illustrate the concept of orthogonality in a matrix experiment. Levels one and two occur the same number of times in each column in the array. Furthermore, for the four rows with level one in column A; two rows have level one of column B and two rows have level two of column B. The same can be said for the four rows with level two of column A. In fact, the same balance of parameter levels can be found for every pair of columns in the array. This balance is illustrated for the first two columns in Figure 13.4. When comparing A1 to A2 in a response table, the effect of B in A1 is the same as the effect of B in A2.

Orthogonality is a pair-wise property of the columns in the array. This means that some columns can be left empty without destroying the orthogonality, or balance, of the array. In many cases it is better not to saturate the array with parameters. This is not a problem as far as experimental balance is concerned. However, *all* the treatment combinations (rows) must be run to maintain the balance condition and preserve the orthogonality.

Because we are assessing only a fraction of all the possible parameter combinations, there are limitations on what the data from such an experiment can be used to quantify. In a tolerance experiment we are interested only in the main effect of each parameter standard deviation or tolerance, independent of all other parameters, on the variational output of the response. Recall from the worst-case and statistical tolerance analysis that it is a common assumption that all components are independent of one another in their effect on the output response. This is not the case for all complex mechanisms and chemical processes. In many cases an orthogonal array that possesses enough experimental capacity, called *degrees of freedom,* must be used to quantify the effects of parameters that are dependent on one another's set points in how they ultimately affect the output response. This is yet another form of design sensitivity that Taguchi considers a noise factor with respect to the response. These interparameter sensitivities are referred to in statistical terms as *interactions.* Many engineers use tolerances to combat the effects of interactions in a design. This is considered a poor strategy relative to taking steps to minimize the presence of the interaction by methods included in concept design and parameter design. Only after these methods have been applied to minimize interactivity should tolerances be used to minimize interactivity effects that are disruptive to product performance (see [R13.3]).

Degrees of Freedom: The Capacity to Do Experimental Analysis

Degree of freedom (DOF) is a concept used to describe how large an experiment must be and how much information can be extracted from the data. In practical terms, the degrees of freedom for a matrix experiment is one less than the number of runs in the experiment:

$$DOF_{exp} = \text{number of runs} - 1 \tag{13.2}$$

The degrees of freedom needed to describe a parameter's main effect is one less than the number of levels tested for that parameter:

$$DOF_f = \text{number of levels} - 1 \tag{13.3}$$

The problem of solving a set of simultaneous equations for a set of unknowns is a good mathematical analogy for this process. The number of equations is analogous to the degrees of freedom of a matrix experiment. The number of unknowns is analogous to the total degrees of freedom for the factor effects:

$$\text{Total } DOF_f = (\text{number of factors})(DOF_f) \tag{13.4}$$

How does the degrees-of-freedom analysis look for the cases we have seen so far?

Degrees of Freedom for the Full Factorial

In the array shown in Figure 13.2, the $DOF_{exp} = 128 - 1 = 127$. If only the main effects of the factors A through G are being studied, then the total DOF_f being used is 7, and there is a waste of 120 degrees of freedom. This is why the full factorial is so inefficient; it generates much more information than is actually used in the analysis. Even if all the two-way interactions are examined (each requiring one DOF), then the total DOF_f needed is still only 28.

DOF for the Fractional Factorial Orthogonal Array

In the array shown in Figure 13.3, the $DOF_{exp} = 8 - 1 = 7$. The degrees of freedom required to calculate the main effects of each factor are total $DOF_f = 7(2 - 1) = 7$. This is matched for 100 percent efficiency. We have no degrees of freedom left open. The fractional factorial can be easily loaded to capacity.

DOF for Two-Level Full Factorial Arrays

Any experiment in which all of the possible combinations of factor levels are tested is, by definition, a full factorial. One common example is the 2^2 full factorial shown in Figure 13.5.

The orthogonal array shown in Figure 13.5 provides three DOF_{exp} but only requires two DOF_f for quantifying the two parameter effects (A and B). Let us define a metric, the experimental efficiency X, where

$$X = \frac{\text{Total DOF}_f}{\text{DOF}_{exp}} \tag{13.5}$$

For the 2^2 full factorial experiment, the efficiency $X = \frac{2}{3}$, or 67 percent.

Figure 13.6 shows a 2^3 full factorial. It provides seven DOF_{exp} but requires only three DOF_f for the factor effects, $X = 43$ percent efficiency.

The convention for naming the fractional factorial orthogonal arrays is

$$L_a\,(b^c)$$

where:

a = the number of experimental runs
b = the number of levels for each factor
c = the number of columns in each array

Figure 13.5 A two-factor, two-level full factorial array

Run	*A*	*B*
1	1	1
2	1	2
3	2	1
4	2	2

Figure 13.6 A three-factor, two-level full factorial

Run	*A*	*B*	*C*
1	1	1	1
2	1	1	2
3	1	2	1
4	1	2	2
5	2	1	1
6	2	1	2
7	2	2	1
8	2	2	2

The L_4 is shown in Figure 13.7. It can handle three factors at two levels. It is the smallest of the two-level orthogonal arrays. The L_4 $DOF_{exp} = 3$. The first two columns of the L_4 are the 2^2 full factorial.

The L_8 is shown in Figure 13.8. It can handle seven factors at two levels. The L_8 $DOF_{exp} = 7$. The first, second, and fourth columns of the L_8 are the 2^3 full factorial.

The L_{12} is shown in Figure 13.9. It can handle 11 factors at two levels. The L_{12} $DOF_{exp} = 11$. The L_{12}, along with the L_{18} (a three-level array), has some special properties with respect to interactions that make it uniquely useful in tolerance analysis.

The L_{12} is a unique array with respect to the effect of interactions, between any two or more factors, on its analysis. Interactions are illustrated later in this chapter using a simple example. In general, an interaction between factors means that the effect of one factor depends upon the level of another. Thus, factor A has a negative slope when factor B is at level one, and factor A has a positive slope when factor B is at level two (see Figure 13.11). In most of the fractional factorial orthogonal arrays, the mathematical effect of an interaction between two factors is blended with at least one other factor effect. This is referred to as *confounding* because the analysis of the means (Chapter 14) result is a numerical sum of the factor effect blended with the interaction effect. These effects cannot be separately evaluated in an array when all of the columns are used, or the array is *saturated.* It is safe to saturate an array when interactions between all parameters are relatively weak; otherwise the interactive effects may cloud the quantification of the main effects coming from each parameter. When severe interactions are suspected, they must be investigated properly.

Figure 13.7 $L_4(2^3)$ orthogonal array

Run	*1*	*2*	*3*
1	1	1	1
2	1	2	2
3	2	1	2
4	2	2	1

Figure 13.8 $L_8(2^7)$ orthogonal array

Run	*1*	*2*	*3*	*4*	*5*	*6*	*7*
1	1	1	1	1	1	1	1
2	1	1	1	2	2	2	2
3	1	2	2	1	1	2	2
4	1	2	2	2	2	1	1
5	2	1	2	1	2	1	2
6	2	1	2	2	1	2	1
7	2	2	1	1	2	2	1
8	2	2	1	2	1	1	2

Figure 13.9 $L_{12}(2^{11})$ orthogonal array

Run	*1*	*2*	*3*	*4*	*5*	*6*	*7*	*8*	*9*	*10*	*11*
1	1	1	1	1	1	1	1	1	1	1	1
2	1	1	1	1	1	2	2	2	2	2	2
3	1	1	2	2	2	1	1	1	2	2	2
4	1	2	1	2	2	1	2	2	1	1	2
5	1	2	2	1	2	2	1	2	1	2	1
6	1	2	2	2	1	2	2	1	2	1	1
7	2	1	2	2	1	1	2	2	1	2	1
8	2	1	2	1	2	2	2	1	1	1	2
9	2	1	1	2	2	2	1	2	2	1	1
10	2	2	2	1	1	1	1	2	2	1	2
11	2	2	1	2	1	2	1	1	1	2	2
12	2	2	1	1	2	1	2	1	2	2	1

Figure 13.10 $L_{16}(2^{15})$ orthogonal array

Run	*1*	*2*	*3*	*4*	*5*	*6*	*7*	*8*	*9*	*10*	*11*	*12*	*13*	*14*	*15*
1	1	1	1	1	1	1	1	1	1	1	1	1	1	1	1
2	1	1	1	1	1	1	1	2	2	2	2	2	2	2	2
3	1	1	1	2	2	2	2	1	1	1	1	2	2	2	2
4	1	1	1	2	2	2	2	2	2	2	2	1	1	1	1
5	1	2	2	1	1	2	2	1	1	2	2	1	1	2	2
6	1	2	2	1	1	2	2	2	2	1	1	2	2	1	1
7	1	2	2	2	2	1	1	1	1	2	2	2	2	1	1
8	1	2	2	2	2	1	1	2	2	1	1	1	1	2	2
9	2	1	2	1	2	1	2	1	2	1	2	1	2	1	2
10	2	1	2	1	2	1	2	2	1	2	1	2	1	2	1
11	2	1	2	2	1	2	1	1	2	1	2	2	1	2	1
12	2	1	2	2	1	2	1	2	1	2	1	1	2	1	2
13	2	2	1	1	2	2	1	1	2	2	1	1	2	2	1
14	2	2	1	1	2	2	1	2	1	1	2	2	1	1	2
15	2	2	1	2	1	1	2	1	2	2	1	2	1	1	2
16	2	2	1	2	1	1	2	2	1	1	2	1	2	2	1

Consider how the subsystems and components in an actual product interact with one another. Rarely does one interaction put all of its energy into a single place. In fact, complex designs and processes can have myriad interactions, some large and some small, all working together. When they work together harmoniously, the effect can be very helpful, but when the interactive effects are harmful they can interfere with the main functionality or effect of one other design element or many. Some designs have parameters that are redundant—they both try to contribute to the functional response in the same manner. If one parameter undoes the effect of another by means of co-contribution effects, the interaction can be very disruptive. This problem can be driven by one parameter tolerance being at the low limit while another co-contributing parameter tolerance resides at the high limit.

The interactive effects within the L_{12} array are distributed more or less evenly so that they all partially confound with one another, as is true in many physical designs. We refer to the L_{12}, the L_{18}, and the L_{36} arrays (the L_{18} and L_{36} have similar interaction properties) as the *engineering arrays.* They are not suitable for modeling main effects *and* interactions; however, they provide a realistic simulation of the actual interplay of parameters that occurs in engineering systems. The distributed effects of the interactions act in a manner similar to noise; after all, undesired interactions in a design are a nuisance.

The main effects that emerge from a properly prepared study using the engineering arrays are strong enough to rely on despite noise and uncontrolled interactions. For that reason, the L_{12} is an excellent experiment for optimizing physical designs that are engineered for *additivity,* or low interactivity between parameters. It is particularly suitable for doing experimental evaluations involving noise factors where two-level factors are studied (see Chapter 14).

The L_{16} is shown in Table 13.10. It can handle 15 factors at two levels. The L_{16} $DOF_{exp} = 15$. Columns 1, 2, 4, and 8 of the L_{16} are the 2^4 full factorial.

There are larger two-level arrays. The L_{32} can hold up to 31 parameters, and the L_{64} can hold up to 63 parameters. It is rare that one would have to evaluate so many critical parameters in a product.

Methods to Account for Interactions within Tolerance Design Experiments

As previously mentioned, in some cases we need to run an experiment to provide information about how one parameter's variability contribution to the output response changes when another parameter's variability level changes. We can think of this as a dependent sensitivity analysis (for more information see Evans). When factors act in an additive fashion, then we can refer to the study as an independent sensitivity analysis.

Interactions Defined

As we have discussed, an interaction between two design parameters means that the effect of one parameter on the response depends upon the level of another parameter. An example of an interaction is shown in Figure 13.11. When parameter B is low, changing parameter A from low to high causes the response to get smaller. When parameter B is high, changing parameter A from low to high causes the response to get bigger. Thus, the effect of parameter A on the response depends upon the level of parameter B. This dependency is reciprocal. That is, the effect of parameter B on the response also depends upon the level of parameter A.

An interaction such as is shown in Figure 13.11 is referred to as *antisynergistic,* since the slope of the parameter effect plots changes sign depending upon parameter B's level. If one parameter changes, then the effect of both parameters on the response changes. The antisynergistic aspect means that if one of the parameters is changed, the other one can end up at its nonoptimum level, *even though it is not changed.* As an engineering issue, this leads to the need to reoptimize the rest of the system after a change occurs in just one parameter. This is a nonrobust condition, since it implies that the system is very sensitive to changes in one of the parameters. In an additive (low-interactivity) situation,

Figure 13.11 An interaction between two parameters

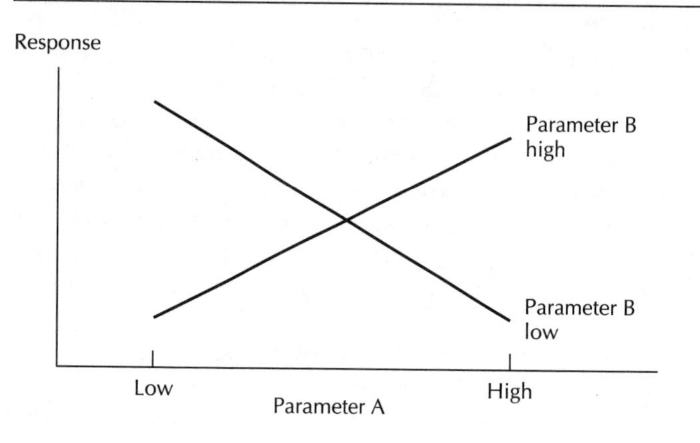

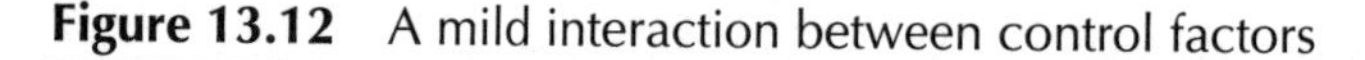
Figure 13.12 A mild interaction between control factors

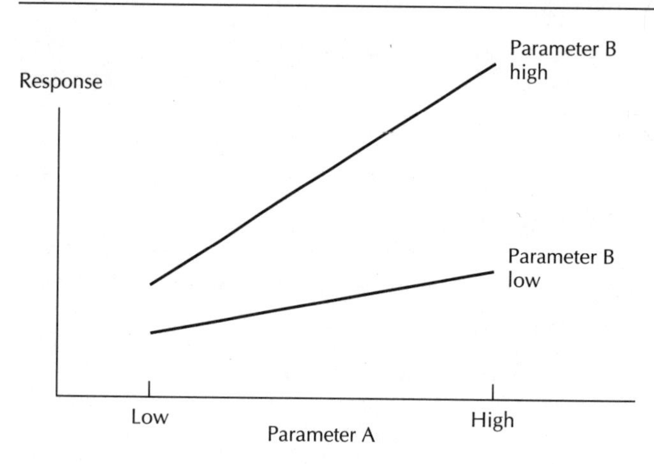

the effect of a change in one parameter is primarily limited to that parameter's main effect. Reoptimization of the other parameters is not needed, since their effect is relatively stable in the absence of strong interactions.

A mild interaction between parameters, as shown in Figure 13.12, is a tolerable situation. Here, although there is a measurable interaction, it is referred to as *synergistic* because the parameter levels do not heavily depend on each other.

The optimum level for control factor A is high, regardless of the level of factor B. Similarly, the optimum level for control factor B is high regardless of the level of factor A. The *magnitude* of the effect of changing factor A depends on the level of factor B, and vice versa. This situation is described as *monotonic;* that is, directionality is consistent between the two slopes. Since the optimal levels stay the same, reoptimization is not needed. Nevertheless, the situation is not ideal. A relatively small loss in robustness, such as might result from a factor-level change, can be amplified by the interactions as their influence is applied as well. In this case the parameters can fluctuate within their tolerances and will not grossly affect functional performance.

Many times the measured interactivity is not a function of a physical phenomenon but rather a function of the response metric or design parameters selected by the engineer. Thus, interactions can be induced by human error in selecting poor representations of what a critical response variable or design parameter actually is. These are sometimes referred to as *virtual* interactions.

When a response variable is selected on the basis of being a direct engineering unit that is expressive of the direct physical activity occurring within the design, the likelihood of an artificial or virtual interaction occurring is low. When a direct engineering unit is used as a credible response variable and an interaction shows up in the quantitative data analysis, it is very likely to be real. A designed experiment must be specifically loaded to provide the proper degrees of freedom to be allocated to facilitate the analysis of both individual main effects and interactive effects. One way of dealing with interactive parameters that happen to have a detrimental effect on the response of a product is to tighten the tolerances on both interactive parameters so as to limit the contribution of the effect. This, of course, is expensive and sometimes necessary.

Quantifying the Effect of an Interaction

Interactions can be estimated using certain types of fractional factorial orthogonal arrays that possess the appropriate degrees of freedom, such as the L_4 orthogonal array shown in Figure 13.13.

This array has the property that column 3 can be used to represent the column 1 and column 2 interaction (A × B). Thus, if factors A and B are assigned to columns 1 and 2, respectively, and column 3 is left empty, then the calculation of the effect of column 3 can be used to study the A × B interaction. Column 3 has one degree of freedom available to account for either the A × B interaction or the effect of a third, new (additive) factor C. Putting a parameter in column 3 would saturate the array and consume the degree of freedom available to quantify the interaction between A and B. This is a good illustration of how degrees of freedom in a two-level design either get used for parameter main effect quantification or for quantification of an interaction between two other parameters.

Things can get confusing when a factor is assigned to a column that is capable of quantifying both an interactive effect and the effect of a factor. The way to avoid this problem is to refrain from using interaction columns that are associated with parameters that may possess strongly interactive relationships with one or more other parameters.

When studying interactions, it is best to use fractional factorial arrays that have enough degrees of freedom to separate the main effects from the two-way interactions, and all the two-way interactions from one another (called a *resolution V design*; see [R13.2 and 3]). If you are justified in believing that just two factors are interacting, you can isolate them and run them through an L_4 array with column 3

Figure 13.13 The L_4 array

	1	*2*	*3*
Run	*A*	*B*	$C = A \times B$
1	1	1	1
2	1	2	2
3	2	1	2
4	2	2	1

left empty. In this way, you can obtain a very clear picture of the two-way interaction effect between the factors.

An approach for choosing the proper column assignments to study interactions is needed. If all the columns of an array are used, as in the L_4 shown in Figure 13.13, then there is some uncertainty as to how the results are to be interpreted. Is the analysis-of-means value due to the factor assigned to that column (the main effect) or to the interactions of other parameter interdependencies also attributed to that column? There is a tradeoff to be made here. The number of degrees of freedom required to study the main effects and interactions increases quickly with the number of factors, as is shown by the column assignments and the number of parameters each orthogonal array can accept without confounding

Figure 13.14

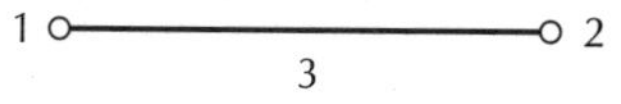

1 o———————o 2
3

Figure 13.15 L_8 linear graphs

Figure 13.16 L_{16} linear graphs*

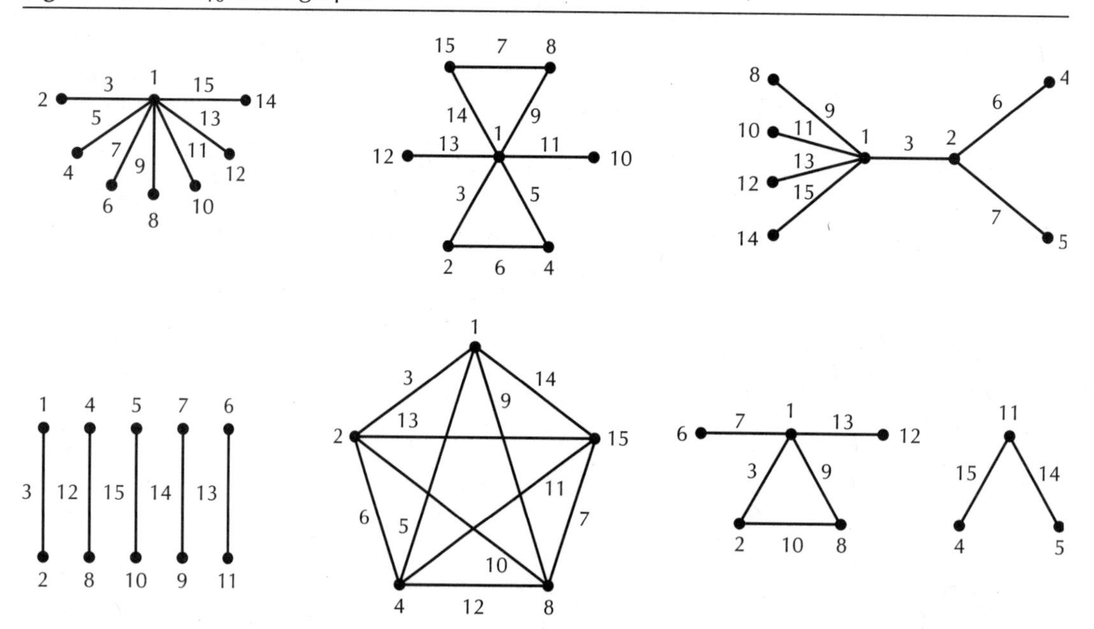

*All linear graphs are used with permission from the American Supplier Institute.

main effects with two-way interactions (that is, interactions between two factors). We will use a linear graph [R13.4] to show where interactive parameters should be assigned for the two-level arrays.

The linear graph [R13.4] for the L_4 array is shown in Figure 13.14.

This linear graph states that to study the interaction between parameters 1 and 2 you must leave column 3 empty. If column 3 is occupied with a third parameter and there is an interaction between parameters 1 and 2 it will be confounded with the response due to parameter 3.

The linear graphs [R13.4] for the L_8 are shown in Figure 13.15.

The linear graphs for the L_8 show where four parameters can be placed, in columns 1, 2, 4, and 7, to allow the three interactions, in columns 3, 5, and 6, to be safely evaluated.

The linear graphs for the L_{16} [R13.4] are shown in Figure 13.16.

Defining the Degrees of Freedom Required for Evaluating Interactions

One other issue to explore is how many degrees of freedom are required to study, or model, an interaction. Remember, for each unknown quantity that is calculated in a matrix experiment, one treatment combination must be run. To describe the interaction between two two-level factors, it is sufficient to find the difference in slope between the two main effects, as shown in Figure 13.17.

In the L_4 example, one two-level column (column 3), with $DOF_f = (2-1) = 1$, is adequate to estimate the interaction between factors A and B. In a two-level array, a column has the same DOF whether it is used to study a parameter or the interaction between two other parameters.

We now have enough background established so that if the team has to, it can set up an experiment to evaluate interactive parameters at two levels. Again, tolerance design is not the best place to study and correct for problematic interactive parameters.

The analysis of the data from fractional factorial orthogonal array matrix experiments will be thoroughly discussed in Chapters 14 and 16.

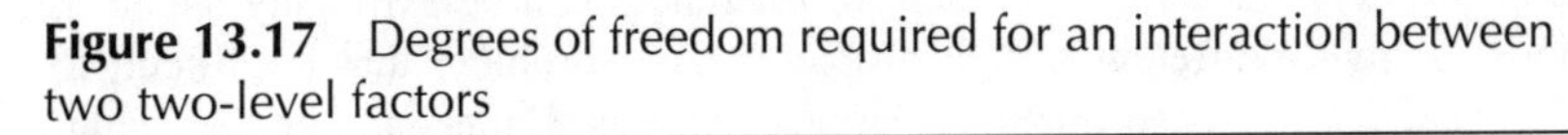

Figure 13.17 Degrees of freedom required for an interaction between two two-level factors

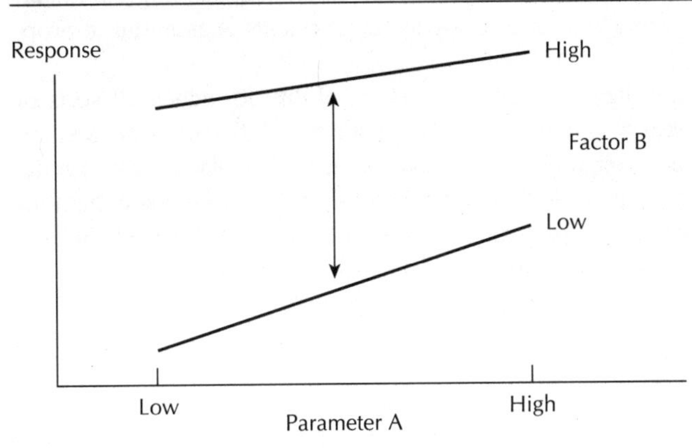

Summary

Chapter 13 is an introduction to the discipline of *designed* experiments. The topics discussed were tailored for their specific application to tolerance development. Several approaches to experimentation have been discussed. The power and efficiency of fractional factorial orthogonal arrays has been illustrated.

The topic of degrees of freedom has been developed to show how the size of an experiment is related to the ability to quantify the main effect of numerous design parameters. We have also discussed the nature of parameter interactions and how to approach their quantification. We have shown how to make room in an array to evaluate interactions through the proper allocation of experimental degrees of freedom.

The following outline reviews the material covered in Chapter 13:

Chapter 13: General Review of Orthogonal Array Experimentation for Tolerance Design Applications

- Developing Tolerances Using a Designed Experiment
- Use of Orthogonal Arrays in Tolerance Design
- The Build-Test-Fix Approach
- Introduction to Full Factorial Experiments
- Designed Experiments Based on Fractional Factorial Orthogonal Arrays
- Degrees of Freedom: The Capacity to Do Experimental Analysis
- Degrees of Freedom for the Full Factorial
- DOF for the Fractional Factorial Orthogonal Array
- The DOF for Two-Level Full Factorial Arrays
- Methods to Account for Interactions within Tolerance Design Experiments
- Interactions Defined
- Quantifying the Effect of an Interaction
- Defining the Degrees of Freedom Required for Evaluating Interactions

The use of designed experiments in the development of tolerances can be extremely useful in optimizing the specifications for many different design problems. Any problem that is difficult to model analytically is a good candidate for experimental tolerance analysis. Understanding the fundamentals of planning, constructing, running, and analyzing designed experiments is essential to properly developing tolerances in modern industry.

A unique feature of experimental optimization techniques is the ability to induce all sorts of noise in the experimental runs. This provokes data that is extremely realistic relative to what can happen in the manufacturing and customer-use environments. The next chapter will introduce the specific methods available to properly identify significant noise factors, and will show how to prepare them for use in a tolerance experiment. Tolerances that help keep a design robust for extended periods of time have a direct impact on reliability, as can be seen in the final case study in Chapter 18.

References

R13.1: *Statistics for Experimenters;* Box, Hunter & Hunter; J. Wiley and Sons, 1978, pg. 478

R13.2: *Quality by Experimental Design* 2nd Ed.; Thomas B. Barker; Marcel Dekker, 1994, pg. 478

R13.3: *Engineering Methods for Robust Product Design;* W. Y. Fowlkes & C. M. Creveling; Addison-Wesley, 1995, pg. 480

R13.4: *Orthogonal Arrays and Linear Graphs;* G. Taguchi and S. Konishi; ASI Press, 1987, pg. 493

Additional references for experimental design topics can be found in Appendix D.

CHAPTER 14

Introducing Noise into a Tolerance Experiment

One of the most powerful features of the Taguchi design process is the introduction of stressful noise into the tolerance analysis experiments. There is no substitute for actually forcing what *can* happen in the manufacturing and use environment to *happen* in the product-development hardware. This is done to aid in the reduction of sensitivity to sources of variation during a product's development. This chapter is concerned with properly identifying and applying stressful noise into the tolerance optimization experiments. All the *critical* noise factors for a component, subsystem, and product need to be identified and scrutinized for their strength and directional effect on the output response. Once this is completed, they can be grouped into two categories so that the design can be thoroughly and bidirectionally stressed in an efficient manner. This technique for putting stress into an experiment is called the *compound noise factor* approach. Tolerance optimization needs to be done while the critical noises are present so that the tolerances are developed to maintain proper output response values. This is another strategy for keeping customer responses on target.

Defining Noises and Creating Noise Diagrams and Maps

Noise, as we know, is any source of variability. The first step in properly stressing a design is to identify the many noise factors that fall under the following categories:

1. Environmental noise
2. Unit-to-unit noise
3. Deterioration noise
4. Surrogate noise

The first category is *external noise,* which refers to sources of variability that are external to the system; that is, completely outside of our control when the product is used as intended. External noises include:

- Environment; temperature, relative humidity, and altitude.
- Duty cycle or load; that is, how much and how often the customer uses the product.
- Set-up conditions, or what optional choices in a complex device are selected by the customer.
- Consumables chosen by the customer for input to the system, such as different papers used in a copier or different fuels, lubricants, or coolant used in an engine.
- Foreseeable misuse and abuse.

These are all examples of factors that can be controlled and simulated in the lab, but that cannot be controlled in actual use without adversely impacting customer satisfaction or adding cost. Many customers do not like to be told what special conditions need to be followed to have proper functionality from their new purchase.

The subsystems in a complex system such as office copier machines and automobiles have another form of environmental noise referred to as *proximity noise.* In this case the noises are external to the subsystem, such as the unwanted effect of one subsystem on another. For example, the heat emitted from the subsystem in a high-speed copier that fuses toner to paper by the application of heat and pressure can influence the performance of neighboring subsystems. While not external to the system, the proximity effect of other subsystems is an important issue each subsystem team should consider when developing a list of noises.

The second category is *unit-to-unit variation,* an internal noise often mislabeled as simply tolerance variation (remember, a tolerance is not a measure of variation, it is a measure of how much variation is allowed before a part is rejected or adjusted). This is the inevitable variation caused by noises in the manufacture and assembly of the component, subassembly, or total product. There is always some variation in parts and manufacturing processes that can cause the performance of the final product to vary. Additionally, there are going to be variations induced by how an assembler or service technician puts components and subassemblies together and then makes adjustments. Some examples are:

- Dimensional variation in mechanical parts, which can stack up in an assembly.
- Electrical component variation.
- Process parameter variation (temperature, speed, time, pressure, and so forth) in setting up a manufacturing device such as a mold injector or chemical reaction.
- Variation in rheological properties of polymers used in a product.
- Variations in product-controlled parameters, such as hydraulic pressure in a machine or ignition timing in an engine.

These are all parameters that vary from unit to unit, often in a way that can be described probabilistically by some statistical distribution (see Chapter 8). Generally, the range of variation cannot be reduced without adding cost, in the form of tighter tolerances, lengthier assembly processes, or applications of process control.

The third category is *deterioration noise.* Also an internal noise, it is the inevitable degradation of equipment components and subsystems over time. Some examples are:

- Corrosion, fatigue, wear, and physical degradation.
- The buildup of impurities in recycled materials such as oil in an engine, or chemicals (including mold injection materials) in a manufacturing process.

- Loss of volatile materials over time such as plasticizers in an automobile dashboard, or low molecular-weight fractions in a polymer.
- The disordering effects of entropy, such as loss of magnetization in an audio- or videotape recording.

These are all sources of variation that increase over time. We can sometimes influence the rate of degradation (see Chapter 12), but eventually all systems suffer from this type of loss. It is this type of noise that is absolutely critical when using the $r_{S/N}$ factor [R14.1] to relate signal-to-noise metrics to reliability metrics (see Chapter 2). Without the inclusion of time-based deterioration noises in robustness and tolerance experiments, the use of $r_{S/N}$ factor methods is inappropriate. The basis for legitimate $r_{S/N}$ factor analysis comes from very thorough FMECA (failure modes effects and criticality analysis) [R14.2] evaluations that are then used to aid in the construction of component and subsystem noise diagrams and system level noise maps. The noise diagram must contain the time-based deterioration factors that directly affect reliability. We often refer to reliability as robustness that lasts over the life of the product, or at least as robustness that deteriorates at an acceptable rate. In fact, in Chapter 12 we reviewed how to tolerance factors that control the rate of deterioration for a product or subassembly.

The fourth category is *surrogate noises* [R14.1], which can be external or internal. Surrogate noise refers to the measurement of variability without necessarily controlling the cause of the variation. This is similar to experimental error but does not refer to measurement error. For surrogate noises, the process data are varying in a manner that is not directly controlled in the experiment. For example, surrogate noises could include:

- Repeat measurement over time during a process run,
- Sampling from different locations where position can dictate variation, and
- Picking off only the high/low values from a strip-chart recording.

In each case, it may be difficult to specify the noise factor, but reducing the variability would help improve quality.

Surrogate noises can be classified into three groups [R14.4]. Such classification can be useful when trying to be thorough in noise analysis and brainstorming. The three groups are:

1. Positional: unit-to-unit, within a unit, operator-to-operator, machine-to-machine, plant-to-plant.
2. Cyclical: lot-to-lot, batch-to-batch, including hysteresis effects.
3. Temporal: hour-to-hour, shift-to-shift, day-to-day, week-to-week.

These classifications can be used as guidelines for thinking about additional sources of variation.

Selecting Noise Factors for Inclusion in a Tolerance Experiment

The choice of the noise factor treatment is a major part of the tolerance experimental planning process. The design is toleranced only against the types of noises that are chosen or against noises that are related to those chosen. If there are sources of variability that affect the system in ways that are not part of the experiment, the design may be susceptible, even after tolerance optimization. Leaving critical noises out of an experiment can allow for very disturbing results later in the design's life cycle.

The first step an engineering team should take is to identify all the noise factors that can introduce variability in the function of the design. This is best done through brainstorming sessions that thoroughly define and prioritize the noises [R14.3]. The goal is to pick a set of noises that are representative of all the sources of variation that can affect the design's performance. The team needs to interview customers, service personnel, suppliers, manufacturing representatives, and assembly workers. A good brainstorming session generates a long list of noises. Remember, there are many sources of variability. Try to identify the various energy transformations in the system, and make sure you have noises that act as sources of variability for each of them.

Use design input energy transformation considerations to guide your selection. The ideal function of the design should have been reviewed and discussed. Consider all the factors affecting the energy transformations that occur in the system, as well as any possible sources of variation *within* each energy transformation, and what other effects can superimpose themselves on the desired transformation, thus introducing variability. Variability may be a result of an inefficient energy transformation, but it could also be a result of an undesired secondary transformation that superimposes on the primary transformation. When a design converts energy, some of the transformation is going to go into nonuseful forms. Look for the mechanisms that facilitate this inefficiency.

Be sure to consider effects that are internal to the product as well as customer observable effects. For example, photographic film handling is accomplished by pushing or pulling a film strip with friction rollers. The energy transformation is from the roller (rotational kinetic energy) to the film (linear kinetic energy). Variation in the coefficient of friction, roller hardness, or nip pressure can all impact this energy transformation.

List the four primary categories of noise. When the initial discussion begins to wane and no one is producing new noise candidates, step back and look at the list. Classify the noises to see if all the categories are represented. Do this with the entire engineering team; it usually gets people thinking again and additional noises will emerge.

As a final strategy, look for surrogate noise factors—ways to find or measure variability other than directly controlling the sources of variability. Surrogate noises can be very useful and are a practical way to capture variation when the noise factors are hard to control in the lab. The surrogate noise should be chosen to allow measured variation with little additional experimental effort. Avoid using surrogate noises, such as replication, that result in additional experimental runs; it is not efficient. Use the first three forms of noise factors, when possible, to replace random variation with *induced* variation.

Noise Diagrams and System Noise Maps

A helpful technique for identifying noise factors and their relationships to the components, subsystems, and product system architecture is found in the construction of noise diagrams and noise maps. A noise diagram is a representation of how the three major forms of noise (external, unit-to-unit, and deterioration) are associated with a design element. A system noise diagram illustrates all the variational interrelationships that exist between the numerous design elements comprising an entire product system. Figure 14.1 shows a noise diagram and Figure 14.2 shows a system noise map. These documents are often preceded by and subsequently supplemented by failure modes, effects and criticality analysis, fault tree analysis, and Ishikawa diagrams.

Noise diagrams and system noise maps are excellent ways to store the information developed around noise factors and their complex effects on the entire product. When a new version of the

Figure 14.1 Noise diagram for an underwater camera enclosure

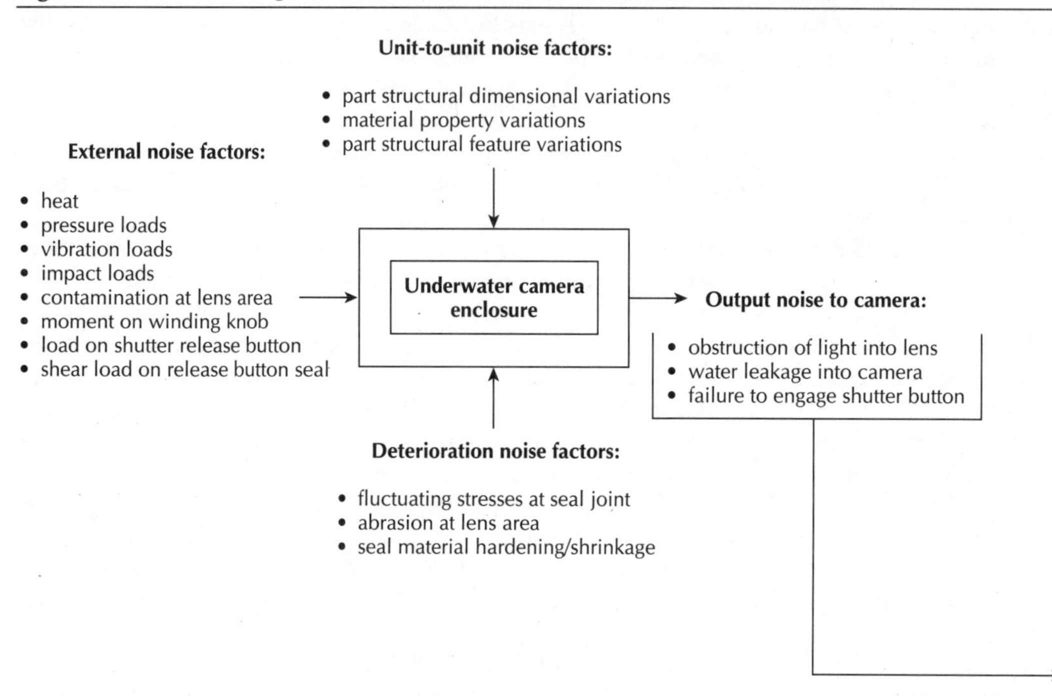

product is developed, these documents can be extremely helpful in the analysis and experimental evaluation activities. Once an engineering organization has worked with a particular technology through numerous commercialization projects, the engineering team will become very familiar with the major sources of noise at work in the product and its manufacturing processes. This familiarity will evolve into a mature capacity to *rapidly* develop new products with appropriate stress testing. Noise diagrams and system noise maps can be upgraded and used repeatedly. They will continually evolve and become a strategic source of engineering information that will provide a competitive advantage over companies that react to noise through ad hoc and iterative build-test-fix methods.

Every link between the components *within* the camera system noise map (Figure 14.2) indicates the presence of unit-to-unit (UU) noise in the form of part dimensional variations, material variations, or part feature variations. Every one of these links represents a required tolerance-analysis area. The system noise map can be very helpful in assessing the specific tolerance-analysis activities required across the entire system. The noise diagram (Figure 14.1) shows the various types of noise that can affect the camera enclosure. For an underwater camera, the enclosure must be designed to be robust against many sources of variation. When parameter design has done all it can do for making the enclosure insensitive to noise, tolerance design must be used to design in the remaining quality through constrained variance reduction for the component parts. Noise diagrams and maps are also of great value in guiding the team when setting up the noise and tolerancing experiments.

A very good source of how to induce noise in experiments can be found in the symposia notes and case studies from all of the past Taguchi symposiums. This information is available from the American Supplier Institute (see Appendix D), and contains the world's best source of examples of noise selection.

Figure 14.2 System noise map for an underwater camera

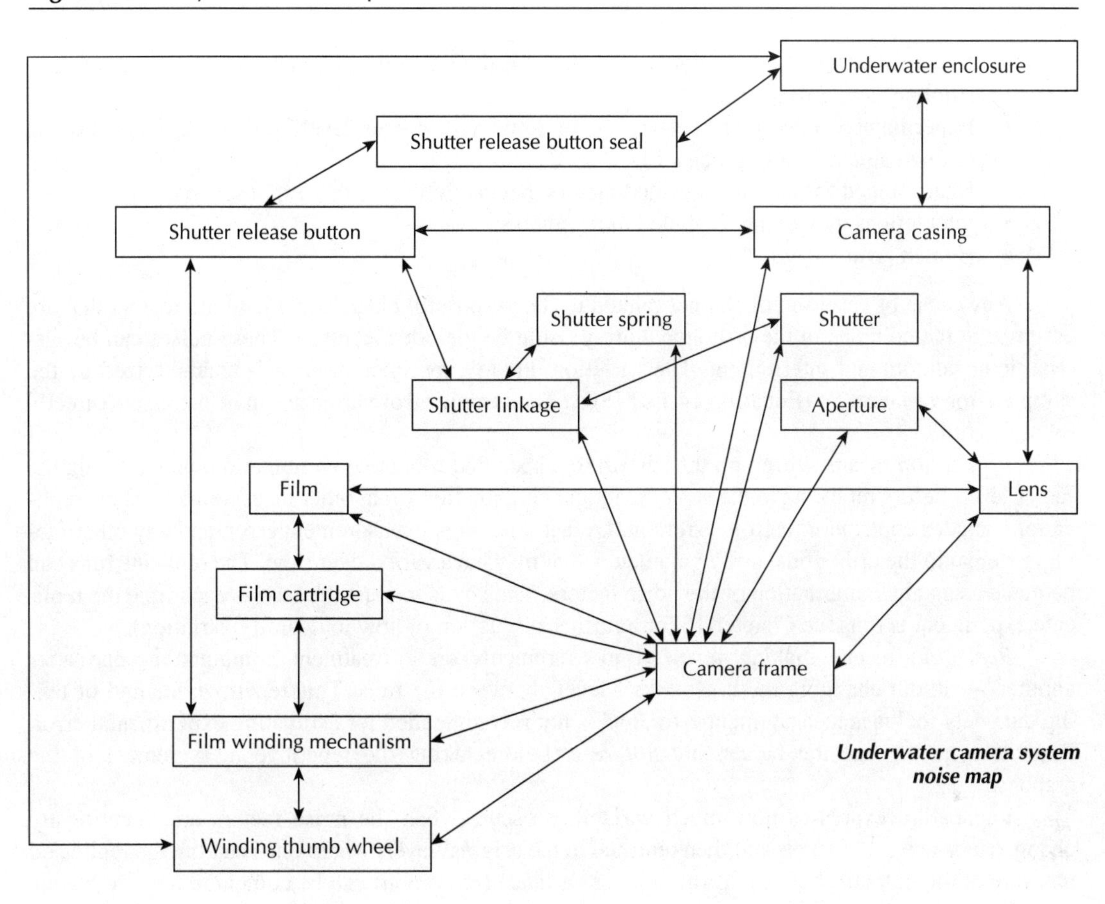

Experimental Error and Induced Noise

Experimental error is referred to by several other names. Statisticians often refer to it as the residual of the experimental data, meaning that portion of the data variation that is not attributable to the effect of any of the main factors in the experiment. Experimental variation in data is quantified using the *mean square deviation due to error* (MSD_e), or the error variance. The concept of the variance is discussed in Chapter 3. The *sample* variance due to experimental error is given by the sum of squares from replicate data points divided by the degrees of freedom allocated (the number of replicates) to quantify variation due only to experimental effects. A replicate will typically be a replicated run from an orthogonal array experiment:

$$S^2_{\text{experimental error}} = \frac{SS}{\text{DOF}_{\text{allocated}}} = \frac{\sum_{i=1}^{n} (y_i - \bar{y})^2}{\text{DOF}_{\text{allocated}}} \tag{14.1}$$

Experimental error is the variation in the data that occurs due to any of the following uncontrolled causes:

- Data measurement noise; that is, variation due to meter error, transducer error, or data-acquisition-system error.
- Experimental noise due to variation in setting the factor levels for an experimental run (human adjustment imprecision).
- Unaccounted for and uncontrolled factors that can influence the response data.
- Interactions that are not included in the analysis.
- Human error.

Any cause of variation that is not related to the purposeful changes made to the factors that are assigned to the columns in the orthogonal array could be included as noise. These noises can be systematic or random and intermittent. The variation due to experimental error is characterized by the sample error variance, S_e^2 (Equation 14.1), MSD_e. The sample error variance can be estimated directly by replication.

Replication means setting up the system to a specified treatment combination, then altering the factor levels before returning to the same treatment combination for another measurement. Thus, replication includes contributions from errors in the factor settings, measurement error, and any other factors external to the orthogonal array parameters that may have varied over time. The replicate runs can be made using any combination of the noise factors. The key is to reset the factors each time the replicate experiment is run (see Chapter 16 for a further discussion of how to quantify variation).

Repetition means making repeated measurements on a treatment combination, one after another—without changing the noise-factor levels between the runs. This repetitive method of taking data only includes measurement error, and is not recommended for estimating experimental error. Repetition means that the factors are *not reset* before taking the repetitive measurement of the response.

Replication expresses how much variability occurs when the noise factors are intentionally changed to a new set of levels and then returned to the original levels where one is taking the replicated measure of the output. The static form of error variance (repetition) can be compared to the dynamic variation (replication) that occurs when noise factors are changed. This comparison is helpful when evaluating the contribution to error due only to measurement instruments.

To identify significant noise factors, one can perform an analysis of variance calculation on the experimental data (see Chapter 16). This analysis produces statistics that can be used to make a comparison between the signals (changes *between* noise factors) and the constant noise *within* the entire data set due to human beings using equipment that is sensitive to noise in a changing environment. It is very important to recognize that the experimental error is assumed to be approximately constant throughout the experimental sessions. The changes between the noise factors are held in contrast to this constant source of variation. This contrast can be expressed in a ratio, using a term called the MSD_{factor} due to the noise factor effects and the term called MSD_{error}. MSD is called the Mean Square Deviation (see Chapter 16) and is a form of variation used in the ANOVA calculation process. The F ratio quantifies the strength of the induced noises relative to the experimental sources of noise:

$$F = MSD_{factor} / MSD_{error}$$

Figure 14.3 illustrates how we seek to distinguish between induced noise in an experiment and random or uncontrollable sources of variation using the MSD_{error}.

Figure 14.3 Forms of error and methods of detection

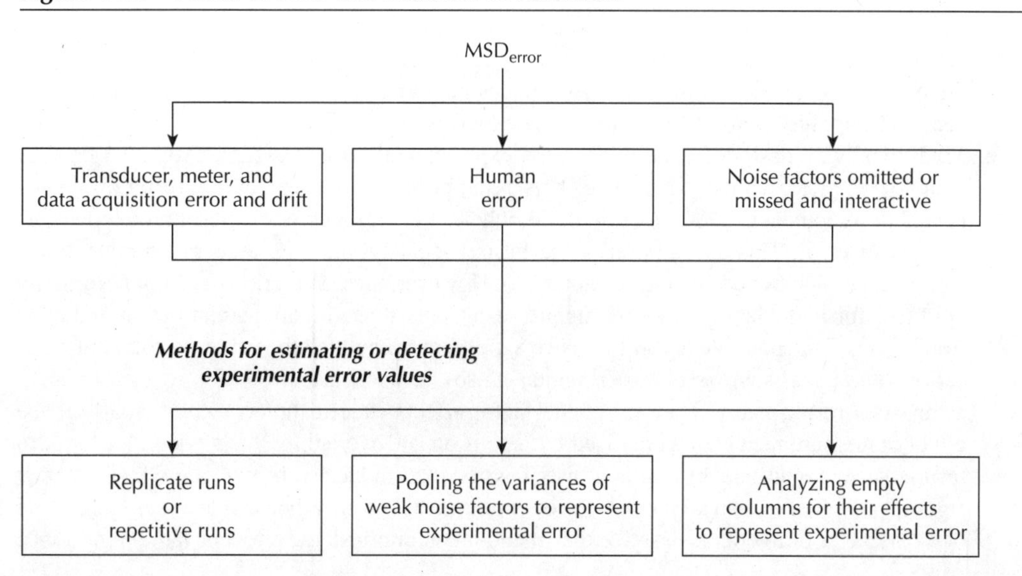

Figure 14.4 Experimental error independent of the factor effects

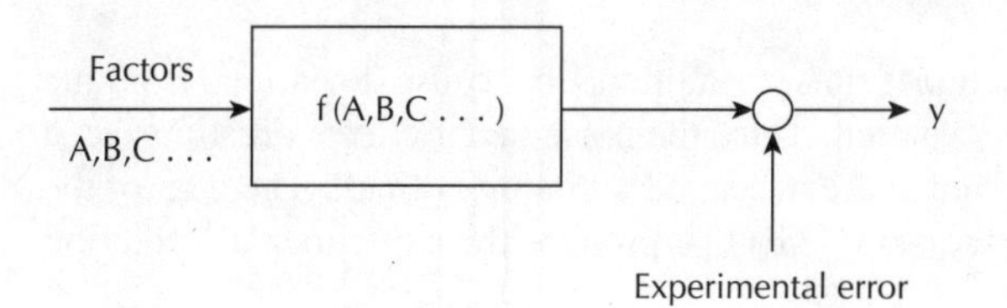

The *F*-ratio, the mean square deviation, and other methods of estimating the error variance are discussed more fully in Chapter 16, on the analysis of variance.

A fundamental assumption of classical DOE is that the same error distribution (variance) applies for the entire experimental space. The situation can be characterized by Figure 14.4, which shows the addition of experimental error as independent of the factor levels and underlying the functional relationship.

Accurate tolerance analysis is achieved through the following approach:

1. Noise factors are used to induce a controllable level of system variation that is separate and distinct from the experimental error discussed so far.
2. Design parameters that are found to be significant in improving robustness during parameter design reduce the level of variation due to noise factors. Unit-to-unit variation is considered a noise in parameter design, and is now considered the main focus of tolerance design; there will still be variation associated with the design parameters. Tolerance design seeks to further enhance quality by adding specific constraints to the design parameter values through

the application of tolerance limits. Thus, one must use the metrics of design parameter variance (in the form of standard deviations) as the experimental control factors in a tolerance design experiment, hence removing them from the category of a noise factor and placing them in category of a control factor for the design. Other noise factors are still in force and need to be applied as stress to the tolerance experiments.

3. Traditionally, in designed experiments, the experimental error is assumed to be larger than many of the factors in the experiment. This assumption, to some extent, is a result of the historical development of DOE in agriculture, where it is impossible to control all the possible sources of error. This assumption is the basis for many analyses used to estimate factor significance such as pooling (see Chapter 16). However, the assumption does not necessarily hold true for many laboratory experiments, especially those involving equipment and electronics. It is quite possible to run extensive experiments in which the variation found in replication (due to *all* sources of experimental noise) is, in fact, entirely due to measurement error. As a result, many if not all of the factor effects exceed the error, particularly if the effect of measurement error is reduced by repetition and averaging. It is a good idea to try to insure such conditions by using noise factors that induce variation at a level that is significantly greater than all the sources of experimental error. When this is done for each run of a parameter or tolerance design experiment, it can diminish the effect of random nuisance factors.

A major goal of the noise experiment is to know what the level of experimental error is and, if possible, find induced noise treatment combinations that can cause variation at a much greater level. Consider Figure 14.5, which compares field variation to laboratory variation (experimental error).

In this typical situation, it is possible to choose noise conditions that cause deviations from the target that are comparable to the range of field variation. Thus, the noise factor effects greatly exceed the experimental error so the contribution of error to the measured variation is small. The goal of the noise experiment is to find compounded noise factors (CNF) that produce the experimental condition pictured in Figure 14.6.

Figure 14.5 Comparison of field distribution and lab distribution

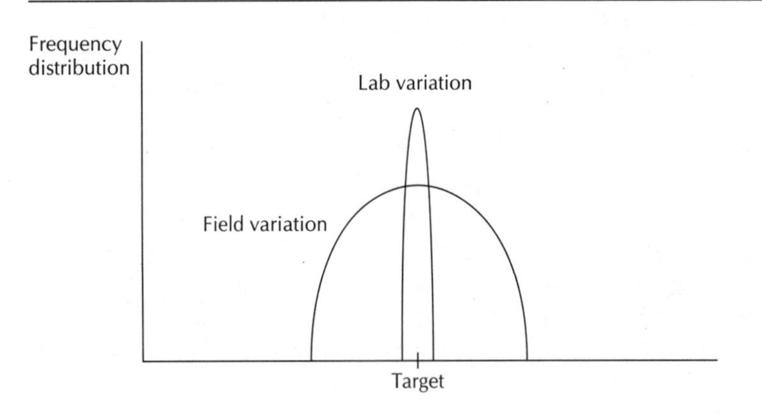

Figure 14.6 Comparison of field distribution and lab distribution with the application of compound noise factors

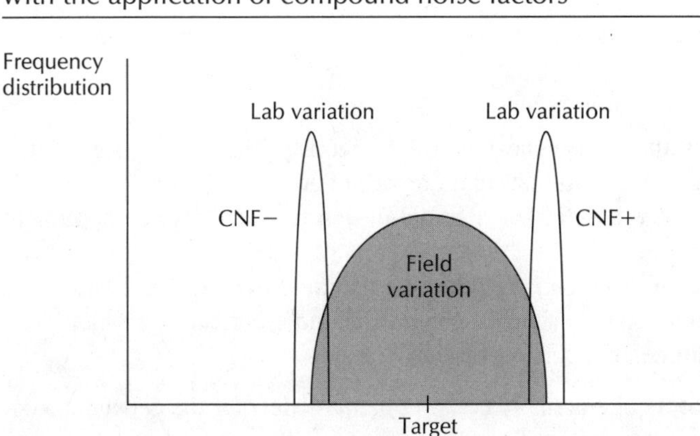

The Noise Factor Experiment

The decision to study noise factors is a serious matter when you are faced with intense pressure to release a design or when you're spending limited resources on testing and experimentation to correct a known problem. Considerable time will need to be spent planning and analyzing the proper engineering output response and understanding the customer observable response and many related noise factors. There is often an urge to rush into testing possible solutions using ad hoc scenarios aimed at getting right to the "obvious" solution. It is hard to justify an effort that does not appear to be leading directly to a solution—but without a disciplined, methodical engineering process, the search for optimal performance is going to be almost totally based on good fortune rather than true insight. The old cliché "the harder I work the luckier I get" is true in design optimization. The more discipline and rigor the engineering team applies to the process steps they employ, the more likely they will develop a truly optimal design. Consider the consequences of using only experience, intuition, and historical skill to speculate on a quick solution that can be built and tested. Anecdotal data and historical experience are only a very small part of the basis upon which a good design must evolve. This approach is not systematic; it sometimes depends on luck to succeed. It does give the appearance of quick response, but even when it is successful, there is no way to know if the solution is optimum. Even more critical is the issue of keeping tolerance problems from recurring. Rework, in the form of having to solve the same basic problems over again, is a significant cause of schedule delay in product development. In fact, it is not uncommon that many of the problems troubling a product at shipping approval have been known and worked on ("solved") throughout the product development process. The use of inefficient approaches such as build-test-fix to solve problems is a major cause of the inability to eliminate problems completely. Inefficient approaches are ones that require an inordinately large number of tests to improve the reliability of the product when the conclusions prove to be inconsistent or fleeting.

The situation is similar with respect to the identification and application of noise. Brainstorming to produce a comprehensive list of noise factors relies on engineering knowledge and experience to predict what the major sources of variability are. While this is a good start, it is inadequate. It is not feasible, due to cost or lack of control, to consider all the noise factors.

Carefully designed noise experiments, which are used to develop the right information in as few tests as possible, are invaluable at preventing oversights and proving assumptions. Only by the application of a disciplined experimental process can one avoid the need to reoptimize due to noises that may have been overlooked.

The noise factor experiment has at least four goals:

1. Identify the few noise factors that cause most of the variability. Minimize effort in the tolerance-optimization experiment by eliminating unimportant noises.
2. Benchmark the performance of the baseline or nominal design (this can also be applied to concept selection).
3. Perfect the experimental procedure by trying it out during the noise experiment. This results in a smoother tolerance optimization experiment, which is the more critical experiment.
4. Define and quantify the magnitude of experimental error.

The goals of the experiment do not require actually creating math models for the dependence of the output response on the noises. Rather, they seek to find the relative significance (magnitude) of the noise factors and the directionality of their effect. The selection of noise factors and levels for the experiment should be guided by the following rule: Do *not* test what you know for sure, but *do* test what you are sure you don't know.

Creating Compound Noise Factors

The brainstorming session can result in a large list of noise factors. It simply is not feasible to run even a fractional factorial (orthogonal array) experiment on a very large number of noise factors. However, existing knowledge of the system may be sufficient to allow some prioritization and grouping of noise factors. Once an exhaustive list of noises exists, the team should rank order the list. The initial list will surely contain some redundancy and separately named noise factors that in fact affect the system in a similar manner.

Factors with known interrelationships can be *compounded* by grouping noise factors that are proved to have a similar effect on the system. For example, if temperature, material, and pressure all can affect the critical dimension of a manufactured part, then those noises may not need to be tested separately. All that is needed is a *magnitudinal* and *directional* knowledge of how they affect the dimension. If high temperature, polymer 1, and high pressure are all known to produce oversized parts,

Figure 14.7 Compounding noises

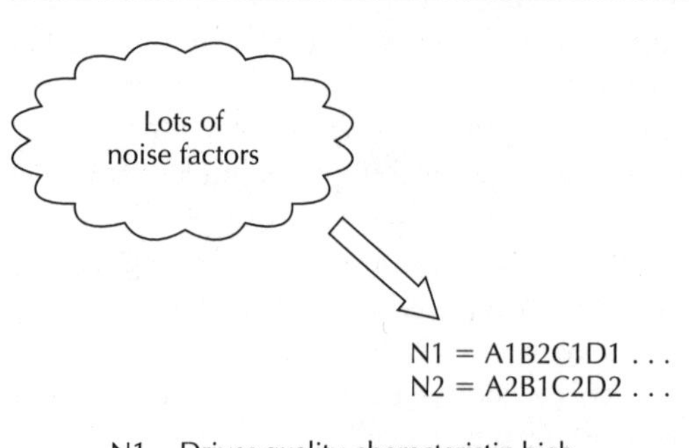

then test those levels together. For undersized parts test low temperature, polymer 2, and low pressure together. Thus, the compound noise factor (CNF), with two levels, can be defined, as shown in Figure 14.7.

$$\text{CNF1 (High)} = \text{temperature (high)} + \text{polymer (1)} + \text{pressure (high)}$$

$$\text{CNF2 (Low)} = \text{temperature (low)} + \text{polymer (2)} + \text{pressure (low)}$$

Setting Up and Running the Noise Experiment

The question of how to examine the effects of noise is best approached by first realizing that the product, subsystem, or component parameters must be set at their *nominal* values during the entire noise factor experimentation process. The noise factor experiment begins with a list of noise factors obtained by applying the process discussed in the previous section. The results of the brainstorming should be collected to produce a candidate list of up to 11 noises. More may require too much experimental effort for studying noises, too few and there is a risk of neglecting important sources of variability. The 11 noise factors can be studied using the Plackett-Burmann array, known as the L_{12} orthogonal array, which allows us to study the noise factors in just 12 experiments. This array is shown in Figure 14.8.

The quantitative effect of any two-way interaction between the factors tested in an L_{12} is more or less uniformly distributed to all columns, as described in Chapter 13. This property makes the L_{12} array ideal for a study of the main effects of the noise factors. The uniform distribution of the interaction effects increases the ability of the main noise effects to stand out without worrying about confounding in particular columns. This acts as a powerful interaction countermeasure because the main effect of any one noise, which was chosen for its presumed large effect on the system under study, is very likely to overwhelm the interactive effects from the other noises. Thus, the noise experiment can be analyzed simply by finding the average response value, through analysis of means, corresponding to each column used in the array.

Figure 14.8 The L_{12} array used for a noise experiment

Run	*1*	*2*	*3*	*4*	*5*	*6*	*7*	*8*	*9*	*10*	*11*
1	1	1	1	1	1	1	1	1	1	1	1
2	1	1	1	1	1	2	2	2	2	2	2
3	1	1	2	2	2	1	1	1	2	2	2
4	1	2	1	2	2	1	2	2	1	1	2
5	1	2	2	1	2	2	1	2	1	2	1
6	1	2	2	2	1	2	2	1	2	1	1
7	2	1	2	2	1	1	2	2	1	2	1
8	2	1	2	1	2	2	2	1	1	1	2
9	2	1	1	2	2	2	1	2	2	1	1
10	2	2	2	1	1	1	1	2	2	1	2
11	2	2	1	2	1	2	1	1	1	2	2
12	2	2	1	1	2	1	2	1	2	2	1

The noise factors chosen by the team, following the process described in the previous section, are assigned to the L_{12}— one to each column. Since there are only 11 columns available, the maximum number of distinct factors to be tested is 11. If necessary, some of the noise factors can be compounded into a single column. Thus, it is actually possible to have many more than 11 factors in the experiment. Compounding results in a loss of information about the *individual* noise factor effects. For that reason, it is important to choose groupings on the basis of common physical effects so that the factors included in a compound noise are synergistic. It is, of course, always possible to leave some columns empty and use less than 11 factors, as illustrated in the next example. For each noise factor, a high and low level must be chosen. Three levels are not necessary because curvature effects are not required, nor do we care about the nominal noise effect in a tolerance experiment. The goal is to find the magnitude of the influence of a noise factor or compound noise factor on the output response, including the direction of that influence.

The team must also consider how the noise factors are to be controlled in the lab. Noise factors should represent conditions that are not controlled in actual product usage but that are controllable in lab testing. This is very important for obtaining reproducible results with a minimum of experimental effort. For example, the operating environment is a common uncontrolled external noise. In lab tests, however, environment can be controlled in an environmental chamber. In this way extreme combinations of humidity and temperature can be used that stress the system and simulate the field variation, but do so in a controlled and reproducible manner.

Analysis of Means for Noise Experiments

The noise factor effects from an L_8 (though you will more often use an L_{12}) experiment are studied using the analysis of means. This reveals which noise factors have a large effect on the value of the output response characteristic. We also seek to identify the *directionality* effects—which noise factor levels make the output response high and which make it low.

Calculating the analysis of means for an L_8 experiment is very easy to do. Figure 14.9 shows an L_8 array. We have chosen to evaluate just five noise factors in this experiment.

Figure 14.9

Run	*1* *A*	*2* *B*	*3* *C*	*4* *D*	*5* *E*	*6*	*7*	Output response values
1	1	1	1	1	1	1	1	y_1
2	1	1	1	2	2	2	2	y_2
3	1	2	2	1	1	2	2	y_3
4	1	2	2	2	2	1	1	y_4
5	2	1	2	1	2	1	2	y_5
6	2	1	2	2	1	2	1	y_6
7	2	2	1	1	2	2	1	y_7
8	2	2	1	2	1	1	2	y_8
	2×3 4×5	1×3	1×2	1×5	1×4	2×4 3×5	2×5 3×4	

The eight output responses ($y_1 - y_8$) contain all the information we need to quantify the average effect for each factor at its high and low levels. To evaluate the eight output responses we follow these simple steps for each level of each factor: For factor A at level 1 we look at the array and find that experiments 1 through 4 contain the four response values ($y_1 - y_4$) that represent what factor A at level 1 is contributing to the overall average response (the sum of y_{1-8} divided by 8). So, we will sum the values of y for experiments one through four and then divide this sum by 4:

$$\text{Average response for } y_{A1} = (y_1 + y_2 + y_3 + y_4)/4$$

The same process is repeated for noise factor A at level 2.

$$\text{Average response for } y_{A2} = (y_5 + y_6 + y_7 + y_8)/4$$

This process is then repeated for each level for each remaining noise factor.

$$\text{Average response for } y_{B1} = (y_1 + y_2 + y_5 + y_6)/4$$

$$\text{Average response for } y_{B2} = (y_3 + y_4 + y_7 + y_8)/4$$

$$\text{Average response for } y_{C1} = (y_1 + y_2 + y_7 + y_8)/4$$

$$\text{Average response for } y_{C2} = (y_3 + y_4 + y_5 + y_6)/4$$

$$\text{Average response for } y_{D1} = (y_1 + y_3 + y_5 + y_7)/4$$

$$\text{Average response for } y_{D2} = (y_2 + y_4 + y_6 + y_8)/4$$

$$\text{Average response for } y_{E1} = (y_1 + y_3 + y_6 + y_8)/4$$

$$\text{Average response for } y_{E2} = (y_2 + y_4 + y_5 + y_7)/4$$

This is all there is to calculating the analysis of means, or ANOM. It works the same way for two- or three-level experiments. The results from the ANOM can be shown in a factor effect plot such as is shown in Figure 14.10.

The ANOM results can be interpreted by inspection of the factor effects plot. The low slope of the factor effects for noise factors B and D indicate that they are not very significant. Factors A, C, and E, however, appear to be important. The compounding of the noise factors is determined by the direction of the noise factor effect *slope*. Thus, in this case, we see that for CNF_+ factor levels A high, C low, and E low (written more compactly as $CNF_+ = A2C1E1$) all work together to cause the output response to deviate in the positive direction. Similarly for CNF_-: factor levels A low, C high, and E high ($CNF_- = A1C2E2$) all work together to drive the output response low.

An *additive model* can be used to predict the level of deviation expected from the compound noise factors. Thus, for the case illustrated here, the predictive equations would be given by:

$$\begin{aligned} y_+ &= \bar{y} + (\bar{y}_{A2} - \bar{y}) + (\bar{y}_{C1} - \bar{y}) + (\bar{y}_{E1} - \bar{y}) \\ y_- &= \bar{y} + (\bar{y}_{A1} - \bar{y}) + (\bar{y}_{C2} - \bar{y}) + (\bar{y}_{E2} - \bar{y}) \end{aligned} \qquad \textbf{(14.2)}$$

Where $\bar{y}$ is the average of all eight of the response values (the sum of y_{1-8} divided by 8) from the experimental runs. We use this additive model to mathematically quantify the magnitude and directionality of a set of compounded noise factors.

Figure 14.10 Compounding of noise factors by ANOM factor effects plot

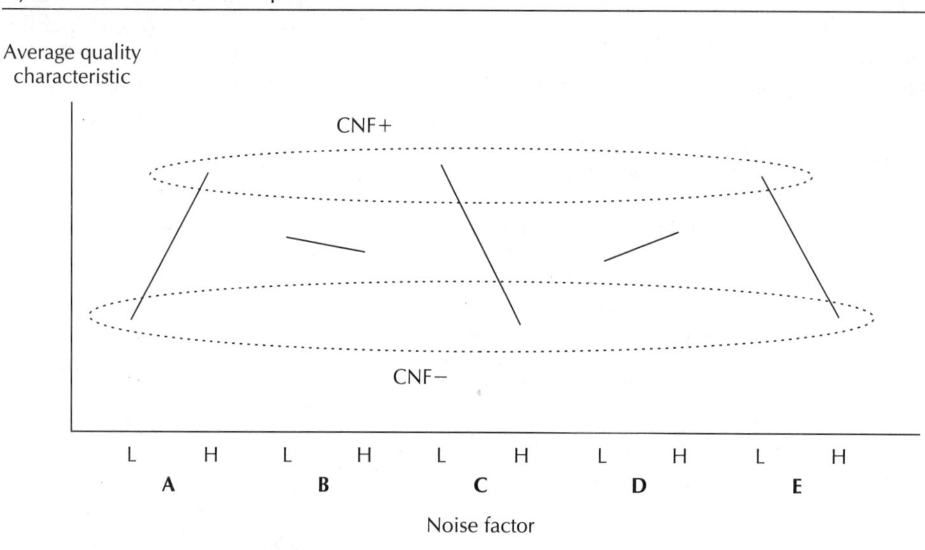

Verification of the Predicted Response

It is important to verify the results of the compounding analysis by testing the two compound noise factors. This is because of the possibility that significant interactions are neglected in the additive model (Equation 14.2). The L_{12} array can act as a countermeasure, but requires more experimental runs than necessary in our case. Nevertheless, if a large number of noise factors are chosen initially, the compounding is likely to be very effective. Remember, the noise factors are chosen because the engineering team predicts that these factors have a significant effect all by themselves. The likelihood of a few interactive effects overwhelming several main effects compounded together is low.

The verification test is done by running the same nominal parameters used for the L_8 experiment at the compound noise factor combinations. If one has not yet performed the parameter design process, the combined effect of all the noises compounded together may cause the subsystem to fail. This is some cause for concern, since it is impossible to analyze the tolerance experiment if several runs do not produce data. The compound noises represent possible combinations that the design will encounter in actual use; if the optimized design cannot handle the noise, there will be a quality issue. Either the design concept must be altered or the product requirements must be reconsidered. Typically, the parameter design process will have settled all the issues around noise susceptibility. If the engineering team has performed parameter design, the team will just be reusing the same noises, except for the unit-to-unit (tolerance) noise.

Usually, the verification test does produce useful data where the output response level for CNF_+ is significantly higher than that for CNF_-. That result alone is adequate, but the values should be checked against the predictive equation (see Equation 14.2). The comparison need not be exact, but if the results compare favorably with the prediction then one can proceed with confidence that interactions between the noise factors are not significant.

Now that the noise factors are understood and compounded, we can focus on how to set up a tolerance experiment.

Summary

In Chapter 14, we have introduced the methods available to the engineering team to drive noise into the designed experiments that will be used to develop tolerances. Without the stressful presence of noise in these experiments, the tolerances will not be developed to support the robustness required to satisfy customer expectations.

The major types of noise have been defined and examples provided. A strategic process has been developed to quantify the magnitude and directionality of the critical noises for any given case. Designed experiments can be used to develop compounded noise factors that can be used to stress a design element undergoing tolerance evaluation. Compounded noise factors are very efficient at inducing stress in a tolerance experiment.

The following outline illustrates the main topics covered in Chapter 14:

Chapter 14: Inducing Noise in a Tolerance Experiment

- Defining Noises and Creating Noise Diagrams and Maps
    - Selecting Noise Factors for Inclusion in a Tolerance Experiment
    - Noise Diagrams and System Noise Maps
    - Experimental Error and Induced Noise
    - The Noise Factor Experiment
    - Creating Compound Noise Factors
    - Setting Up and Running the Noise Experiment
    - Analysis of Means for Noise Experiments
    - Verification of the Predicted Response

The benefits of inducing noise in a tolerance experiment are all centered around the need for creating realistic conditions during the development of tolerances. One of the benefits of using quality engineering from the very beginning of product development is that the noise factors and their compounding structure will be done during parameter design. This means tolerance design can be expedited since the noise experiments will have already been conducted.

Now that we have an appreciation for how to induce stress in a tolerance experiment with a high degree of efficiency, we can move on to discuss the options available to the team for setting up a tolerance experiment.

References

R14.1: *Quality Engineering Using Robust Design;* M. Phadke; Prentice Hall, 1989, pg. 502

R14.2: *Failure Mode, Effects, and Criticality Analysis (FMECA);* R. Borgovini, S. Pemberton and M. Rossi; Reliability Analysis Center; ITT Research Institute, 1993, pg. 503

R14.3: *Taguchi Methods;* G. S. Peace; Addison-Wesley, 1993, pg. 505

R14.4: *World Class Quality: Using Design of Experiments to Make It Happen;* K. R. Bhote; AMACOM, 1991, pg. 504

Additional references pertaining to noise factors can be found in Appendix D.

CHAPTER 15

Setting up a Designed Experiment for Variance and Tolerance Analysis

Generally speaking, tolerance experiments are set up much in the same way as robustness and noise experiments. One similarity is that parameter set points under evaluation are placed into the columns of an appropriate orthogonal array. A difference is that the levels of the parameters are set at the initial *one standard deviation limit* around the nominal set point (we will call this a *statistical variance experiment*). Another, less frequently mentioned and often misunderstood, approach in current experimental tolerance literature is to set the critical parameters at their actual tolerance limits (a worst-case tolerance experiment) that are identified using one or more of the techniques defined in Chapters 4 through 12. In this case, the parameter limits could be the three-sigma limits around the nominal set point.

Statistical variance experiments are reviewed in Taguchi's book, *Introduction to Quality Engineering* [R15.1], as well as in a few other books, the most notable being *Engineering Quality By Design* [R15.2], by Thomas Barker of the Rochester Institute of Technology. Barker presents a very clear description of the proper way to conduct a statistical variance experiment. Neither he nor Taguchi call it by this name, but the process they discuss in their writing is the same process we will be discussing. We will call it a statistical variance experiment because we are experimenting with a statistic that is representative of the manufacturing or assembly variance for each parameter.

In the experimental setup we do not use the variance but rather the *standard deviation,* since it is in the direct units of the parameter being evaluated. In analytical tolerance development the variance *must* be used when adding the effects of variability between numerous design parameters. The same is true in experimental tolerance analysis. Variances are arithmetically additive, while standard deviations are not.

We must use the variance and standard deviation in tolerance experimentation because they are the fundamental physical measures of variation. They are the statistics that are calculated from the measured output responses during the manufacturing or assembly process. They are the statistics that come from the ANOVA data decomposition process. They are the statistics we will use in the variance reduction equations later in Chapter 17. Numerically speaking, everything that is required to work with tolerance analysis is based on the statistics of the variance and the standard deviation.

In a worst-case tolerance experiment, we are using an orthogonal array to search the performance of the design as it is exposed to the various *tolerance* limits. We do not use one standard deviation limits in these experiments—the full tolerance is the object of the evaluation. The experimental runs from the array are put through the rigors of the compounded noise factors from the noise experiment (see Chapter 14) to additionally stress the worst-case conditions for the design. So, in reality, a worst-case tolerance experiment is another form of noise experiment and is to be considered a true *worst-case scenario.* It should be noted that these types of experiments will definitely overstate the variation associated with the output response's distribution if they are run at the three-sigma limits that are traditionally applied as component tolerances (Barker, 1990, p. 115 [R15.2]). A worst-case tolerance experiment can show, much like a traditional worst-case tolerance analysis (Chapter 5), just where the sensitivities lie for each of the parameters and the overall variance of the response. These types of experiments can be conducted *after* the statistical variance experiments are run and adjustments have been made to lower the critical parameter standard deviations as a result of loss function analysis. Thus, they are an option for gaining additional worst-case information once the one standard deviation values for each parameter are placed back into final tolerance form. These kinds of experiments are not valid for refining tolerances, and are not a substitute for a statistical variance experiment. We will discuss worst-case tolerance experiments in more detail later in this chapter.

We are looking for intelligent ways to improve quality while adding cost in a strategic, disciplined manner. In robust design experiments we seek to hold costs constant at an initially determined low level from the output of the concept design process. The premise is to start out with a superior *low-cost* design concept, make it robust through parameter design, and then add cost as necessary at the final stage of development through tolerance design.

Preparing to Run a Statistical Variance Experiment

Statistical variance experiments are normally carried out in two experimental stages. The first experiment is defined by setting each critical design parameter at its $\pm 1\sigma$ values. A two- or three-level array is selected with enough degrees of freedom to hold all the parameters. The experiment need not be saturated. The experimental runs from the orthogonal array are each conducted at each of the two levels of the compound noise factors as defined by the process discussed in Chapter 14.

The second experimental stage employs a verification experiment. It is run to verify the effect of the new *adjusted* sigma levels after they are strategically lowered using loss function analysis.

Selecting the Right Orthogonal Array for the Experiment

Selecting an array for a statistical variance experiment is not difficult. There will be very few times when your selection will be other than a *two-level* array. If the nominal design set point is well characterized at this point in the development cycle, there is no need to include it as a third level. Three-level

Figure 15.1 A tolerance experimental layout with compound noise factors

	Compounded noise factors at two levels	
	N1	N2
Orthogonal array containing parameter levels		

experiments take longer to run and tell you nothing you won't already know about the nominal set point. Special steps must be taken if one chooses to run three levels. We will discuss these shortly. The ANOVA process you will use to evaluate the data is not designed to provide insight into *level* differences (ANOM does this). The ANOVA process can only provide information concerning the effect of changes occurring *between* the individual parameters on the output variation. If there are not any significant interactive relationships between the parameters being toleranced, it is a simple matter of picking a two-level array that has enough degrees of freedom to handle all the parameters you wish to evaluate.

Since a two-level array has one DOF available for each column, it is very simple to match the number of parameters with the number of columns in an array. If you have two or three parameters use an L_4. If you have three to seven use an L_8; eight to 11, an L_{12}; nine to 15, an L_{16}; and for more than 15 use the L_{32}. These recommendations change if you want to study interactive effects. We have already demonstrated how to load an array when it is necessary to evaluate an interaction (Chapter 13). By this point in a design project one really should not be doing interactive evaluations. That is an issue that should be heavily evaluated in concept design with central composite DOE's and resolution-4 or -5 orthogonal arrays (see Box et al. 1978 and Barker 1994 [R15.3 and 4]). Some interaction analysis may be necessary during robustness optimization, but only when you are working with an unknown design with respect to interactivity. Tolerance design is not a good place for a first encounter with interaction evaluations. If it is the first time a product has been put through the rigors of a designed experiment, you will have a good reason to look for the interactions—better safe than sorry. Again, the most effective place to study interactivity between parameters is in concept design.

If the team is using a designed experiment to evaluate a set of parameters for the first time, and they are going to use the one standard deviation limits around the nominal as the parameter-level set points, then it is appropriate to consider using a three-level design. This will enable the team to learn the most from the experiment. In this case you would want to perform two types of analysis. The first type is called the analysis of means (ANOM), as shown in Figure 15.2. It allows you to make plots based on the mean output response relative to each parameter level, and will show any curvature in the parameter effects on the output (see Chapter 14).

The second type of analysis is ANOVA (see Chapter 16). This data analysis procedure, as we have stated, will show the relative contribution each parameter has on the variation in the output

Figure 15.2 Plots from an analysis of means

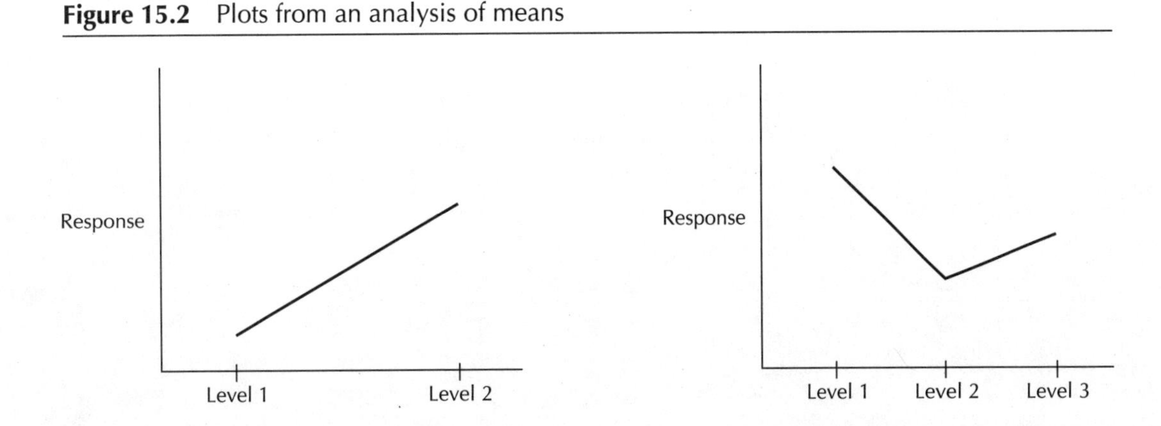

response. Some experiments can be quite large and take significantly more effort to develop and run when including parameter interactions. To evaluate an interaction between two parameters in a three-level design, you must sacrifice (leave empty) *two* columns as defined by the interaction table or linear graph for the array (see Fowlkes and Creveling [R15.5] for a complete explanation of studying interactions in a three-level experiment). There are some special precautions you will need to take to use a three-level design to evaluate tolerance parameters set at some value related to the standard deviation. We will present them in the next section.

Special Instructions for Running Three-Level Statistical Variance Experiments

Statistical variance experiments are analogous to statistical RSS tolerance analysis, because both are based upon the use of the variational statistic sigma. For a statistical variance experiment, when the $\pm 1\sigma$ parameter set points and the nominal set point are assigned to a three-level array, the standard deviation values *must be modified.* This portion of Chapter 15 is considered a special topic and is presented for those interested in the deeper topics available in designed experiments as applied to tolerance analysis. The material from this point on is optional and is included to make the text thorough as a handbook for the practicing engineer, designer, or technician.

The three-sigma paradigm states that the parameter tolerances are to be set at plus and minus *three* standard deviations away from the nominal target set point of the parameter. If you have broken out of the three-sigma paradigm and have tolerances based on Cp's greater than one, or perhaps based on the loss function, then the two extremes of the three levels still need to be assigned based on an adjusted value of the standard deviation associated with the nominal parameter set point of the component. The engineering team must be very clear on how a designed experiment impacts tolerance analysis in an experimental context.

When we are running a designed experiment we are sampling, in a specific and limited manner, from a nonnormal distribution. If we are not careful, we will make our data appear as if it came from a broader distribution than might otherwise be expected with real production processes. This can make the values appear much different than they really are.

If we take a sample of components from a real manufacturing process, it should be drawn at random so that the parts we measure are independent from one another. Let's say we draw 50 parts from a

Figure 15.3

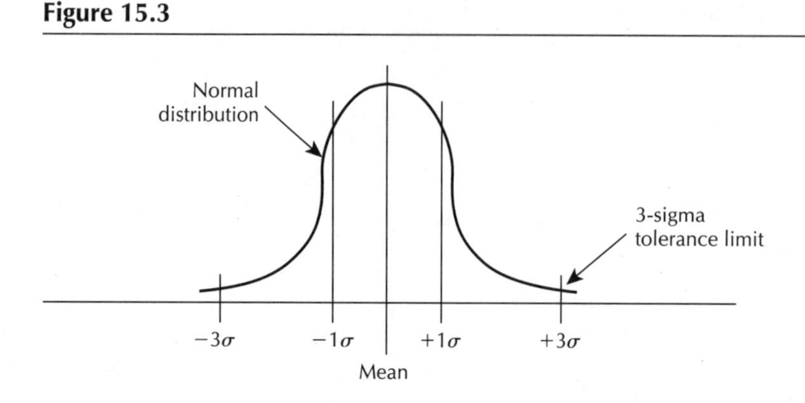

machining process and measure a critical dimension that is supposed to be one inch (the nominal target). We will find that the 50 part dimensions actually form a normal distribution (assuming the machining process produces parts that are normally distributed about an average value). We can obtain two important statistics from the sample: the mean and the standard deviation around the mean.

Figure 15.3 displays these statistics along with another very important set of values—the three-sigma limits that are normally assigned as the part tolerances. If we conducted a tolerance analysis out in the real world, it is this process of randomly sampling from a normally distributed group of components and the quantification of the sample statistics that would guide us. In this book, we have covered using Monte Carlo simulations (Chapter 8) and designed experiments (Chapter 17) to speed up the process of tolerance analysis through the use of computer-aided tolerance analysis. If we know the standard deviation and type of output distribution (usually normal) associated with critical dimensions on various components and the assembly math model, then we can go directly to Monte Carlo simulation and simulate many thousands of random samples of the output values for the assembly of the various parts. From this simulation we can create an assembly output *distribution* to discover output statistics about the variability in the assembly.

Sometimes the Monte Carlo approach is not a viable option because we don't have an accurate math model to use in the simulation. In these cases we must use a designed experiment to do the simulation. So, Monte Carlo and designed experiment simulations are very much related. The problem is that running a Monte Carlo is straightforward because we can make it sample randomly from whatever distribution we choose, but a designed experiment cannot sample randomly, nor can it use any other distribution but a biased version of the uniform distribution that is designed into the array by its orthogonal structure. The very thing that makes a balanced array efficient and capable of disciplined data accumulation makes it very constrained in the nature of its output distribution characteristics. Barker (1990) [R15.2], describes the situation as follows:

> *. . . the designed experiment has a structure and each of the levels or conditions for each factor appears systematically. We need this system for the analysis , but such a systematic variation in the factors will never take place in reality. We would expect a random, possibly normal distribution of conditions. . . . These normally distributed inputs would then lead to the desired normal distribution of the output.*

When a three-level experiment is set up with the $\pm 1\sigma$ values and the nominal set point for the parameters used as the input, the situation is not representative of an input from a normal distribution.

In fact, the input is representative of a biased form of a uniform distribution, because the array forces the occurrence of the three input levels (nominal $-\ \sigma$, nominal, and nominal $+\ \sigma$) to be *equally* likely. An unbiased uniform distribution would be composed of an unlimited range of equally likely values, *not just three,* as we see in the use of a three-level array. Another way to handle such a problem would be to use a discrete distribution to analyze the effect of using three levels in a designed experiment. We will see an example of this later in this chapter.

The three values assigned to the levels in the orthogonal array are the *only* values that the design parameters can assume during the experiment. In the real manufacturing and assembly processes the parameters assume a continuum of values that are going to follow a normal distribution (see Figure 15.4). We can make an adjustment to the input standard deviations to make them create output data that appears as though it was created from normally distributed inputs rather than inputs from uniform distributions. Applying this adjustment corrects for the structure and constraining nature of using a designed experiment to efficiently generate tolerance data. Remember that any value from a uniform distribution is just as likely to occur as any other. A normal distribution has a very distinct, symmetrical, bell-shaped bias associated with the random likelihood of any value occurring. Because the likelihood of a parameter assuming a value away from the targeted nominal *diminishes* with a normal distribution, but not with a uniform distribution, we have to make an adjustment to the standard deviation used in a three-level experiment. We do this so that the output distributions from either a uniformly distributed set of parameters or a normally distributed set of parameters generate essentially the same output distribution and statistics. We will illustrate how this works in detail shortly. It is important that you understand how employing designed experiments *correctly* can greatly aid in the development of tolerances.

Figure 15.4 shows how the uniform and normal distribution are constrained by one-sigma limits and three-sigma production tolerance limits, respectively, for what range of values can be used as input to a designed experiment or the Monte Carlo simulation.

When we use a $\pm 1\sigma$ limit on a uniformly distributed input parameter we are constraining the input values to the values at or inside of the -1σ and $+1\sigma$ lines shown in Figure 15.4. In a designed experiment, only three discrete (nonrandom) values from within this range are allowed to occur (mean $-\sigma$, mean, and mean $+\sigma$). A production sample or a Monte Carlo simulation will draw random values

Figure 15.4

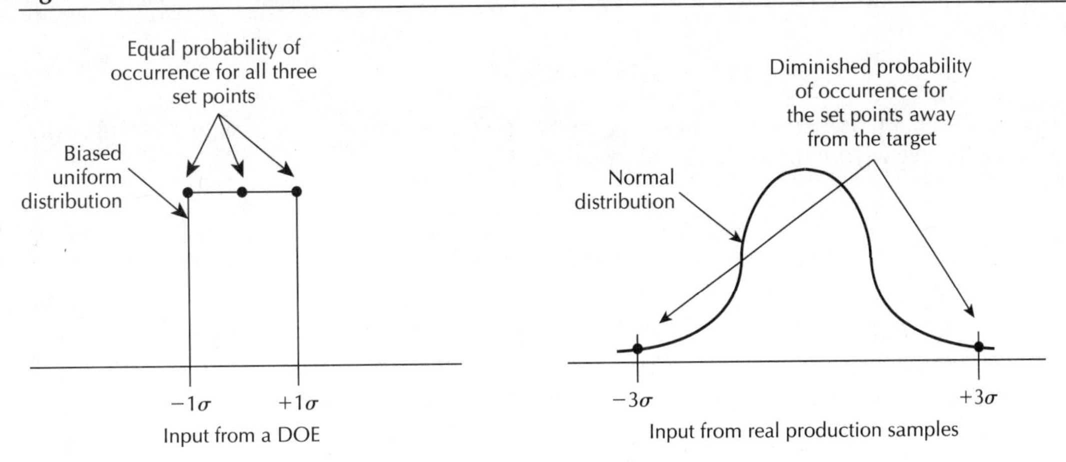

from a continuum *within* the three-sigma limit range. If you want to force the uniformly distributed parameters, limited to just three levels, to produce output responses that looked like those coming from normal, continuous distributions, then the uniform distribution must be expanded into a broader range of allowable values. Only then can it supply the appropriate input values that will mimic normally distributed inputs.

Just how much wider should we make the uniform distribution? What mechanism and metric do we use to do this? We are going to use the standard deviation as the metric for adjusting the set points of the uniform distribution. The mathematical explanation of how to apply an expansion factor to the parameter standard deviation to make the output distribution appear as if it came from a normal input distribution is shown below [R15.2]:

The population variance is defined by:

$$\sigma^2 = \frac{\sum (x - \mu)^2}{N} \quad \textbf{(15.1)}$$

We can define any deviation from the mean (μ) as Δ. This will represent the limits, in units of standard deviations, for the parameter set points in the three-level designed experiment. Equation 15.2 shows the three levels from the array as they form an expression for the variance.

$$\sigma^2 = \frac{\sum ((\mu - \Delta) - \mu)^2 + (\mu - \mu)^2 + ((\mu + \Delta) - \mu)^2}{3} \quad \textbf{(15.2)}$$

Condensing the expression by canceling out the mean terms we get:

$$\sigma^2 = \frac{(-\Delta)^2 + (+\Delta)^2}{3} \quad \textbf{(15.3)}$$

The variance is now shown to be 2/3 of the square of the tolerable limits:

$$\sigma^2 = \frac{2(\Delta)^2}{3} \quad \textbf{(15.4)}$$

Taking the square root of both sides we find that the value for the standard deviation must be the square root of 2/3 times the tolerable limit.

$$\sigma = \sqrt{\frac{2}{3}}\,(\Delta) \quad \textbf{(15.5)}$$

Putting the expression in terms of Δ, with respect to the expanded number of standard deviations, we find the required expanded standard deviation to be . . .

$$\Delta = \sqrt{\frac{3}{2}}\,\sigma \quad \textbf{(15.6)}$$

Thus, the deviation from the mean for the two levels away from the nominal target must be represented by **1.2247**$\boldsymbol{\sigma}$—not 1σ. This is the appropriate correction we must use to force the uniformly distributed input parameter levels to emulate inputs from normal distributions in three-level experiments.

For a two-level experiment ($N = 2$) we can repeat the math and find that no correction is necessary:

$$\sigma^2 = \frac{\sum ((\mu - \Delta) - \mu)^2 + (\mu - \mu)^2 + ((\mu + \Delta) - \mu)^2}{2} \tag{15.7}$$

Again, condensing the expression by canceling out the mean terms we get:

$$\sigma^2 = \frac{(-\Delta)^2 + (+\Delta)^2}{2} \tag{15.8}$$

$$\sigma^2 = \frac{2(\Delta)^2}{2} \tag{15.9}$$

$$\Delta = \sqrt{\frac{2}{2}}\,\sigma \tag{15.10}$$

For a two-level experiment, the output distribution will look as though it came from a series of parameters that are varying according to a normal distribution. We do not need to adjust the standard deviation values for parameters run in two-level experiments. In the three-level case, if one does not apply the 1.2247 correction value to the input standard deviations for the parameters, the same probability is being assigned to the extremes that are being assigned to the nominal. This, we know, is simply not how the real world works. For a one-sigma uniform distribution case, this means one would *underestimate* the actual variability that would be expected from normal input distributions. For the three-sigma case, where the tolerances are based on three standard deviations, one would greatly *overestimate* the nature of the input variability but would have a real simulation of a worst-case condition (which we know almost never happens).

We will use *Crystal Ball* Monte Carlo simulation software to demonstrate (approximately) the effect on an output distribution created by three inputs from biased uniform distributions versus three inputs from normal distributions. The following three-output distribution graphs are created by three input parameter distributions being constrained by uniformly distributed inputs at one-sigma limits to *approximate* a designed experiment, uniformly distributed inputs set at 1.2247σ limits to make the uniform distribution from a designed experiment emulate a normal distribution, and normally distributed inputs set at three-sigma limits to emulate a random production sample. We will then conduct two of the same three simulations using a three-level L_9 orthogonal array experiment. We will compare the results to illustrate exactly what is going on in both the Monte Carlo and the designed experimental simulation. You will then see first-hand why the 1.2247σ conversion must be done with three-level experiments.

The example will focus on three blocks that are being machined such that the measured standard deviation is 0.000333 inch. The three input parameters each have nominal target dimensions of 1 inch. The three-sigma tolerances are plus or minus 0.001 inch. The real production output response is the sum of the three input parameters.

The three output distributions, which will be shown shortly, are the outcome of various changes made to the constraints (1σ, 1.2247σ, and 3σ) limiting the values that the three parameters can attain during the running of a 5,000-sample *Crystal Ball* Monte Carlo simulation. Figure 15.6 demonstrates how the process of using an L_9 designed experiment will generate data that will appear as a normal output distribution.

Figure 15.5 Real production output distribution

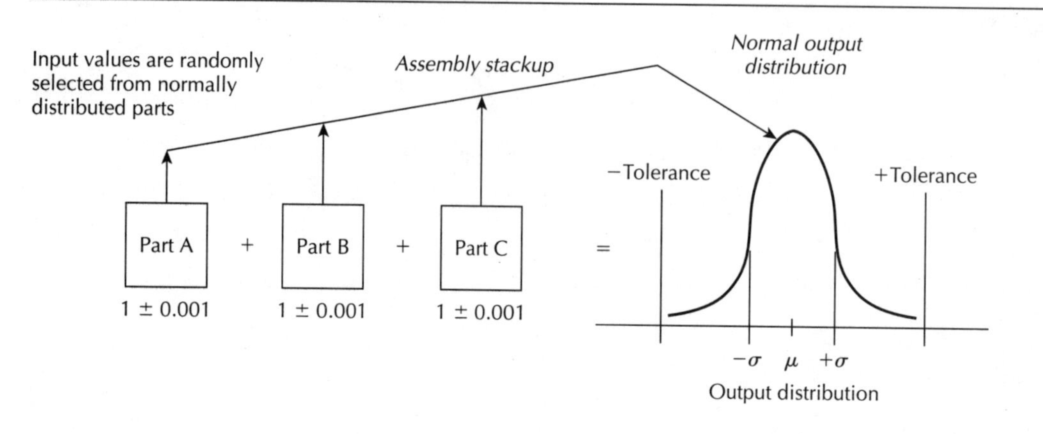

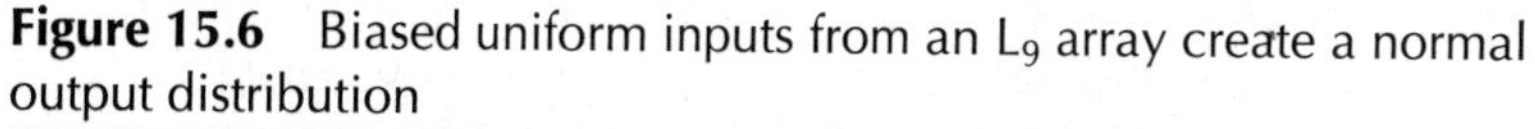

Figure 15.6 Biased uniform inputs from an L_9 array create a normal output distribution

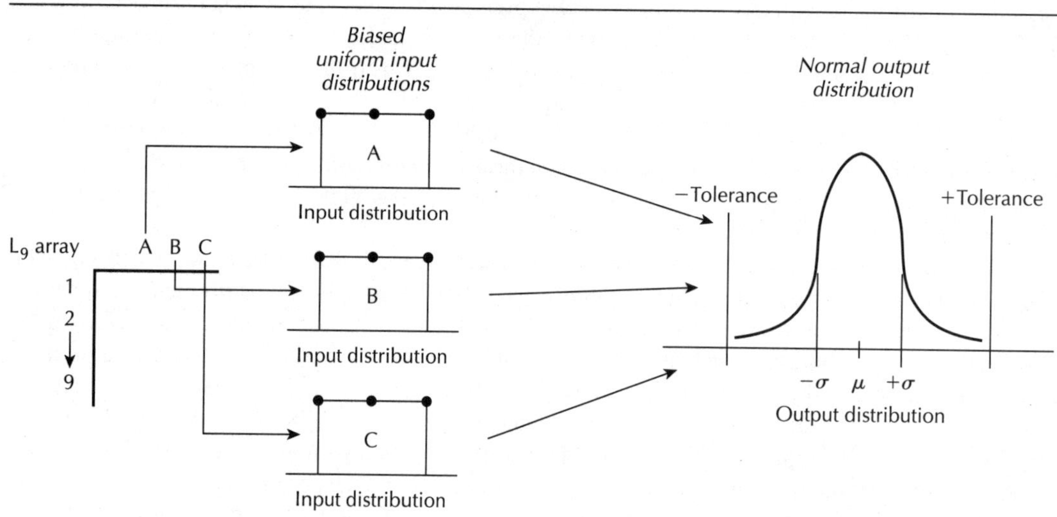

Let's compare how Monte Carlo and designed experimental simulations create statistics that characterize the nature of variation in a tolerance example. The following distribution illustrates the output from a Monte Carlo simulation for three parameters (A, B, and C) that are input as *normal* distributions set at their three-sigma limits: The part tolerance = .001 inch and sigma = .000333 inch for each part.

The statistics associated with the Monte Carlo output distribution are as shown in Figure 15.8.

This data is representative of what we would expect from a manufacturing process capability analysis. The manufacturer would make 5,000 parts according to the specification limits (1.000 inch $\pm 3\sigma$ = .001 inch). We know the manufacturer screens 100 percent of the parts so we would find no parts beyond the three-sigma tolerance limits. This is exactly what the distribution in Figure 15.8 shows. Can we run a designed experiment to emulate this same scenario?

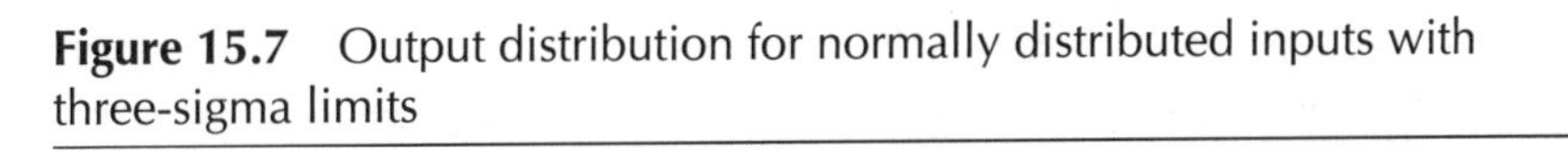

Figure 15.7 Output distribution for normally distributed inputs with three-sigma limits

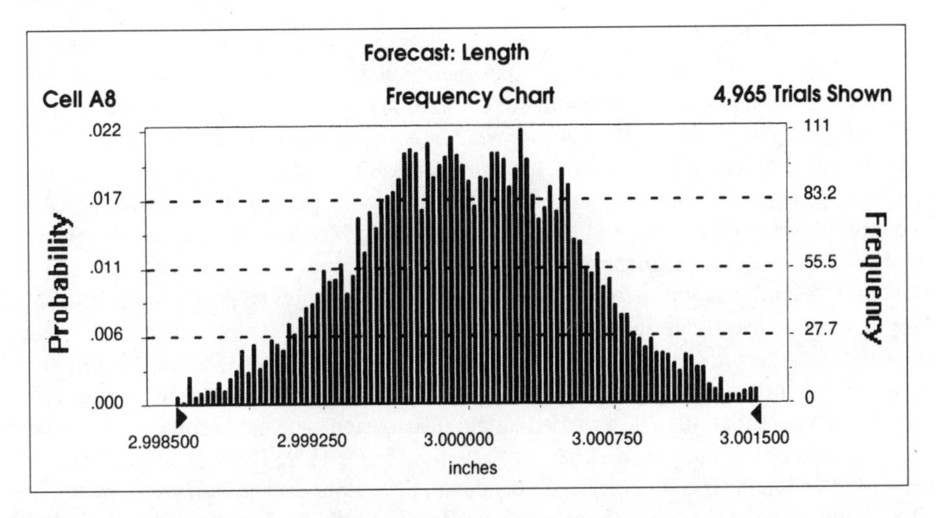

Figure 15.8 Output response statistics for normally distributed inputs with three-sigma limits

Forecast: Length

Edit Preferences View Run Help

Cell A8 Statistics

Statistic	Value
Trials	5,000
Mean	3.000000
Median (approx.)	2.999997
Mode (approx.)	3.000285
Standard Deviation	0.000573
Variance	0.000000
Skewness	-0.01
Kurtosis	2.94
Coeff. of Variability	0.00
Range Minimum	2.997778
Range Maximum	3.002249
Range Width	0.004471
Mean Std. Error	0.000008

It is not possible to run a designed experiment that is comprised of random samples from a normal distribution, as we did in this Monte Carlo simulation. The L_9 experiment is capable only of providing nonrandom inputs from a uniform distribution. We will have to move on to the next Monte Carlo simulation, where we will use *uniform* distributions set at one-sigma limits to further develop this illustration. We will then set up an L_9 designed experimental simulation for uniformly distributed input values that are set at one-sigma limits to compare to the Monte Carlo simulation output.

The next Monte Carlo output distribution shows what happens when the three input parameters are represented by biased *uniform* distributions set at one-sigma limits. We are doing this with the Monte Carlo simulation to approximate how a designed experiment will simulate the sampling process and construct a distribution with its associated output statistics. The Monte Carlo simulation cannot exactly duplicate how a designed experiment will select and deploy sample values in the simulation. Monte Carlo simulations sample randomly from a *continuum of infinite values* from whatever distribution they are programmed to use. A designed experiment uses nonrandom samples that are usually set at *either two or three discrete levels.* Designed experiments can represent only biased uniform distributions because of their orthogonal (balanced) structure. These are very big differences in how samples are deployed into a simulation to describe an output distribution. The next Monte Carlo simulation we will use to emulate a designed experiment is shown in Figure 15.9. The one-sigma limits = 0.999667 inch and 1.000333 inch. The statistics associated with the output distribution are as shown in Figure 15.10. The output statistics show that the mean values for the normally distributed input case (at three-sigma limits) and the uniformly distributed input case (at one-sigma limits) are the same. The standard deviations for the two distributions are quite different:

$$\sigma_{\text{normal inputs}} = 0.000573 \qquad \sigma_{\text{uniform inputs}} = 0.000337$$

This indicates that the uniformly distributed inputs are *understating* the true output-response variability. This shows the fundamental problem of working with designed experiments to simulate the effect of varying tolerances that have been reduced to the one standard deviation limits. The Monte

Figure 15.9 Output distribution for uniformly distributed inputs with one-sigma limits

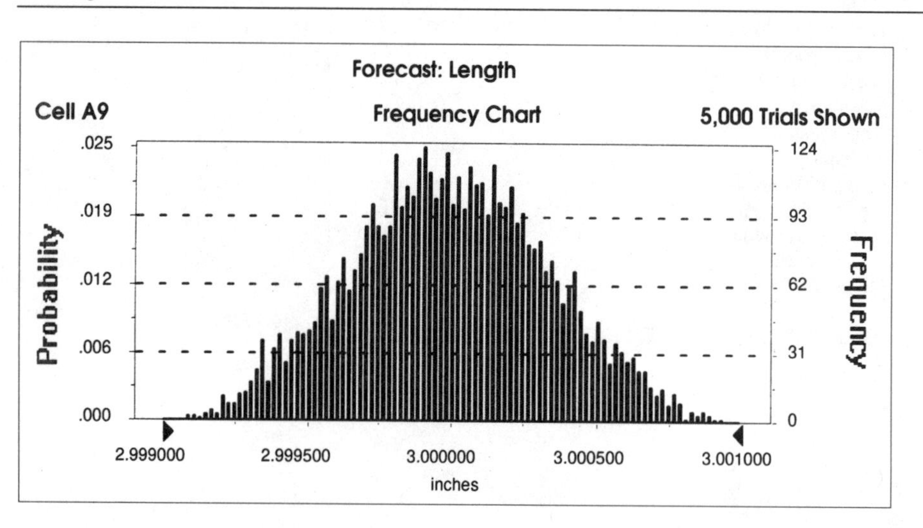

Figure 15.10 Output response statistics for uniformly distributed inputs with one-sigma limits

Forecast: Length

Edit Preferences View Run Help

Cell A9 Statistics

Statistic	Value
Trials	5,000
Mean	2.999998
Median (approx.)	2.999993
Mode (approx.)	2.999910
Standard Deviation	0.000337
Variance	0.000000
Skewness	0.02
Kurtosis	2.61
Coeff. of Variability	0.00
Range Minimum	2.999083
Range Maximum	3.000927
Range Width	0.001845
Mean Std. Error	0.000005

Carlo simulation only approximates the actual output distribution a three-level designed experiment will produce. The Monte Carlo randomly samples across the continuum of uniformly distributed input values for each part. A three-level designed experiment, such as an L_9, can input only nonrandom values. In this case they are (mean -1σ), (mean), and (mean $+ 1\sigma$). The Monte Carlo simulation produces an output standard deviation that is *smaller* than what the L_9 designed experiment produces. This means that the designed experiment, which we will look at shortly, will produce a normal distribution with a standard deviation statistic that will be different than the Monte Carlo simulation but will still *understate* the actual output variability that is to be expected from real samples of the three parts. Running the L_9 simulation shown in Figure 15.11 will prove this point.

Summing the three part values, set at the one-sigma limits, for the nine runs of the L_9 array produces the output values shown in Figure 15.11.

The output distribution statistics are calculated to be:

$$\text{Mean} = 2.999963$$

$$\text{Standard deviation} = 0.000484$$

$$\text{Skew} = -0.70$$

$$\text{Kurtosis} = 4.07$$

Comparing the Monte Carlo simulation to the designed experiment simulation we find:

$$\sigma_{\text{Monte Carlo } 3\sigma \text{ normal inputs}} = 0.000573$$

$$\sigma_{\text{Monte Carlo } 1\sigma \text{ uniform inputs}} = 0.000337$$

$$\sigma_{L_9\ 1\sigma \text{ uniform inputs}} = 0.000484$$

Figure 15.11 The L_9 experiment output for uniformly distributed inputs at one-sigma limits

Run	1 *Part A*	2 *Part B*	3 *Part C*	4	*Output values*
1	0.999667	0.999667	0.999667	1	2.999001
2	0.999667	1.000000	1.000000	2	2.999667
3	0.999667	1.000333	1.000333	3	3.000333
4	1.000000	0.999667	1.000000	3	2.999667
5	1.000000	1.000000	1.000333	1	3.000000
6	1.000000	1.000333	0.999667	2	3.000000
7	1.000333	0.999667	1.000333	2	3.000333
8	1.000333	1.000000	.0999667	3	3.000000
9	1.000333	1.000333	1.000000	1	3.000666

This data suggests that the Monte Carlo simulation actually drew many more samples that were closer to the mean than the L_9 was "allowed" to draw. The L_9 could not sample values *inside* of the one-sigma limits, but the Monte Carlo could. Remember, the L_9 has only three levels to include in the sample input data (see Figure 15.11). The Monte Carlo can select, at random, from an infinite number of values between -1σ and $+1\sigma$. This explains why the sigma value for the Monte Carlo is smaller than the L_9.

The standard deviation at the one-sigma limits is too low in comparison to reality, as defined by the Monte Carlo output standard deviation obtained by inputs drawn from a normal distribution set at three-sigma limits. We must take corrective action to force the uniformly distributed inputs to produce output statistics that look like they came from normally distributed inputs sampled at three-sigma limits. We can now apply the 1.2247-correction multiplier to the input standard deviation values to accomplish this goal. First, we will rerun the Monte Carlo simulation with each part still represented by a uniform distribution but now limited by plus and minus $\mathbf{1.2247\sigma}$.

As shown in Figure 15.12, the 1.2247 Sigma limits = 0.999592 inch and 1.000408 inches. The statistics associated with the output distribution are as shown in Figure 15.13. Summing the three part values, set at the 1.2247σ limits, for the nine runs of the array produces the output values shown in Figure 15.14.

The output distribution statistics are calculated to be:

$$\text{Mean} = 3.000000$$

$$\text{Standard Deviation} = 0.000612$$

$$\text{Skew} = -0.86$$

$$\text{Kurtosis} = 3.83$$

Comparing the Monte Carlo simulations to the designed experiment simulation we find:

$$\sigma_{\text{Monte Carlo } 3\sigma \text{ normal inputs}} = 0.000573$$

$$\sigma_{\text{Monte Carlo } 1.2247\sigma \text{ uniform inputs}} = 0.000409$$

$$\sigma_{L_9 \; 1.2247\sigma \text{ uniform inputs}} = 0.000612$$

Figure 15.12 Output distribution for uniformly distributed inputs with 1.2247σ limits

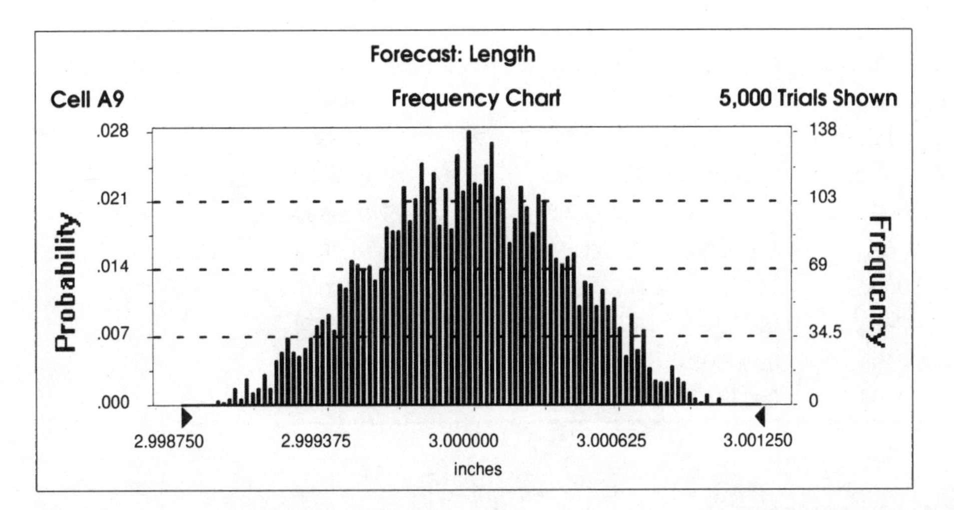

Figure 15.13 Output response statistics for uniformly distributed inputs with 1.2247σ limits

Forecast: Length

Edit Preferences View Run Help

Cell A9 **Statistics**

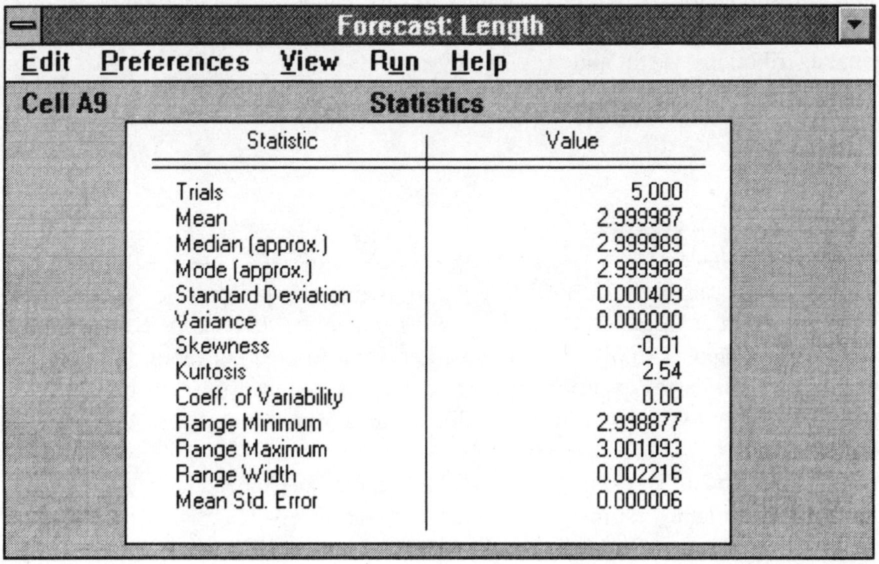

Statistic	Value
Trials	5,000
Mean	2.999987
Median (approx.)	2.999989
Mode (approx.)	2.999988
Standard Deviation	0.000409
Variance	0.000000
Skewness	-0.01
Kurtosis	2.54
Coeff. of Variability	0.00
Range Minimum	2.998877
Range Maximum	3.001093
Range Width	0.002216
Mean Std. Error	0.000006

Figure 15.14 The L_9 experiment output for uniformly distributed inputs at 1.2247σ limits

Run	1 Part A	2 Part B	3 Part C	4	Output values
1	0.999592	0.999592	0.999592	1	2.998776
2	0.999592	1.000000	1.000000	2	2.999592
3	0.999592	1.000408	1.000408	3	3.000408
4	1.000000	0.999592	1.000000	3	2.999592
5	1.000000	1.000000	1.000408	1	3.000408
6	1.000000	1.000408	0.999592	2	3.000000
7	1.000408	0.999592	1.000408	2	3.000408
8	1.000408	1.000000	0.999592	3	3.000000
9	1.000408	1.000408	1.000000	1	3.000816

We find that the 1.2247 transformation is moving the standard deviation back to values that are more closely aligned to the output created by the Monte Carlo simulation, which uses normally distributed input values at the three-sigma limits. When looking at the L_9 statistics constrained by 1.2247σ limits, we find that the standard deviation is no longer understating the standard deviation found in the Monte Carlo simulation used to emulate the real case. In fact, it is a little bit bigger than the simulated "real-world" standard deviation. Notice that the Monte Carlo simulation, for 1.2247σ limited uniform input distributions, is still understating the variability.

Can we do better than this? Is there a way to a get a closer match from designed experiment statistics to "real-world" statistics?

Using the $\sqrt{3}\sigma$ Transformation

There is an improvement that can be made to the $\sqrt{3/2}\ \sigma$ transformation process for statistical variance experiments being run at three levels. In the paper *Statistical Tolerance Analysis Using a Modification of Taguchi's Method* [R15.6], an alternative approach to creating more accurate output response data from three-level array experiments is presented. As we have stated, using the $\sqrt{3/2}$ σ transform helps make the input to the experiment look like it came from normal rather than uniform distributions.

Errico and Zaino have shown that using the $\sqrt{3/2}\ \sigma$ transform does indeed help, and provides response data output that reasonably estimates the mean and variance—but it does not provide very reasonable statistics for skew and kurtosis. In their work, Errico and Zaino found that using the $\sqrt{3/2}\ \sigma$ transform in an experiment, in comparison to output statistics from a 250,000-trial Monte Carlo simulation, produced means that were within 0.027 percent of each other, and standard deviations that were within 0.03 percent of each other. This is comforting until one looks into the skew and kurtosis values that this transformed input produces. They found that the differences in the skew were more than 36 percent. The differences in the measures of kurtosis were a whopping 103 percent. So there can be real errors resident in the data produced using the $\sqrt{3/2}\ \sigma$ transform when it comes to the

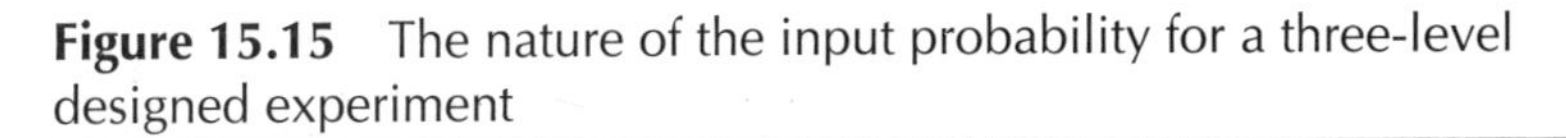

Figure 15.15 The nature of the input probability for a three-level designed experiment

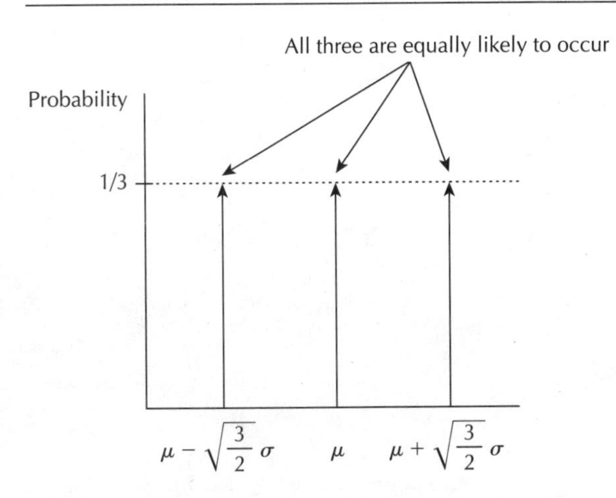

skew and kurtosis. This can lead to errors in tolerance allocation in certain critical assemblies where understanding the degree of normality of the output distribution is important. In most cases the error will not matter.

The $\sqrt{3/2}\ \sigma$ transform is based on the concept of a uniform distribution, as shown in Figure 15.15, where the probability of occurrence is broken down into three blocks, each associated with one of the three levels of the experimental parameters. Thus, the probability of a random variable assum-ing either the mean or plus or minus the $\sqrt{3/2}\ \sigma$ is one-third, or one in three. This is what we mean when we say the uniform distribution represents *equally likely chances* of a value occurring. An orthogonal array forces the randomness out of the experimental order and replaces it with the pattern of values found in the body of the array as ones, twos, and threes.

Errico and Zaino define a similar approach to transforming the standard deviations for three-level statistical variance experiments. They suggest using the $\sqrt{3}\ \sigma$ as the transformation value.

The idea behind their approach is to redefine the weighting of the input variable probabilities (1/3) by dividing the probability of an occurrence of a random variable, such as a *level* in a tolerance experiment, into six "probable blocks" of occurrence (see Figure 15.16). A random variable having the value of the mean is given a four-in-six chance of occurring; the probability of a random variable assuming the value of the mean plus or minus the $\sqrt{3}\ \sigma$ is assigned a one-in-six chance of occurring. This is how the probabilities are assigned in the $\sqrt{3}\ \sigma$ transformation case.

The $\sqrt{3}\ \sigma$ transform has probability weighting based on Figure 15.16.

Here we see how the mean is weighted with a probability of 4/6. The $\mu \pm \sqrt{3}\sigma$ values are each given 1/6 weights. The fact that the weights are uneven between the mean and the transformed standard deviation values must be taken into account when calculating the mean, standard deviations, skew, and kurtosis values. This weighted calculation process is relatively easy to accommodate, as we shall see shortly.

The weighting factors, w_i, provide a lot more resolution in the calculations, thus minimizing the propagation of error in the estimate of the statistics—particularly in the skew and kurtosis. The weighting process works in the following manner.

Figure 15.16 The new probability weighting for a three-level experiment

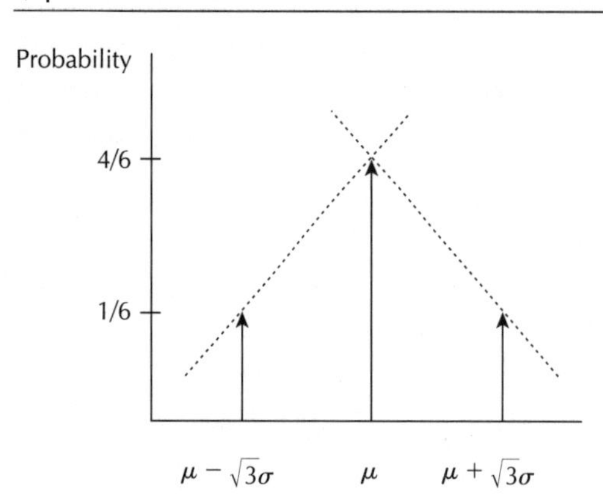

In an L_{27} for three parameters, each at three levels, the weights are: $w_i = 1/6$ for the standard deviations at levels 1 or 3, and 4/6 for the nominal values at level 2.

So, whenever the array states, for a given run, that the pattern of ones, twos, and threes are in a certain order, we can assign 1/6 for the ones and threes, and 4/6 for the twos. In the three-part case we have been working with the L_9 array; but because we must use unequal weights for the three-level experiment we can no longer use the L_9. We must use a full factorial array so that each possible combination of the three parts at three levels is run in the experiment. Only in this way can we use 100 percent of the weighting values. If we were to use a fractional factorial experiment such as the L_9, we would not be able to properly calculate the weighted statistics. This is true because we are obtaining only a fraction of the probable combinations. All combinations for the three parts are required to use the weighted statistical calculations correctly. The L_{27} is the smallest full factorial array that can handle three factors (columns 1, 2, and 5 make up the full factorial array for three factors at three levels). The weighting values are calculated in Figure 15.17.

The probabilities for each part value occurring at each level (1/6 for level 1, 4/6 for level 2, and 1/6 for level 3) are shown in Table 15.1, the L_{27} full factorial array for the three-part case.

The formulas for calculating *weighted* means, standard deviations, skews, and kurtosis values are as follows:

$$\bar{y}_{weighted} = \frac{(weight_1 \times y_1 + weight_2 \times y_2 + \ldots + weight_n \times y_n)}{(weight_1 + weight_2 + \ldots + weight_n)} = \frac{\sum_{i=1}^{n} w_i y_i}{\sum_{i=1}^{n} w_i} \quad \textbf{(15.11)}$$

$$\sigma^2_{weighted} = \sum_{i=1}^{n} w_i (y_i - \bar{y}_{weighted})^2 \quad \textbf{(15.12)}$$

$$Skew_{weighted} = \sum_{i=1}^{n} w_i (y_i - \bar{y}_{weighted})^3 \quad \textbf{(15.13)}$$

Table 15.1

Run number	*Part A*	*Part B*	*Part C*	*Weight factor*
1	1/6	1/6	1/6	1/216
2	1/6	1/6	4/6	4/216
3	1/6	1/6	1/6	1/216
4	1/6	4/6	1/6	4/216
5	1/6	4/6	4/6	16/216
6	1/6	4/6	1/6	4/216
7	1/6	1/6	1/6	1/216
8	1/6	1/6	4/6	4/216
9	1/6	1/6	1/6	1/216
10	4/6	1/6	1/6	4/216
11	4/6	1/6	4/6	16/216
12	4/6	1/6	1/6	4/216
13	4/6	4/6	1/6	16/216
14	4/6	4/6	4/6	64/216
15	4/6	4/6	1/6	16/216
16	4/6	1/6	1/6	4/216
17	4/6	1/6	4/6	16/216
18	4/6	1/6	1/6	4/216
19	1/6	1/6	1/6	1/216
20	1/6	1/6	4/6	4/216
21	1/6	1/6	1/6	1/216
22	1/6	4/6	1/6	4/216
23	1/6	4/6	4/6	16/216
24	1/6	4/6	1/6	4/216
25	1/6	1/6	1/6	1/216
26	1/6	1/6	4/6	4/216
27	1/6	1/6	1/6	1/216

$$Kurtosis_{weighted} = \sum_{i=1}^{n} w_i(y_i - \bar{y}_{weighted})^4 \quad \textbf{(15.14)}$$

These modified formulas (where multiplying by $(1/N)$ is replaced by the weighting factor) must be used to calculate the *weighted* output statistics, since we have assigned two different probabilities (1/6 and 4/6) to the experimental input levels in the three-level L_9 array.

For an L_9 setup based on the $\sqrt{3/2}\ \sigma$ transformation we see the following weights:

Rows 1 – 9: 1/3 1/3 1/3 . . . the weight multiplier for the response = $(1/3)^3$

All nine of the rows have the *same* probability due to the use of uniformly distributed (equally likely) set point probabilities of one in three chances. Therefore, they do not require the added step of calculating weighted statistics. The statistics for the equally distributed probability case are all calculated with a weighting factor of 1. The 1.2247 transform greatly simplifies the process, and also works with any fractional factorial array. The real problem is that one cannot study very many tolerance parameters with the 1.732σ approach, since it requires full factorial designs at three levels. This will greatly limit its application in many instances, and is a big problem when considering many, if not most, tolerance problems. Typical designs have more than three critical parameters. An L_{27} can handle only three parameters in a full factorial setup.

A Comparison of Output Statistics

Figures 15.17 and 15.18 show the Monte Carlo simulation output distribution and statistics for the three parts input as uniformly distributed values constrained by 1.732σ limits.

The 1.732-sigma limits = 0.999423 inch and 1.000577 inches.

The statistics associated with the output distribution are as shown in Figure 15.18.

The statistics in this Monte Carlo output are biased on the fact that we used a uniform distribution to emulate the output of an L_{27} experiment but did not apply the weighting factor to the data. The way to correct for this is to define a custom distribution in *Crystal Ball* to account for the 1/6 probabilities for the 1.732's limits, and the 4/6 probability for the mean value occurring.

The results for each experimental run are shown in Table 15.2 for the L_{27} full factorial array for the three-part case. Summing the three part values, set at the 1.732σ limits, for the 27 runs of the array produces the output values shown.

Figure 15.17 Output distribution for uniformly distributed inputs with 1.732σ limits

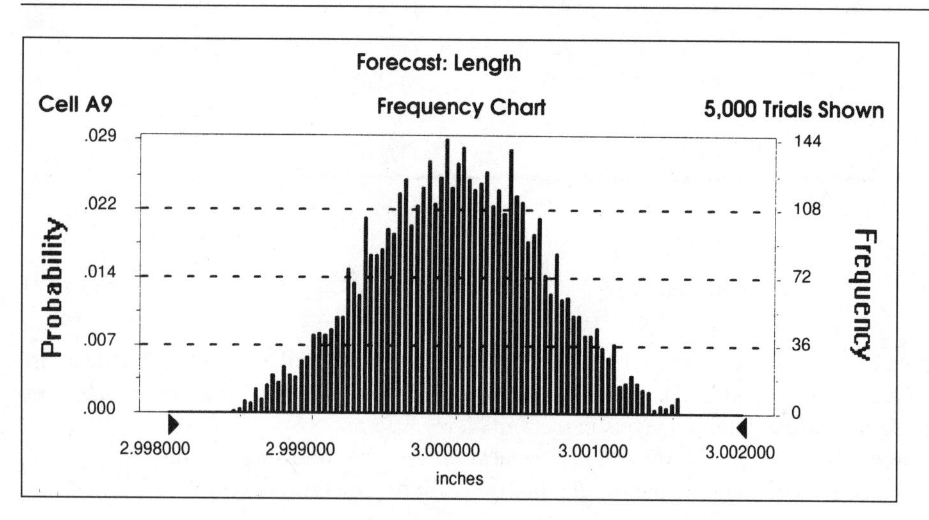

Figure 15.18 Output response statistics for uniformly distributed inputs with 1.732σ limits

Forecast: Length

Edit Preferences View Run Help

Cell A9 Statistics

Statistic	Value
Trials	5,000
Mean	3.000010
Median (approx.)	3.000016
Mode (approx.)	2.999940
Standard Deviation	0.000584
Variance	0.000000
Skewness	-0.03
Kurtosis	2.59
Coeff. of Variability	0.00
Range Minimum	2.998398
Range Maximum	3.001569
Range Width	0.003171
Mean Std. Error	0.000008

Table 15.3 displays how the probability weights are to be assigned to each response so that we can calculate the weighted statistics.

The weighted output distribution statistics are calculated to be:

$$\text{Mean} = 3.000000$$

$$\text{Standard Deviation} = 0.000577$$

$$\text{Skew} = 0.000$$

$$\text{Kurtosis} = 3.000$$

Comparing the Monte Carlo simulations to the designed experiment simulation we find:

$$\sigma_{\text{Monte Carlo } 3\sigma \text{ normal inputs}} = 0.000573$$

$$\sigma_{\text{Monte Carlo } 1.732\sigma \text{ uniform inputs}} = 0.000584$$

$$\sigma_{L_{27}\ 1.732\sigma \text{ uniform inputs}} = 0.000577$$

Here we see the two Monte Carlo simulations coming fairly close together, and the standard deviation for the nine data points from the L_{27} equaling 0.000577. Table 15.4 is a comparison chart for all the experimental and Monte Carlo cases we have simulated. The most realistic simulation is *underlined* (normally distributed inputs at three-sigma limits).

The 1.732σ transform definitely produces data and output statistics that come the closest to the most realistic simulation (normally distributed inputs set at three-sigma limits—such as one would get from a 100 percent inspected sample of production parts). The unfortunate reality is that this 1.732σ

Table 15.2 The results from running the L_{27} full factorial array

Run	*1* Part A	*2* Part B	*5* Part C	*Response values*
1	.999423	.999423	.999423	2.99827
2	.999423	.999423	1.00000	2.998846
3	.999423	.999423	1.000577	2.999423
4	.999423	1.00000	.999423	2.998846
5	.999423	1.00000	1.00000	2.999423
6	.999423	1.00000	1.000577	3.000000
7	.999423	1.000577	.999423	2.999423
8	.999423	1.000577	1.00000	3.000000
9	.999423	1.000577	1.000577	3.000577
10	1.00000	.999423	.999423	2.998846
11	1.00000	.999423	1.00000	2.999423
12	1.00000	.999423	1.000577	3.000000
13	1.00000	1.00000	.999423	2.999423
14	1.00000	1.00000	1.00000	3.000000
15	1.00000	1.00000	1.000577	3.000577
16	1.00000	1.000577	.999423	3.000000
17	1.00000	1.000577	1.00000	3.000577
18	1.00000	1.000577	1.000577	3.001154
19	1.000577	.999423	.999423	2.999423
20	1.000577	.999423	1.00000	3.000000
21	1.000577	.999423	1.000577	3.000577
22	1.000577	1.00000	.999423	3.000000
23	1.000577	1.00000	1.00000	3.000577
24	1.000577	1.00000	1.000577	3.001154
25	1.000577	1.000577	.999423	3.000577
26	1.000577	1.000577	1.00000	3.001154
27	1.000577	1.000577	1.000577	3.001731

transform can be applied only to full factorial experiments, which greatly limits its applicability to practical tolerance problems.

This exercise sends a strong message to both the analytical and experimental engineering communities: both Monte Carlo and designed experiment methods work and should be used to confirm each other's validity. When we determine that a given set of parameters will perform in some statistically significant manner using analytical methods, we should always seek an experimental verification

Table 15.3 The L_{27} experiment output for uniformly distributed inputs at 1.732σ limits

Run	*Part A*	*Part B*	*Part C*	*Weight*	*Response*
1	1/6	1/6	1/6	1/216	2.99827
2	1/6	1/6	4/6	4/216	2.998846
3	1/6	1/6	1/6	1/216	2.999423
4	1/6	4/6	1/6	4/216	2.998846
5	1/6	4/6	4/6	16/216	2.999423
6	1/6	4/6	1/6	4/216	3.000000
7	1/6	1/6	1/6	1/216	2.999423
8	1/6	1/6	4/6	4/216	3.000000
9	1/6	1/6	1/6	1/216	3.000577
10	4/6	1/6	1/6	4/216	2.998846
11	4/6	1/6	4/6	16/216	2.999423
12	4/6	1/6	1/6	4/216	3.000000
13	4/6	4/6	1/6	16/216	2.999423
14	4/6	4/6	4/6	64/216	3.000000
15	4/6	4/6	1/6	16/216	3.000577
16	4/6	1/6	1/6	4/216	3.000000
17	4/6	1/6	4/6	16/216	3.000577
18	4/6	1/6	1/6	4/216	3.001154
19	1/6	1/6	1/6	1/216	2.999423
20	1/6	1/6	4/6	4/216	3.000000
21	1/6	1/6	1/6	1/216	3.000577
22	1/6	4/6	1/6	4/216	3.000000
23	1/6	4/6	4/6	16/216	3.000577
24	1/6	4/6	1/6	4/216	3.001154
25	1/6	1/6	1/6	1/216	3.000577
26	1/6	1/6	4/6	4/216	3.001154
27	1/6	1/6	1/6	1/216	3.001731

that is also statistically valid. We have just demonstrated how to do this check-and-balance procedure. We have also shown that there is nothing to fear from running only 9 or 27 experiments. The statistics show that 5,000 Monte Carlo simulations and 9 or 27 orthogonal array experimental runs all yield remarkably similar results.

So, we have shown what is probabilistically going to happen in a statistical variance experiment and a Monte Carlo simulation—but what about a worst-case scenario?

Table 15.4

$\mu \pm 1\sigma$	$\mu \pm 1\sigma$	$\mu \pm 3\sigma$	$\mu \pm \sqrt{\frac{3}{2}}\sigma$	$\mu \pm \sqrt{\frac{3}{2}}\sigma$	$\mu \pm \sqrt{3}\sigma$	$\mu \pm \sqrt{3}\sigma$
L_9 array	Monte Carlo	Monte Carlo	L_9 array	Monte Carlo	L_{27} array	Monte Carlo
$\mu = 3.0$	$\mu = 3.0$	$\mu = 3.0$	$\mu = 3.0$	$\mu = 3.0$	$\mu = 3.0$	$\mu = 3.0$
$\sigma = .000484$	$\sigma = .000337$	$\sigma = .000573$	$\sigma = .000612$	$\sigma = .000409$	$\sigma = .000577$	$\sigma = .000584$
skew = −.70	skew = 0.02	skew = −.01	skew = −.86	skew = −.01	skew = .00	skew = −.03
kurt. = 4.07	kurt. = 2.61	kurt. = 2.94	kurt. = 3.83	kurt. = 2.54	kurt. = 3.00	kurt. = 2.59

Conducting a Tolerance Experiment for Worst-Case Conditions

Running a worst-case tolerance experiment can be of use in some circumstances. This approach is analogous to performing a traditional worst-case tolerance analysis, as discussed in Chapter 5. It is to be done with rigorous attention to detail. The team will be gathering data created in a noisy environment of its own making, and forcing what *can happen to happen* while doing their best to keep experimental error to an absolute minimum. It is desirable to have the signal from the parameter variations, as assailed by additional sources of noise, stand out above the noise that is induced by just running the experiment. When loading the array with the parameter variations to be evaluated, with each parameter set point placed at the high and low level of the tolerances, the team must conduct each experimental run exactly as it is stated in the pattern defined by the array. Table 15.5 shows an L_8 loaded and ready to be run.

This example illustrates how one can study mixtures of technology. There are mechanical, electrical, fluid dynamic, and magnetic technologies used in the function of this product. All elements must be evaluated at their one-sigma limits (using a two-level experiment). Then we can study the worst-case scenario at three- or six-sigma limits. The experimental evaluation of these parameters will provide the sensitivity and variability of performance data required to make rational decisions concerning optimal tolerances. Each of the eight runs must be exposed to the compound noises defined by the noise experiment and the analysis of its output. The results of such an experiment can then be analyzed using ANOVA (Chapter 16).

It is crucial that the tolerance limits used in the experiment have some degree of integrity. That is why it is so important to understand all the tolerancing methods discussed in this book. The initial tolerances can come from lookup tables, supplier Cp or Cpk values, worst-case analysis, statistical analysis, Monte Carlo analysis, or loss function-based analysis. No matter where the initial tolerance originated, the Taguchi tolerance design process is going to scrutinize the initial tolerance in light of the quality-loss function. The process for analyzing the data from a tolerance experiment is exactly the same as for analyzing a statistical variance experiment.

Let's walk through the same process we used to analyze the $\mu \pm 3\sigma$, $\mu \pm \sqrt{3/2}\ \sigma$ and $\mu \pm \sqrt{3}\sigma$ cases, only now we will analyze the $\mu \pm 3\sigma$ tolerance levels for the four-factor assembly case from Chapter 5.

Table 15.5

Run	1 *Magnitude strength*	2 *Air-flow rate*	3 *Voltage bias*	4 *Spacing gap*	5 *Shaft rpm*	6 *Surface flatness*	7 *Axial skew*
1	900 gauss	10 cfm	−300 volts	.020 in.	100 rpm	.0000 in.	−.005 in.
2	900 gauss	10 cfm	−300 volts	.030 in.	102 rpm	.0025 in.	+.005 in.
3	900 gauss	14 cfm	−340 volts	.020 in.	100 rpm	.0025 in.	+.005 in.
4	900 gauss	14 cfm	−340 volts	.030 in.	102 rpm	.0000 in.	−.005 in.
5	1100 gauss	10 cfm	−340 volts	.020 in.	102 rpm	.0000 in.	+.005 in.
6	1100 gauss	10 cfm	−340 volts	.030 in.	100 rpm	.0025 in.	−.005 in.
7	1100 gauss	14 cfm	−300 volts	.020 in.	102 rpm	.0025 in.	−.005 in.
8	1100 gauss	14 cfm	−300 volts	.030 in.	100 rpm	.0000 in.	+.005 in.
	2 × 3	1 × 3	1 × 2	1 × 5	1 × 4	1 × 7	1 × 6
	4 × 5	4 × 6	4 × 7	2 × 6	2 × 7	2 × 4	2 × 5
	6 × 7	5 × 7	5 × 6	3 × 7	3 × 6	3 × 5	3 × 4

An L_9 Experiment and Monte Carlo Simulation Using $\mu \pm 3\sigma$ Levels Assuming Uniform Distributions

Table 15.6 shows the L_9 array loaded with the $\mu \pm 3\sigma$ set points.

We can manually calculate the gap dimension by using the following gap model:

$$\text{Gap} = \text{Part D} - (\text{Part A} + \text{Part B} + \text{Part C})$$

The gap dimensions for each row of the L_9 array are as follows:

Row 1: 0.007

Row 2: 0.006

Row 3: 0.005

Row 4: 0.007

Row 5: 0.003

Row 6: 0.005

Row 7: 0.004

Row 8: 0.006

Row 9: 0.002

Table 15.6

Run	1 Part A	2 Part B	3 Part C	4 Part D
1	.999	1.999	1.499	4.504
2	.999	2.000	1.500	4.505
3	.999	2.001	1.501	4.506
4	1.000	1.999	1.500	4.506
5	1.000	2.000	1.501	4.504
6	1.000	2.001	1.499	4.505
7	1.001	1.999	1.501	4.505
8	1.001	2.000	1.499	4.506
9	1.001	2.001	1.500	4.504

Running the nine data points through a statistical analysis we find the following information:

L_9 mean = 0.005

L_9 standard deviation = 0.0017

L_9 skew = 0.000

L_9 kurtosis = 2.93

These statistics represent the nature of the distribution of the output gap dimensions as constrained by the input component dimensions. These values can be compared to the same sort of data being generated in a 5000-trial Monte Carlo simulation with the same constraints on the input values—except that the sample inputs are not constrained by the patterns of the orthogonal array matrix values.

The output distribution from the Monte Carlo simulation is shown in Figure 15.19.

The statistical data, describing the distribution from the Monte Carlo is shown in Table 15.7.

Comparison of L_9 gap statistics to the Monte Carlo gap statistics shows the following results.

L_9 mean = 0.005	Monte Carlo mean = 0.005
L_9 standard deviation = 0.0017	Monte Carlo standard deviation = 0.0012
L_9 skew = 0.000	Monte Carlo skew = −0.03
L_9 kurtosis = 2.935	Monte Carlo kurtosis = 2.70

We can now compare the three-sigma results to the one-sigma results, as shown in Table 15.8. We have now quantified the distinctions between a tolerance experiment and a statistical variance experiment.

There is a 194 percent difference between the L_9 standard deviation data, and a 200 percent difference between the Monte Carlo standard deviation data. These differences are minimal. The differ-

Figure 15.19

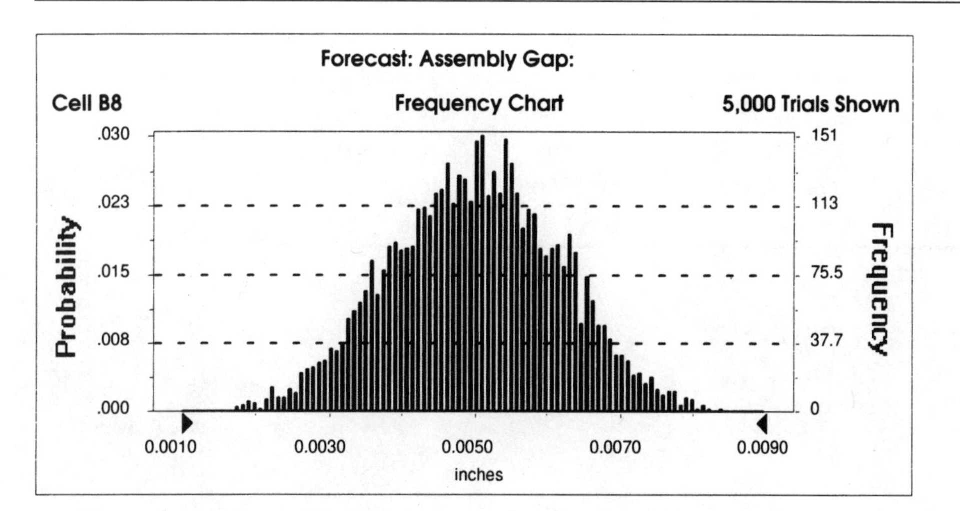

Table 15.7 Forecast: Assembly gap

Statistic	*Value*
Trials	5,000
Mean	0.0050
Median (approx.)	0.0050
Mode (approx.)	0.0051
Standard deviation	0.0012
Variance	0.0000
Skewness	−0.03
Kurtosis	2.70
Coefficient of variability	0.23
Range minimum	0.0014
Range maximum	0.0085
Range width	0.0071
Mean standard error	0.0000

ences between the one- and three-sigma cases is large and significant for how we evaluate setting up tolerance limits. This illustrates the difference between a worst-case analysis and a statistical analysis. The key question to ask is what is the acceptable limit of the gap as far as the customer loss (A_0 and Δ_0) is concerned. We must now use the quality-loss function to establish the economical internal manufacturing limits (A and Δ) and the cost the company can bear in maintaining them.

Table 15.8

$\mu \pm 1\sigma$	$\mu \pm 1\sigma$	$\mu \pm 3\sigma$	$\mu \pm 3\sigma$
L_9 Array	Monte Carlo	L_9 Array	Monte Carlo
$\mu = 0.005$	$\mu = 0.005$	$\mu = 0.005$	$\mu = 0.005$
$\sigma = .000579$	$\sigma = .0004$	$\sigma = .0017$	$\sigma = .0012$
skew = .023	skew = .02	skew = .000	skew = −.03
kurt. = 2.32	kurt. = 2.63	kurt. = 2.935	kurt. = 2.70

Metrology and Experimental Technique

A last word concerning how physical performance measurements are to be taken, recorded, and processed is in order. All experiments should be planned well in advance. The reduction of measurement error as an unwanted source of noise is of paramount importance. The transducers and data-acquisition systems used to actually gather and record the physical performance of the design should be of high quality, known capability for precision and accuracy, and in calibration. The quality of the data is the basis for the integrity of the tolerances that will eventually come from the analysis. We have listed some excellent sources for experimental metrology in Chapter 2 (Block 4 of the system reliability growth discussion).

Summary

This chapter has focused on the topic of properly setting up an experiment to evaluate standard deviations and tolerances for specific elements in a design. Two distinct types of experimental approaches have been discussed. Statistical variance experiments work with the manufacturing or assembly standard deviations associated with design parameters. The experimental data from such an approach provides a basis to do a tolerance analysis very much like an RSS analysis (Chapter 6). The tolerance experiment is a form of worst-case analysis that is less realistic in comparison to the statistical variance experiment. Ample warning is given for the restricted use of this approach (see Chapter 18 for a case study employing a form of this approach).

We have reviewed the way to prepare for conducting both two-level and three-level orthogonal array experiments. The simplest approach is to use a two-level array. Employing a three-level array requires a unique transformation of the component standard deviation values before they can be safely loaded into the experimental array. Three-level arrays require more experimental runs and may not add a lot of additional benefit to the analysis. A special section of the chapter was dedicated to this optional topic.

Chapter 15: Setting up a Designed Experiment for Variance and Tolerance Analysis

- Preparing to Run a Statistical Variance Experiment
  - Selecting the Right Orthogonal Array for the Experiment
  - Special Instructions for Running Three-Level Statistical Variance Experiments
- Using the $\sqrt{3}\sigma$ Transformation
- A Comparison of Output Statistics
- Conducting a Tolerance Experiment for Worst-Case Conditions
- An L_9 Experiment and Monte Carlo Simulation Using $\mu \pm 3\sigma$ Levels Assuming Uniform Distributions
- Metrology and Experimental Technique

The use of designed experiments in the statistical analysis of tolerances is most helpful when working with designs that are difficult to model analytically. The experimental approach allows one to develop data that are easily analyzed for parameter sensitivities with respect to the response variable.

You will enjoy the benefits of this approach if the simple rules contained in this chapter are followed for properly preparing the level set points for the experimental runs. Once the data are obtained from the experiments, the ANOVA data processing method can be used to quantify the main effect of each parameter on the response under evaluation. This is the topic of the next chapter.

References

R15.1: *Introduction to Quality Engineering;* G. Taguchi; Asian Productivity Organization, 1986, pg. 531

R15.2: *Engineering Quality By Design;* T. Barker; Marcel Dekker, 1990, pg. 531

R15.3: *Statistics for Experimenters;* Box, Hunter and Hunter; J. Wiley and Sons, 1978, pg. 535

R15.4: *Quality by Experimental Design* 2nd Ed.; Thomas B. Barker; Marcel Dekker, 1994, pg. 535

R15.5: *Engineering Methods for Robust Product Design;* W. Y. Fowlkes and C. M. Creveling; Addison-Wesley, 1995, pg. 537

R15.6: "Statistical Tolerance Analysis Using a Modification of Taguchi's Method"; J. R. Errico and N. A. Zaino; *Technometrics,* November 1988, Vol. 30, No. 4, pg. 557

CHAPTER 16

The ANOVA Method

Analysis of variance (ANOVA) is a computational technique that quantitatively estimates the relative contribution each parameter variation makes to the overall response variation. The contribution of each parameter is typically expressed as a percentage of the overall measured response. For example: four parameters are governing a product's response. Parameter A is contributing 36 percent of the response, parameter B is contributing 12 percent, parameter C 40 percent, parameter D 7 percent, and the remaining 5 percent of the response is attributed to random experimental effects (sometimes called *experimental error*). Thus, 95 percent of the measured response is due to the main effects attributable to the four design parameters.

ANOVA uses a mathematical technique known as the sum of squares to quantitatively examine the deviation of the parameter effect values from the overall experimental mean response. This is referred to as the variation *between* the design parameters. The magnitude of the significance of the individual parameters is quantified by comparing the variation between the parameter effects against the variation in the experimental data due to random experimental error and the effects of unrepresented interactions. The variation of the data due to random experimental variability and interactions is referred to as the variation *within* the control factors that make up the experiment. In classical DOE, the interactions are treated as additional significant effects and are kept strictly separated from the random experimental variability. Equation 16.1 is referred to as the *F-ratio.* It can be formed by relating the parameter variation (also called the *mean square due to a parameter*) in the numerator and the variation due to experimental error (the *mean square due to experimental error*) in the denominator.

$$F = \frac{MS}{MS_e} = \frac{\textit{Mean square due to a design parameter}}{\textit{Mean square due to experimental error}} \tag{16.1}$$

The ANOVA process uses the F ratio to help the engineering team to gain insight into which parameter tolerances are the most and least important. In fact, if a comparison between the effect of parameter D (7 percent) and the random experimental effects (5 percent) is made, it is not at all clear that parameter D is truly contributing to the response in any significant way beyond what is attributable to random experimental effects. This kind of information can be very helpful when decisions have to be made about which parameter tolerances are worth tightening for the sake of balancing cost-versus-customer quality expectations.

Accounting for Variation Using Experimental Data

The ANOVA process is used to decompose the main parameter effects and the effects due to error from the experimental data. Everything the engineering team needs to know about the relative contribution due to each design parameter is hidden within the data gathered for each row of the array. The orthogonal array has a distinct structure that produces data the ANOVA process can use to separate all the effects into distinct and useful data structures for each parameter. ANOVA has the mathematical capability to process the experimental data into sums of squares that hold the key to the quantification of the main effect contribution for each design parameter.

We can illustrate the three forms of analysis that are used to quantify the variation resident in the experimental parameters and the experimental error effects.

1. The first form is the sum of squares. It quantifies the differences associated with a parameter or error replicates from the overall mean of the data set:

$$SS\ parameter\ or\ error = \sum_{i=1}^{n} (y_i - \bar{y})^2$$

2. The second form is the mean of the sum of squares. It quantifies the mean value of the sum of squares associated with a parameter or with error replicates by dividing by the DOF allocated to the parameter or set of error replicates. The DOF for each parameter will either be one (for two-level arrays) or two (for three-level arrays):

$$MS = \frac{SS_{parameter\ or\ error}}{\text{DOF}\ for\ the\ parameter}$$

3. The third form is the mean square deviation, or true variance. It quantifies the true variance for a parameter or for error replicates by dividing *MS* by the DOF-1 for the entire experiment:

$$S^2 = MSD = \frac{MS_{parameter\ or\ error}}{\text{DOF}\ for\ the\ entire\ experiment}$$

These three forms represent the mechanics of the analysis of variance. ANOVA is used to help make decisions during the tolerance design phase of product development (Chapter 17). The parameters that contribute high percentages to the measured response variation are the focus of the cost versus performance balancing activities that take place during the tolerance design process.

Since ANOVA can be viewed as a procedure that decomposes the variation for each parameter relative to the overall mean response, it is important to remember that the variance is a statistic that measures the width of a distribution of data about the mean value. In the Taguchi approach, ANOVA can be applied to two forms of data. First, it can be applied to the data as measured in *engineering units.* This is the way it is used in the tolerance design process. ANOVA is used in tolerance design to identify those critical parameters that are worth considering for tolerance tightening, variance reduction, improving material grades, or some other means of improving quality. It can also be used to justify loosening tolerances for product cost reduction purposes. Second, it can be applied to the data after it has been transformed into signal-to-noise ratios. This is done in the parameter design process. ANOVA is used in parameter design to aid in identifying the strongest contributing control factors so that an accurate predictive model can be constructed and used during the optimum nominal set point verification process.

A Note on Computer-Aided ANOVA

This book uses two ANOVA software packages to illustrate how this process can be carried out on a personal computer. We have chosen two software packages for two reasons. First, because this text can be used as a follow-up to Fowlkes and Creveling (1995) [R16.1]. That text is based on the use of WinRobust software, which includes a basic ANOVA capability. Many readers will have WinRobust and therefore will prefer to see the material in this text illustrated with WinRobust examples. The second reason to include two ANOVA packages is that many practitioners of Taguchi methods are using ANOVA-TM software package from Advanced Systems & Designs, Troy, Michigan, as recommended by the American Supplier Institute. This book is intended to facilitate the standard application of Taguchi's approach to tolerance design as recommended by the American Supplier Institute (Dr. Taguchi is the Executive Director of ASI). ANOVA-TM is an advanced ANOVA package written explicitly for performing designed experiments for parameter design and tolerance design. It has a stronger and more fully featured ANOVA capability than WinRobust. We will show how two software packages solve the same ANOVA problems; the distinctions between the two will help clarify many of the details of computer-aided analysis of variance.

An Example of the ANOVA Process

We will use a simple belt-drive system as an example of a parameter tolerance experiment. Table 16.1 shows the L_9 matrix experiment and the responses obtained from it.

WinRobust and ANOVA-TM use fairly simple calculations to process the response data from the experiments. The following example demonstrates the sum of squares on the response values from the experimental runs.

First, the overall mean ($\overline{y}$), from which all the variation is calculated, is given by:

$$\overline{y} = \frac{1}{9}\sum_{i=1}^{9} (13.47 + 31.97 + \ldots + 20.09) = 22.60 \qquad \textbf{(16.2)}$$

The ANOVA process proceeds by determining the *grand total sum of squares* (GTSS):

$$\begin{aligned} GTSS &= \sum_{i=1}^{9} y_i^2 \\ &= (13.5)^2 + (32)^2 + \ldots + (20.1)^2 \\ &= 4{,}961.6 \end{aligned} \qquad \textbf{(16.3)}$$

Table 16.1 Data for a belt-drive system

A	*B*	*C*	*D*	*Response*
1	1	1	1	13.5
1	2	2	2	32.0
1	3	3	3	29.0
2	1	2	3	22.2
2	2	3	1	15.8
2	3	1	2	27.4
3	1	3	2	15.4
3	2	1	3	29.0
3	3	2	1	20.1

The GTSS can be decomposed into two parts. The first is the sum of the squares due to the overall experimental mean:

$$\begin{aligned} \textit{SS due to the overall mean} &= (\textit{\# of experiments}) \times \bar{y}^2 \\ &= 9 \times (22.60)^2 = 4{,}596.8 \end{aligned} \qquad \textbf{(16.4)}$$

The second part is the sum of the squares due to variation about the mean, referred to as the total sum of the squares (TSS):

$$\begin{aligned} TSS &= \sum_{i=1}^{9} (y_i - \bar{y})^2 \\ &= (13.5 - 22.60)^2 + (32 - 22.60)^2 + \ldots + (20.1 - 22.60)^2 \\ &= 364.3 \end{aligned} \qquad \textbf{(16.5)}$$

Note that the GTSS = TSS + SS due to overall mean.

The technique of summing squares can be used to define the contribution of each individual parameter within the total sum of squares. The sum of squares method is based on numerically quantifying the variation that is induced by the parameter effects around the overall experimental mean response ($\bar{y}$). This is why it is called the analysis of *variance.* The process uses the building blocks of the variances (sum of squares and mean squares) to quantify the strength of the parameter main effects. The ANOVA relationships can be summarized by Figure 16.1.

The L_9 experiment is structured so that the orthogonal (balanced) nature of the designed experiment produces data that must be gathered in the following format:

- three runs (1, 2, and 3) for A at the low level (-1σ),
- three runs (4, 5, and 6) for A at nominal level, and
- three runs (7, 8, and 9) for A at high level ($+1\sigma$)

Therefore, for parameter A, the sum of the squares due to variation about the mean is:

$$SS_A = (\textit{\# of exp. at A1})(\bar{y}_{A_1} - \bar{y})^2 + (\textit{\# of exp. at A2})(\bar{y}_{A_2} - \bar{y})^2 + (\textit{\# of exp. at A3})(\bar{y}_{A_3} - \bar{y})^2$$

Figure 16.1 Decomposition of the sums of squares

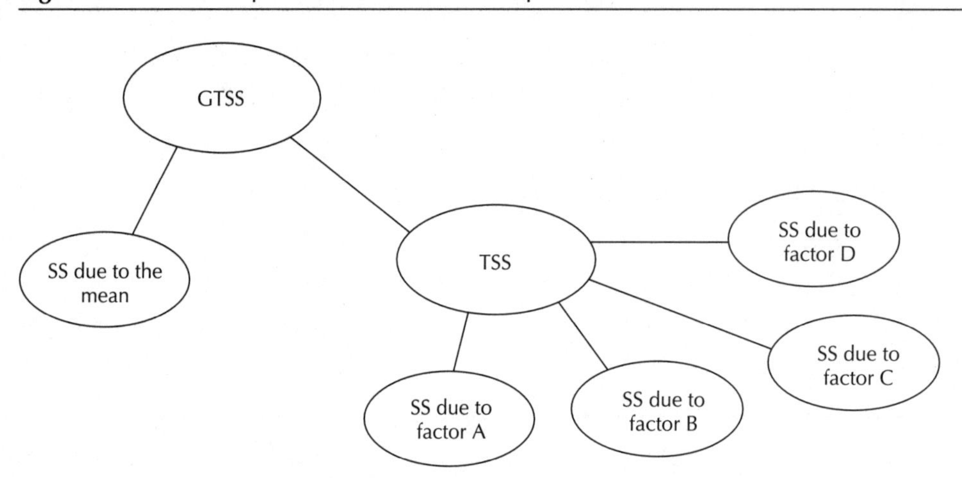

$$= 3(\bar{y}_{A_1} - \bar{y})^2 + 3(\bar{y}_{A_2} - \bar{y})^2 + 3(\bar{y}_{A_3} - \bar{y})^2$$
$$= 3(24.49 - 22.60)^2 + 3(21.80 - 22.60)^2 + 3(21.51 - 22.60)^2$$
$$= 16.2 \qquad \textbf{(16.6)}$$

where $\bar{y}_{Ai}$ is the average of the three response samples for each parameter level (i = 1, 2, or 3). When this procedure is repeated for factors B, C, and D, the sums of squares are as follows:

$SS_B = 140.3$
$SS_C = 30.6$
$SS_D = 177.4$

These values represent a rough measure of the relative importance of each parameter in the variation in the output response. If the components are added (16.2 + 140.3 + 30.6 + 177.4 = 364.5), the total SS as found in equation 17.5 is obtained (ignoring rounding errors).

The percentage contribution of the relative effect each parameter has on the output response is found using the following formula:

$$\textit{Percentage contribution} = (SS_{factor}/\textit{Total SS}) \times 100$$

Factor A: (16.2/364.5) × 100 = 4.4 percent

Factor B: (140.3/364.5) × 100 = 38.5 percent

Factor C: (30.6/364.5) × 100 = 8.4 percent

Factor D: (177.4/364.5) × 100 = 48.7 percent **(16.7)**

Degrees of Freedom in ANOVA

Note that no percentage has been assigned to account for experimental error contributions. Does this mean that there is no error in this experiment? No! The reason no sum of squares is assigned to account for experimental error can be understood by looking at the degrees of freedom available from the L_9 experiment.

A degree of freedom is an *independent* parameter associated with either:

- A matrix experiment,
- A design parameter, or
- A sum of squares computation.

For example, an L_9 orthogonal array has nine experiments, so it has nine degrees of freedom, and its grand total sum of squares has nine degrees of freedom. Calculating the overall mean for any orthogonal array takes up one degree of freedom, as does the sum of squares due to the overall experimental mean. As a result:

$$\text{Total SS} = (\text{GTSS} - \text{SS due to the overall mean})$$

$$\text{Total DOF} = (\text{\# of experiments} - 1)$$

The total sum of squares can have only eight degrees of freedom. Similarly, for parameters in an orthogonal array, one degree of freedom is allocated for each *level* of parameter. For the overall experimental mean of each parameter, subtract one degree of freedom. Therefore, for each three-level factor, there are $3 - 1$, or two degrees of freedom. In general, the degrees of freedom associated with a parameter is one less than the total number of levels for that parameter.

In this L_9 example, there are eight DOF available to estimate parameter tolerance effects. All eight are employed in evaluating the four parameters. No DOF is available to be allocated to quantifying experimental error effects. No experimental energy is spent on accounting for experimental error; this would incur the cost of running larger experiments, consuming time, effort, and money. Let's discuss what options are available to quantify experimental error.

Figure 16.2 Decomposition of the degrees of freedom

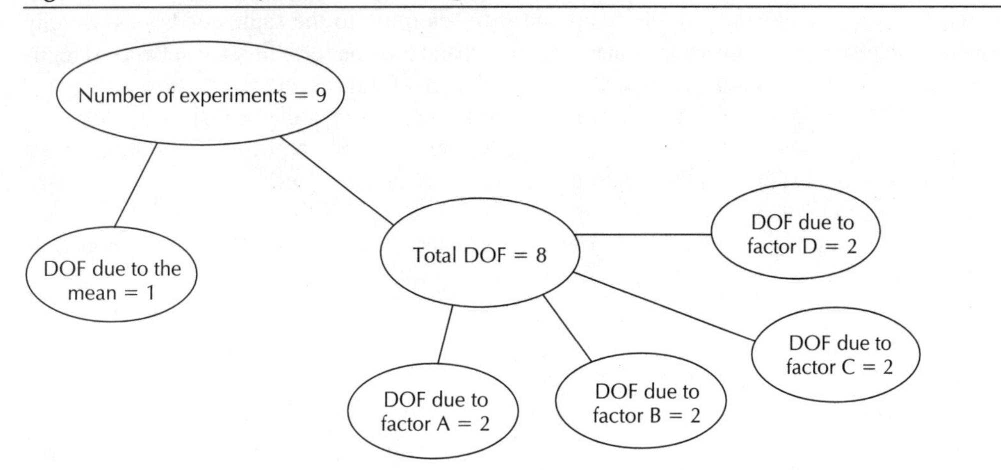

Error Variance and Pooling

In order to calculate the F ratio, experimental error must be estimated—here's one way to do it. The mean square for a parameter is defined by:

$$MS = \frac{\textit{sum of the squares due to a parameter}}{\textit{degrees of freedom for the parameter}} \tag{16.8}$$

The error mean square (MS_{error}) for the parameters is defined by:

$$MS_{error} = \frac{\textit{sum of the squares due to experimental error}}{\textit{degrees of freedom for experimental error}} \tag{16.9}$$

In the interest of seeking the most information possible from an orthogonal array matrix experiment, *all of the available degrees of freedom* are used to focus on the study of the parameters. That's why the mean square of the error in the belt-drive example is not estimated—at least not directly. A method of estimating the MS due to error does, however, exist. The method is called *pooling* the sum of the squares of the parameters that have a small contribution to the overall response.

A reasonable guideline is to pool the sum of squares of up to half of the parameters (the half defined by their percent contribution to be least significant). That makes about half of the degrees of freedom available to estimate the MS due to error. In this instance, the MS from weak parameters are made to stand in as an estimate of the experimental error. Note that this is only a guideline; in some experiments, all the parameters may be highly significant, in others few parameters may be significant. Therefore, the method of pooling typically provides an exaggerated estimate of experimental error, and is viewed as a conservative approach. An example of pooling follows shortly.

Error Variance and Replication

A more direct way to estimate the error variance is by repeating three to five times the measurement of the response at the same parameter nominal set point. This means setting up the same parameter levels and noise factor levels to replicate the response measurement three to five times. Remember, a true replicate requires physically changing the setup and then returning to the same conditions. Simply repeating measurements on a given setup is mere repetition, and does not give the same effects as replication. Repetition simply measures the variation in your system of transducers and meters.

One common strategy for the replicate case is to do a repeat setup at the beginning, middle, and end of the overall experiment. Since the parameters are the same for each replicated measurement, all of the variability can be attributed to the experimental error variance. This technique, however, misses any error due to parameter interactions.

The sum of squares due to the experimental error determined by replication is calculated as follows:

$$MS_{error} = \frac{1}{(r-1)} \sum_{j=1}^{r} (y_j - \overline{y}_R)^2 \tag{16.10}$$

where r = number of replicate measurements, y_j = individual response values for the replicates, and $\overline{y}_r$ = average of the replicate response values. The value of r will usually be between three and five replicates.

This error estimate can now be used in the same way as the one obtained by pooling. In WinRobust, this value must be manually calculated and entered in the space provided as an external prediction or estimate of error. ANOVA-TM has a feature in its menu that allows one to automatically include replicates in the design so that the software will automatically calculate the MS due to error.

Error Variance and Utilizing Empty Columns

Another common technique for estimating experimental error is to gather data on it—including interactive effects (often done with an L_{12} or an L_{18} array)—using any empty columns in the matrix experiment. One or more empty columns can be analytically processed just as if there were a parameter being studied. The numerical values produced in the analysis of variance for the empty columns are expressing the random experimental variability and any interaction effects that show up in the empty columns (for an L_{12} or an L_{18} only). This technique then automatically includes interactions as a source of experimental error. WinRobust and ANOVA-TM ANOVA tables display the numerical values for the sum of squares for any empty columns not assigned to a parameter. Be sure any columns not being occupied by a parameter are left unassigned when setting up the experimental array.

The *F*-Test

The ANOVA table will also include the *F*-ratio, also referred to as the variance ratio. This ratio, actually a ratio of mean squares, is used to test for the significance of parameter effects. The *F*-ratio is given by Equation 16.1. When F is much greater than 1 the effect of the parameter is large compared to the variance due to experimental error and interaction effects.

Here are some suggested rankings (generalized values) for F ratios:

- $F < 1$: the experimental error outweighs the parameter effect; the parameter is insignificant and indistinguishable from the experimental error.
- $F \approx 2$: the parameter has only a moderate effect compared to experimental error.
- $F < 4$: the parameter is strong compared to experimental error and is clearly significant.

Use these general guidelines when deciding where to tighten tolerances, reduce variation, or upgrade an engineering material specification. When it is necessary to be exact in your knowledge of significance of the F ratio, the use of an *F*-ratio table is recommended. The top row of an *F*-ratio table contains the degrees of freedom used in the numerator, and the left column of the table contains the degrees of freedom used to estimate the experimental-error term. For a detailed explanation of the formal statistical ANOVA process, see Freund (1988), Custer et al. (1993), and Ross (1988) [R16.2, 3, and 4].

In summary, the F ratio is the statistical analog to Taguchi's signal-to-noise ratio for the design parameter effect versus the experimental error.[1] The *F*-ratio uses information based on mean squares to define the relationship between the power of the parameter effects (a type of signal) and the power of

1. Do not confuse the signal-to-noise ratio used in the Taguchi parameter design approach to analyzing experimental data with the ratio used in defining the F statistic in the ANOVA process. While they are generally describing the same things, they are mathematically different.

the experimental error (a type of noise). Taguchi's signal-to-noise ratios depend on the physical case at hand and come in many forms, depending upon the type of response being measured (STB, NTB, and so forth).

A WinRobust ANOVA Example

Performing an ANOVA is simplified by using WinRobust. The same example will be repeated shortly using the more advanced ANOVA-TM software package. The completed L_9 array (from the belt-drive example), loaded and ready for data entry, is shown in Figure 16.3.

To enter the data, open the Response menu, set up the response using the Define Response window, shown in Figure 16.4.

Figure 16.3 WinRobust main screen showing belt-drive L_9 array

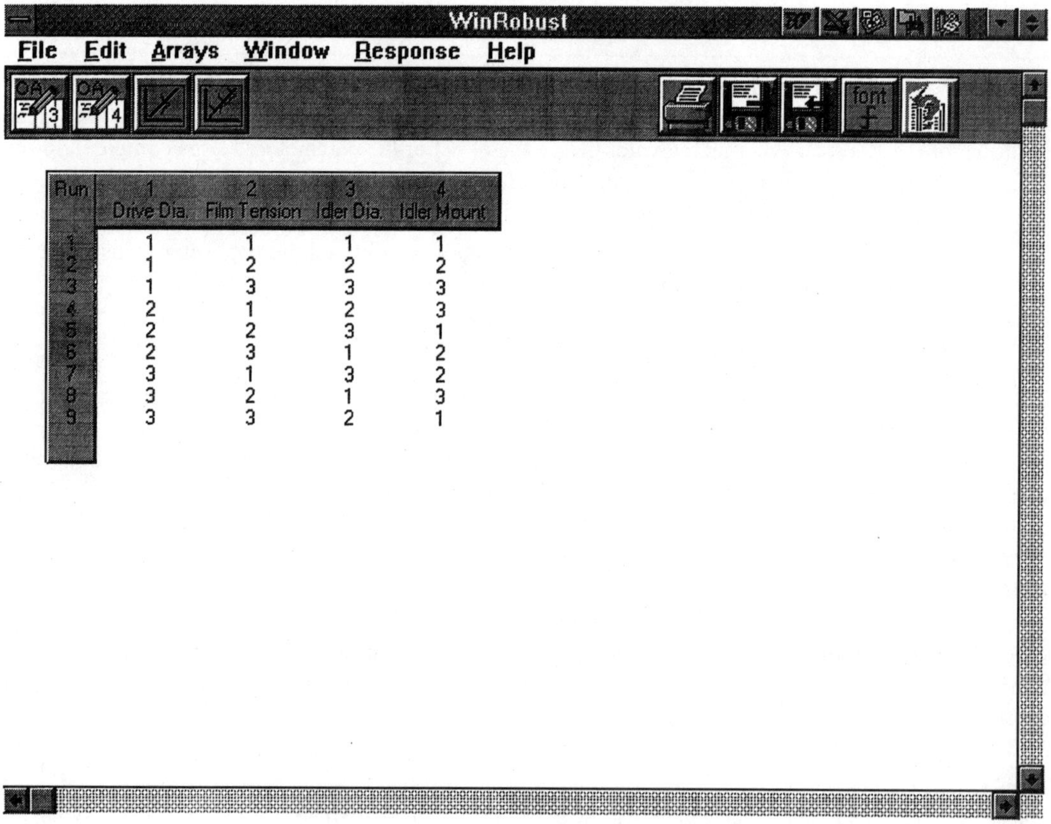

WinRobust

File Edit Arrays Window Response Help

Run	1 Drive Dia.	2 Film Tension	3 Idler Dia.	4 Idler Mount
1	1	1	1	1
2	1	2	2	2
3	1	3	3	3
4	2	1	2	3
5	2	2	3	1
6	2	3	1	2
7	3	1	3	2
8	3	2	1	3
9	3	3	2	1

Figure 16.4 The WinRobust Define Response window

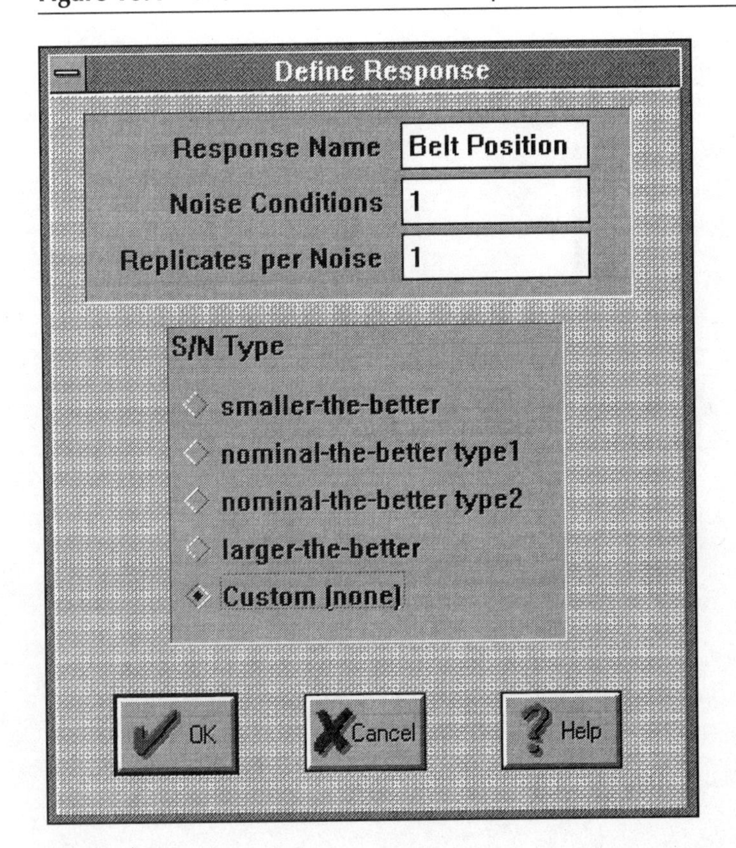

Use the Define Response window to set up WinRobust for proper data entry. The response is "belt position." Select Custom (none) to leave the data in standard engineering units; otherwise WinRobust will transform it into some form of signal-to-noise ratio.

The nine responses are shown entered into the Response window, shown in Figure 16.5. Click "Crunch" to tell the program to perform the ANOVA calculations.

Figure 16.6 displays the results of the ANOVA calculations performed by WinRobust. The ANOVA data seen here are for the main parameter effects only. In order to compute *F* ratios, you must choose an option for estimating the error. The parameters are rank-ordered according to the magnitude of their mean squares. Factors D and B contribute far more than half the total variation. Therefore, the pooling guideline for using about half of weakest parameters certainly applies here. Parameters C and A should have their sums of squares (SS) and mean squares (MS) pooled together and attributed to experimental error. By doing this, the effects of drive roller diameter and idle roller diameter are being considered indistinguishable from experimental noise (random variation).

The window shown in Figure 16.7 contains WinRobust's three options for estimating experimental error. This window is opened by selecting the Options menu from the main menu. Figure 16.7 shows the pooling option chosen. The value 4 has been entered to tell WinRobust to use the four

Figure 16.5 The Response data-entry window

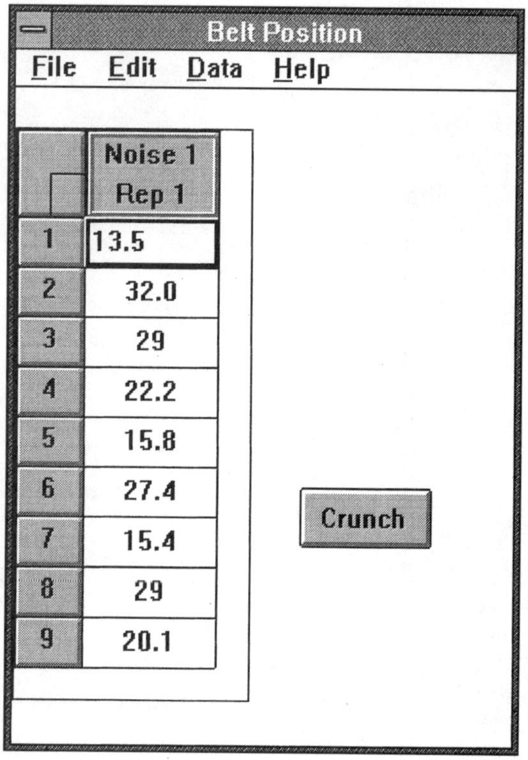

Belt Position

File Edit Data Help

	Noise 1 Rep 1
1	13.5
2	32.0
3	29
4	22.2
5	15.8
6	27.4
7	15.4
8	29
9	20.1

Crunch

degrees of freedom associated with the two parameters that have relatively weak responses compared with the two others. WinRobust is programmed to select the *weakest* parameters for pooling when commanded to do so.

After performing the pooling operation, the ANOVA table is modified as shown in Figure 16.8. The F ratios appear indicating the level of significance for each of the parameters. Consulting an F-ratio table for two DOF for the numerator and four DOF for the denominator at the 90 percent confidence level shows that 4.32 is the critical F value above which the control factors can be considered statistically significant in their contribution to the response performance of the design. At the 95 percent confidence level, the critical F value is 6.9. By this standard, the parameters are marginally significant with respect to the estimated experimental error. This is one reason that pooling for error is somewhat imprecise in comparison to direct measures of experimental error through taking actual replicates.

Here it is shown that the two remaining parameters are moderately active in provoking a significant response with respect to the output variation. They are good candidates for tolerance tightening, variance reduction, or material upgrading. This is how ANOVA can make it quite clear what the critical-to-function (CTF) parameters really are. The other parameters now become candidates for tolerance relaxation for economic balance within the design.

Figure 16.6 Results of the ANOVA

WinRobust - C:\WINROB\L9DRV.RDE - ANOVA

Print Edit Options Help

Belt Position-Means

Factor	SS	d.o.f.	mean sq	F
Idler Mount	180.33	2	90.16	
Film Tension	145.08	2	72.54	
Idler Dia.	34.70	2	17.35	
Drive Dia.	20.40	2	10.20	

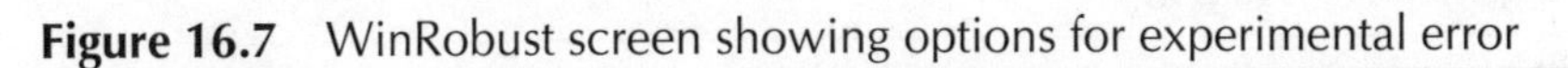
Figure 16.7 WinRobust screen showing options for experimental error

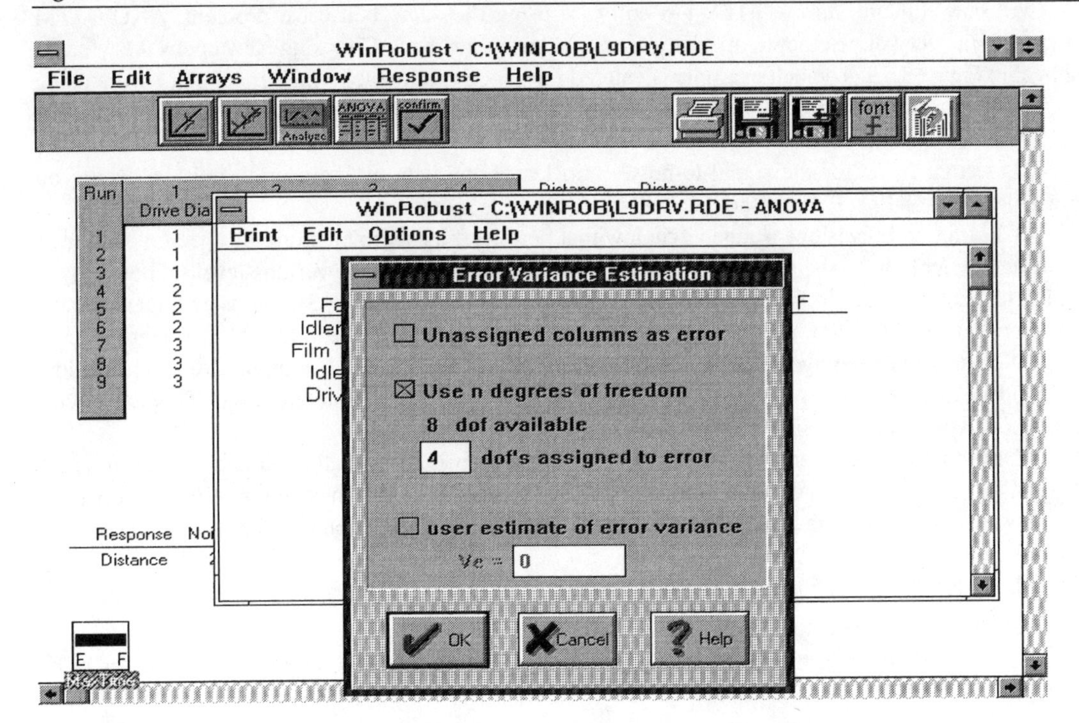

Figure 16.8 Results of the ANOVA after choosing pooling option

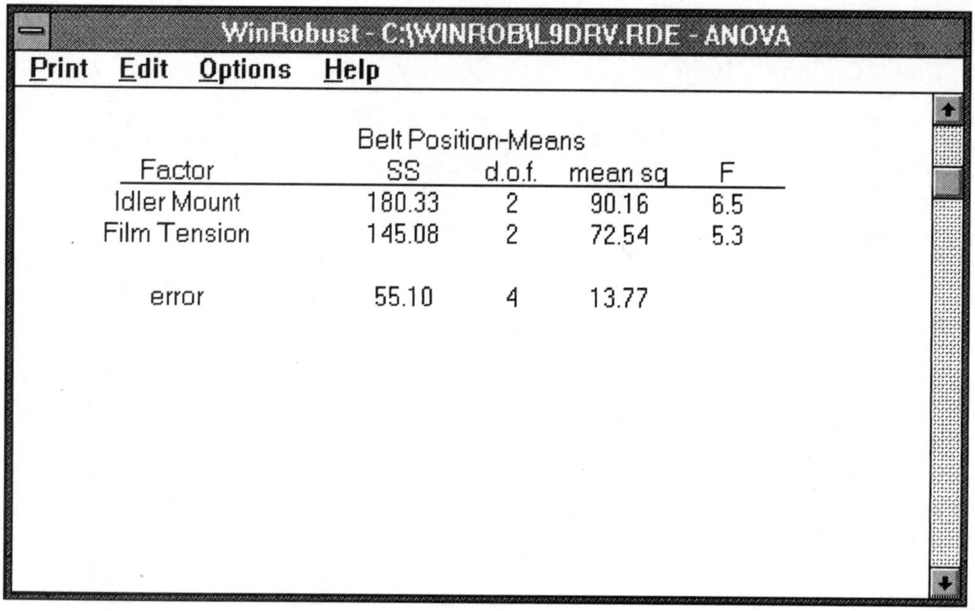

WinRobust - C:\WINROB\L9DRV.RDE - ANOVA

Print Edit Options Help

Belt Position-Means

Factor	SS	d.o.f.	mean sq	F
Idler Mount	180.33	2	90.16	6.5
Film Tension	145.08	2	72.54	5.3
error	55.10	4	13.77	

An ANOVA-TM Example [R16.5]

We will now illustrate the ANOVA-TM software using the same belt-drive problem. ANOVA-TM starts with a query of whether to open a new or existing experiment. Choosing New opens the window shown in Figure 16.9. The belt example is entered into the Experiment box. The fact that we are using continuous variables is entered into the Data box. Under Group One we have selected an L_9 orthogonal array.

Selecting None for the signal-to-noise ratio leaves the data in engineering units. Clicking on Variable tells ANOVA-TM that the data will be continuous in nature.

Clicking on Labels opens a new window that permits the user to enter all the specific parameter information, including labels (A, B, C, and D), parameter names, and the various-level set points (typically, nominal $- 1.2247\sigma = 1$, nominal $= 2$, and nominal $+ 1.2247\sigma = 3$). In this example we have left them set at 1, 2, and 3, even though the data were collected with the 1.2247 adjustment in force.

Now that the labeling is complete the Input menu can be opened. The Input/Edit Data window will appear, as shown in Figure 16.11 with nine data points for the L_9 entered into it. This is the window used to place the data into ANOVA-TM.

Once the data is entered in the window, the next step is to open the Output menu and select ANOVA table. The basic ANOVA Table window will appear, as shown in Figure 16.12. Notice no *F* ratios are calculated. The program is waiting for specific instructions on pooling to estimate experimental error.

Let's define each column for ANOVA-TM's table.

- Source is the name of the parameter.
- Pool is the column with buttons to select which parameters to pool or unpool for error estimation.

Figure 16.9 The main Experimental Layout window

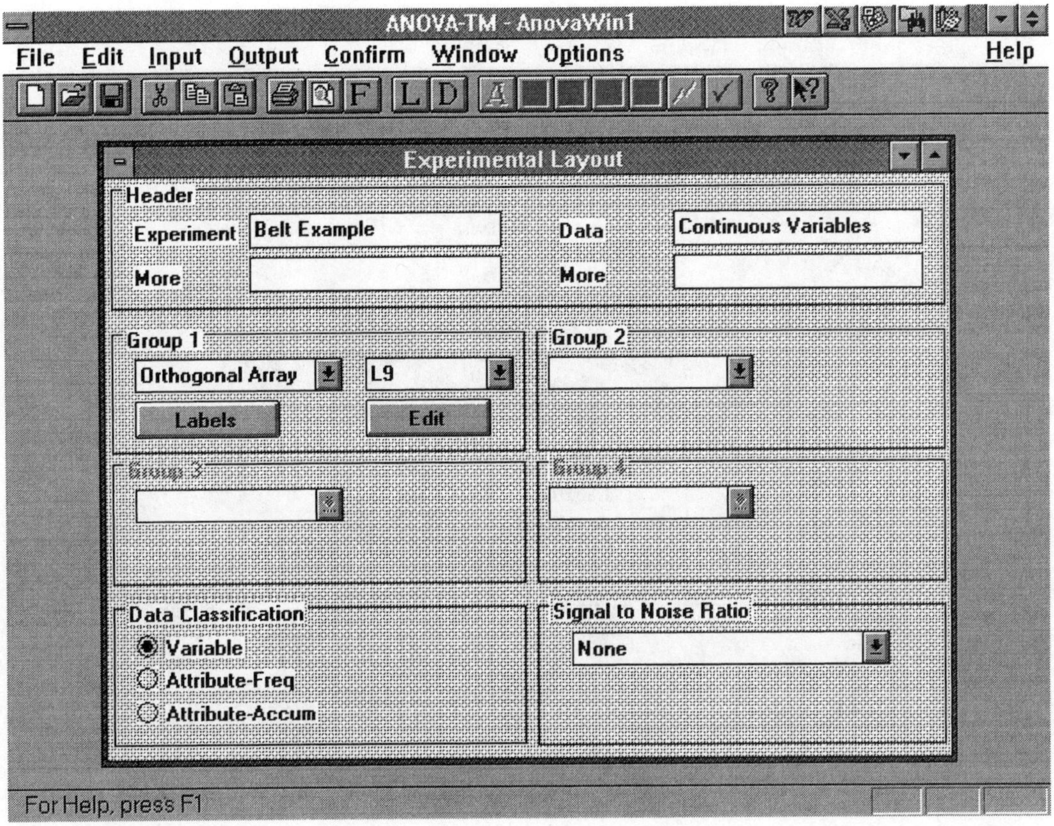

- DF is the parameter's degree of freedom (one DOF for a two-level array and two DOF for a three-level array).
- S is the parameter's sum of squares value.
- V is the parameter's mean square value (the ANOVA-TM manual calls it the variance, where $V = S/DF$—but it is not the true parameter variance).
- F is the F ratio. $F = V/V_e$, where V_e is the pooled values of V for weak parameters.
- S' is the parameter *pure variation,* $S' = S - [(V_e \times (\mathrm{DF})]$.
- ρ is the parameter percent contribution ratio. $\rho = S'/S_t(100)$, where S_t is the sum of all S values in the experiment.

For the rows, e_1 is the *between-experiment* error, e_2 is the error due to taking replicates, and (e) is the pooled error. The Total row is the sum of the column values.

This completes the basic introduction to the two software packages. ANOVA-TM has many additional features to help further analyze the data, including many that generate plots and perform sensitivity analysis. While ANOVA, in and of itself, is a form of sensitivity analysis, ANOVA-TM can perform a special form of sensitivity analysis that requires the use of data that is transformed

Figure 16.10 Parameter labeling window

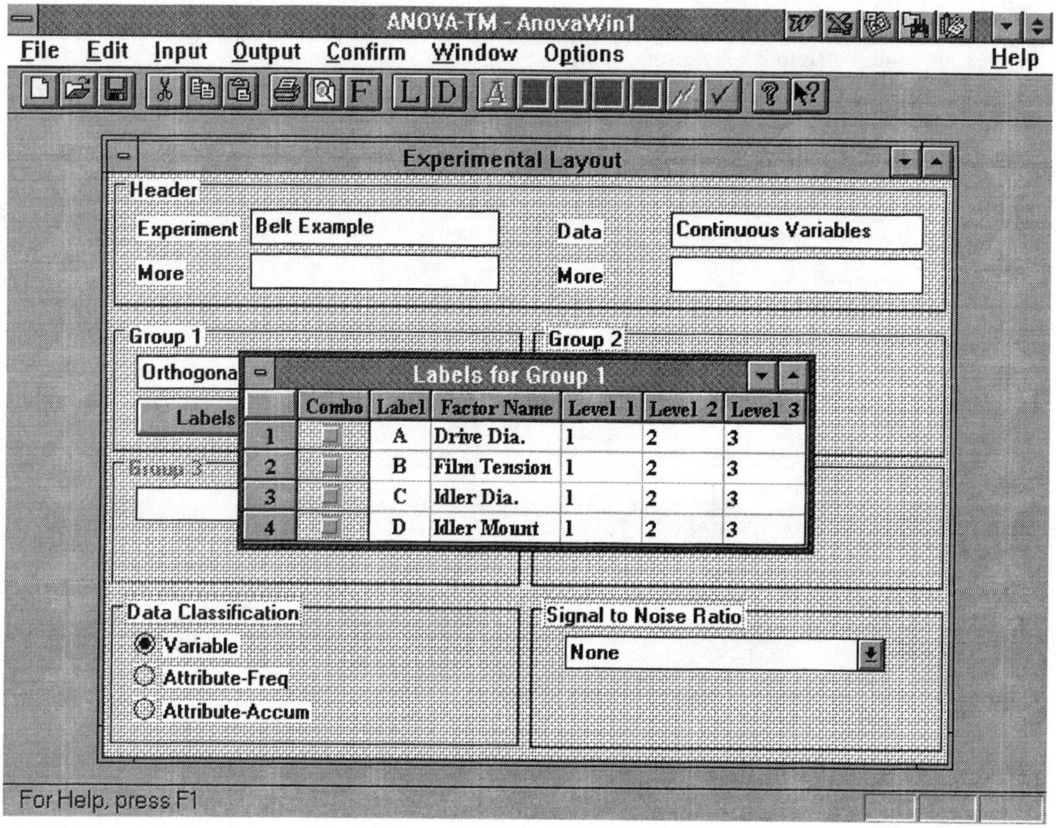

ANOVA-TM - AnovaWin1

File Edit Input Output Confirm Window Options Help

Experimental Layout

Header

Experiment: Belt Example

Data: Continuous Variables

More

More

Group 1

Orthogona

Labels

Group 2

Group 3

Labels for Group 1

	Combo	Label	Factor Name	Level 1	Level 2	Level 3
1		A	Drive Dia.	1	2	3
2		B	Film Tension	1	2	3
3		C	Idler Dia.	1	2	3
4		D	Idler Mount	1	2	3

Data Classification

- Variable
- Attribute-Freq
- Attribute-Accum

Signal to Noise Ratio

None

For Help, press F1

Figure 16.11 The Input/Edit Data window

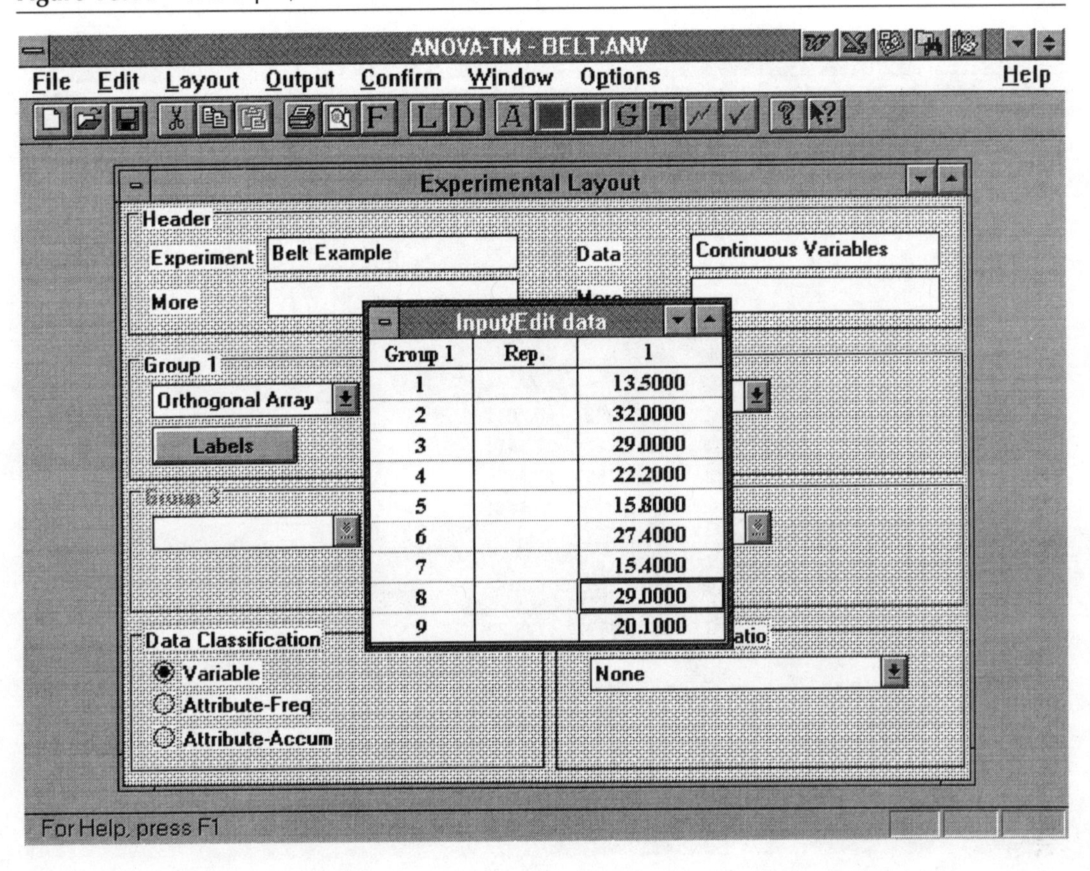

Figure 16.12 The ANOVA table

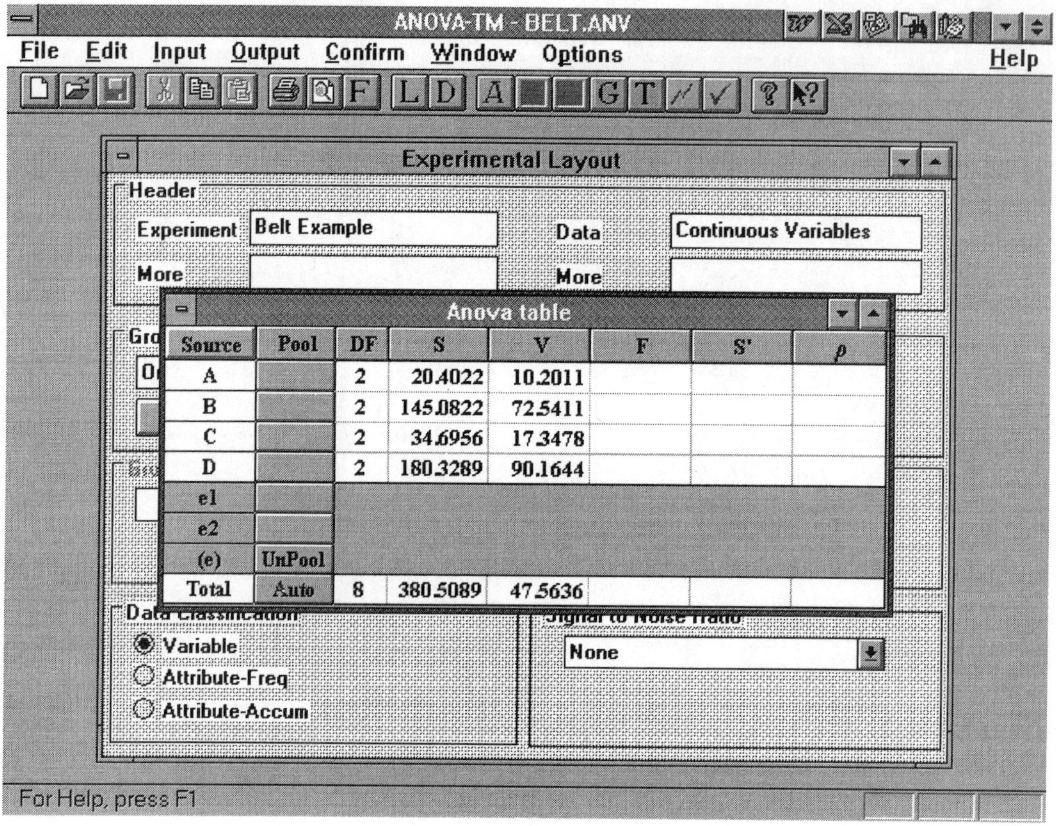

ANOVA-TM - BELT.ANV

File Edit Input Output Confirm Window Options Help

Experimental Layout

Header

Experiment: Belt Example

Data: Continuous Variables

More

More

Anova table

Source	Pool	DF	S	V	F	S'	ρ
A		2	20.4022	10.2011			
B		2	145.0822	72.5411			
C		2	34.6956	17.3478			
D		2	180.3289	90.1644			
e1							
e2							
(e)	UnPool						
Total	Auto	8	380.5089	47.5636			

Data Classification

Variable

Attribute-Freq

Attribute-Accum

Signal to Noise Ratio

None

For Help, press F1

Figure 16.13 The ANOVA table with pooling

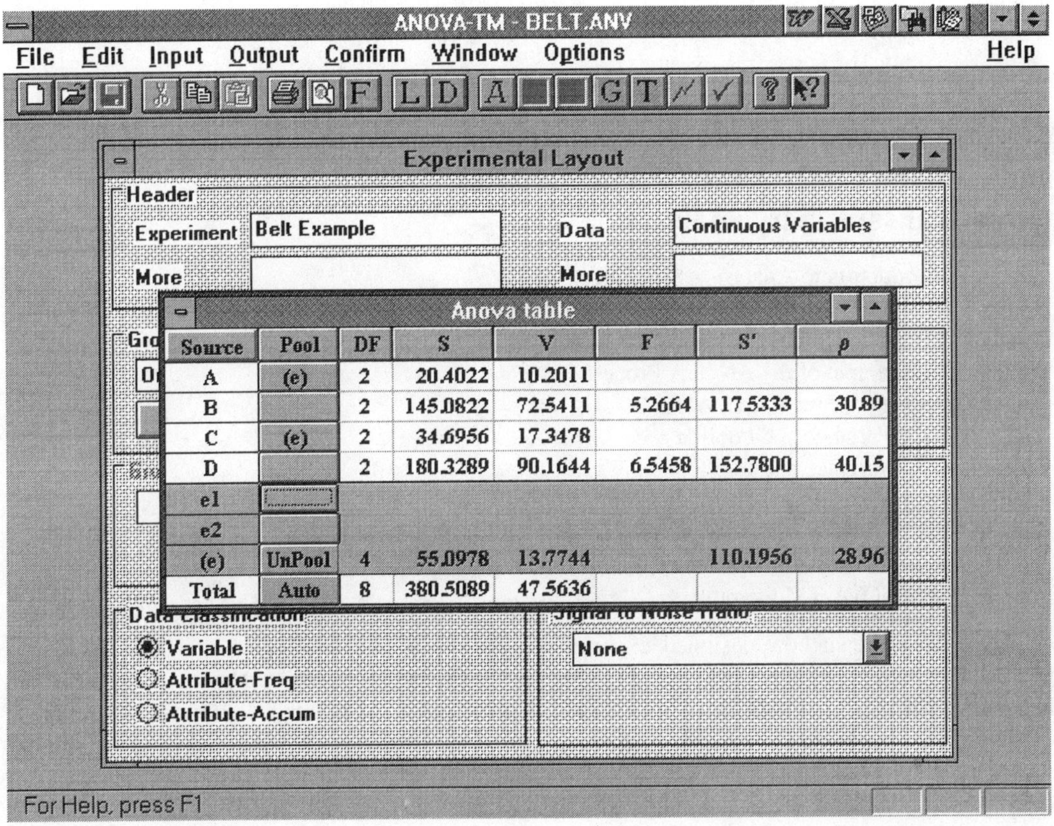

Source	Pool	DF	S	V	F	S'	ρ
A	(e)	2	20.4022	10.2011			
B		2	145.0822	72.5411	5.2664	117.5333	30.89
C	(e)	2	34.6956	17.3478			
D		2	180.3289	90.1644	6.5458	152.7800	40.15
e1							
e2							
(e)	UnPool	4	55.0978	13.7744		110.1956	28.96
Total	Auto	8	380.5089	47.5636			

into specific signal-to-noise units (dB). Consult the ANOVA-TM manual for a detailed explanation of this form.

Whether you prefer to use WinRobust, ANOVA-TM or both, the mechanics underlying the ANOVA calculation process are the same.

Summary

Analysis of variance is central to the statistical analysis of designed experiments. ANOVA can decompose the variation in a set of data into the main parameter effects, including interactions (if enough experimental DOF are allocated and the experimental columns are loaded properly per Chapter 13), with the remaining variation assignable to experimental error. ANOVA is based on the principle that the sums of squares are additive. It is a mathematical treatment of the data that gives quantitative information about the significance of the design parameters. Such information can be used to formulate and answer, with statistical rigor, hypotheses concerning the probability that a factor judged significant is indeed active, and vice versa—that a factor judged *not* significant is indeed negligible. Such information is highly useful for identifying tolerance sensitivity in a design parameter.

The contents of Chapter 16 are outlined below:

Chapter 16: The ANOVA Method for Identifying and Ranking Control Factor Sensitivity to Variability (empirical sensitivity analysis)

- Introduction to Analysis of Variance
- Accounting for Variation Using Experimental Data
- A Note on Computer-Aided ANOVA
- An Example of the ANOVA Process
- Degrees of Freedom in ANOVA
- Error Variance and Pooling
- Error Variance and Replication
- Error Variance and Utilizing Empty Columns
- The *F*-Test
- A WinRobust ANOVA Example
- An ANOVA-TM Example

We have shown how designed experiments can aid in the development of the disciplined approach to tolerance analysis introduced in Chapter 13. Chapter 14 illustrated how to force what can happen to happen through the development and deployment of compounded stressful noise factors in the designed experiment. In Chapter 15 we discussed the two types of experimental approaches to setting up a designed experiment to evaluate standard deviations or tolerances as they affect a measured response variable. This chapter has introduced more tools required to use designed experiments by reviewing the data-analysis process called ANOVA. We are now ready to explore a comprehensive case study that will demonstrate the detailed steps of the final process in off-line quality engineering.

References

R16.1: *Engineering Methods for Robust Product Design;* W. Y. Fowlkes and C. M. Creveling; Addison-Wesley, 1995, pg. 583

R16.2: *Modern Elementary Statistics;* J.E. Freund; Prentice Hall, 1988, pg. 594

R16.3: *Analysis of Variance;* L. Custer, D. R. McCarville, M. Harry and J. Prins; Addison-Wesley, 1993, pg. 594

R16.4: *Taguchi Techniques for Quality Engineering;* P. Ross; McGraw Hill, 1988, pg. 594

R16.5: *ANOVA-TM Users Manual;* Advanced Systems and Designs, Inc., 1996, pg. 601

See Appendix E for information about ANOVA-TM and WinRobust.

CHAPTER 17

The Tolerance Design Process: A Detailed Case Study

We will bring this book to a close by thoroughly reviewing an experimentally based tolerance analysis. The case study [R17.1] we will review and expand was presented by Alan Wu at the Eleventh Annual Taguchi Symposium in Costa Mesa, California, in 1993. Alan Wu is an instructor and expert in quality engineering at the American Supplier Institute, Taguchi's base of operation in the United States, in Allen Park, Michigan. ASI is the premier organization in the West for the dissemination of quality engineering methodologies and training materials, and it is *the* place to go for comprehensive training in concept design, parameter design, and tolerance design methodologies. ASI has this author's gratitude for granting permission to use this case study as a teaching mechanism. The case presented in this chapter covers all the elements of a proper tolerance design process.

The Steps for Performing the Tolerance Design Process

1. Define the engineering design parameters to be evaluated. This includes their optimized nominal target set points from the parameter design process. The output response we are seeking to keep on target will carry over from the work done in the parameter design process. It is imperative that the target value be defined at this point. If you are still developing for the optimum *nominal* parameter set points, the parameters are not ready to move forward into the tolerance design process.
2. Develop the quality-loss function for the output response (see Chapter 11). The customer loss, A_0, needs to be defined at this point. The target for the output response, along with A_0, is used in the definition of the customer tolerance limit Δ_0. The type of the problem (NTB, STB, LTB, or asymmetric) should be known from the parameter design; if it is not defined, then it must be determined at this point in the process. If the customer tolerance is not in units

that can be measured by the engineering team, the appropriate conversion must be made to align the output response used in the tolerance experiment with the customer's observed response. In Chapter 13, this was called the intermediate high-level tolerance (Δ_I). A key skill to develop is the ability to translate nontechnical customer terms into engineering terms upon which A_0 and Δ_0 are based.

3. Define the manufacturing or assembly standard deviation (1σ) above and below the mean set point for each critical parameter under evaluation. These will serve as the parameter level set points when loaded into an orthogonal array. These values are usually determined by looking at the $\pm 3\sigma$ tolerances assigned to each parameter nominal set point. The manufacturer of the components should provide an estimate or direct measure of the standard deviations for the critical parameters from the components currently being made (see Chapter 14).
4. Select the appropriate orthogonal array with enough degrees of freedom to facilitate the proper evaluation of the effect of the one standard deviation set points for each parameter on the output response (see Chapter 13). Determine whether the analysis requires two or three levels. Two levels are usually sufficient, and are more economical to run (see Chapter 15).
5. Align the appropriate compound noise factors with the orthogonal array experiment. If necessary, conduct appropriate noise experiments to identify the severity (magnitude) and directionality of noise factor effects on the output. Develop compound noise factors if this was not done during parameter design (see Chapter 14).
6. Set up and run the experiment. Collect the data under rigorous conditions with controlled experimental techniques to minimize experimental error in the data. If you are not pooling weak parameter effects to represent experimental error, take three to five replicates for one experiment to have enough data to quantify the experimental error variance. If not all the columns of the array are used, the degrees of freedom from these unassigned columns can be used to quantify the experimental error (see Chapter 16).
7. Conduct a formal analysis of variance on the data using a personal computer. Identify the overall mean square deviation (MSD) for the entire experiment, as well as the sum of squares, mean sum of squares, MSDs and F ratios for each parameter [$F = (MS_{parameter})/(MS_{error})$].
8. Calculate the total loss for the baseline design using the measure of true variance (total $MSD = (\sigma^2)$) from the ANOVA data in the quality-loss function $(L) = k(\sigma^2)$.
9. Quantify the loss due to each parameter in the baseline design. This is done by using the mean square deviation ($MSD = \sigma^2_{parameter}$) for each parameter, from the ANOVA data, by plugging the parameter MSD's into the quality-loss function $(L) = k(\sigma^2_{parameter})$.
10. Calculate the percentage ($\%_{parameter} = [(MS_{parameter})/(MS_{total})]100$) each parameter contributes to the overall variability in the output response. This is the MS for each individual parameter divided by the MS for the entire experiment (see Chapter 16). This information will be required for tolerance reallocation through tightening or loosening tolerances.
11. Calculate the process capability (Cp) for the design using the upper and lower customer tolerance limits from the loss function and the standard deviation from the entire experiment (take the square root of the MSD for the entire experiment). The Cp will be $(USL\text{-}LSL)/6\sigma$ (see Chapter 3).
12. Construct the variance equation. This is used to govern the mechanics of changing the parameter variances. It is based on cost balancing, from customer losses to company costs down to the actual costs associated with making each manufacturing process capable of stable output of the *required* variance. Here is an introduction to how it works.

The design will have a *current* quality-loss value defined from the loss function and the data from the ANOVA. The *required* quality-loss value is the reduced level of loss that is required to stay well below customer loss limits, which can be economically justified as an expenditure by the company. A loss-reduction ratio can be defined as follows:

$$\frac{Loss_{required}}{Loss_{current}} = loss\ reduction\ ratio \tag{17.1}$$

If you know the required or desired amount of loss that your business can afford to absorb in reducing customer loss (as opposed to funding warranty costs), the loss reduction ratio will be very helpful in allocating the overall variance reduction requirements down to the parameters. The loss terms can be converted to mean square deviation terms:

$$\frac{Loss_{required}}{Loss_{current}} = \frac{k(MSD_{required})}{k(MSD_{current})} \tag{17.2}$$

The MSD terms are the same as variance terms. Since the *k* values are the same they cancel out of the equation:

$$= \frac{k(MSD_{required})}{k(MSD_{current})} = \frac{\sigma^2_{required}}{\sigma^2_{current}} = variance\ reduction\ ratio \tag{17.3}$$

The variance equation simply states how the overall variance ratio relates to the reduction in individual parameter variances. We have two options on how to apply the variance equation [R17.2]. They are as follows:

Option 1: A Company Cost-Driven Process

This process is driven by how a company can afford to control loss with respect to its own internal financial performance requirements and its customer's financial loss minimization requirements. Here we know the *required* variance, and we're capable of allocating that overall new variance down to each parameter variance, as shown in Figure 17.1.

The term *variance adjustment factor* (VAF) needs to be determined to create new parameter MSD values ($VAF^2 \times \sigma^2_{parameter}$). VAF is tactically developed for each parameter by working with the

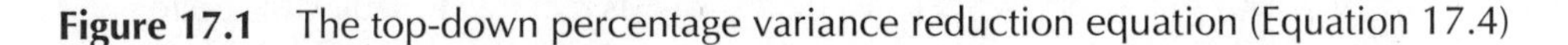

Figure 17.1 The top-down percentage variance reduction equation (Equation 17.4)

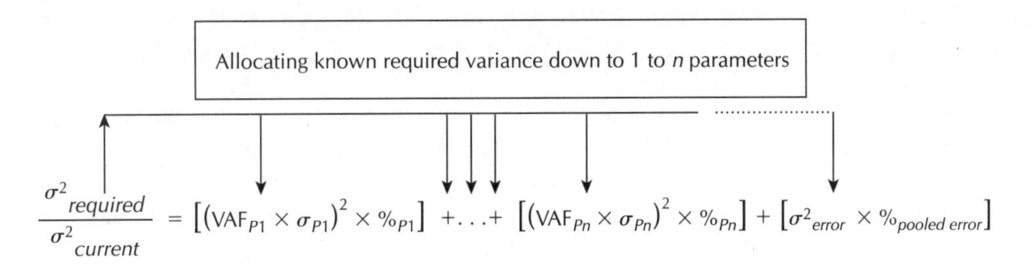

manufacturing engineer. The variance for each parameter is defined by $\sigma^2_{\text{parameter}}$. The $\%_{\text{parameter}}$ term represents the percent contribution each parameter makes to the overall output response from the ANOVA data. Note that the error variance term is included at the end of the equation. This experimental variation "constant" must be subtracted from the variance reduction ratio (left-hand side of the equation), since it has no role to play in the reduction process.

Each parameter is allocated a *reasonable* portion of the total variance reduction requirement. This process puts the burden of variance reduction on the individual parameter manufacturing engineers. If one parameter bottoms out on variance reduction, another will have to pick up the difference. If a few of the parameters can economically absorb the total reduction requirement and perhaps some extra, then all the remaining parameter variances may be relaxed as determined by manufacturing costs for each parameter. It won't be very often that one knows the overall required variance based on what the company can afford to "invest" in loss reduction. It is much more likely that there may be a target variance for the output response that can be allocated down to the parameters. By far, the most likely cases will require a bottom-up approach to overall variance reduction. In this case, you will have to accept whatever overall reduction can be obtained after summing up all the parameter variance reductions and the costs accrued to obtain them. This is the focus of option 2.

Option 2: A Manufacturing Capability and Cost-Driven Process

What is the manufacturing process economically capable of producing? Sometimes allocating variance reduction down to the parameters looks feasible until the manufacturing engineer has a chance to provide the facts on just how low the variance for a particular parameter can be driven. Each manufacturing process has a variance output that the current state of the art can support, and no more—*at any price.* Thus, the term $(\text{VAF}_{P_1} \times \sigma_{P_1})^2$ can go only so far down and that is all, even with 100 percent inspection and part sorting. We are left to consider a different approach—one that generates lower variances on a parameter-by-parameter basis as the respective manufacturing process capabilities dictate.

Option 2 reverses the flow of the variance reduction process. Instead of allocating variance reduction down to the parameters, the process is begun at the parameter level, where each manufacturing process is optimized, and the resulting variance reduction is delivered to sum with all the other variance reductions. A new, lower overall variance is produced by the best each manufacturing process has to offer from an economic perspective. If cost is no object—a rare case in modern industry—then the lowest *feasible* variance can be obtained from the manufacturing process. If the economics of the business, based on loss-function analysis, allow for this rare case, then it can be employed; at least one has a credible metric to steady one's nerves when this radical case is required to satisfy the customers' needs.

Option 2 takes on the characteristics shown in Figure 17.2. Both top-down and bottom-up variance reduction equations have the same terms and form. How the *process* of variance reduction is carried out is different, as we have illustrated.

The unknown value that we usually seek to accurately define in either approach is the individual parameter term—VAF. This value is a fraction (less than 1) that expresses a portion of the overall variance reduction that must come from a *specific* parameter. Typically, each VAF will be different unless similar components are being used more than once in a design (such as circuit components, for instance).

Figure 17.2 The bottom-up variance reduction equation (Equation 17.5)

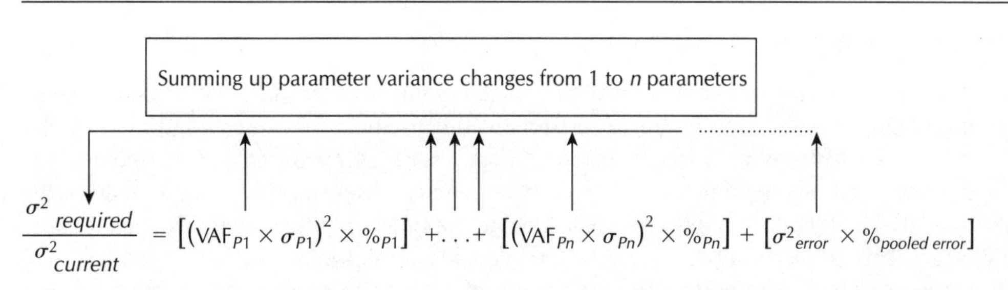

$$\frac{\sigma^2_{required}}{\sigma^2_{current}} = \left[(\text{VAF}_{P1} \times \sigma_{P1})^2 \times \%_{P1}\right] + \ldots + \left[(\text{VAF}_{Pn} \times \sigma_{Pn})^2 \times \%_{Pn}\right] + \left[\sigma^2_{error} \times \%_{pooled\ error}\right]$$

13. Use the new, lower parameter MSD values in the loss function to calculate the new, lower loss values. Recall from Chapter 11 that when the variance drops in value, losses to the customer will follow. Once this is done, the cost differential values (increase in cost for each parameter to attain the required variance reduction) for each parameter can be calculated to show the gain in quality through the reduction in loss. In these cases quality gains are quantified through reduction in quality-loss values. Note that the loss values go down while component costs go up. Ideally, losses will go down at a greater rate than the rate of increase in the manufacturing costs. This is the beginning of the cost balancing process.
14. Meet with the manufacturing engineers responsible for the production processes that make the components to define the increase in component costs associated with the reduction in each parameter MSD. This new, higher cost is called the *upgrade cost.* The difference between the reduction in quality loss (in dollars) and the increase in component cost is called the *net gain* or *net loss.* If there is a net gain in quality through loss reduction, we proceed with the upgrade; if the cost of upgrading exceeds the reduction in quality loss, we do not proceed. The new, lower total loss can be calculated from the net positive quality gains from the components selected to be upgraded.
15. Obtain the new, upgraded components and rerun the parameters in the orthogonal array experiment at the new standard deviation limits as defined by the reduction in parameter MSDs. Use the original compound noises in this verification experiment. Rerun the ANOVA on the new set of experimental data points. Calculate the new signal-to-noise ratio for the experiment to compare in a new $r_{S/N}$ factor comparison (see Chapter 2).
16. Calculate the new Cp from the Customer tolerance limits and the new standard deviation coming from the upgraded design data (using the new ANOVA values) from the verification experiment. Display the old and new values relative to the design's cost and quality:
    - Old and new quality loss (in dollars)
    - Upgrade costs and net gain in loss reduction (in dollars)
    - Original CP and new CP
    - Old signal/noise ratio from gate 1 and new signal-to-noise ratio for gate 3
    - Old $r_{S/N}$ factor from gate 1 and new $r_{S/N}$ factor for gate 3 (from Chapter 2)

This process can be used on either statistical variance experiments or tolerance experiments. Adjustments have to be made when working with tolerance experiments since they are not built on direct changes in the standard deviations of the parameters. The recommended approach is the statistical variance experimental approach because it deals directly with standard deviations and variances.

The ASI Circuit Case Study [R17.1]

Identifying the Parameters for Tolerancing

The case we are about to analyze is based on an electrical circuit that is made up of 23 components.

The control circuit is designed to use 14 resistors, two capacitors, two diodes, and four transistors to convert a single voltage input into a specific output frequency measured in Hertz (cycles per second). From analytical engineering analysis and experimental parameter design data analysis the team finds that only 12 of the 23 components are critical to the function of the circuit from a tolerancing point of view. This is not to say that the other components are not important—just that they are known to be secondary in their importance to the physics of actively controlling the variation for output of the circuit. They have undergone analysis previous to tolerancing and have been found to be weak compared to the other components with respect to their tuning effect on the *average* output response. They have also been found to be weak in how they help control *variability* in the response. So, the design has already undergone a first iteration of a rough form of sensitivity analysis during parameter design.

When a design undergoes the rigorous evaluation of the parameters design process it is scrutinized for parameter sensitivity to noise factors. This has two benefits: First is that the parameters can be sized for robustness against noise; second is that the ranking of parameter contribution to output-response variation is quantified such that a screening process can be employed to rationalize what parameters are worth putting into tolerancing experiments. Some factors will not have a very strong effect on the average output response.

Figure 17.4 illustrates a parameter that is *not* a good candidate for tolerance analysis.

Figure 17.5 is representative of a parameter that *is* a good candidate for experimental tolerance analysis. The reason is that this parameter exhibits sensitivity to level fluctuations in *both* mean performance output and signal-to-noise output for the product response characteristic. Figure 17.4 shows that

Figure 17.3*

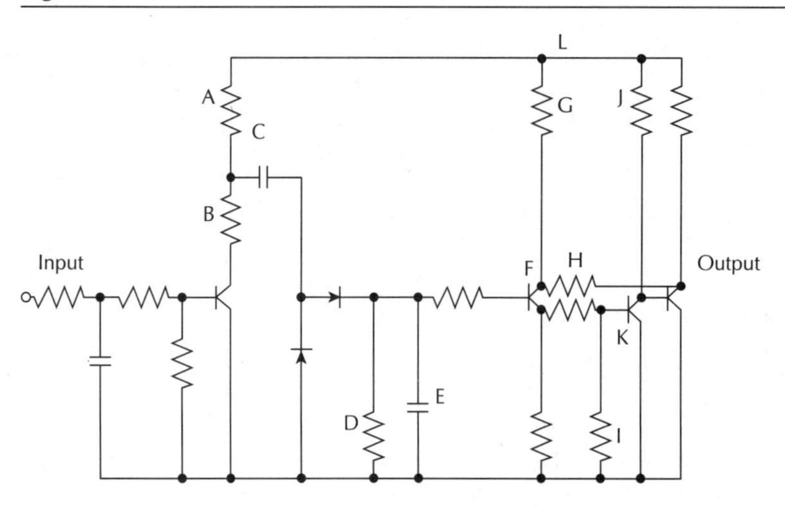

*Reprinted by permission of ASI.

Figure 17.4

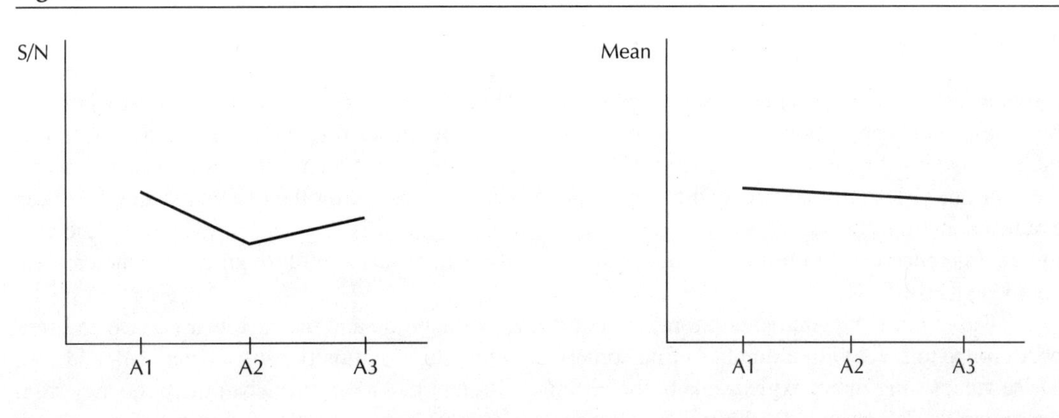

Figure 17.5

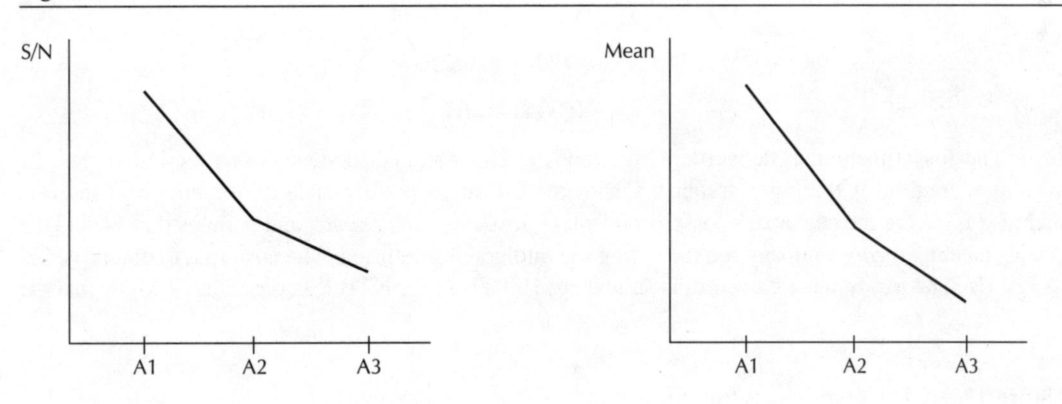

once the parameter is set at the optimum level or region for maximum signal-to-noise output (to reduce sensitivity to noise), the mean response is stable even with fluctuations around the optimum nominal set point. Minor changes in noise sensitivity may occur, but nominal output should be very stable. This is a case where wider tolerances may be economically acceptable. If the slope for Figure 17.4 were very steep, then it too would be a candidate for tolerance analysis.

Improving Quality in Cases Where Parameter Design Was Not Done

If this design had not been taken through parameter design but is just being leveraged for a new application, then all 23 components would probably have to be included in the initial experiment to rank sensitivities. Engineering analysis may suffice to eliminate some of the parameters, but the team would have nowhere near the information they would have had through the application of parameter design. Many companies do not yet habitually practice robust design, and therefore will have to use a longer sensitivity-analysis process for tolerance experimentation.

Constructing the Loss Function

The circuit is intended to produce a targeted output frequency of 570 Hz. As is always the case with the Taguchi approach, we will focus our efforts on driving the distributed performance of the output response as close to the targeted set point as possible. This circuit is a current design that is being reused in a new application. The management of the company decided to minimize the risk of performance problems by insisting on the use of loss function-based tolerance design, as opposed to taking the time and risk of developing a whole new circuit. Unfortunately, after the existing design had been evaluated and put through robustness optimization, it still lacked the capability required by the customer. Tolerance design must be done to systematically reduce variation through careful increases in component costs.

The engineering team joined forces with the marketing group and the customer service and support groups to meet with a number of customers to define the loss function parameters (A_0 and Δ_0). These values were direct expressions of the cost the customer had to absorb when the frequency from the circuit drifted away from the target. After meeting with a substantial cross-section of customers, the corporate team developed the following loss-function parameters:

$$A_0 = \$100$$

$$(\Delta_0)^2 = (720 - 420/2)^2$$

$$k = 0.00444\ (\$/\text{Hz}^2)$$

The loss function is defined in Figure 17.6. The nominal-the-best loss function represents an approximation of the costs associated with the off-target performance of the circuit. The costs included in A_0 are the customer's losses in business, losses in repair costs, and perhaps the costs to the circuit-manufacturing company for correcting the intolerable condition. The customer tolerance is 420 to 720 Hz. The loss at these extremes is valued at \$100. From the NTB loss function we know that the

Figure 17.6 The quality loss function

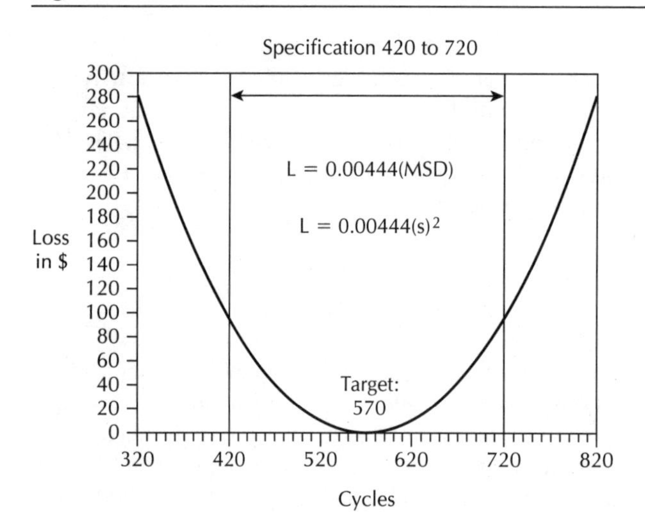

economic coefficient will equal $100/(570−420)^2. Thus, $k = 0.00444(\$/\text{Hz}^2)$. Since we are interested in analyzing a number of circuits and not just one, we must use the average loss function for n samples:

$$L(y) = k[\sigma^2] \tag{17.6}$$

The loss function becomes:

$$L(y) = 0.00444[\sigma^2]$$

With the loss function defined we are ready to study the components that control the physical output of the circuit. Each circuit component will contribute, in a unique and quantifiable manner, to the variation of the output frequency. The mechanism we will use to methodically gather data concerning the frequency output is the L_{16} orthogonal array. Taguchi has referred to the orthogonal array as a *design inspection device.* The orthogonal array is structured so that each circuit component is given an equal and balanced opportunity to exhibit its behavior on the frequency. Table 17.1, which lists components and their parameter set points, will guide in the selection of an appropriate orthogonal array.

The $\pm 1\sigma$ limits are defined for each component of the circuit. If we were conducting a tolerance experiment, we would be using the actual production tolerance estimates, which would be substantially wider than one standard deviation; typically they are three standard deviations on either side of the mean. Under six-sigma paradigm conditions they would be set at six standard deviations on either side of the mean, because the width of the distribution is substantially more narrow than is found in the three-sigma paradigm (see Figure 17.7). When sufficient work is put into optimizing the robustness of the manufacturing process—taking it into a six-sigma paradigm—the resulting output distribution can be quite helpful. One sigma still should be used when the component has parameters with a manufacturing Cpk of 1.5 or greater, because the ANOVA process is based on the calculation of parameter variances.

Table 17.1

Component	*Units*	*Target*	*Target + 1σ*	*Target − 1σ*
Resistor A	Kilo ohms	2.2	2.31	2.09
Resistor B	Kilo ohms	470	493.5	446.5
Capacitor C	Micro farads	0.68	0.71	0.646
Resistor D	Kilo ohms	100	105	95
Capacitor E	Micro farads	10	12	8
Transistor F	HFE	180	270	90
Resistor G	Kilo ohms	10	10.5	9.5
Resistor H	Kilo ohms	1.5	1.575	1.425
Resistor I	Kilo ohms	10	10.5	9.5
Resistor J	Kilo ohms	10	10.5	9.5
Transistor K	HFE	180	270	90
Input voltage L	Volts	6.5	6.8	6.2

Figure 17.7 The three- and six-sigma paradigms

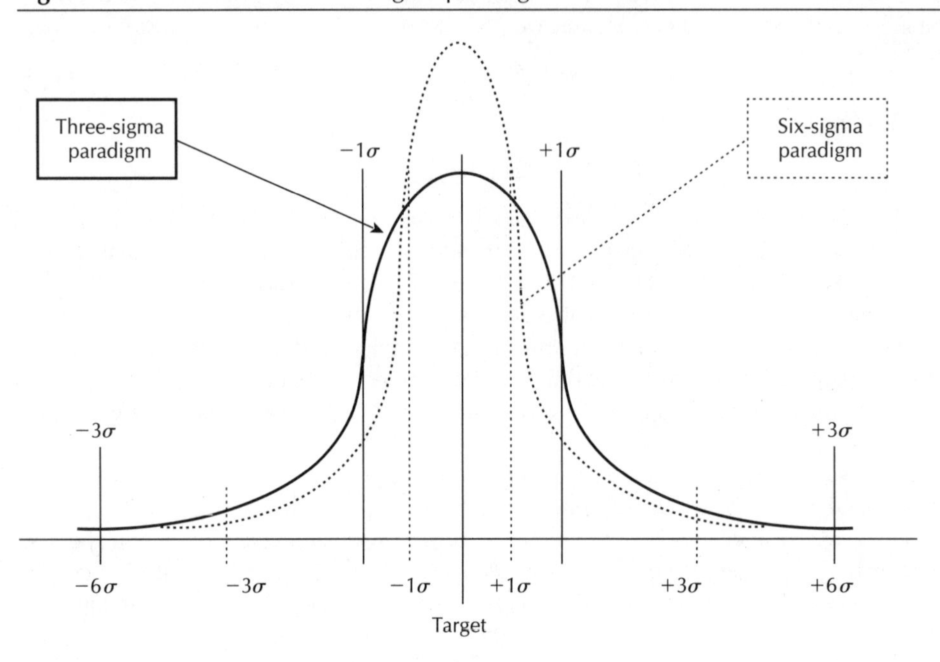

Our case is based on a design from the three-sigma paradigm with the experimental parameter limits set at the one-sigma limits.

Setting Up and Running the Experiment

Now that we have identified the parameters for the experiment and their levels, we can select an appropriate array. Because we have 12 parameters and have concluded that there are no significant interactions to worry about, we find that an L_{16} array has enough degrees of freedom for this case. We need just 12 DOF, and the L_{16} provides 15. We will demonstrate the capabilities of WinRobust and ANOVA-TM software in this case study. After opening WinRobust, open the Select Orthogonal Array box under Array menu, as shown in Figure 17.8. You will find the two-level and three-level arrays. Click the L_{16} diamond, then click OK.

WinRobust will immediately display an empty L_{16} array, as shown in Figure 17.9. You need to load it with the parameters and standard deviation set points previously defined.

To load the first parameter click on the letter A in the array column. After clicking on the letter Figure 17.10 will appear. Type in the name of the parameter and then enter each set point, as shown in Figure 17.10. The lower set points (1s) will be the parameter nominal minus the standard deviation, and the upper set points (2s) will be the parameter nominal plus the standard deviation.

Click OK when you have entered the name and set points. This process is repeated for all 11 of the remaining parameters. Figure 17.12 shows the L_{16} with all 12 parameters properly loaded. Note we have left the last three columns empty. Do not open these columns. If you so, you must unassign them so that their degrees of freedom are left open to be pooled to represent experimental error (see Chapter 16).

Figure 17.8

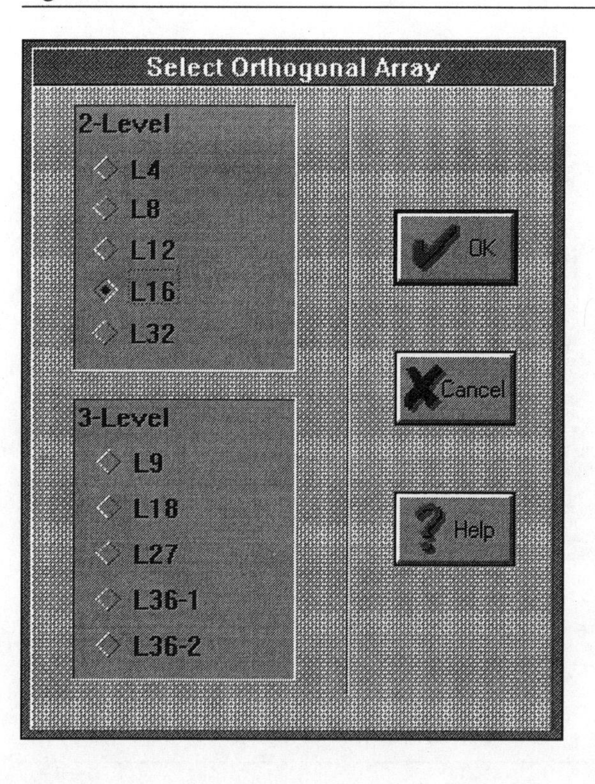

Figure 17.9 The L_{16} orthogonal array

Run	*1*	*2*	*3*	*4*	*5*	*6*	*7*	*8*	*9*	*10*	*11*	*12*	*13*	*14*	*15*
1	1	1	1	1	1	1	1	1	1	1	1	1	1	1	1
2	1	1	1	1	1	1	1	2	2	2	2	2	2	2	2
3	1	1	1	2	2	2	2	1	1	1	1	2	2	2	2
4	1	1	1	2	2	2	2	2	2	2	2	1	1	1	1
5	1	2	2	1	1	2	2	1	1	2	2	1	1	2	2
6	1	2	2	1	1	2	2	2	2	1	1	2	2	1	1
7	1	2	2	2	2	1	1	1	1	2	2	2	2	1	1
8	1	2	2	2	2	1	1	2	2	1	1	1	1	2	2
9	2	1	2	1	2	1	2	1	2	1	2	1	2	1	2
10	2	1	2	1	2	1	2	2	1	2	1	2	1	2	1
11	2	1	2	2	1	2	1	1	2	1	2	2	1	2	1
12	2	1	2	2	1	2	1	2	1	2	1	1	2	1	2
13	2	2	1	1	2	2	1	1	2	2	1	1	2	2	1
14	2	2	1	1	2	2	1	2	1	1	2	2	1	1	2
15	2	2	1	2	1	1	2	1	2	2	1	2	1	1	2
16	2	2	1	2	1	1	2	2	1	1	2	1	2	2	1

Figure 17.10 Entering parameter set points into the L_{16}

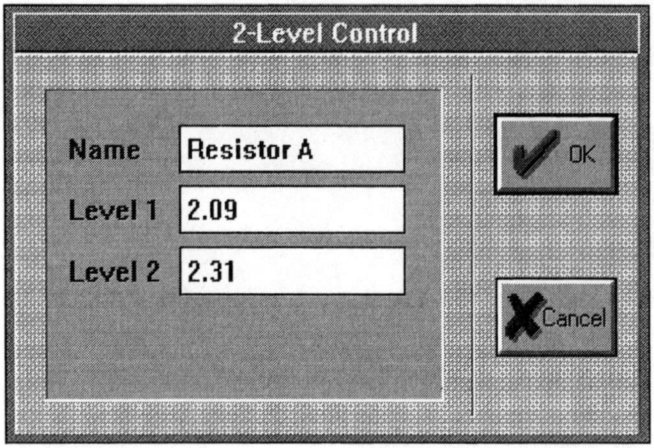

2-Level Control

Name Resistor A

Level 1 2.09

Level 2 2.31

OK

Cancel

Figure 17.11 The L_{16} array loaded with the 12 parameters

Run	1 Res. A	2 Res. B	3 Cap. C	4 Res. D	5 Cap. E	6 Trans. F	7 Res. G	8 Res. H	9 Res. I	10 Res. J	11 Trans. K	12 Voltage L	13	14	15
1	2.09	446.5	.646	95	8	90	9.5	1.425	9.5	9.5	90	6.2	1	1	1
2	2.09	446.5	.646	95	8	90	9.5	1.575	10.5	10.5	270	6.8	2	2	2
3	2.09	446.5	.646	105	12	270	10.5	1.425	9.5	9.5	90	6.8	2	2	2
4	2.09	446.5	.646	105	12	270	10.5	1.575	10.5	10.5	270	6.2	1	1	1
5	2.09	493.5	.714	95	8	270	10.5	1.425	9.5	10.5	270	6.2	1	2	2
6	2.09	493.5	.714	95	8	270	10.5	1.575	10.5	9.5	90	6.8	2	1	1
7	2.09	493.5	.714	105	12	90	9.5	1.425	9.5	10.5	270	6.8	2	1	1
8	2.09	493.5	.714	105	12	90	9.5	1.575	10.5	9.5	90	6.2	1	2	2
9	2.31	446.5	.714	95	12	90	10.5	1.425	10.5	9.5	270	6.2	2	1	2
10	2.31	446.5	.714	95	12	90	10.5	1.575	9.5	10.5	90	6.8	1	2	1
11	2.31	446.5	.714	105	8	270	9.5	1.425	10.5	9.5	270	6.8	1	2	1
12	2.31	446.5	.714	105	8	270	9.5	1.575	9.5	10.5	90	6.2	2	1	2
13	2.31	493.5	.646	95	12	270	9.5	1.425	10.5	10.5	90	6.2	2	2	1
14	2.31	493.5	.646	95	12	270	9.5	1.575	9.5	9.5	270	6.8	1	1	2
15	2.31	493.5	.646	105	8	90	10.5	1.425	10.5	10.5	90	6.8	1	1	2
16	2.31	493.5	.646	105	8	90	10.5	1.575	9.5	9.5	270	6.2	2	2	1
	2×3	1×3	1×2	1×5	1×4	1×7	1×6	1×9	1×8	1×11	1×10	4×8	1×12	2×12	3×12
	4×5	4×6	4×7	2×6	2×7	2×4	2×5	2×10	2×11	2×8	2×9	5×9	4×9	4×10	4×11
	6×7	5×7	5×6	3×7	3×6	3×5	3×4	3×11	3×10	3×9	3×8	6×10	5×8	5×11	5×10
	8×9	8×10	8×11	8×12	9×12	10×12	11×12	4×12	5×12	6×12	7×12	7×11	6×11	6×8	6×9

Figure 17.12 The Labeling window

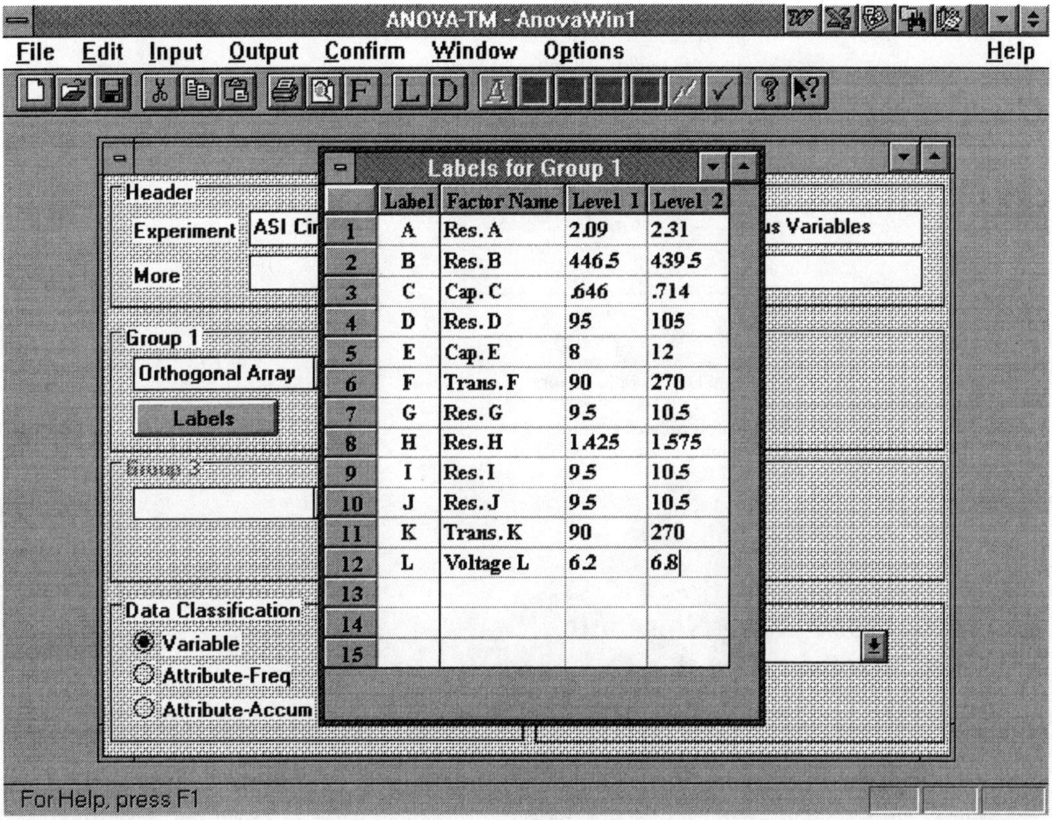
ANOVA-TM - AnovaWin1
File Edit Input Output Confirm Window Options Help
Labels for Group 1
Header
Experiment ASI Ci
s Variables
More
Group 1
Orthogonal Array
Labels
Data Classification
Variable
Attribute-Freq
Attribute-Accum
Label Factor Name Level 1 Level 2
1 A Res. A 2.09 2.31
2 B Res. B 446.5 439.5
3 C Cap. C .646 .714
4 D Res. D 95 105
5 E Cap. E 8 12
6 F Trans. F 90 270
7 G Res. G 9.5 10.5
8 H Res. H 1.425 1.575
9 I Res. I 9.5 10.5
10 J Res. J 9.5 10.5
11 K Trans. K 90 270
12 L Voltage L 6.2 6.8
13
14
15
For Help, press F1

We will use the numerical values for these three unassigned columns in the ANOVA to quantify the mean square due to experimental error value, MS_{error}. We will not take replicates in this case study; if the array were saturated, we would probably take three replicates for each of the experimental runs. The other option is to pool the weak contributors to the overall response. We will pool the DOF from the weak factors and the DOF from the empty columns in this case study.

For software comparison purposes, ANOVA-TM is used to set up the experiment in the following manner. Under the File menu, open a new experiment. Select an L_{16} Orthogonal Array from Group One. Using the Labeling process, defined in Chapter 16, the window shown in Figure 17.12 can be constructed. ANOVA-TM is now ready for data input from the L_{16} experiment.

Two- versus Three-Level Experiments

Notice that we have selected a two-level array. Since the circuit has already been through parameter design, we have no real need to expend the experimental energy (DOF) or expense to evaluate the target level of the parameters. We already know they are where we want them to be from a robustness standpoint.

Figure 17.13 Noise diagram for a circuit

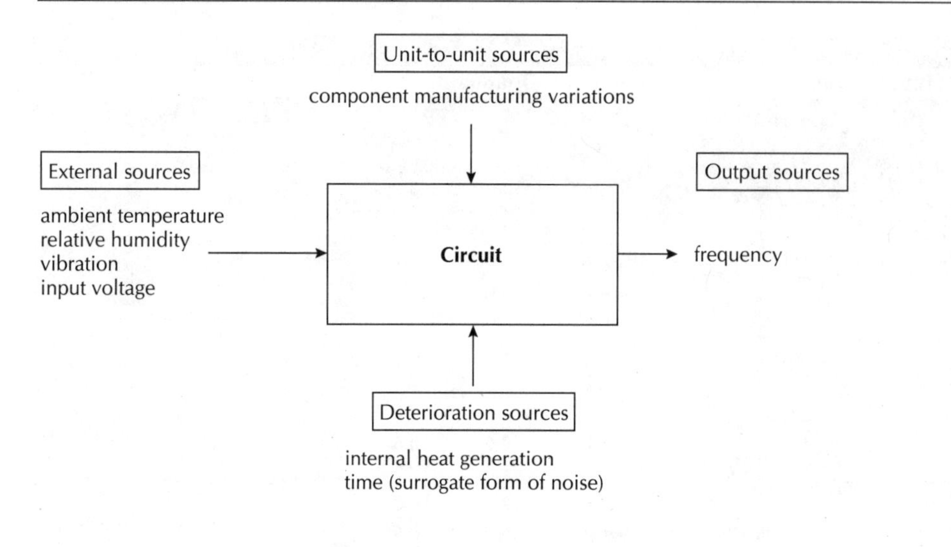

Techniques for Putting Noise into the Tolerance Experiment

In this case we are working with an electronic circuit. The sources of noise for such a device may be defined in a noise diagram such as is shown in Figure 17.13.

In this case, we find that the input voltage can be included as a tolerance parameter in the main orthogonal array; the unit-to-unit noises are the standard deviations used to set the parameters. The external noises are set at worst-case levels of 90°F and 75 percent relative humidity. The circuit was subjected to a shock test prior to testing to emulate shipping conditions. The deterioration noise was induced by building the circuit from pre-aged components that have been intentionally stressed at accelerated temperatures. Prestressing components is a common strategy for inducing noise in an experiment.

Running the Experiment

The test plan called for taking only one frequency reading (no replicates) due to time constraints. To enter the response WinRobust will be working with, one must open the Response menu. Select Add Static Characteristic. Figure 17.14 will appear. In the Response box enter Frequency. In the Noise Conditions box enter 1. In the Replicates per Noise box enter 1. In the S/N Type section click on Custom (none); the S/N ratios are used in parameter design, not tolerance design. This leaves the response in the engineering units of frequency.

ANOVA-TM prepares for data entry as follows.

The Data Classification is input as Variable. The Signal-to-Noise is selected as None. This leaves the data untransformed and in standard engineering units.

Figure 17.14

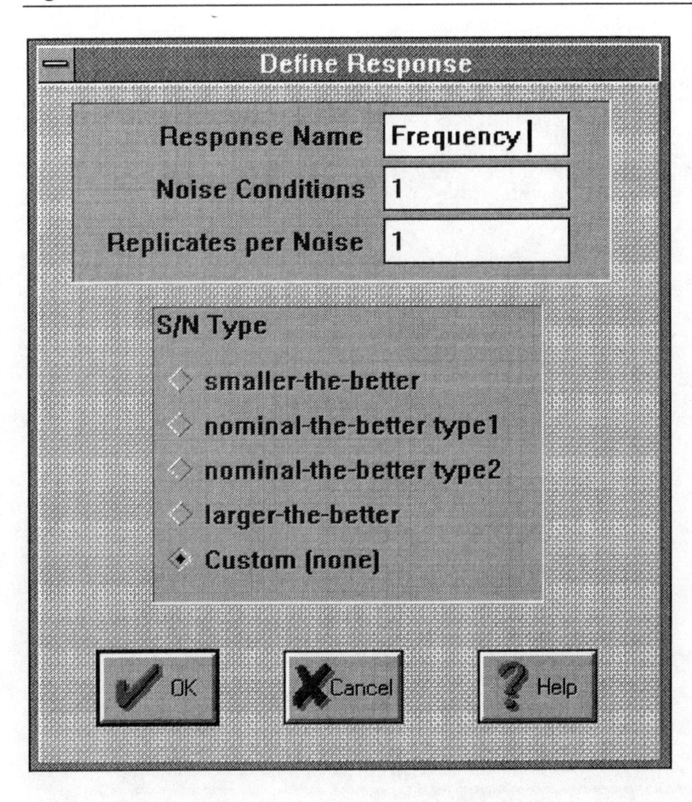
Define Response
Response Name
Frequency
Noise Conditions
1
Replicates per Noise
1
S/N Type
smaller-the-better
nominal-the-better type1
nominal-the-better type2
larger-the-better
Custom (none)
OK
Cancel
Help

Figure 17.15 The ANOVA-TM main experimental layout window

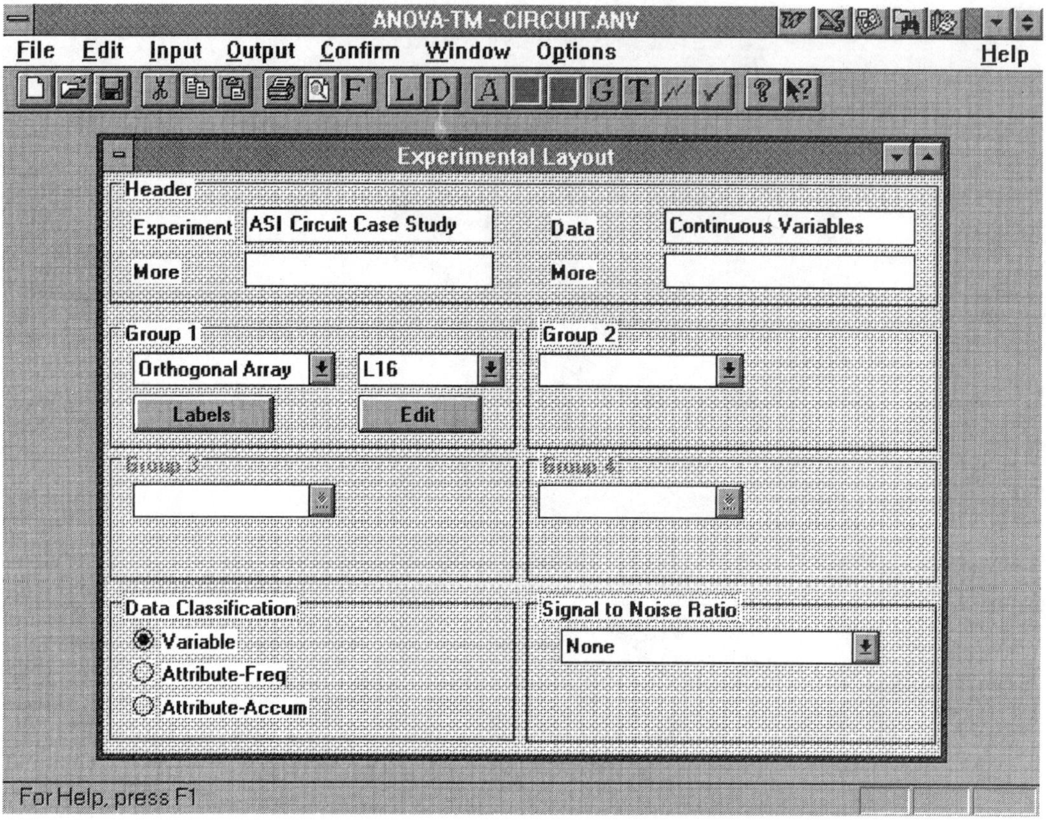

Data Entry

Now that the WinRobust software knows what the response is and how many data points are going to be entered you can click OK. The frequency data entry window shown in Figure 17.16 will appear. These are the spaces where you will enter the 16 data points from each of the experimental runs.

Once the data is entered, click on Crunch to enter the data into WinRobust.

Figure 17.17 shows the results.

We did not need to take replicates in this experiment because (1.) we have three empty columns with three DOF available to quantify experimental error, and (2.) we plan to pool the weak parameter data to add additional degrees of freedom to estimate experimental error.

ANOVA-TM inputs the data as follows.

Open the Input menu and the Input/Edit Data window shown in Figure 17.18 will appear. The data points can then be entered according to the order from the L_{16} array.

Both software packages are now loaded with the data and ready to output their respective ANOVA tables.

Figure 17.16 The Frequency data entry window for the L_{16}

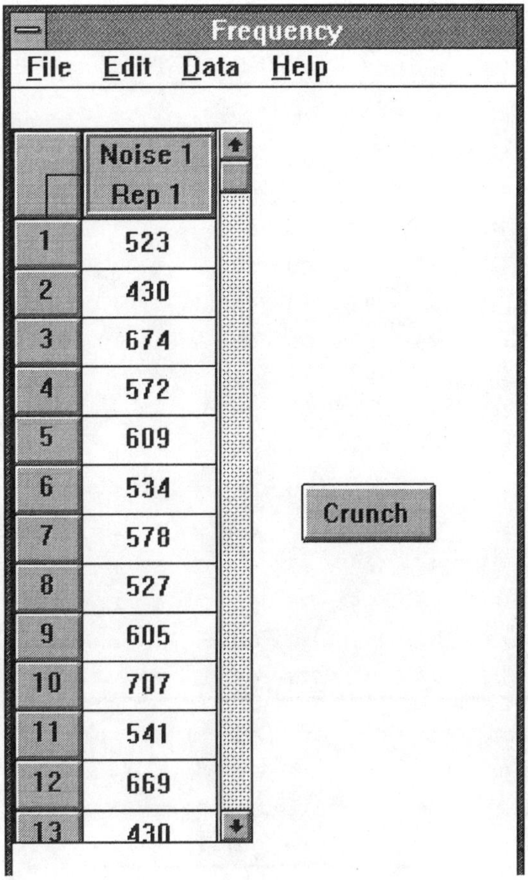

Frequency

File Edit Data Help

	Noise 1 Rep 1
1	523
2	430
3	674
4	572
5	609
6	534
7	578
8	527
9	605
10	707
11	541
12	669
13	430

Crunch

Figure 17.17 The L_{16} experimental results

Run	1 Res. *A*	2 Res. *B*	3 Cap. *C*	4 Res. *D*	5 Cap. *E*	6 Trans. *F*	7 Res. *G*	8 Res. *H*	9 Res. *I*	10 Res. *J*	11 Trans. *K*	12 Voltage *L*	13	14	15	Frequency S/N
1	2.09	446.5	.646	95	8	90	9.5	1.425	9.5	9.5	90	6.2	1	1	1	523.00 dB
2	2.09	446.5	.646	95	8	90	9.5	1.575	10.5	10.5	270	6.8	2	2	2	430.00 dB
3	2.09	446.5	.646	105	12	270	10.5	1.425	9.5	9.5	90	6.8	2	2	2	674.00 dB
4	2.09	446.5	.646	105	12	270	10.5	1.575	10.5	10.5	270	6.2	1	1	1	572.00 dB
5	2.09	493.5	.714	95	8	270	10.5	1.425	9.5	10.5	270	6.2	1	2	2	609.00 dB
6	2.09	493.5	.714	95	8	270	10.5	1.575	10.5	9.5	90	6.8	2	1	1	534.00 dB
7	2.09	493.5	.714	105	12	90	9.5	1.425	9.5	10.5	270	6.8	2	1	1	578.00 dB
8	2.09	493.5	.714	105	12	90	9.5	1.575	10.5	9.5	90	6.2	1	2	2	527.00 dB
9	2.31	446.5	.714	95	12	90	10.5	1.425	10.5	9.5	270	6.2	2	1	2	605.00 dB
10	2.31	446.5	.714	95	12	90	10.5	1.575	9.5	10.5	90	6.8	1	2	1	707.00 dB
11	2.31	446.5	.714	105	8	270	9.5	1.425	10.5	9.5	270	6.8	1	2	1	541.00 dB
12	2.31	446.5	.714	105	8	270	9.5	1.575	9.5	10.5	90	6.2	2	1	2	669.00 dB
13	2.31	493.5	.646	95	12	270	9.5	1.425	10.5	10.5	90	6.2	2	2	1	430.00 dB
14	2.31	493.5	.646	95	12	270	9.5	1.575	9.5	9.5	270	6.8	1	1	2	480.00 dB
15	2.31	493.5	.646	105	8	90	10.5	1.425	10.5	10.5	90	6.8	1	1	2	578.00 dB
16	2.31	493.5	.646	105	8	90	10.5	1.575	9.5	9.5	270	6.2	2	2	1	668.00 dB
	2×3	1×3	1×2	1×5	1×4	1×7	1×6	1×9	1×8	1×11	1×10	4×8	1×12	2×12	3×12	
	4×5	4×6	4×7	2×6	2×7	2×4	2×5	2×10	2×11	2×8	2×9	5×9	4×9	4×10	4×11	
	6×7	5×7	5×6	3×7	3×6	3×5	3×4	3×11	3×10	3×9	3×8	6×10	5×8	5×11	5×10	
	8×9	8×10	8×11	8×12	9×12	10×12	11×12	4×12	5×12	6×12	7×12	7×11	6×11	6×8	6×9	

Figure 17.18 The ANOVA-TM Input/Edit Data window

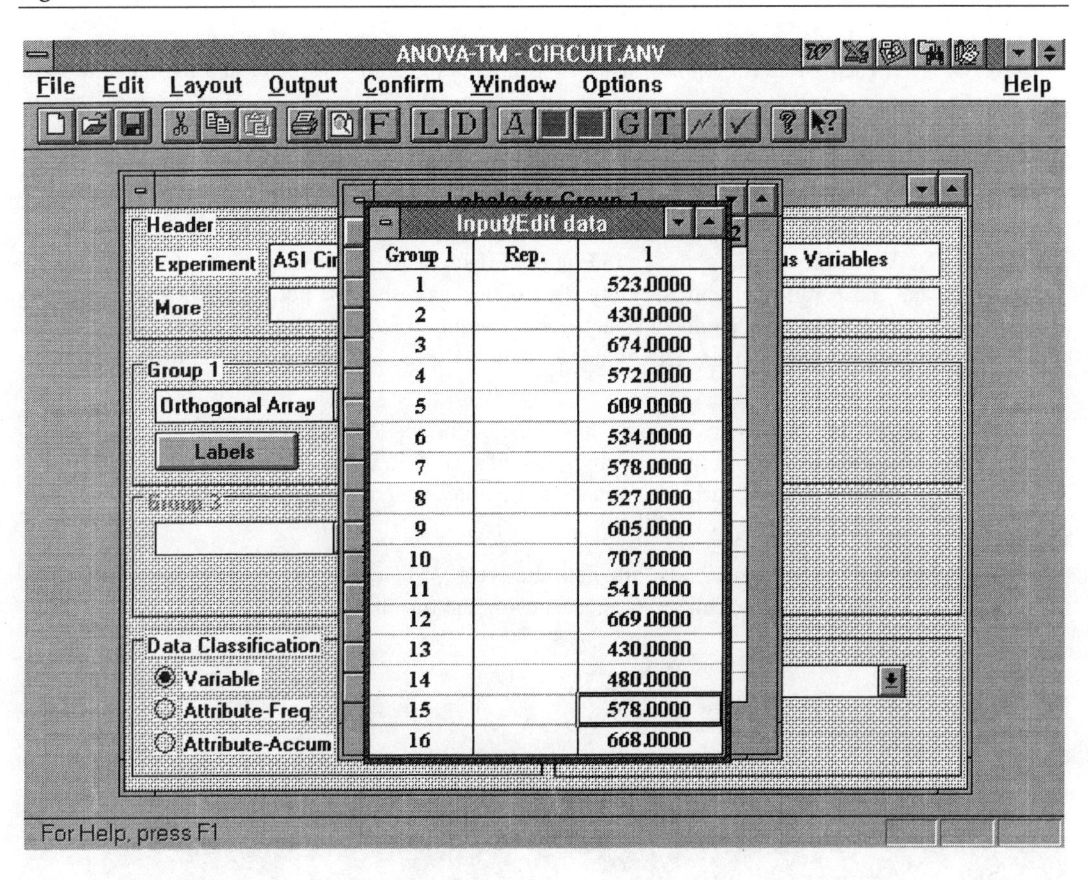

Measuring the Proper Response Characteristic

A direct physical measure of output can be readily gathered with an oscilloscope, which can be calibrated and is very easy to evaluate for accuracy and precision. Since our response variable is continuous and is quantified in Hertz, there is little concern that the data is of high quality. Since we have forced what can happen to happen in this experiment, there is very low risk that variability from random effects is going to outweigh the effect of the design parameters being changed. Such an engineered experiment takes teamwork and skill to put together and execute. There is little room for sloppy thinking or workmanship in such an endeavor.

Interactions in Tolerance Experiments

The engineering team that developed this circuit determined that the parameter interactivity in its design was at a low enough level that there were no degrees of freedom allocated to study such effects. The interactive effects will essentially act as a noise in this experiment.

Analyzing the Data

We have entered the 15 data points into WinRobust for analysis. It has a good capability for performing ANOVA on Taguchi-style orthogonal array experiments for both parameter and tolerance design.

Applying ANOVA

Let's walk through how to conduct an ANOVA on the data step by step. The first step is to look under the Window menu. Select Results and Analysis or click on the ANOVA box just below the menu bar to display the ANOVA results window, shown in Figure 17.19.

Notice that the F ratio column is empty; this is because we have not told WinRobust how to account for experimental error. To do so select *Error Pooling* under the Options menu. The window shown in Figure 17.20 will appear. This window contains three options to tell WinRobust how to account for experimental error.

- unassigned columns as error (pools unassigned columns to represent error),
- use n degrees of freedom (pools your choice of n DOF to represent error), and
- user estimate of error variance (you must manually calculate and enter the error variance from replicate runs).

WinRobust will not automatically calculate the error variance due to replicate runs. You must manually calculate the variance of all the replicates using the formula:

$$V_e = \frac{1}{n-1} \sum (y_i - \bar{y})^2$$

Figure 17.19 The ANOVA window prior to estimation of error

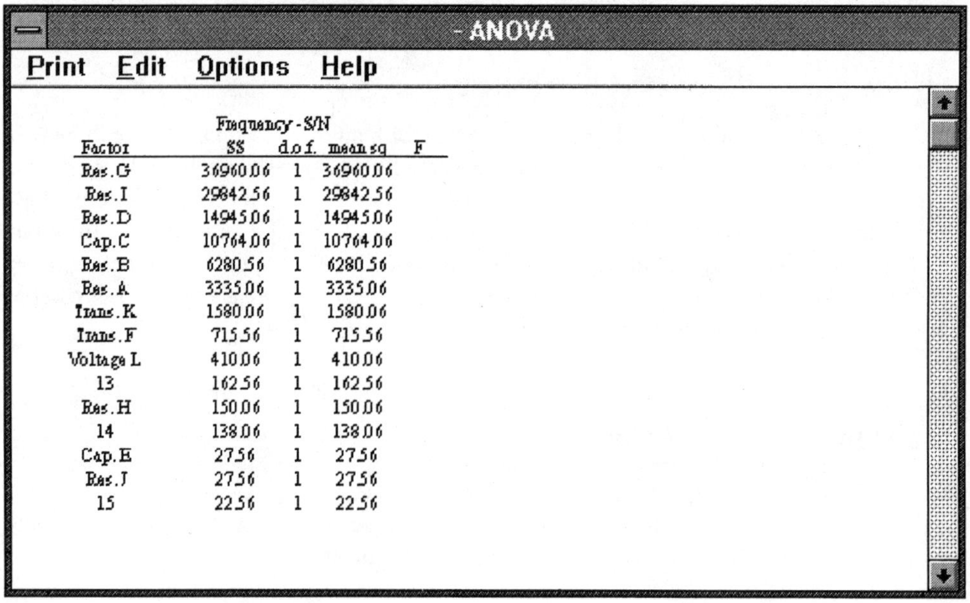

- ANOVA

Print Edit Options Help

Frequency - S/N

Factor	SS	d.o.f.	mean sq	F
Res.G	36960.06	1	36960.06	
Res.I	29842.56	1	29842.56	
Res.D	14945.06	1	14945.06	
Cap.C	10764.06	1	10764.06	
Res.B	6280.56	1	6280.56	
Res.A	3335.06	1	3335.06	
Trans.K	1580.06	1	1580.06	
Trans.F	715.56	1	715.56	
Voltage L	410.06	1	410.06	
13	162.56	1	162.56	
Res.H	150.06	1	150.06	
14	138.06	1	138.06	
Cap.E	27.56	1	27.56	
Res.J	27.56	1	27.56	
15	22.56	1	22.56	

Figure 17.20 The Error Variance Estimation window

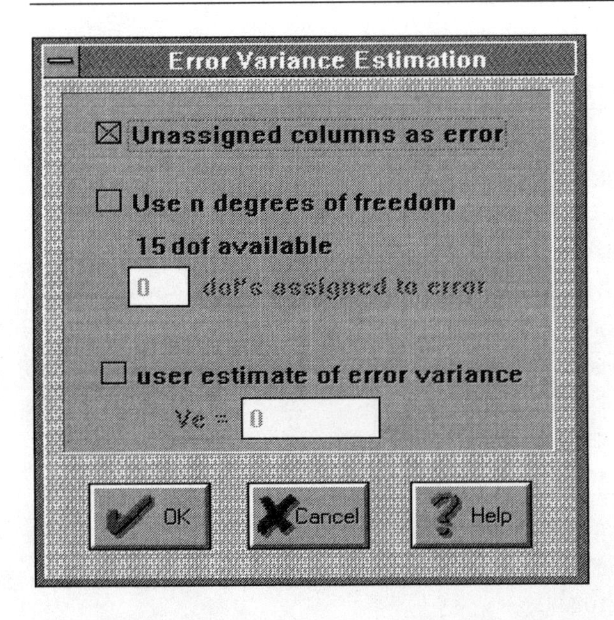

This value can also be calculated with a calculator with statistical functions.

First, we will just pool the unassigned columns as estimates of error variance; just click on the box for this assignment and an X will appear. The ANOVA table appears as shown in Figure 17.21.

Now we can see what the *F* ratio values are for the 12 parameters. We have pooled the empty columns to represent the experimental error. The parameters Resistor H, Capacitor E, and Resistor J are indistinguishable from experimental error and are additional candidates to include in the estimate of experimental error. Let's include their three DOFs as additional sources of error. We must tell Win-Robust we want to use n = DOF.

We return to the Error Variance Estimation window, click on the Use n Degrees of Freedom box, type in 6, and click OK. The new ANOVA table shown in Figure 17.23 is produced.

Now we see the top nine significant parameters, which are the ones we will consider in the loss function-based tolerance analysis. We can now calculate the percentage each parameter contributes to the overall response. The total sum of mean squares is 104,920.67. The total DOF equals 15. The percentages are as follows:

Resistor G	[36,960/104,920.67] × [100]	= 35.23%
Resistor I	[29,842.5/104,920.67] × [100]	= 28.44%
Resistor D	[14,945/104,920.67] × [100]	= 14.24%
Capacitor C	[10,764/104,920.67] × [100]	= 10.26%
Resistor B	[6,280.5/104,920.67] × [100]	= 5.99%
Resistor A	[3,335/104,920.67] × [100]	= 3.18%
Transistor C	[1,580/104,920.67] × [100]	= 1.50%
Transistor F	[715.5/104,920.67] × [100]	= 0.68%
Voltage L	[410/104,920.67] × [100]	= 0.39%

Figure 17.21

Factor	Frequency-Mean SS	d.o.f.	mean sq	F
Res. G	36960.00	1	36960.00	342.2
Res. I	29842.50	1	29842.50	276.3
Res. D	14945.00	1	14945.00	138.4
Cap. C	10764.00	1	10764.00	99.7
Res. B	6280.50	1	6280.50	58.2
Res. A	3335.00	1	3335.00	30.9
Trans. K	1580.00	1	1580.00	14.6
Trans. F	715.50	1	715.50	6.6
Voltage L	410.00	1	410.00	3.8
Res. H	150.00	1	150.00	—
Cap. E	27.50	1	27.50	—
Res. J	27.50	1	27.50	—
error	324.00	3	108.00	

Figure 17.22

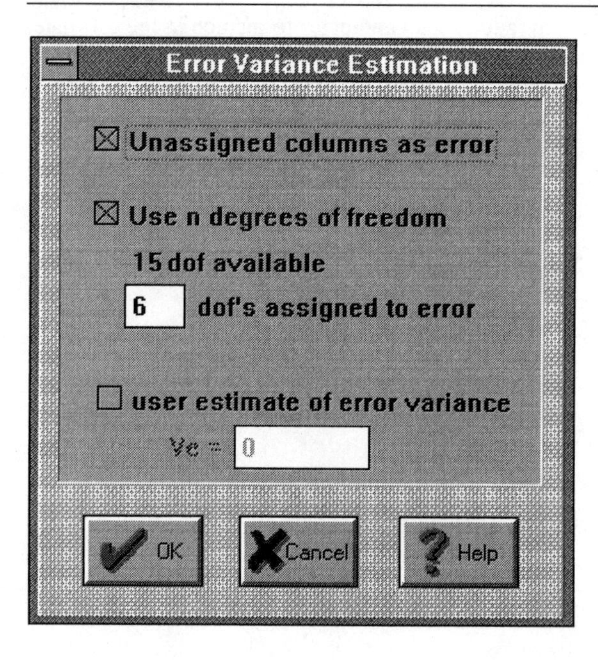

Figure 17.23 The ANOVA table for the top nine significant parameters

Factor	*SS*	*d.o.f.*	*mean sq*	*F*
	Frequency-Mean			
Res. G	36960.00	1	36960.00	419.2
Res. I	29842.50	1	29842.50	338.5
Res. D	14945.00	1	14945.00	169.5
Cap. C	10764.00	1	10764.00	122.1
Res. B	6280.50	1	6280.50	71.2
Res. A	3335.00	1	3335.00	37.8
Trans. K	1580.00	1	1580.00	17.9
Trans. F	715.50	1	715.50	8.1
Voltage L	410.00	1	410.00	4.7
error	529.00	6	88.17	

Figure 17.24 Plot of parameter contributions

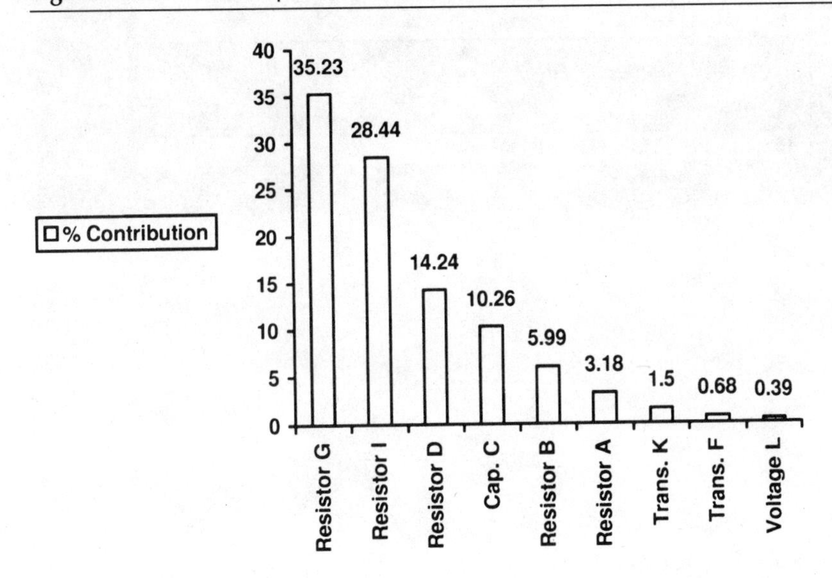

Figure 17.24 shows a plot of the percent contribution each parameter provides to the variation in the output response.

We can now see quite clearly what parameters are most significant in their influence on the output response. We will use this ranking information to aid in the determination of the *critical-to-function* (CTF) parameters.

Figure 17.25 ANOVA-TM ANOVA table window prior to pooling

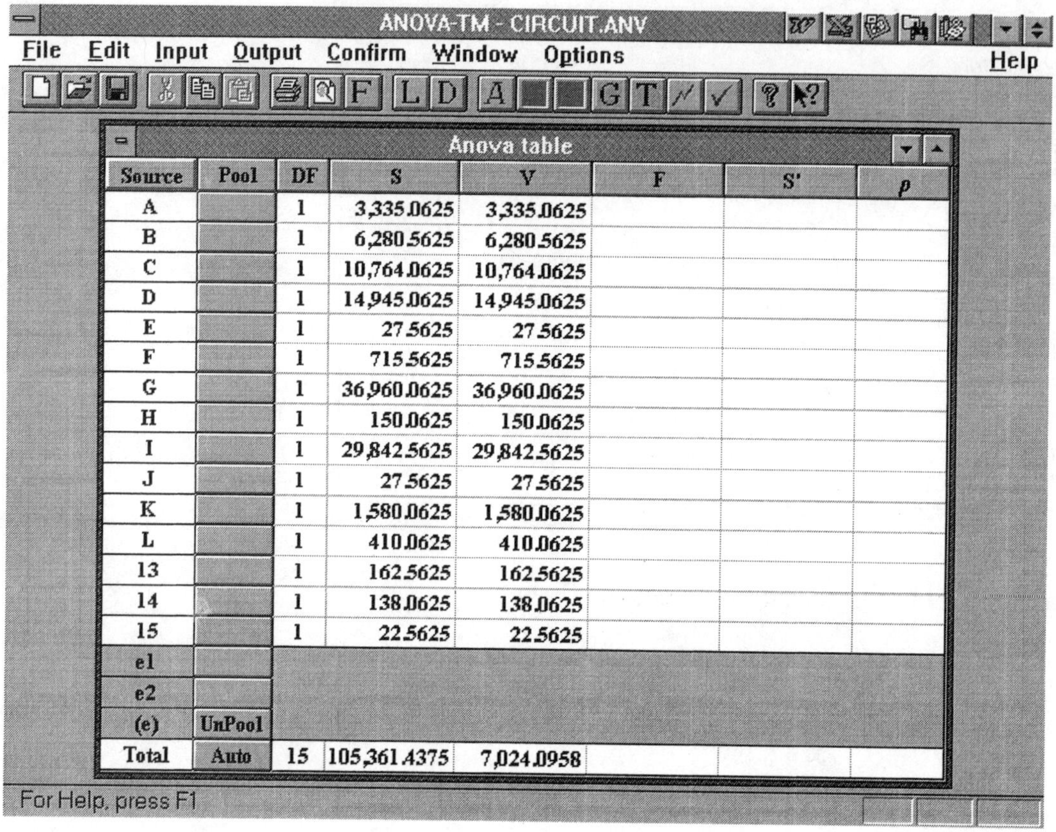

ANOVA-TM - CIRCUIT.ANV

File Edit Input Output Confirm Window Options Help

Anova table

Source	Pool	DF	S	V	F	S'	ρ
A		1	3,335.0625	3,335.0625			
B		1	6,280.5625	6,280.5625			
C		1	10,764.0625	10,764.0625			
D		1	14,945.0625	14,945.0625			
E		1	27.5625	27.5625			
F		1	715.5625	715.5625			
G		1	36,960.0625	36,960.0625			
H		1	150.0625	150.0625			
I		1	29,842.5625	29,842.5625			
J		1	27.5625	27.5625			
K		1	1,580.0625	1,580.0625			
L		1	410.0625	410.0625			
13		1	162.5625	162.5625			
14		1	138.0625	138.0625			
15		1	22.5625	22.5625			
e1							
e2							
(e)	UnPool						
Total	Auto	15	105,361.4375	7,024.0958			

For Help, press F1

ANOVA-TM processes the L_{16} data as follows.

Choose the Output menu, select ANOVA, then select Table to display the window shown in Figure 17.25.

The ANOVA table shows all the sum of squares and mean square values prior to any pooling of empty columns or low-value design parameters.

Selecting the three empty columns and the three weakest circuit components as estimates for experimental error produces the changes in ANOVA-TM's table shown in Figure 17.26.

Figure 17.26 The ANOVA table after pooling

ANOVA-TM - CIRCUIT.ANV

File Edit Input Output Confirm Window Options Help

Anova table

Source	Pool	DF	S	V	F	S'	ρ
A		1	3,335.0625	3,335.0625	37.8715	3,247.0000	3.08
B		1	6,280.5625	6,280.5625	71.3194	6,192.5000	5.88
C		1	10,764.0625	10,764.0625	122.2321	10,676.0000	10.13
D		1	14,945.0625	14,945.0625	169.7097	14,857.0000	14.10
E	(e)	1	27.5625	27.5625			
F		1	715.5625	715.5625	8.1256	627.5000	0.60
G		1	36,960.0625	36,960.0625	419.7026	36,872.0000	35.00
H	(e)	1	150.0625	150.0625			
I		1	29,842.5625	29,842.5625	338.8793	29,754.5000	28.24
J	(e)	1	27.5625	27.5625			
K		1	1,580.0625	1,580.0625	17.9425	1,492.0000	1.42
L		1	410.0625	410.0625	4.6565	322.0000	0.31
13	(e)	1	162.5625	162.5625			
14	(e)	1	138.0625	138.0625			
15	(e)	1	22.5625	22.5625			
e1							
e2							
(e)	UnPool	6	528.3750	88.0625		1,320.9375	1.25
Total	Auto	15	105,361.4375	7,024.0958			

For Help, press F1

Relating the ANOVA Data to the Loss Function and Process Capability (Cp)

Recall that the quality-loss function for the circuit was determined to be:

$$L(y) = 0.00444\sigma^2$$

We can calculate σ^2 for the circuit by dividing the total mean square value by the total DOF:

$$\sigma^2 = 104{,}920.67/15$$

$$\sigma^2 = 6994.71$$

The total quality loss for the circuit is:

$$\text{Loss} = 0.00444 \times 6994.71$$

$$\text{Loss} = \$31.05$$

Table 17.2

Parameter	*% Contribution*	*MSD = MS/15*	*Parameter Loss*
Resistor G	35.23	2,464	$10.94
Resistor I	28.44	1,989.5	$8.83
Resistor D	14.24	996.3	$4.42
Capacitor C	10.26	717.6	$3.19
Resistor B	5.99	418.7	$1.86
Resistor A	3.18	222.3	$0.99
Transistor K	1.50	105.3	$0.47
Transistor F	0.68	47.7	$0.21
Input Voltage L	0.39	27.3	$0.12
Experimental Error	.084	5.88	$0.03
Totals	99.92	6,994.71	$ 31.05

The process capability index (Cp) based on the current circuit tolerances and the standard deviation from the square root of the total experimental variance, σ^2, is:

$$Cp = (USL - LSL)/6\sigma$$

$$Cp = (720 - 420)/6(83.6)$$

$$Cp = .60$$

We must now focus on the individual parameters, their mean square deviation values, and their individual contributions to the quality loss. Each parameter column has 16 runs that cut through it, and each parameter has two levels assigned to it; therefore, each parameter has 16 opportunities to express its influence on the response. This means that $n - 1$ DOF are available to express the contribution each parameter is making. We can take the mean square values for the parameters and divide them by 15 to provide the measure we call the mean square deviation ($\text{MSD}_{\text{parameter}} = \text{MS}/n - 1\ \text{DOF} = \sigma^2$). For the nine parameters that are found to be significant ($F > 4$), we quantify the $\text{MSD}_{\text{parameter}}$ and the quality loss [$\text{Loss} = \text{k}(\text{MSD}_{\text{parameter}})$], as shown in Table 17.2.

We have developed a great deal of quantitative information concerning the parameter $\pm 1\sigma$ contributions to variation. Now we are poised to take specific action to balance the quality (output variability) and costs (on lower σ values).

Defining the Critical to Function (CTF) Factors

Recall that the parameters that rank high in F ratio value and high in percentage contribution to the overall output response variation are the *critical-to-function* parameters. These are parameters that have a *significant* effect on the response; they are to be the focus of the cost and quality balancing activities. There are three quantifiable measures that define which parameters are CTF and which are not:

- *F* ratio,
- Percentage contribution, and
- Parameter quality loss in dollars.

One must use engineering judgment along with these quantitative measures to guide the team in the task of selecting which parameters to invest money in to reduce their standard deviations. In fact, the lowest ranking parameters may be good candidates to consider for cost reduction by *loosening* their tolerances. The CTF parameters should be used to shrink the output variability. A new parameter can be defined to account for cost reduction. This is the topic of the next section.

Defining the Cost Improvement Parameters (CIP)

A few parameters can be identified that are potential cost improvement parameters. As mentioned, these will be the low-ranking parameters from the ANOVA table. Often one can use a CIP to help offset the cost increases that must be made to drive the output variation to an acceptable point. Figure 17.27 illustrates the quality-loss costs associated with the current circuit.

Before we can begin to make judgments on what parameters to target for variance reduction, we need to look at the costs associated with such reductions. We should take a moment to reflect back on the $r_{S/N}$-factor analysis discussed in Chapter 2. We are now down at the component level, where we are ready to shrink variability that was identified as the mechanism behind true reliability growth—*if* the reduction in variance is based on data generated in the presence of deterioration noise factors. This circuit data has deterioration noise resident within its values. Thus, as we selectively reduce parameter variation, by tightening tolerances, we build in the $r_{S/N}$-factor contributions that will build up to meet

Figure 17.27 Plot of parameter quality loss in dollars

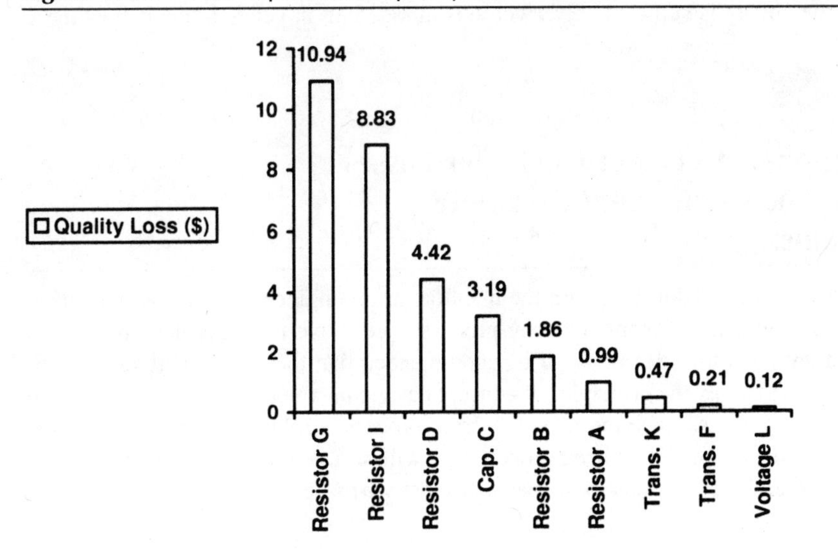

the overall reliability goals for the circuit. We not only balance costs and increase quality, but we strategically build in reliability that can be traced back up through the system. You now are working with an engineering process that habitually produces the following measurable items:

- Cost balance,
- Quality that is focused on customer expectations, and
- Reliability growth

Let's move on to see just how we identify the cost to the company in upgrading parameters through variance reduction.

Identifying and Quantifying the Costs Associated with Improving Quality

As we have discussed, there are two ways to work with the component manufacturing engineer relative to variance reduction:

- Allocated reduction targets: Enter the process with a predetermined, required level of variance reduction that *must* be attained, and
- Upgrade based on manufacturing process capability: Enter the process looking for the lowest possible variance reduction that is economically feasible for the individual component-manufacturing processes.

When pressed to the limit, these approaches are both going to run into the wall of physical law as it applies to manufacturing technology. The first approach has a definite target to go after—the quality loss. The second approach does not have a specific quality-loss target, but seeks to get the best variance reduction that is economically possible. The reduction in quality loss is going to be whatever falls out of the process relative to the output-response variation reduction. Practically speaking, the rigid allocation approach makes it harder to reduce company costs since one is forcing each critical parameter to its lowest limit until the quality is sufficient. Every case is unique but, in general, this is how these two approaches work.

Working with Suppliers to Lower Customer Losses through Reducing the Component Parameter Standard Deviations

A critical condition one will need to develop, with the manufacturing engineer, is the manufacturing process robustness optimization. If the manufacturing process is sensitive to noise, it is difficult to achieve the variance required at a minimum cost. It is recommended that the engineering team work with the manufacturing team to apply the principles of concept design, parameter design, and tolerance design to the manufacturing process. Once the manufacturing process is robust and under statistical control, one can be assured that the manufacturing process Cpk will be at a maximum. When the manufacturing Cpk is maximized, the conditions exist that will greatly help the team get the design Cpk maximized.

Calculating New Variances and MSD Values Using the Variance Equation

The circuit components were all toleranced initially by using $\pm 1\sigma$ about the nominal set points, which corresponded to approximately ± 5 percent of the nominal set point. The team decided to look at *uniformly* improving each circuit component (except for the input voltage, which is an external noise over which they have no direct control) by lowering the variability around the nominal to approximately ± 1 percent. This means that the standard deviation must be reduced by a factor of 5. In the variance equation, modified to state its terms in MSD form, the VAF (variance reduction factor) will have to be $(1/5)^2$.

$$\frac{\sigma^2_{required}}{\sigma^2_{current}} = [(VAF^2_{P1} \times MSD_{P1}) \times \%_{P1}] + \ldots + [(VAF^2_{Pn} \times MSD_{Pn}) \times \%_{Pn}]$$

Note that the $MSD_{\text{experimental error}}$ has been left out of the equation in this case study because its value is so small. Normally, it is subtracted from the ratio on the left-hand side of the equation.

$$\text{Uniform VAF} = 0.04$$

We can now return to the nine parameters, apply the VAF to each of them, and calculate the new mean square deviation values $[(VAF^2_{Pn} \times MSD_{Pn}) \times \%_{Pn}]$, including their percent contribution to the overall response. Table 17.3 shows the old MSD values, the adjusted MSD values, the old loss, the new loss, and the net reduction in loss.

The loss values in the table are calculated by using $0.00444(MSD_{old})$ and $0.00444(MSD_{new})$. The reduction in loss is the difference between the old and new losses. Notice the dramatic reduction in quality loss (−\$29.69) when we reduce the MSD by a factor of 5 ($.04 \times MSD_{old}$). The new total loss

Table 17.3 Comparison table for MSDs and losses

Parameter	*Percent Contribution*	*MSD_{old}*	*MSD_{new} .04(MSD_{old})*	*$Loss_{old}$*	*$Loss_{new}$*	*Net Δ in Loss*
Resistor G	35.23	2,464	98.56	\$10.94	\$0.44	−\$10.50
Resistor I	28.44	1,989.5	79.58	\$8.83	\$0.35	−\$ 8.48
Resistor D	14.24	996.3	39.85	\$4.42	\$0.18	−\$ 4.24
Capacitor C	10.26	717.6	28.70	\$3.19	\$0.13	−\$ 3.06
Resistor B	5.99	418.7	16.75	\$1.86	\$0.07	−\$ 1.79
Resistor A	3.18	222.3	8.89	\$0.99	\$0.04	−\$ 0.95
Transistor K	1.50	105.3	4.21	\$0.47	\$0.02	−\$ 0.45
Transistor F	0.68	47.7	1.90	\$0.21	\$0.01	−\$ 0.20
Input Voltage L	0.39	27.3	1.09	\$0.12	\$0.005	−\$0.015
Experimental Error	.084	5.88	5.88	\$0.03	–	–
Totals	99.92	6,994.71	285.41	\$ 31.05	\$1.36	−\$29.69

Figure 17.28 Plot of new quality loss in dollars

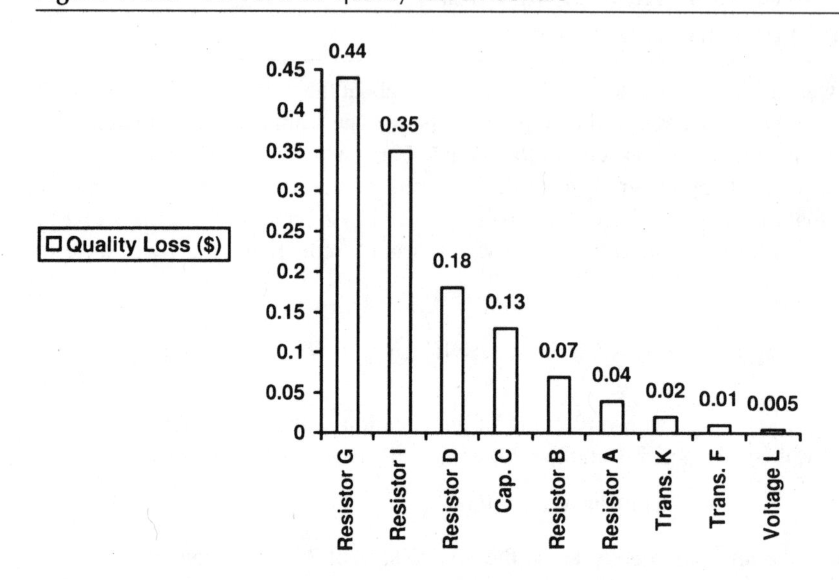

for the improved circuit is $1.36 (down from $31.05). These are remarkable results. Figure 17.28 displays the new quality-loss graph for the circuit.

We must now face the fact that it costs money to get these results. We are now ready to find out what the costs will be to attain the factor-of-5 reduction in variability.

Quantifying the Cost of Reducing the Parameter Standard Deviations

We have identified that we need to lower each parameter's MSD by a factor of 5. This will have to be verified by working with the suppliers of the nine circuit components. Each component will have a cost associated with the required reduction in output variability. For resistors, the cost increase is $0.06 each; the capacitor upgrade will cost $1 each; the transistor upgrades $2. Table 17.4 shows the upgrade costs and the new net change in the loss values.

We can see from the totals that the upgrade costs are small relative to the savings gained through the upgrade process. We still see an improvement in loss of $24.39 per circuit after paying $5.30 to upgrade each.

Figure 17.29 displays the new net change in loss including the upgrade costs for each component.

Notice that the two transistors end up creating *negative* quality-loss gains. This illustrates what happens when the cost to upgrade exceeds the quality-loss gain achieved by parameter variance reduction. We can now decide whether to include all the upgraded parameters or just those that deliver positive changes in net quality loss. Now we can see the power the analysis of variance and the quality-loss function deliver to the variance-reduction-versus-cost decision-making process. These are the data we need to make cost balancing decisions in tolerance development.

Table 17.4 Comparison table for MSDs and losses

Parameter	*MSD_{new} .04(MSD_{old})*	*$Loss_{new}$*	*Net Δ in Loss*	*Cost of Upgrade*	*New Net Δ in Loss*
Resistor G	98.56	$0.44	−$10.50	$0.06	−$10.44
Resistor I	79.58	$0.35	−$8.48	$0.06	−$8.42
Resistor D	39.85	$0.18	−$4.24	$0.06	−$4.18
Capacitor C	28.70	$0.13	−$3.06	$1.00	−$2.06
Resistor B	16.75	$0.07	−$1.79	$0.06	−$1.73
Resistor A	8.89	$0.04	−$0.95	$0.06	−$0.89
Transistor K	4.21	$0.02	−$0.45	$2.00	+$1.55
Transistor F	1.90	$0.01	−$0.20	$2.00	+$1.80
Input Voltage L	1.09	$0.005	−$0.015	–	–
Experimental Error	5.88	–	–		–
Totals	6,994.71	$1.36	−$29.69	$5.30	−$24.39

Figure 17.29 Plot of new net change in loss after upgrading costs are included

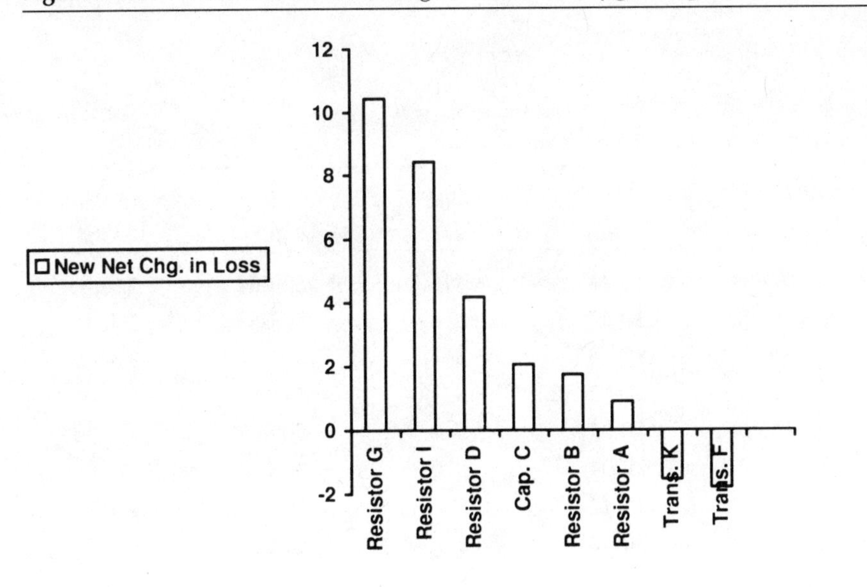

Identifying and Quantifying the Opportunities for Lowering Costs

To decide what to do about the transistors, we need to look at the relative contribution they make to overall variation reduction in the output. Their relative contributions are 1.50 percent and 0.68 percent. These are very weak contributors to the reduction in variation when compared to resistors G, I, and D and capacitor C, which together make up more than 88 percent of the contribution to the output-response variation. We conclude that it is not worth the cost to upgrade the transistors; we will leave them at their current levels and cancel their consideration in the circuit output variance reduction.

Relaxing Tolerances and Material Specifications of CIPs to Balance Cost and Quality

A cost improvement parameter is one that is not contributing a significant percentage to the output response. In our case, we find that transistor F can be downgraded in quality by selecting a low-cost transistor that has a standard deviation of ±10 percent. The new, lower cost of this transistor is $1; so we cut a dollar out of the circuit cost and pay a very small penalty in the overall quality loss for the circuit. Table 17.5 illustrates the final circuit quality loss improvements and MSD values for upgraded and downgraded components. We increase transistor F's MSD by a factor of two and incur greater quality loss in doing so. We decrease transistor F's cost by $1. We are leaving transistor K as it was originally designed. Note the changes highlighted for transistors K and F.

Table 17.5 Comparison table for final MSDs and losses

Parameter	MSD_{new}	$Loss_{new}$	*Net Δ in Loss*	*Cost of Upgrade*	*New Net Δ in Loss*
Resistor G	98.56	$0.44	−$10.50	$0.06	−$10.44
Resistor I	79.58	$0.35	−$ 8.48	$0.06	−$ 8.42
Resistor D	39.85	$0.18	−$ 4.24	$0.06	−$ 4.18
Capacitor C	28.70	$0.13	−$ 3.06	$1.00	−$ 2.06
Resistor B	16.75	$0.07	−$ 1.79	$0.06	−$ 1.73
Resistor A	8.89	$0.04	−$ 0.95	$0.06	−$ 0.89
Transistor K	105.3	$0.47	$ 0.00	$0.00	$ 0.00
Transistor F	95.4	$0.42	+$ 0.30	(−$1.00)	−$ .70
Input Voltage L	1.09	$0.005	−$ 0.015	–	–
Experimental Error	5.88	–	–	–	–
Totals	480	$2.11	−$28.74	$.30	−$28.44

New Loss and Cp after Upgrading and Downgrading

The new value for quality loss is \$2.11 versus \$31.05.

The new process capability index is:

$$Cp = (USL - LSL)/6\sigma \qquad (\sigma = \text{sqr. root of } MSD)$$

$$Cp = (USL - LSL)/6(21.9)$$

$$Cp = 2.28$$

Using Tolerance Design to Help Attain Six-sigma Quality Goals

We have taken a circuit with 0.60 process capability and improved it to one that possesses a Cp of 2.28. This means we went from very poor capability (less than three sigma) to a capability in excess of six sigma. Recall from Chapter 3 that six-sigma quality is equal to a Cp of 2.0. We have demonstrated how a statistical variance experiment can be used to drive a six-sigma quality growth process during the design phase (Phase 3) of product development.

Using Tolerance Design to Improve System Reliability

Not only have we shown how to create six-sigma quality in a circuit, we have also demonstrated how to develop a resistance to noises based on external and degradation sources. Tolerancing a design in the context of these reliability-depleting mechanisms will aid in the length of time it takes for the design to move off target. The variance reductions we have forced into the circuit lie at the heart of the physics of reliability. The *r*-factor goals for each component are met through variance reduction. The circuit's *r* factor is greatly improved—MSD_{old} versus MSD_{new} is:

$$\Delta\, MSD = 6{,}994.71 - 480$$

$$r \text{ factor} = 6994.71/480 = 14.57X$$

This corresponds to a little over a 14.5X improvement in the circuit output variance. This is a very helpful change to deliver to the overall product reliability goals. If we believe we have properly stressed the circuit with noises that strongly influence reliability—namely external and deterioration noises—the *r* factor improvement for the circuit is somewhere in the vicinity of 14X. This value can be confirmed through a stress test of the design to calculate the new S/N value for the response. This will produce a new $r_{S/N}$ factor by comparing the final design S/N ratio to the baseline S/N ratio (see Chapter 2). Using this process we can see how, through tolerance analysis, an engineering team can help grow reliability by working directly on the mechanisms that control it.

Summary

Chapter 17 brings the material contained in Section 3 together into one comprehensive case study. There are a number of helpful tools that can be used in the tolerance design process. We have shown that there are at least 16 steps that can be followed to properly conduct a tolerance analysis based on a designed experiment.

The chapter has demonstrated how the quality-loss function can be deployed in the development of specifications in a six-sigma context. Each parameter can be scrutinized for its contribution to output-response variation. Then each parameter can be used in the cost-versus-quality balancing process where each parameter variance can be considered for reduction. The process can also produce data that can be used to loosen tolerances, or enlarge the manufacturing variance, on parameters that do not prove to be critical to the function of the design.

Chapter 17 is outlined as follows:

Chapter 17: The Tolerance Design Process: A Detailed Case Study

- The Steps for Performing the Tolerance Design Process
  - Option 1: A Company Cost-Driven Process
  - Option 2: A Manufacturing Capability and Cost-Driven Process
- The ASI Circuit Case Study
  - Identifying the Parameters for Tolerancing
  - Improving Quality in Cases Where Parameter Design Was Not Done
  - Constructing the Loss Function
  - Setting Up and Running the Experiment
- Two- versus Three-Level Experiments
- Techniques for Putting Noise into the Tolerance Experiment
  - Running the Experiment
- Data Entry
  - Measuring the Proper Response Characteristic
- Interactions in Tolerance Experiments
  - Analyzing the Data
  - Applying ANOVA
- Relating the ANOVA Data to the Loss Function and Process Capability (Cp)
- Defining the CTF Factors
- Defining the Cost Improvement Parameters (CIP)
- Identifying and Quantifying the Costs Associated with Improving Quality
- Working with Suppliers to Lower Customer Losses through Reducing the Component Parameter Standard Deviations
- Calculating New Variances and MSD Values Using the Variance Equation
- Quantifying the Cost of Reducing the Parameter Standard Deviations
- Identifying and Quantifying the Opportunities for Lowering Costs
- Relaxing Tolerances and Material Specifications of CIPs to Balance Cost and Quality
- New Loss and Cp after Upgrading and Downgrading

- Using Tolerance Design to Help Attain Six-sigma Quality Goals
- Using Tolerance Design to Improve System Reliability

The tolerance design process is very helpful when attempting to develop tolerances on subsystems that contain mechanical assemblies, electrical and electronic assemblies, and mixtures of various technologies that perform a targeted physical process that must be constrained. Tolerance analysis that is facilitated by designed experiments is most useful when a mathematical model for the design function is difficult or impossible to create. Many of the risky assumptions present in complex models are removed when the engineering team works with experimental data derived from designed experiments that have been performed in the presence of realistic noise factors. It is in this stressful environment that tolerance data can be produced that is capable of enhancing the reliability of the design.

It is very important to recognize the benefits of using disciplined engineering analysis prior to the planning of designed experimentation. The data from such experimentation, in conjunction with the analysis from the application of the quality-loss function, will greatly expand the engineering team's ability to rapidly design products for optimized quality, balanced cost, and enhanced reliability.

Chapter 18 contains five real case studies from Kodak. The last of these case studies demonstrates a unique application of the use of a designed experiment during system tolerance development.

References

R17.1: *Tutorial C: Tolerance Design;* Bob Moesta and Alan Wu; Case Studies and Tutorials, 11th Annual Taguchi Symposium; American Supplier Institute, 1993, pg. 609

R17.2: *Engineering Quality By Design;* T. Barker; Marcel Dekker, 1990, pg. 614

For additional references on the topic of Tolerance Design see Appendix D.

SECTION IV

Industrial Case Studies

Section IV contains Chapter 18: Drive-System Case Studies. These five case studies focus on applying many of the principles taught in this book. Rather than discussing disconnected cases from separate designs, this section concludes the book by showing how to develop tolerances for a real mechanical drive system and its subassemblies and components. In this way the reader will see a practical case from the bottom up.

Chapter 18 draws on the material presented in Chapters 4, 5, 6, 7, 8, 13, 15, and 17. The cases have been presented in as simple a manner as possible to illustrate the general tolerance-development procedures while keeping the flavor of real cases from industry. The highlight of Chapter 18 is the use of both traditional tolerancing methods and experimentally based tolerance methods being popularized by Taguchi.

CHAPTER 18

Drive-System Case Studies

This chapter contains five actual case studies that demonstrate how to apply many of the methods taught in this text. As in all real cases, the execution of the tolerance-development process must accommodate unique and sometimes unusual circumstances. The five case studies are all related to developing specifications for a mechanical drive system for an office copier. They follow the sequential development of tolerances for the components and ultimately the integration of all the subassemblies and components into the drive system for a high-volume copy machine. This case study is from Eastman Kodak Company.[1]

The first case study begins with standard purchased components that are found in four of the drive system's subassemblies. These subassemblies are called *drive modules.* The drive system is made up of four drive modules that transmit power through a flexible, toothed drive belt connected to an a.c. drive motor. Each of the four drive modules is essentially the same in construction and application. The loads on the modules do vary since they have a variety of inertial startup loads, steady-state torque loads, and fluctuating torque loads that they must sustain during operation. The loads on the drive system are at their highest when starting the motion of all the masses. These are inertial and frictional loads that uniquely stress the drive system whenever the copy machine is started. The steady-state loads are present whenever the drive system reaches constant velocity; that is to say, *after* all the transient startup loads have been brought into equilibrium. The fluctuating loads can occur during steady-state operation when one or more loads are "clutched" in or out of service.

This series of case studies will demonstrate the difference between tolerancing for fit and assembleability and tolerancing for balancing dynamic loads and wear of materials. Essentially, we will demonstrate how to make components and assemblies fit together properly and then how to make them transmit kinetic energy efficiently in the presence of debilitating noise factors.

Cases 1 through 4 will focus on the components of a single drive module. They deal with the development of geometric tolerances that primarily have ramifications for proper assembly relationships. They contain no direct tolerance analysis related to drive loads or functional dynamics—these

1. Special thanks to Don Hensel and Greg Mahoney for helping with these case studies.

issues will be covered in the fifth case. All of the design issues related to properly engineering the drive module components for strength were done prior to the tolerance analysis (load-bearing and power-transmission capacity).

Case 1 will focus on tolerancing issues relative to the purchase of standard, off-the-shelf components. This material was introduced in Chapter 4. This first case study will illustrate one practical way to apply this material.

Cases 2 through 4 will focus on the *assembly* of both standard purchased components and custom manufactured components that are designed and toleranced by the drive-system engineering team. This drive-module assembly tolerance analysis will be demonstrated in three sequential case studies. Case 2 will reflect the material discussed in Chapter 5 (worst-case tolerance analysis); case 3 will reflect the RSS approach to tolerance analysis as discussed in Chapter 6; and the final drive-module case study, Case 4, will demonstrate how the computer-aided tolerance method of Monte Carlo simulation, from Chapter 8, can be used to analyze the assembly.

We will then turn our attention to a final case study on the complete drive system. To integrate the four drive modules—the drive belt, idler rollers and the vertical plate that geometrically and structurally supports them—we must employ some of the experimental methods of Taguchi's tolerance design process (Section III). This is necessary because the traditional, analytical methods of tolerance analysis are not fully capable of facilitating the development of tolerances for this complex system of components acting in their dynamic environment (various loads *and* dimensional variation due to mounting on a separate supporting surface). In addition, the case further develops the system tolerances for their contribution to reducing variability in the presence of environmental and deterioration noise factors. The first four cases are very helpful for analyzing the effects of unit-to-unit variability. The additional environmental and deterioration factors include worn and contaminated drive belts and drive-module components, as well as the varying loads they encounter. Simulating these in mathematical or graphical models can be very difficult. In fact, attempting to simulate these sources of variability rather than physically experimenting with them can be somewhat risky.[2] This is so because the results of the simulation can be difficult to verify without actually running the real hardware in the presence of real noises. The results of stressful experiments on real hardware are reasonably indisputable when conducted properly (see Chapters 13 through 17).

Let's move on to defining the system and drive modules before we proceed with the case studies.

The Drive System

Figure 18.1 illustrates the complete drive system. The complete system contains a drive motor, two drive belts, five drive modules, two idler rollers, and one tensioning pulley. The focus of this case study is on the four drive modules and drive belt located on the right-hand side of the figure.

Each drive module provides power to a different copier subsystem. In Figure 18.2 we illustrate the rear of the drive-system test fixture. Here we see that there are four electronically controlled brakes, used to induce the various loads that are possible during the operation of the copier. The test fixture is called a *robustness test fixture.* Text fixtures such as this are used in industry to evaluate production like components and assemblies for optimum nominal and tolerance set points.

2. See chapter 1 in *Modeling Engineering Systems,* by J. W. Lewis, HighText Publications, Inc., 1994.

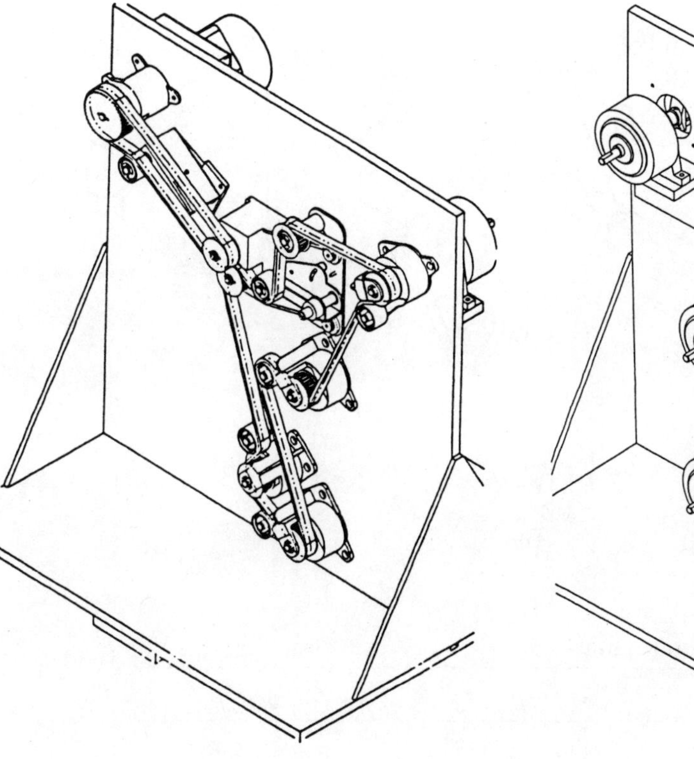

Figure 18.1 Drive-system test fixture

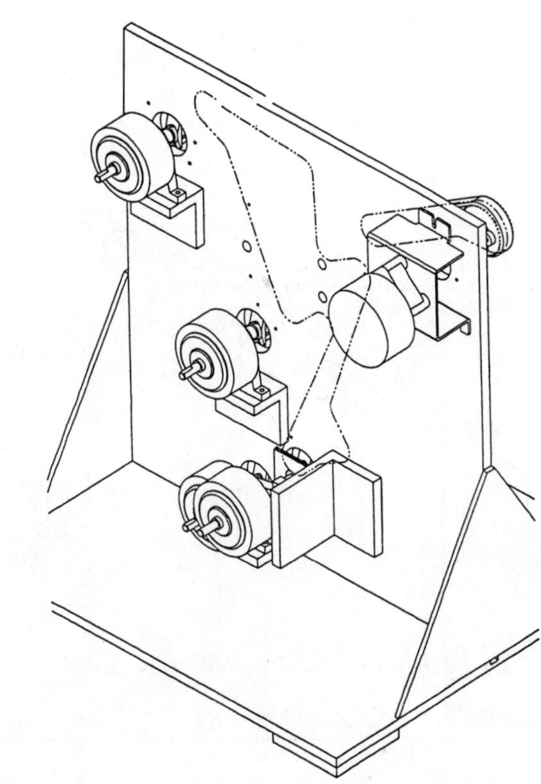

Figure 18.2 The rear view of the drive-system test fixture

The Drive Module

Figure 18.3 displays an exploded view of one of the drive modules. A drive module consists of 14 individual parts. Some parts are used twice; thus, the bill of material (Table 18.1) counts only 11 total parts. The drive module takes input power from the belt, transmits it through the pulley to a pin, then through the shaft to another pin, and finally through the coupling to a mating coupling located on an adjoining subsystem. All the other components serve supporting functions to constrain these power transmission components. Tolerancing the drive module has two priorities: constraining the geometry of the components to facilitate efficient power transmission, and constraining the geometry of the components so the drive module can be assembled efficiently from differing lots of parts.

The bill of material defines each part by name, whether it is a purchased (standard) component or a custom part being designed by the drive-system engineering team. The bill also lists the number of times the part is used and its function.

Figure 18.3 A drive module

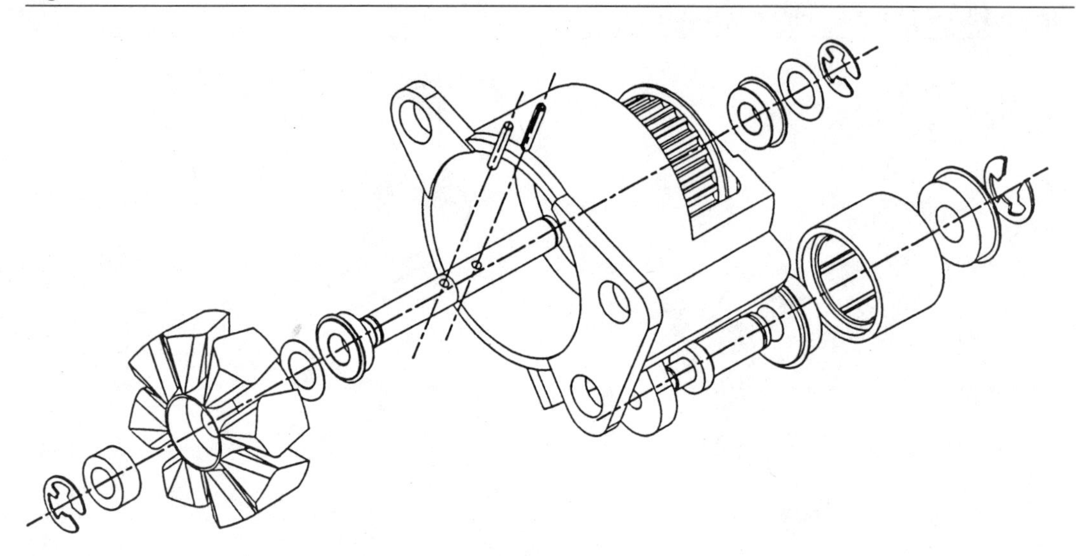

Table 18.1 Drive-module bill of material

Item #	*Description*	*Type of Part*	*Quantity*	*Function*
1	Shaft	Custom	1	Power transmission
2	Drive coupling	Custom	1	Power transmission
3	Dowel pin	Purchased	1	Power transmission
4	Spacer	Purchased	1	Support
5	E-ring	Purchased	2	Support
6	Washer	Purchased	2	Support
7	Bearing (8 mm)	Purchased	2	Support
8	Spring pin	Purchased	1	Power transmission
9	Housing	Custom	1	Support
10	Pulley (24 groove)	Custom	1	Power transmission
11	Idler assembly (flat)	Custom	1	Support

A completely assembled drive module is shown in Figure 18.4.

Now that we have illustrated the drive system and a drive module, let's move on to the case studies.

Figure 18.4 A completely assembled drive module

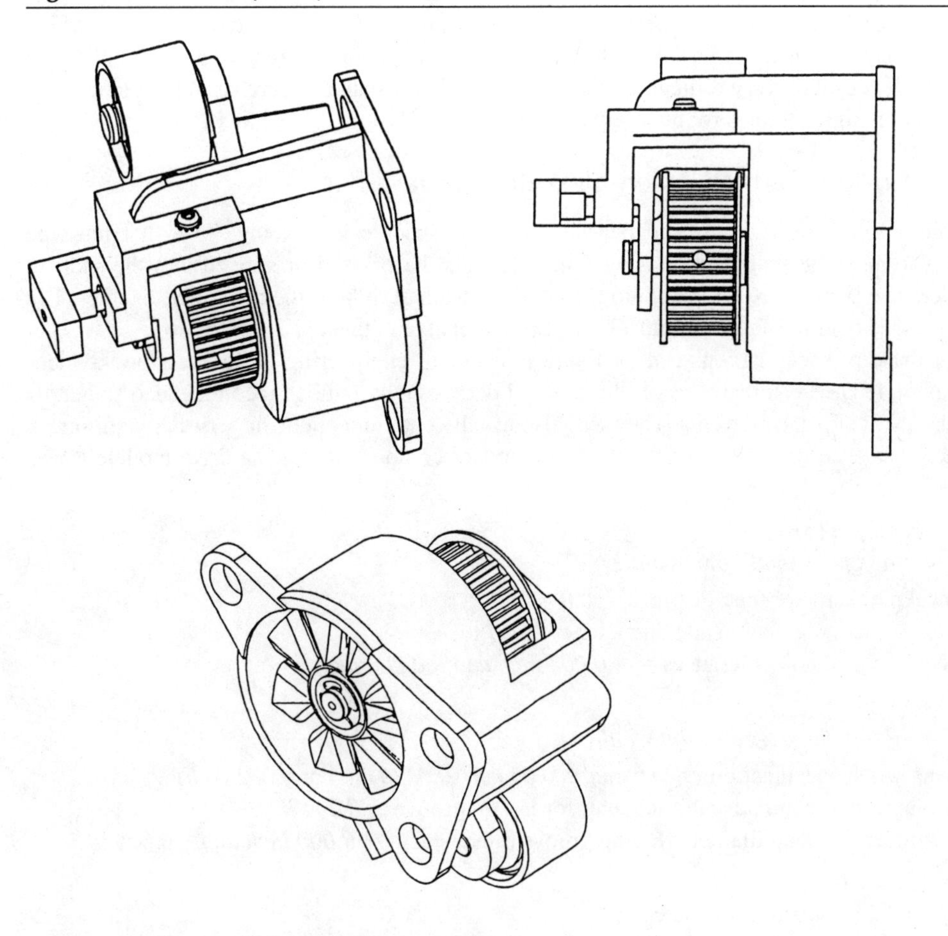

Case 1: Defining Tolerances for Standard Drive-Module Components

Every component in the drive module, whether it functions as a critical power-transmission element or a support mechanism, must be specified with nominal dimensions and their attendant tolerances. This case study addresses all the components in the drive module that are standard parts purchased from some outside source. The processes for working with outside parts suppliers (vendors) are discussed in Chapter 4.

The parts in the drive module that will be purchased are as follows:

Item #3 Dowel pin (1)

Item #5 E-rings (2)

Item #6 Washer (2)

Item #7 8mm bearings (2)

Item #8 Spring pin (1)

These five components will already possess nominal dimensions with standard tolerances. The design task is to size these components correctly with the rest of the drive module components, which will be *custom* designs. It is not unusual to alter the custom component designs to accommodate the purchased parts. It is usually very difficult and expensive to alter a standard purchased part, since they are manufactured in high volumes for a variety of uses.

Tolerancing Custom Parts to Assemble with Standard Parts

The dowel pin, spring pin, E-rings, and bearings all have specific interfacing "fits" that must be specified in relation to the shaft and housing. The pins must be pressed or slipped into the shaft to transmit power. The E-rings are clipped onto the ends of the shaft to hold the assembly together. The bearings are pressed onto the shaft and must have slip fits relative to the housing.

Each of these fit specifications can be found in standard engineering design handbooks mentioned in Chapter 4. The manufacturers of the standard parts usually publish recommended standards for various classes of slip and press fits. They will also usually entertain phone or personal inquiries on special applications. The specifications for the five standard components for the drive module are as follows:

Item #3 dowel pin (1)
Slip/light press fit in shaft hole at pulley

- Dowel pin diameter specifications = *0.094 inch ±0.001*
- Shaft pin holc nominal inside diameter = *0.094 inch*
- Recommended hole tolerances = *+0.003 inch and −0.000 inch*

Item #5 E-rings (2)
Shaft E-ring slot diameter and slot width

- E-ring width specifications = *0.6 mm ± 0.06 mm* (*0.0236 inch ± 0.002 inch*)
- Recommended nominal shaft diameter for E-ring groove = *0.252 inch*
- Recommended shaft diameter E-ring groove tolerances = *+0.000 inch and −0.004 inch*

Item #6 Washer (2)
Loose fit on shaft outside diameter:

- Washer outside diameter specifications = *0.640 inch ± 0.003 inch*
- Washer inside diameter specifications = *0.320 inch ± 0.002 inch*
- Washer width = *0.010 inch ± 0.001 inch*
- Shaft nominal diameter = *0.3150 inch* (one place) and *0.3140 inch* (one place)
- Shaft diametrical tolerances = *+0.0004 inch −0.0000 inch* (one place) and ± *0.0005 inch* (one place)

Item #7 Bearings 8mm (2)
Press fit onto the outside diameter of the shaft
Recommended shaft outside diameter specifications for a press fit:

- Nominal shaft diameter = *0.3150 inch*
- Shaft diametrical tolerances = *+0.0004 inch −0.0000 inch*
- Bearing width specifications = *0.1969 inch ± 0.0005 inch*

Recommended slip fit for bores in housing:

- Recommended nominal bore diameter for slip fit = *0.6304 inch* (two places)
- Bore diametrical tolerances = ± *0.0005 inch* (two places)

Item #8 Spring pin (1)
Press fit in shaft hole at couple

- Spring pin diameter specifications = *0.101 inch* ±*0.002 inch*
- Shaft pin nominal hole inside diameter = *0.094 inch*
- Recommended press fit hole tolerances = *+0.003 inch −0.000 inch*

Concluding this case study, we find that all of the specifications for the standard parts are easily obtained. The washer width can be specified either now (as shown above) or later in the assembly analysis case studies if a different thickness washer is deemed necessary. There are numerous widths of washers — one can be chosen initially and adjusted later.

Each of the components we have listed in this case is a standard item used commonly in commercial designs. Virtually all of the specifications are obtained through catalogues, common design literature developed and maintained within a company's design organization, or direct inquiries to the component supplier. Many corporations have organizations that provide design-specification information for standard components; typically, these organizations are called group technology centers. They specialize in promoting the use of a common standardized design elements so that the corporation can enjoy the benefits of purchasing these items in bulk quantities at a great cost savings.

Case 2: Drive Module for Worst-Case Assembly Analysis

Once the standard parts are defined and their specifications documented, the effect of assembly requirements can be synthesized with the load-bearing requirements of the custom components. The custom components have already been designed for strength and power-transmission characteristics. We can now perform a worst-case assembly stackup tolerance analysis including all of the components, both standard and custom. Figure 18.5 illustrates the stackup of components in the assembly.

The assembly illustrated in Figure 18.5 shows the dimensional steps (vectors) from point A to point L that will be included in the stack analysis. We are choosing to show the gap as L to A. Actually, there is nothing in the design to prevent the gap from appearing on both sides of the bearing support. The pins that transmit the power are dimensioned and toleranced such that they do not limit the axial travel of the pulley or the couple. The dowel pin is specified as a slip fit to facilitate ease of assembly. The inner bearing race constrains the axial motion of the couple and the pulley. This is fine because the race turns right along with the pulley and couple. If contact was induced by very large lateral belt forces, the bearings would have to be selected to handle these loads (thrust-bearing application). We will study the effects of positional tolerances of the module and its support plate in the presence of stressful noise factors as they relate to lateral belt forces in the final, Taguchi tolerance design case study. For now let's proceed with the worst-case tolerance stackup analysis. The measured response in this stack analysis is the gap.

Worst-Case Linear Stack Analysis

We would like the total gap dimension to be no smaller than *0.050 inch* (L to A). Our sign convention is positive to the right and negative to the left. The standard tolerances are obtained from the component suppliers' recommendations. The four component's tolerances are obtained by blending spacial requirements with economic manufacturing tolerances for similar parts.

Figure 18.5 Drive module assembly (not to scale)

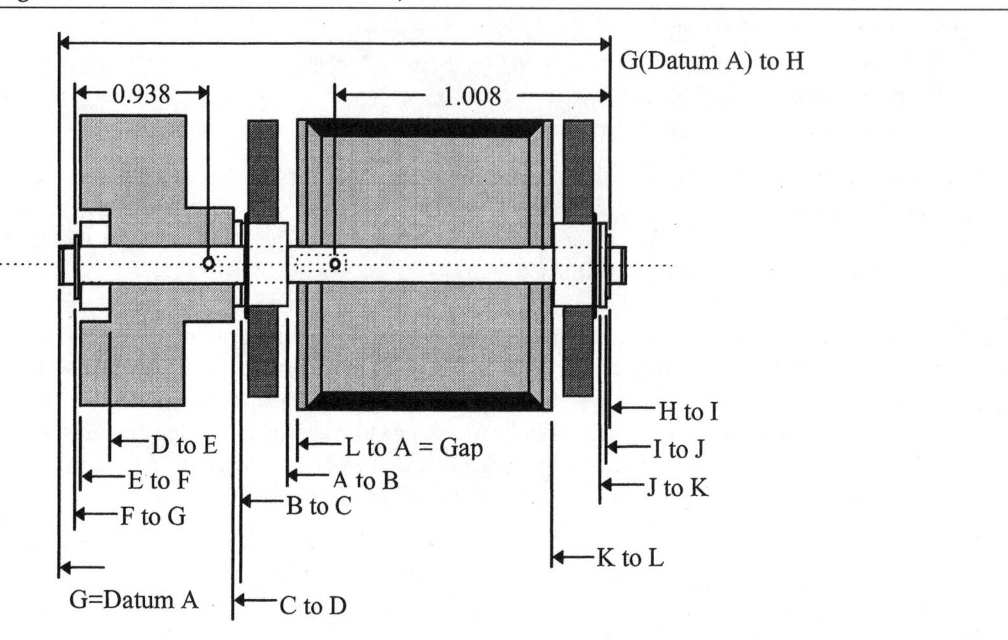

Table 18.2

	Nominal	*Tolerance*	*Worst-case dimension*
A to B = bearing width	− 0.1969 inch	+ 0.0005 inch	− 0.1974 inch
B to C = washer width	− 0.0100 inch	+ 0.0010 inch	− 0.0110 inch
C to D = couple bore width	− 0.7260 inch	+ 0.0070 inch	− 0.7330 inch
D to E = spacer width	− 0.2400 inch	+ 0.0050 inch	− 0.2450 inch
E to F = E-ring width	− 0.0236 inch	+ 0.0020 inch	− 0.0256 inch
F to G = E-ring to datum A	− 0.0630 inch	+ 0.0050 inch	− 0.0680 inch
G to H = datum A to E-ring	+ 2.4550 inch	− 0.0050 inch	+ 2.4500 inch
H to I = E-ring width	− 0.0236 inch	+ 0.0020 inch	− 0.0256 inch
I to J = washer width	− 0.0100 inch	+ 0.0010 inch	− 0.0110 inch
J to K = bearing width	− 0.1969 inch	+ 0.0005 inch	− 0.1974 inch
K to L = pulley width	− 0.8710 inch	+ 0.005 inch	− 0.8760 inch
L to A = Gap			+ **0.0600 inch**

The worst-case gap (smallest) is found to be 0.010 inch larger than the minimum. This minimum was selected to allow the pulley and the couple to move about the shaft axially to promote freedom of rotation. The minimum gap is also present to assure that thermal expansion of all the parts will not create a binding situation.

The method discussed in Chapter 5 for linear worst-case tolerance analysis has been demonstrated in this case study. Every component in the stack analysis has been placed at its worst-case dimension to force the gap to be at its very smallest allowable dimension. We have shown that even in a worst-case scenario, the minimum gap requirement is not surpassed. We are left to conclude that the drive module is going to assemble and function in an acceptable manner. We cannot speak with any statistical rigor with respect to the gap analysis except to say that it is not very likely that the drive module is going to experience assembly problems due to the component tolerances we have developed. The next case study will make use of statistical principles to aid in the gap analysis.

Case 3: Drive Module for RSS (Statistical) Assembly Analysis

Moving on to the study of variation in an assembly while taking the laws of probability into account, we will apply the root sum of squares approach to the stack analysis of the drive module. Because each of the components is likely to possess a distribution of values that will contribute to the variation in the assembly stack dimension, we must make some statistical assumptions about the nature of the component dimensional variation. The dimensional variation we care about is directly related to the component dimensions that factor into the overall stackup dimension.

For this case, the three-sigma paradigm is assumed. This means that each component tolerance is assumed to be set at plus or minus three standard deviations about the nominal specification (based on a normal distribution of data about the mean). If the true derivation of the tolerance is known, including the nature of the statistical distribution underlying its variability, this information should be used. The best way to find this information is to call or visit the component manufacturer directly. If the information is unknown, then using the three-sigma paradigm is a reasonable approximation you will have to consider. Will this approach place the design in jeopardy? It depends. If you are working with critical assemblies with very demanding tolerance constraints to promote proper performance, it is strongly recommended that you spend the time and resources to determine the true nature of variation behind the recommended tolerances. For this case we will use the *variance* values for performing the RSS calculations based on the following assumption:

$$\text{Variance} = (\text{Tolerance}/3Cp)^2$$

where $Cp = 1$ for the three-sigma paradigm.

Figure 18.6 illustrates the stackup of components in the drive module.

RSS Case Linear Stack Analysis

Now, using the root sum of squares approach, we can calculate a new value for the gap. We would like the total gap dimension to be no less than *0.050 inch* (L to A). Our sign convention remains positive to the right and negative to the left.

Based on the assumption of normally distributed component dimensions and three-sigma tolerance limits, we find the stackup standard deviation to be + 0.004262 inch. We multiply the gap standard deviation by three, and calculate the gap tolerances to be ±0.01279 inch. This value must be subtracted from the nominal gap of 0.094 inch. The result is that the gap shrinks to 0.081 inch. This is quite different than the value we calculated for a worst-case scenario (0.060 inch). When the laws

Figure 18.6 Drive-module assembly (not to scale)

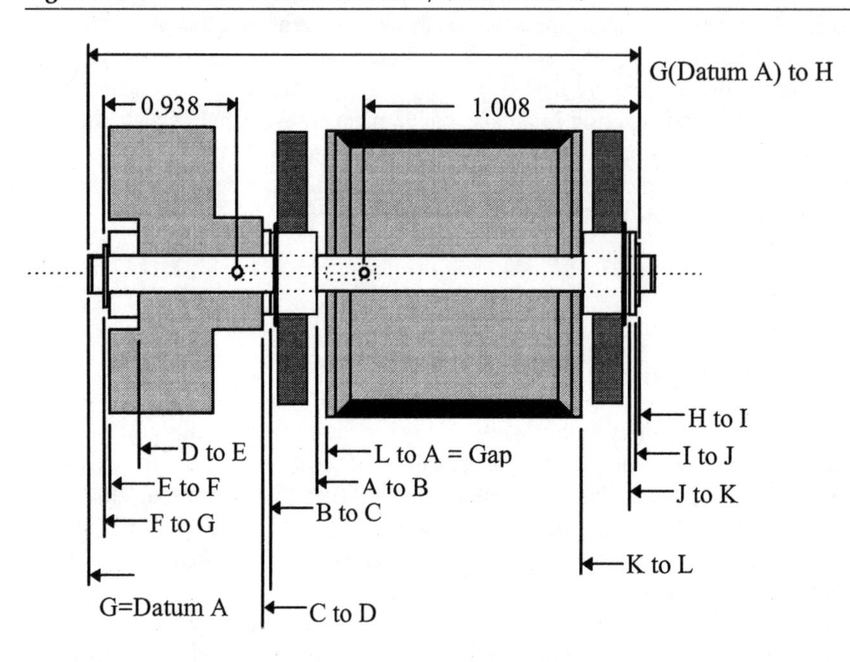

Table 18.3

	Nominal	*(Tolerance/3)*^{2}	*Variance*
A to B = bearing width	−0.1969 inch	+ (0.0005 inch/3)^2	2.777E-8
B to C = washer width	−0.0100 inch	+ (0.0010 inch/3)^2	1.111E-7
C to D = couple bore width	−0.7260 inch	+ (0.0070 inch/3)^2	0.000005
D to E = spacer width	−0.2400 inch	+ (0.0050 inch/3)^2	0.000003
E to F = E-ring width	−0.0236 inch	+ (0.0020 inch/3)^2	4.444E-7
F to G = E-ring to datum A	−0.0630 inch	+ (0.0050 inch/3)^2	0.000003
G to H = datum A to E-ring	+2.4550 inch	− (0.0050 inch/3)^2	0.000003
H to I = E-ring width	−0.0236 inch	+ (0.0020 inch/3)^2	4.444E-7
I to J = washer width	−0.0100 inch	+ (0.0010 inch/3)^2	1.111E-7
J to K = bearing width	−0.1969 inch	+ (0.0005 inch/3)^2	2.777E-8
K to L = pulley width	−0.8710 inch	+ (0.0050 inch/3)^2	0.000003

L to A = gap Taking the square root of the sum of squares (variances) we get +**0.004262 inch**

The nominal gap = 0.094 inch

of probability are taken into account, we see through the three-sigma paradigm and the root sum of squares approach that the gap is not expected to approach the 0.050 inch minimum.

Case 4: Drive Module for Computer-Aided Assembly Analysis

The use of Crystal Ball to perform Monte Carlo analysis (see Chapter 8) on the drive module assembly stackup can greatly aid in determining just how the assembly process will look over many thousands of trials. We can *practice* assembling the drive module over and over again through the computer-aided simulation capabilities resident in Crystal Ball.

The unique requirements for Monte Carlo analysis, beyond what is required to perform worst-case or RSS analysis, are the development of a mathematical model that represents the stackup of component tolerances in the assembly and the definition of individual component variability distributions. These are some of the major topics discussed in Chapter 8.

Fortunately for the drive module case study, the mathematical model is linear. We can simply add or subtract the stackup dimensions to assess what the assembly gap will be. Figure 18.7 is helpful in developing the gap equation we will enter into Crystal Ball in the Forecast cell.

Every time Crystal Ball performs an assembly stackup trial, the math model (forecast) entered is loaded with component values that are selected randomly from the user-assigned statistical distributions (assumptions) that are representative of how the parts will vary naturally in their real production environment.

We will use Figure 18.7 to help illustrate the process of computer-aided assembly analysis.

The assembly gap math model is defined as:

$$\textit{Assembly gap} \doteq [(-\text{bearing width}) + (-\text{washer width}) + (-\text{couple width})$$

$$+$$

$$(-\text{E-ring width}) + (-\text{left side of left E-ring groove to datum A})$$

$$+$$

$$(\text{datum A to right side of right E-ring groove}) + (-\text{E-ring width})$$

$$+$$

$$(-\text{washer width}) + (-\text{bearing width}) + (-\text{pulley width})]$$

(Remember to follow the proper sign convention.)

The individual component specifications are listed below. The one standard deviation values are assumed to be representative of what *could* be occurring as an output from the manufacturing processes being used to make the components. We do not have direct measures of the one-sigma values in this case study. Table 18.4 contains back-calculated one-sigma values based on the tolerances from the component catalogue values and supplier recommendations:

$$\pm 1\sigma = \sqrt{\frac{\textit{Tolerance}}{3}}$$

Figure 18.7 Drive-module assembly (not to scale)

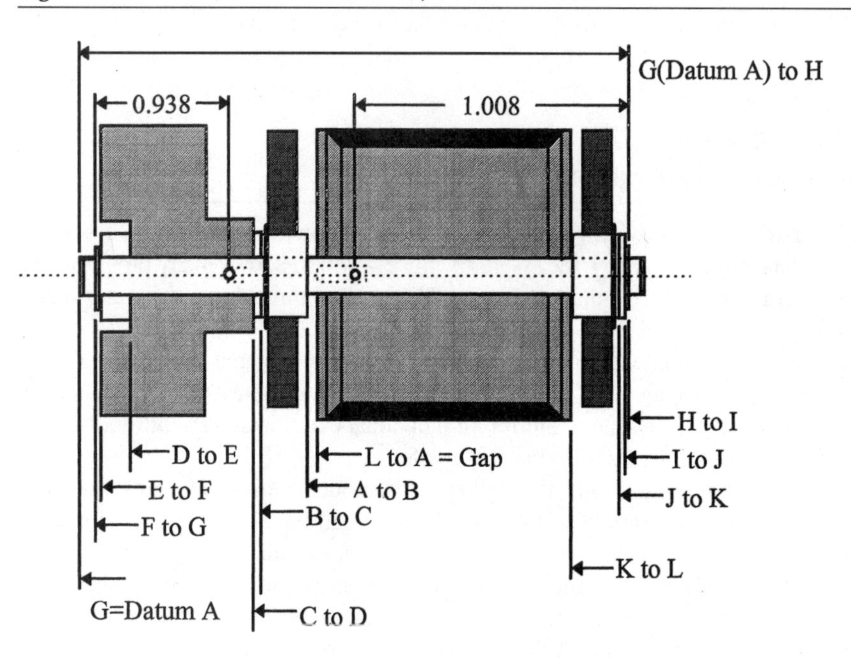

Table 18.4

	Nominal	*Tolerance*	*One standard deviation*
A to B = bearing width	0.1969 inch	± 0.0005 inch	± 0.000167 inch
B to C = washer width	0.0100 inch	± 0.0010 inch	± 0.000333 inch
C to D = couple bore width	0.7260 inch	± 0.0070 inch	± 0.002236 inch
D to E = spacer width	0.2400 inch	± 0.0050 inch	± 0.001732 inch
E to F = E-ring width	0.0236 inch	± 0.0020 inch	± 0.000667 inch
F to G = E-ring to datum A	0.0630 inch	± 0.0050 inch	± 0.001732 inch
G to H = datum A to E-ring	2.4550 inch	± 0.0050 inch	± 0.001732 inch
H to I = E-ring width	0.0236 inch	± 0.0020 inch	± 0.000667 inch
I to J = washer width	0.0100 inch	± 0.0010 inch	± 0.000333 inch
J to K = bearing width	0.1969 inch	± 0.0005 inch	± 0.000167 inch
K to L = pulley width	0.8710 inch	± 0.005 inch	± 0.001732 inch

The Monte Carlo simulation will be run in two stages: once for the assumption that all component dimensions vary according to normal distributions and once for the assumption that certain components vary according to uniform and triangular distributions while some are still assumed to follow normal distributions. This way the engineering team can assess whether the assumption of normal or other distributions has any meaningful affect on the assembly-gap prediction. This is one of the

Figure 18.8 Crystal Ball on Excel worksheet

Microsoft Excel - DS1.XLS

File Edit View Insert Format Tools Data Window Cell Run Help

Component:	Nominal:	Tolerance:	1 Std. Dev.:	USL:	LSL:
Bearing Width	0.19690	0.0005	±0.000167	0.19740	0.19640
Washer Width	0.01000	0.001	±0.000333	0.01100	0.00900
Couple Bore Width	0.72600	0.007	±0.002236	0.73300	0.71900
Spacer Width	0.24000	0.005	±0.001732	0.24500	0.23500
E-Ring Width	0.02360	0.002	±0.000667	0.02560	0.02160
E-Ring to Datum A	0.06300	0.005	±0.001732	0.06800	0.05800
Datum A to E-Ring	2.45500	0.005	±0.001732	2.46000	2.45000
E-Ring Width	0.02360	0.002	±0.000667	0.02560	0.02160
Washer Width	0.01000	0.001	±0.000333	0.01100	0.00900
Bearing Width	0.19690	0.0005	±0.000167	0.19740	0.19640
Pulley Width	0.87100	0.005	±0.001732	0.87600	0.86600
Gap:	0.09400			0.07000	0.11800
			W.C. Gap:	0.06000	0.12800

Sheet1 Sheet2 Sheet3 Sheet4 Sheet5 Sheet6

Ready

strongest benefits of performing Monte Carlo analysis. We can test the impact of our assumptions when we do not know for sure what the nature of the component distributions really are.

Testing assumption 1 (*all* components follow normal distributions)

The tolerance analysis worksheet is the first thing to be constructed, as shown in Figure 18.8. Each component is set up in an Assumption cell as a normal distribution in Crystal Ball. All of the normal distributions are constrained by the three-sigma tolerance limits found in column D. The actual tolerance limits are displayed in columns F and G.

Running the Monte Carlo simulation for 5,000 trials, we find the gap to have the distribution shown in Figure 18.9.

The gap statistics are computed as shown in Figure 18.10. The gap statistics provide a more complete understanding of what is to be expected with respect to gap variability. In this case the gap does not need to be toleranced—but it could be if necessary. Often, a gap tolerance will be derived from the gap standard deviation. Note this gap is the same one we calculated in the RSS case.

Figure 18.9 Gap frequency distribution

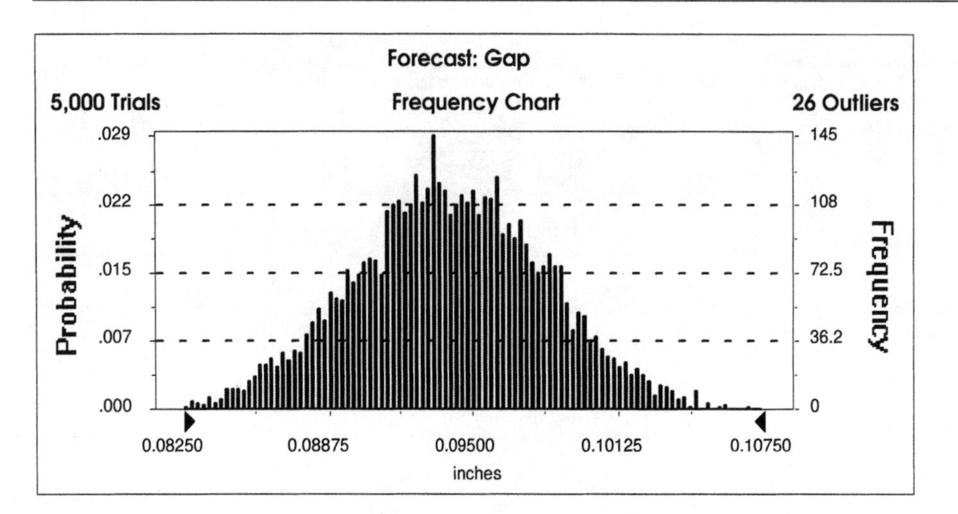

Figure 18.10 Gap tolerance-analysis statistics

Forecast: Gap

Edit Preferences View Run Help

Cell C17 Statistics

Statistic	Value
Trials	5,000
Mean	0.09405
Median	0.09402
Mode	---
Standard Deviation	0.00422
Variance	0.00002
Skewness	-0.04
Kurtosis	3.06
Coeff. of Variability	0.04
Range Minimum	0.07748
Range Maximum	0.10984
Range Width	0.03236
Mean Std. Error	0.00006

The sensitivity analysis for the case where all factors are assumed to follow normal distributions is as shown in Figure 18.11.

The sensitivity analysis reveals that the magnitude of the tolerance sensitivities is driven by the range of the tolerances relative to the nominal dimensions to which they are associated. This was one of the topics discussed in Chapter 7. Here, all of the components have the same sensitivity effect, since

Figure 18.11 Sensitivity analysis

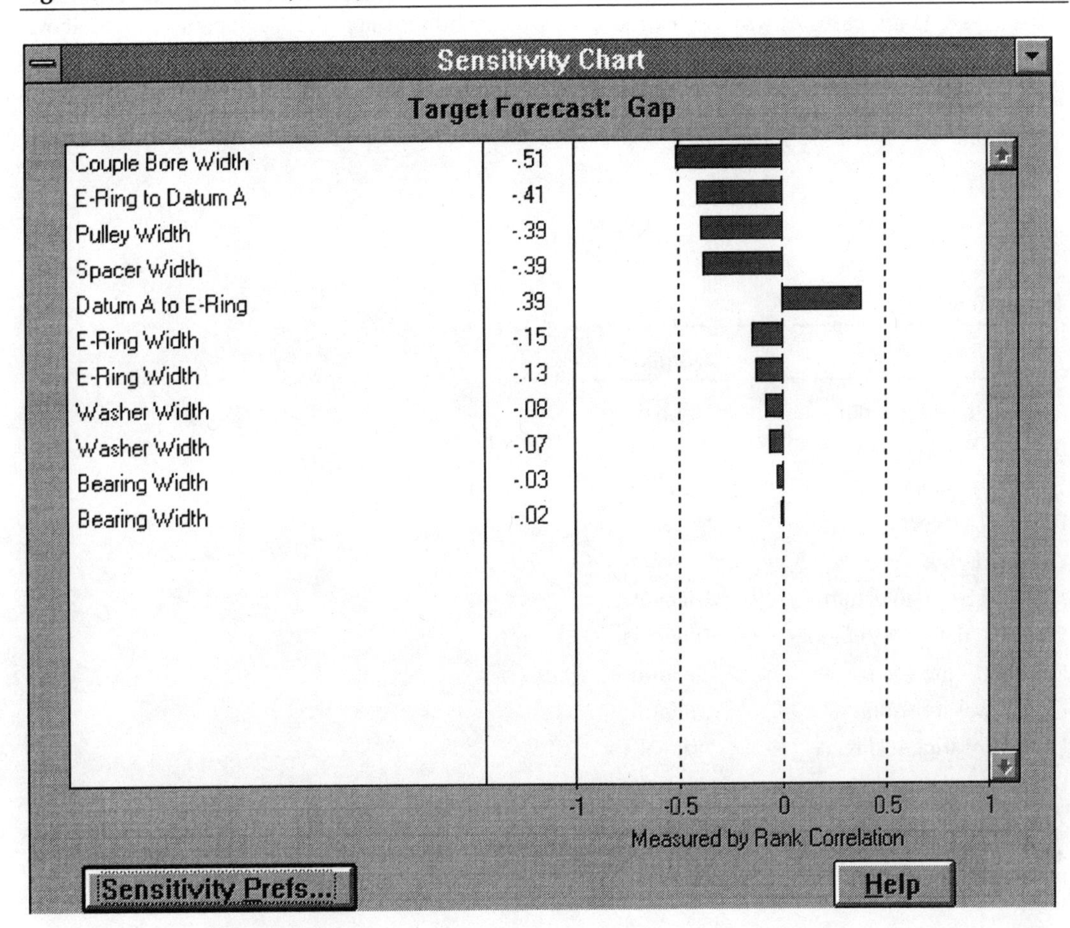

we assume that they all obey normal distributions. The next simulation will show how the effect of different distribution assumptions affect certain component sensitivities.

Testing assumption 2 (all components *do not* follow normal distributions)

The assumption cells in the worksheet are altered to accommodate the changes shown in Table 18.5 in the component distribution assumptions.

We could try every possible (reasonable) type of distribution for every component in the assembly if we chose. In fact, we could use a full factorial orthogonal array to guide the computer-aided experimentation. A fractional factorial array could also be employed to analyze the effect of each type of distribution on the output of the Monte Carlo analysis. This would be fairly easy to do if one limited the evaluations to three distributions—say the uniform, normal, and triangular types. This would require a three-level array. With our 11 factors it would take an L_{27} array to provide a fractional factorial analysis. In this case, we will just use these three distributions run at the specifications listed above. Running the Monte Carlo simulation for 5,000 drive-module assembly trials we find the gap to obtain the distribution shown in Figure 18.12. The gap statistics are computed to be as shown in Figure 18.13.

Table 18.5

	Distribution
A to B = bearing width	Normal
B to C = washer width	Uniform
C to D = couple bore width	Normal
D to E = spacer width	Normal
E to F = E-ring width	Uniform
F to G = E-ring to datum A	Triangular
G to H = datum A to E-ring	Triangular
H to I = E-ring width	Uniform
I to J = washer width	Uniform
J to K = bearing width	Normal
K to L = pulley width	Normal

Figure 18.12 Gap frequency distribution

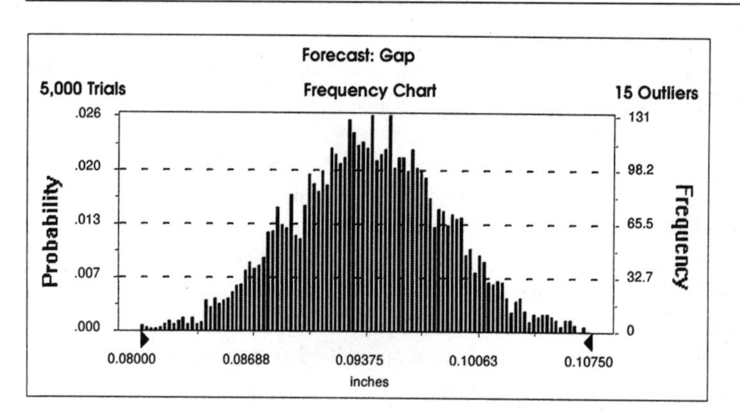

Figure 18.13 Gap tolerance-analysis statistics

Forecast: Gap

Edit Preferences View Run Help

Cell C17 Statistics

Statistic	Value
Trials	5,000
Mean	0.09397
Median	0.09398
Mode	---
Standard Deviation	0.00469
Variance	0.00002
Skewness	0.02
Kurtosis	2.94
Coeff. of Variability	0.05
Range Minimum	0.07655
Range Maximum	0.11054
Range Width	0.03400
Mean Std. Error	0.00007

Notice that the statistics vary slightly from the case where all distributions are assumed to be normal. As should be expected, the mean is smaller and the standard deviation a bit larger (mainly due to the assumption of uniform distributions for some components).

The sensitivity analysis for this multidistributional evaluation is as shown in Figure 18.14.

The Monte Carlo simulations for components that all follow normal distributions and the simulations for certain components following various other distributions provided further insight into the nature of the resultant average gap dimension and the amount of variation we can expect. Even if our assumption of all components following normal distributions is wrong, according to the alternative assumption, the assembly is still going to fit together as required. Monte Carlo analysis is extremely helpful in exploring "what-if" scenarios when you don't know the underlying distributions for the manufactured-component dimensions. In addition, we viewed the sensitivities that were resident in the components due to their tolerance ranges and their assigned distributions. This allowed the team to assess the criticality of each component in promoting the proper assembly of the drive module.

We have briefly discussed some ancillary design issues concerning design elements that are not directly involved in the stack analysis. Design is part science, requiring expertise in statics, dynamics, strength of materials, machine design, vibration analysis, and numerous other topics. It is also partially an art, requiring a sense of space and form that facilitates assembleability and synthesis with other product components and assemblies. It is commonplace that design is the balance of form, fit, and function. The drive module has many elements that require detailed engineering analysis to assure that it can sustain the various loads it will encounter over its design life. So far we have concerned ourselves only with the portion of the design task that promotes assembleability and freedom of rotational motion.

Figure 18.14 Sensitivity analysis

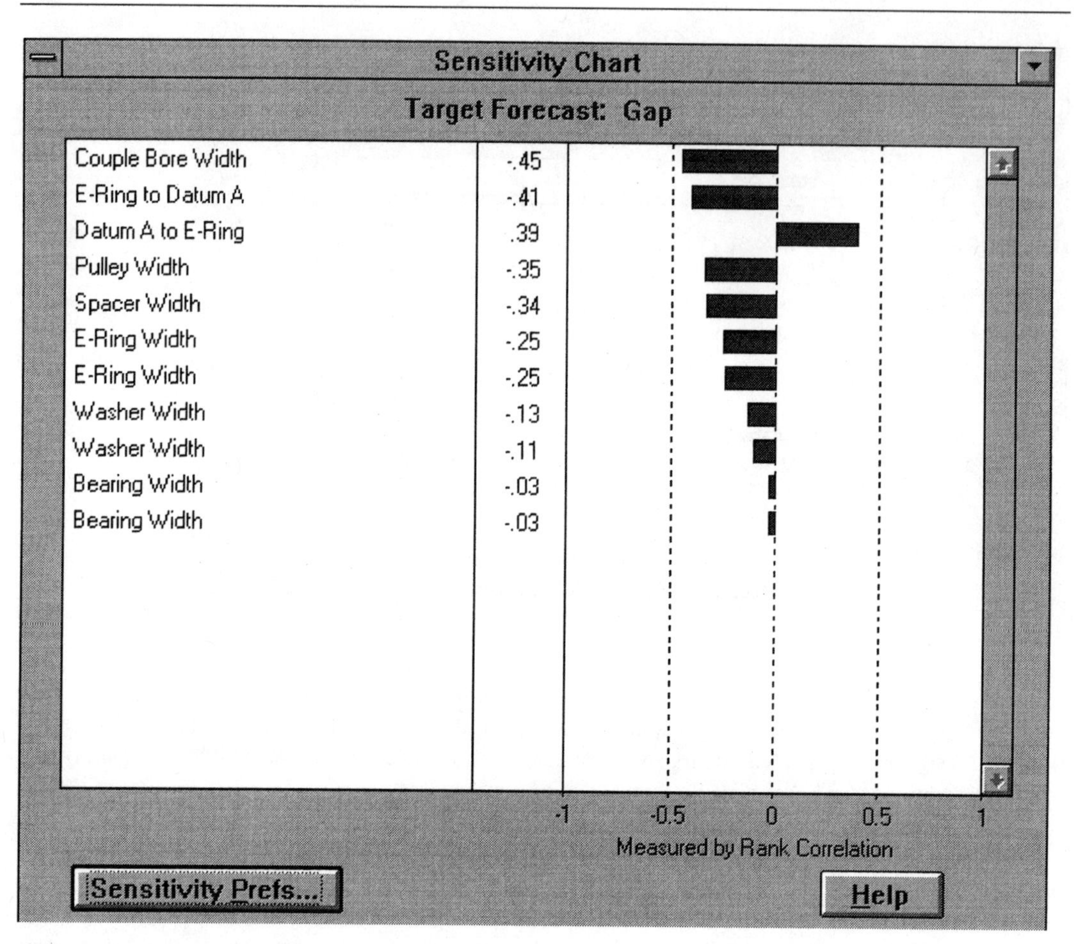

Case 5: Drive System Aided by the Use of a Designed Experiment

The drive module tolerance analysis was readily handled through the application of traditional one-dimensional analytical stackup analysis. After the engineering team assessed the integration of all four drive modules into a drive system, the emphasis changed from one of assembleability to one of drive-belt lateral forces. Minimizing lateral belt tracking forces is one of the major factors that maximizes drive belt life—a reliability issue.

The mounting holes for the assembly of the drive modules to the support plate had to be specified and toleranced prior to the belt force evaluation. The drive modules were properly located and toleranced with respect to one another and the support they are mounted on by the methods of graphical 2-D CAT analysis, using proprietary software. In this case, the computer does all the math while the designer specifies the geometries that establish the relationships in the tolerance problem. Our intent is not to minimize or neglect the importance of this form of tolerance analysis, but we choose to move on to provide a unique exposure to a real case study involving the use of a designed experiment. Taguchi's approach can be applied to a designed experiment in the initial determination of tolerances as they relate to a critical-to-function parameter (lateral belt forces). This case utilizes the worst-case approach since the engineering team was exploring the system's sensitivity to angular deviations (with unknown standard deviations) in the presence of serious noise factors. Here, the tolerance development is not so concerned with assembly issues as it is with drive-system reliability. It is important to recognize that tolerance development is just as appropriate for developing reliability into a design as it is for developing assembleability.

Preparing for the Drive System Tolerance Development Experiment

When the drive modules are connected to the vertical supporting plate in the back of the copier, lateral and angular misalignment can occur from variations in the support plate flatness and the flatness of the mounting faces on the back of the drive module. The sources of the misalignment are typically found to be due to manufacturing (unit-to-unit) variability in the stamping and welding of the vertical support plate, the unit-to-unit variability in the drive module mount surfaces, and the loading of the support plate through the drive modules as a result of the tension applied to the drive belt. Additional noise factors believed to affect the variability in drive belt lateral forces are as follows:

1. Aged versus new drive belts
2. Clean versus contaminated drive belts and components (oil, toner, paper dust, etc.)
3. Drive system loads changing during a run versus steady-state drive loads

The tolerance factors in this case study are as follows:

1. Drive module A angular misalignment
2. Drive module A lateral misalignment
3. Drive belt tension

The engineering team decided to simplify the evaluation of the drive system by breaking the problem into reasonable portions. This is a common strategy in developing and optimizing system specifications.

The first step was to construct the system of four drive modules but to only evaluate lateral belt forces at drive module A. This requires the three other drive modules to be set up at their nominal levels. Drive module A was the focal point for the *initial* study of lateral belt forces. The forces were induced by setting the drive module at three levels of angular misalignment, three levels of lateral misalignment, and three levels of belt tension. The team developed a special force-transducer mount to facilitate the measurement of the lateral belt forces for given amounts of belt tension and misalignment at the drive module–mount plate interface. The force transducer sensed the force at the pulley.

The mechanics of a tensioned belt acting on a pulley are such that the coefficient of friction ($\mu = N/F_f$, where N is the normal force due to belt tension and F_f is the frictional force between the pulley and belt) between the belt and the pulley is a factor in determining just when the belt breaks free of the pulley and slides to one side and begins to wear. If the lateral forces induced between the belt and the pulley are bigger than the frictional force governed by the static coefficient of friction and the normal force, then the belt is going to move. The belt will accelerate until it hits the side of the pulley. Once its

Figure 18.15 Drive module to mount plate misalignment

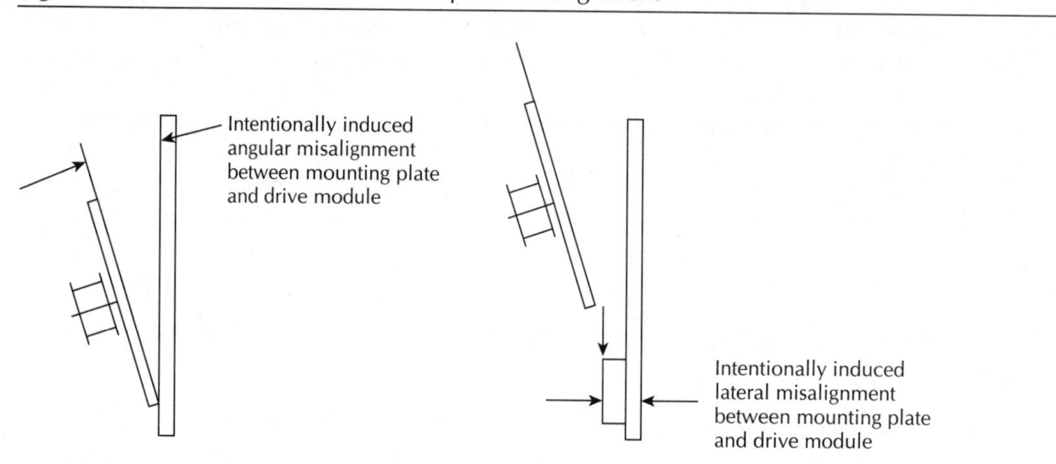

motion is halted by the edge of the pulley, the belt will push against the pulley with whatever force is being induced by the friction between the pulley's and the belt's teeth.

The drive-system team was concerned that the amount of angular and lateral misalignment not be allowed to cause premature belt failures due to excessive wear—thus the need to tolerance (limit) the angular and lateral alignment of the drive modules as well as the belt tension. The use of a designed experiment should always be preceded by an appropriate amount of engineering analysis. Often a designed experiment is used to optimize systems that are very difficult to model. This does not mean that the team should not attempt to model the system to a level that gives insight into how the experiment should be set up. Problems can arise when an engineering team models with the exclusion of experimentation or when the team experiments with the exclusion of modeling. It is wise to always model as far as you can, then move on to *designed* experimentation.

The team constructed the force-vector diagram shown in Figure 18.16 to aid in determining just how to set up the experiment. The drive-belt tension applies a bending force on each pulley shaft. The first thing to happen, under this load, is that the shaft deflects as a cantilevered beam. This induces some angular misalignment. The load is then transmitted to the supporting plate through the bearings and the drive-module housing. The belt tension can be adjusted to be relatively high. This may place enough load on the support plate to elastically "pucker" the support material, thus causing additional angular misalignment of the drive module. The lateral misalignment is mostly due to unit-to-unit manufacturing variability.

Setting Up the Experiment

The engineering team had three factors to evaluate for constraining the individual contributions to lateral belt forces. They also had three noise factors that were of concern. The ideal function of the drive system would be to experience no lateral drive-belt forces at any of the pulleys. Practical experience guided the team to stress the system simultaneously with the development of tolerances to promote reliability in the design.

The engineering team met with the manufacturing group responsible for making the machine frame, which included the support plate for the drive system. No one knew where to limit the amount

Figure 18.16 Force vector diagram

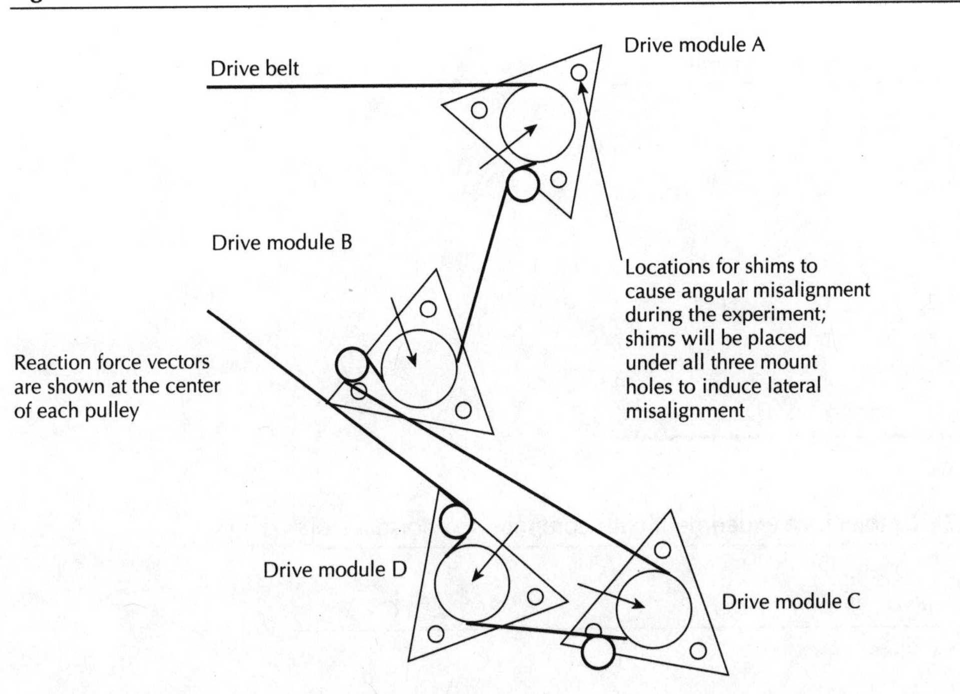

of angular or lateral misalignment. They determined three initial levels of angular and lateral misalignment that were reasonable to explore, and three levels of belt tension that were reasonable. Belt tension can be adjusted by assembly personnel as well as field engineers.

The experiment required the use of an L_9 orthogonal array. They were aware of the risk of some confounding of the results due to tolerance-factor interactivity. The factors were placed in the array to minimize the possibility of confounding between the two misalignment factors. As shown in Table 18.6, columns A and B were loaded with angular misalignment and belt tension, respectively. Column C was loaded with lateral misalignment. Column D was left empty.

The three noise factors were compounded during a brainstorming session as follows:

Compound noise factor at level 1 (*low coefficient of friction case*):

Belt Age = *old*

Contamination = *contaminated*

Loading = *constant*

Compound noise factor at level 2 (*high coefficient of friction case*):

Belt age = *new*

Contamination = *clean*

Loading = *varying*

The L_9 experiment and the compound noise factors combine as shown in Table 18.7.

Table 18.6 L_9 tolerance experiment

Run	*1* *Angular*	*2* *Tension*	*3* *Lateral*	*4*
1	1	1	1	1
2	1	2	2	2
3	1	3	3	3
4	2	1	2	3
5	2	2	3	1
6	2	3	1	2
7	3	1	3	2
8	3	2	1	3
9	3	3	2	1

Table 18.7 L_9 tolerance experiment with compounded noise factors

Run	*1* *Angular*	*2* *Tension*	*3* *Lateral*	*4*	*CNF1*	*CNF2*
1	1	1	1	1		
2	1	2	2	2		
3	1	3	3	3		
4	2	1	2	3		
5	2	2	3	1		
6	2	3	1	2		
7	3	1	3	2		
8	3	2	1	3		
9	3	3	2	1		

After the L_9 experiment has been conducted and analyzed, a more extensive and focused evaluation could be developed if necessary. In WinRobust such a follow-up experiment could take on the form shown in Table 18.8.

In this more extensive L_{12} experiment, *all* of the drive modules could be monitored for the level of lateral force induced during the experiment. Only two levels should be evaluated now that the team has information about the sensitivity of the system. In addition, the manufacturing team may have had time to make enough productionlike frames to begin to get sample measurements of the standard deviation of the flatness of the mounting plate from a coordinate-measurement machine. Then, the more acceptable variance experiment (see Chapter 15) can be conducted with the L_{12}.

Life tests on the belts can now be conducted with known levels of lateral forces to determine how many cycles are capable of being run before the belt fails. The cost of a belt failure can be quantified by the field-service engineers. This would include the cost of the belt, the cost of the service call to

Table 18.8 Possible L_{12} tolerance experiment

Run	1 DM-A	2 DM-B	3 DM-3	4 DM-4	5 Belt Tension	CNF #1	CNF #2
1	1	1	1	1	1		
2	1	1	1	1	1		
3	1	1	2	2	2		
4	1	2	1	2	2		
5	1	2	2	1	2		
6	1	2	2	2	1		
7	2	1	2	2	1		
8	2	1	2	1	2		
9	2	1	1	2	2		
10	2	2	2	1	1		
11	2	2	1	2	1		
12	2	2	1	1	2		

replace it, and the cost of down time for the owner of the copier. The quality-loss function can then be developed for additional tolerance development and cost balancing.

The results from the L_9 experiment for the force at drive module A will be used to set the initial tolerances for all the other drive modules.

Additional forms of stressful noises to the system are differing angular and lateral misalignments and belt tensions at all four drive modules. Initially, the engineering team was searching for the realistic level of force and parameter sensitivities that could be measured and tolerated. The data from the first experiment could be used as necessary to guide in the development of additional experiments such as the aforementioned L_{12}. As you can see from this case study, initial tolerance experiments may be in order long before real manufacturing or assembly-standard deviations can be quantified or a loss function constructed. Preliminary tolerance experiments are very enlightening when working with complex systems that have no previous tolerance data available.

Data Acquisition

Do not be deceived by the seemingly smooth transition from defining the experiment to recording and analyzing the data. The engineering team spent several weeks getting the right transducer selected, ordered, installed, and debugged. They used computer-aided data acquisition and analysis. It is not a trivial matter to get everything set up just right so the speed and efficiency of the computer can be utilized.

Once the data-acquisition equipment was obtained and connected, the team had to be sure the numbers coming out of the trial runs were correct. After several modifications to be sure the signals from the transducer were being processed correctly, the team was ready to perform the L_9 experiment.

Table 18.9 Experimental data

Run #	*Compound Noise Factor 1*	*Compound Noise Factor 2*
1	0.80	0.30
2	2.25	1.40
3	4.00	2.60
4	3.00	2.00
5	5.25	3.40
6	4.60	3.20
7	5.50	3.30
8	6.70	4.00
9	8.00	4.30

Results from the L_9 Experiment

The L_9 experiment yielded the following data (in pounds) as shown in Table 18.9.[3]

The data was processed in WinRobust. Table 18.10, the WinRobust Response table, and Figure 18.17, the WinRobust Response plot, display the results for the individual factors and the ANOVA for the experiment.

WinRobust L_9 ANOVA results are shown in Table 18.11.

WinRobust rapidly analyzes the data and shows the results. The team now has a good deal of insight concerning the individual factor effects as well as the ranked importance of each factor as it relates to the inducement of lateral forces. Notice the size of the error term due to the random effects of just running the experiment.

The angular misalignment had the strongest effect, and belt tension the next strongest. Lateral misalignment and the effects contained in the empty column (interactivity between columns A and B and experimental error) were not very significant.

The magnitude and directionality of the two main factors is quantified for future use in tolerance decision-making activities. The team can immediately get the belt-life tests running, and can hold technical discussions with the belt supplier to determine the risks associated with allowing certain amounts of lateral force to be present during the life of the belt. The team can determine the initial amount of angular and lateral constraint they want to specify as a tolerance range. The belt tension specification can be reviewed and rational limits can be tried on the evolving drive system.

3. The data has been altered to protect the proprietary results that were actually developed.

Table 18.10 Table of L_9 results

Factor	*Level*	*Force (S/N)*
Angle	1	1.89
	2	3.58
	3	5.30
Tension	1	2.48
	2	3.83
	3	4.45
Lateral	1	3.27
	2	3.49
	3	4.01
4	1	3.67
	2	3.38
	3	3.72

Figure 18.17 L_9 individual effects plot

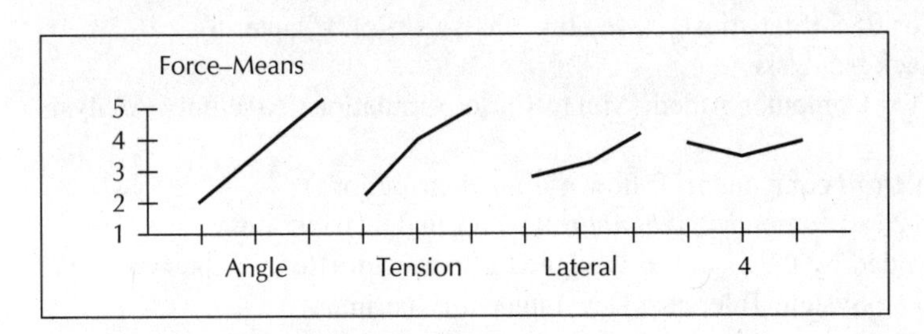

Table 18.11 L_9 ANOVA results

	Force-Mean			
Factor	*SS*	*D.O.F.*	*Mean Sq*	*F*
Angle	34.85	2	17.43	8.9
Tension	12.14	2	6.07	3.1
Lateral	1.74	2	0.87	—
4	0.42	2	0.21	—
Error	17.57	9	1.95	

Summary

This fifth case study is a sample of one of the numerous ways to use a designed experiment in the ongoing development of tolerances in a drive system. The engineering team knew very little about the effect of the changes they were inducing in the drive system. Rather than design a massive, all-inclusive experiment, they chose to run the relatively short exploratory L_9 experiment you have just seen. During the ongoing design evaluations for the copier, if belt life was showing signs of becoming a problem, the team could further explore the tolerancing of the drive system by adding additional force transducers and looking at additional levels of angular misalignment and belt tensions. The knowledge they gained from inducing this variation, in the context of a worst-case study, allowed the team to go forward with reasonable information that could be built upon if the need arose.

Chapter 18 concludes by providing real examples of tolerance-design case studies. The chapter is structured around five cases and is outlined as follows:

Chapter 18: Drive-System Case Studies

- The Drive System
- The Drive Module
- Case 1: Defining Tolerances for Standard Drive-Module Components (Ref. Chapter 4)
  - Tolerancing Custom Parts to Assemble with Standard Parts
- Case 2: Drive Module for Worst-Case Assembly Analysis (Ref. Chapter 5)
  - Worst-Case Linear Stack Analysis
- Case 3: Drive Module for RSS (Statistical) Assembly Analysis (Ref. Chapter 6)
  - RSS Case Linear Stack Analysis
- Case 4: Drive Module for Computer-Aided (Monte Carlo Simulation) Assembly Analysis (Ref. Chapter 8)
  - Testing Assumption 1 (*all* components follow normal distributions)
  - Testing Assumption 2 (all components *do not* follow normal distributions)
- Case 5: Drive System Aided by the Use of a Designed Experiment (Ref. Chapters 13 – 17)
  - Preparing for the Drive System Tolerance Development Experiment
  - Setting Up the Experiment
  - Data Acquisition
  - Results from the L_9 Experiment

Tolerance design is a very broad topic requiring the use of many different techniques and processes. Even though there is substantial diversity in the way an engineering team can conduct the development of tolerances, I hope the eighteen chapters of this text have helped provide a stuctured context and a rational process for you to follow.

APPENDIX A:

The Z Transformation Tables and the t Transformation Table[1]

Enter table with Z value (positive or negative) to two decimal places (hundredths). First decimal-place figure (tenths) is in the first column. Second decimal-place figure (hundredths) is in a column heading. For Z *positive,* tabular area is in the *right tail.* For Z *negative,* tabular area is in the *left tail.*

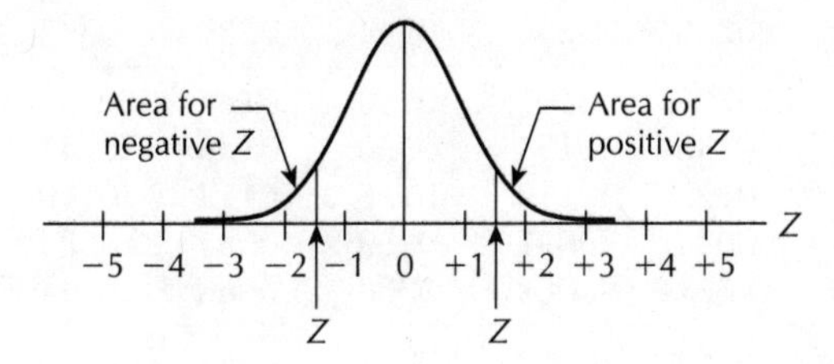

1. Z transformation tables are used with permission from Addison-Wesley Publishing Co. (Z tables are from *SPC for the Rest of Us;* H. Pitt; Addison-Wesley, 1994; t table is from *Six Sigma Producibility Analysis and Process Characterization;* M. J. Harry and J. R. Lawson; Addison-Wesley, 1992.)

Table A.1 Areas under the standard normal curve (*Z = 0.00 to 3.89. Areas to four decimal places*)

	Z (hundredths)									
Z (tenths)	.00	.01	.02	.03	.04	.05	.06	.07	.08	.09
	TAIL AREA									
0.0	.5000	.4960	.4920	.4880	.4840	.4801	.4761	.4721	.4681	.4641
0.1	.4602	.4562	.4522	.4483	.4443	.4404	.4364	.4325	.4286	.4247
0.2	.4207	.4168	.4129	.4090	.4052	.4013	.3974	.3936	.3897	.3859
0.3	.3821	.3783	.3745	.3707	.3669	.3632	.3594	.3557	.3520	.3483
0.4	.3446	.3409	.3372	.3336	.3300	.3264	.3228	.3192	.3156	.3121
0.5	.3085	.3050	.3015	.2981	.2946	.2912	.2877	.2843	.2810	.2776
0.6	.2743	.2709	.2676	.2643	.2611	.2578	.2546	.2514	.2483	.2451
0.7	.2420	.2389	.2358	.2327	.2297	.2266	.2236	.2207	.2177	.2148
0.8	.2119	.2090	.2061	.2033	.2005	.1977	.1949	.1922	.1894	.1867
0.9	.1841	.1814	.1788	.1762	.1736	.1711	.1685	.1660	.1635	.1611
1.0	.1587	.1562	.1539	.1515	.1492	.1469	.1446	.1423	.1401	.1379
1.1	.1357	.1335	.1314	.1292	.1271	.1251	.1230	.1210	.1190	.1170
1.2	.1151	.1131	.1112	.1093	.1075	.1057	.1038	.1020	.1003	.0985
1.3	.0968	.0951	.0934	.0918	.0901	.0885	.0869	.0853	.0838	.0823
1.4	.0808	.0793	.0778	.0764	.0749	.0735	.0721	.0708	.0694	.0681
1.5	.0668	.0655	.0643	.0630	.0618	.0606	.0594	.0582	.0571	.0559
1.6	.0548	.0537	.0526	.0516	.0505	.0495	.0485	.0475	.0465	.0455
1.7	.0446	.0436	.0427	.0418	.0409	.0401	.0392	.0384	.0375	.0367
1.8	.0359	.0351	.0344	.0336	.0329	.0322	.0314	.0307	.0301	.0294
1.9	.0287	.0281	.0274	.0268	.0262	.0256	.0250	.0244	.0239	.0233
2.0	.0228	.0222	.0217	.0212	.0207	.0202	.0197	.0192	.0188	.0183
2.1	.0179	.0174	.0170	.0166	.0162	.0158	.0154	.0150	.0146	.0143
2.2	.0139	.0136	.0132	.0129	.0125	.0122	.0119	.0116	.0113	.0110
2.3	.0107	.0104	.0102	.0099	.0096	.0094	.0091	.0089	.0087	.0084
2.4	.0082	.0080	.0078	.0075	.0073	.0071	.0069	.0068	.0066	.0064
2.5	.0062	.0060	.0059	.0057	.0055	.0054	.0052	.0051	.0049	.0048
2.6	.0047	.0045	.0044	.0043	.0041	.0040	.0039	.0038	.0037	.0036
2.7	.0035	.0034	.0033	.0032	.0031	.0030	.0029	.0028	.0027	.0026
2.8	.0026	.0025	.0024	.0023	.0023	.0022	.0021	.0021	.0020	.0019
2.9	.0019	.0018	.0017	.0017	.0016	.0016	.0015	.0015	.0014	.0014
3.0	.0014	.0013	.0013	.0012	.0012	.0011	.0011	.0011	.0010	.0010
3.1	.0010	.0009	.0009	.0009	.0008	.0008	.0008	.0008	.0007	.0007
3.2	.0007	.0007	.0006	.0006	.0006	.0006	.0006	.0005	.0005	.0005
3.3	.0005	.0005	.0005	.0004	.0004	.0004	.0004	.0004	.0004	.0004
3.4	.0003	.0003	.0003	.0003	.0003	.0003	.0003	.0003	.0003	.0002

3.50 to *3.61:* Area = .0002.
3.62 to *3.89:* Area = .0001.

(*Table continues on next page*)

Example: If $Z = -1.96$, area in *left* tail = .0250.

Table A.1 continued Areas under the standard normal curve (*Z = 3.90 to 4.90. Areas to six decimal places*)

Z (tenths)	*.00*	*.01*	*.02*	*.03*	*.04*	*.05*	*.06*	*.07*	*.08*	*.09*
	TAIL AREA									
3.9	.000048	.000046	.000044	.000043	.000041	.000039	.000038	.000036	.000035	.000033
4.0	.000032	.000030	.000029	.000028	.000027	.000026	.000025	.000024	.000023	.000022
4.1	.000021	.000020	.000019	.000018	.000017	.000017	.000016	.000015	.000015	.000014
4.2	.000013	.000013	.000012	.000012	.000011	.000011	.000010	.000010	.000009	.000009
4.3	.000009	.000008	.000008	.000008	.000007	.000007	.000007	.000006	.000006	.000006
4.4	.000005	.000005	.000005	.000005	.000005	.000004	.000004	.000004	.000004	.000004
4.5	.000003	.000003	.000003	.000003	.000003	.000003	.000003	.000002	.000002	.000002
4.6	.000002	.000002	.000002	.000002	.000002	.000002	.000002	.000002	.000001	.000001
4.7	.000001	.000001	.000001	.000001	.000001	.000001	.000001	.000001	.000001	.000001
4.8	.000001	.000001	.000001	.000001	.000001	.000001	.000001	.000001	.000001	.000001
4.9	ZERO ----------									

Table A.2 Commonly used tail areas and their corresponding values of Z

Area	*Z (+ or −)*
.1000	1.282
.0500	1.645
.0250	1.960
.0100	2.327
.0050	2.576
.0010	3.090

Note: The Z value for any point X on the base scale is the number of standard deviations that the point is away from the mean. In formula form,

Z value for any Point = (Point − Mean)/Standard deviation

or

$$Z_X = \frac{X - \mu}{\sigma_X}$$

If X is to the left of the mean, its corresponding Z value is negative.
If X is to the right of the mean, its corresponding Z value is positive.

Table A.4 Percentiles of the t distributions

v \ $1-\alpha$	0.60	0.70	0.80	0.90	0.95	0.975	0.99	0.995
1	0.325	0.727	1.376	3.078	6.314	12.706	31.821	63.657
2	0.289	0.617	1.061	1.886	2.920	4.303	6.965	9.925
3	0.277	0.584	0.978	1.638	2.353	3.182	4.541	5.841
4	0.271	0.569	0.941	1.533	2.132	2.776	3.747	4.604
5	0.267	0.559	0.920	1.476	2.015	2.571	3.365	4.032
6	0.265	0.553	0.906	1.440	1.943	2.447	3.143	3.707
7	0.263	0.549	0.896	1.415	1.895	2.365	2.998	3.499
8	0.262	0.546	0.889	1.397	1.860	2.306	2.896	3.355
9	0.261	0.543	0.883	1.383	1.833	2.262	2.821	3.250
10	0.260	0.542	0.879	1.372	1.812	2.228	2.764	3.169
11	0.260	0.540	0.876	1.363	1.796	2.201	2.718	3.106
12	0.259	0.539	0.873	1.356	1.782	2.179	2.681	3.055
13	0.259	0.538	0.870	1.350	1.771	2.160	2.650	3.012
14	0.258	0.537	0.868	1.345	1.761	2.145	2.624	2.977
15	0.258	0.536	0.866	1.341	1.753	2.131	2.602	2.947
16	0.258	0.535	0.865	1.337	1.746	2.120	2.583	2.921
17	0.257	0.534	0.863	1.333	1.740	2.110	2.567	2.898
18	0.257	0.534	0.862	1.330	1.734	2.101	2.552	2.878
19	0.257	0.533	0.861	1.328	1.729	2.093	2.539	2.861
20	0.257	0.533	0.860	1.325	1.725	2.086	2.528	2.845
21	0.257	0.532	0.859	1.323	1.721	2.080	2.518	2.831
22	0.256	0.532	0.858	1.321	1.717	2.074	2.508	2.819
23	0.256	0.532	0.858	1.319	1.714	2.069	2.500	2.807
24	0.256	0.531	0.857	1.318	1.711	2.064	2.492	2.797
25	0.256	0.531	0.856	1.316	1.708	2.060	2.485	2.787
26	0.256	0.531	0.856	1.315	1.706	2.056	2.479	2.779
27	0.256	0.531	0.855	1.314	1.703	2.052	2.473	2.771
28	0.256	0.530	0.855	1.313	1.701	2.048	2.467	2.763
29	0.256	0.530	0.854	1.311	1.699	2.045	2.462	2.756
30	0.256	0.530	0.854	1.310	1.697	2.042	2.457	2.750
40	0.255	0.529	0.851	1.303	1.684	2.021	2.423	2.704
60	0.254	0.527	0.848	1.296	1.671	2.000	2.390	2.660
120	0.254	0.526	0.845	1.289	1.658	1.980	2.358	2.617
∞	0.253	0.524	0.842	1.282	1.645	1.960	2.326	2.576

APPENDIX B:

The Adjusted Z Transformation Tables[1]

1. Reprinted with permission from Eastman Kodak Company; generated by Tom Albrecht of Eastman Kodak Company with references to *Non-Normal Data Analysis,* (Multiface Publishing Co., 1990).

Table B.1 Positive skew table: single tail (left-hand side of distribution) probability = 0.00135 or .135 percent

Kurtosis:	*Skew: 0*	*Skew: 0.2*	*Skew: 0.4*	*Skew: 0.6*	*Skew: 0.8*	*Skew: 1.0*
−1.2	−1.727	−1.496	−1.230	−0.975	−0.747	—
−1.0	−1.966	−1.696	−1.384	−1.089	−0.836	—
−0.8	−2.210	−1.912	−1.555	−1.212	−0.927	−0.692
−0.6	−2.442	−2.129	−1.740	−1.348	−1.023	−0.766
−0.4	−2.653	−2.335	−1.930	−1.496	−1.125	−0.841
−0.2	−2.839	−2.522	−2.116	−1.655	−1.235	−0.919
0.0*	*−3.000**	−2.689	−2.289	−1.817	−1.356	−1.000
0.2	−3.140	−2.836	−2.447	−1.976	−1.485	−1.086
0.4	−3.261	−2.966	−2.589	−2.127	−1.619	−1.178
0.6	−3.366	−3.079	−2.716	−2.265	−1.754	−1.277
0.8	−3.458	−3.179	−2.829	−2.396	−1.887	−1.381
1.0	−3.539	−3.267	−2.930	−2.512	−2.004	−1.491
1.2	−3.611	−3.346	−3.020	−2.617	−2.122	−1.602
1.4	−3.674	−3.416	−3.100	−2.713	−2.243	−1.713
1.6	−3.731	−3.479	−3.173	−2.799	−2.345	−1.815
1.8	−3.782	−3.535	−3.238	−2.877	−2.439	−1.916
2.0	−3.828	−3.586	−3.298	−2.949	−2.525	−2.023
2.2	−3.870	−3.633	−3.352	−3.013	−2.604	−2.115
2.4	−3.908	−3.675	−3.401	−3.073	−2.676	−2.202
2.6	−3.943	−3.714	−3.446	−3.127	−2.743	−2.283
2.8	−3.975	−3.750	−3.487	−3.177	−2.805	−2.358
3.0	−4.004	−3.782	−3.525	−3.223	−2.861	−2.428
3.2	−4.031	−3.813	−3.561	−3.265	−2.914	−2.493
3.4	−4.056	−3.841	−3.593	−3.305	−2.962	−2.553
3.6	−4.079	−3.867	−3.624	−3.341	−3.007	−2.610
3.8	−4.101	−3.891	−3.652	−3.375	−3.050	−2.662
4.0	−4.121	−3.914	−3.678	−3.407	−3.089	−2.711

*When kurtosis and skew = 0, we have 3σ quality.

**The values for all upper right-hand table entries are mathematically impossible and are blank in all the tables.

Table B.2 Positive skew table: single tail (right-hand side of distribution) probability = 0.99865 or 99.865 percent

Kurtosis:	*Skew: 0*	*Skew: 0.2*	*Skew: 0.4*	*Skew: 0.6*	*Skew: 0.8*	*Skew: 1.0*
−1.2	1.727	1.871	1.895	1.803	1.636	—
−1.0	1.966	2.134	2.169	2.061	1.856	—
−0.8	2.210	2.400	2.454	2.349	2.108	1.822
−0.6	2.442	2.648	2.726	2.646	2.395	2.052
−0.4	2.653	2.869	2.969	2.926	2.699	2.314
−0.2	2.839	3.060	3.179	3.173	2.993	2.608
0.0*	*3.000**	3.224	3.358	3.385	3.259	2.914
0.2	3.140	3.364	3.510	3.564	3.488	3.206
0.4	3.261	3.484	3.639	3.715	3.681	3.468
0.6	3.366	3.588	3.749	3.837	3.844	3.693
0.8	3.458	3.678	3.844	3.951	3.981	3.883
1.0	3.539	3.757	3.925	4.044	4.044	4.043
1.2	3.611	3.826	3.997	4.124	4.145	4.177
1.4	3.674	3.887	4.060	4.193	4.278	4.290
1.6	3.731	3.941	4.115	4.253	4.351	4.403
1.8	3.782	3.990	4.164	4.307	4.414	4.470
2.0	3.828	4.034	4.208	4.354	4.468	4.539
2.2	3.870	4.073	4.247	4.396	4.517	4.600
2.4	3.908	4.109	4.283	4.434	4.560	4.653
2.6	3.943	4.141	4.315	4.467	4.598	4.700
2.8	3.975	4.171	4.344	4.498	4.632	4.741
3.0	4.004	4.199	4.371	4.526	4.662	4.778
3.2	4.031	4.224	4.396	4.551	4.690	4.810
3.4	4.056	4.247	4.418	4.574	4.715	4.840
3.6	4.079	4.268	4.439	4.595	4.738	4.866
3.8	4.101	4.288	4.458	4.615	4.759	4.890
4.0	4.121	4.307	4.476	4.633	4.778	4.911

*When kurtosis and skew = 0, we have 3σ quality.

Table B.3 Positive skew table: single tail (left-hand side of distribution) probability = 0.0000034 or .00035 percent

Kurtosis:	*Skew: 0*	*Skew: 0.2*	*Skew: 0.4*	*Skew: 0.6*	*Skew: 0.8*	*Skew: 1.0*
1.2	−1.732	−1.496	−1.230	0.975	−0.747	—
1.0	−1.999	−1.706	−1.384	−1.089	−0.836	—
0.8	−2.336	−1.963	−1.562	−1.212	−0.927	−0.692
0.6	−2.761	−2.285	−1.773	−1.349	−1.023	−0.766
0.4	−3.277	−2.686	−2.034	−1.505	−1.125	−0.841
0.2	−3.866	−3.166	−2.359	−1.690	−1.236	−0.919
0.0*	*−4.500**	−3.706	−2.750	−1.914	−1.360	−1.000
0.2	−5.151	−4.283	−3.202	−2.188	−1.502	−1.086
0.4	−5.800	−4.875	−3.696	−2.517	−1.668	−1.179
0.6	−6.431	−5.464	−4.213	−2.895	−1.867	−1.280
0.8	−7.038	−6.038	−4.737	−3.310	−2.105	−1.394
1.0	−7.615	−6.592	−5.257	−3.751	−2.386	−1.524
1.2	−8.161	−7.121	−5.765	−4.203	−2.702	−1.675
1.4	−8.675	−7.624	−6.255	−4.657	−3.049	−1.852
1.6	−9.159	−8.099	−6.725	−5.105	−3.418	−2.059
1.8	−9.614	−8.548	−7.173	−5.541	−3.798	−2.296
2.0	−10.041	−8.972	−7.599	−5.964	−4.183	−2.560
2.2	−10.442	−9.372	−8.003	−6.370	−4.567	−2.848
2.4	−10.819	−9.748	−8.387	−6.760	−4.944	−3.151
2.6	−11.174	−10.104	−8.750	−7.132	−5.312	−3.466
2.8	−11.508	−10.439	−9.094	−7.488	−5.669	−3.785
3.0	−11.823	−10.756	−9.420	−7.827	−6.014	−4.105
3.2	−12.120	−11.055	−9.729	−8.149	−6.347	−4.423
3.4	−12.401	−11.338	−10.022	−8.456	−6.666	−4.735
3.6	−12.666	−11.606	−10.300	−8.749	−6.973	−5.041
3.8	−12.917	−11.861	−10.564	−9.028	−7.267	−5.339
4.0	−13.155	−12.102	−10.815	−9.293	−7.549	−5.628

*When kurtosis and skew = 0, we have 4.5σ quality.

Table B.4 Positive skew table: single tail (right-hand side of distribution) probability = 0.9999966 or 99.99966 percent

Kurtosis:	*Skew: 0*	*Skew: 0.2*	*Skew: 0.4*	*Skew: 0.6*	*Skew: 0.8*	*Skew: 1.0*
−1.2	1.732	1.877	1.897	1.803	1.636	—
−1.0	1.999	2.177	2.190	2.063	1.856	—
−0.8	2.336	2.557	2.557	2.376	2.109	1.822
−0.6	2.761	3.035	3.021	2.765	2.407	2.052
−0.4	3.277	3.609	3.588	3.252	2.770	2.315
−0.2	3.866	4.257	4.244	3.843	3.221	2.625
0.0*	*4.500**	4.946	4.955	4.520	3.771	2.998
0.2	5.151	5.648	5.690	5.252	4.414	3.455
0.4	5.800	6.340	6.423	6.007	5.127	4.005
0.6	6.431	7.011	7.136	6.760	5.879	4.642
0.8	7.038	7.651	7.820	7.494	6.642	5.346
1.0	7.615	8.257	8.469	8.197	7.397	6.089
1.2	8.161	8.828	9.080	8.865	8.127	6.846
1.4	8.675	9.365	9.655	9.495	8.823	7.597
1.6	9.159	9.868	10.193	10.087	9.485	8.330
1.8	9.614	10.340	10.697	10.641	10.110	9.031
2.0	10.041	10.782	11.169	11.161	10.697	9.698
2.2	10.442	11.196	11.610	11.647	11.247	10.331
2.4	10.819	11.584	12.024	12.102	11.764	10.928
2.6	11.174	11.949	12.411	12.528	12.247	11.490
2.8	11.508	12.292	12.775	12.928	12.701	12.018
3.0	11.823	12.615	13.117	13.303	13.126	12.514
3.2	12.120	12.919	13.439	13.656	13.526	12.980
3.4	12.401	13.205	13.742	13.988	13.901	13.418
3.6	12.666	13.476	14.028	14.300	14.254	13.830
3.8	12.917	13.732	14.298	14.595	14.587	14.217
4.0	13.155	13.975	14.553	14.873	14.901	14.582

*When kurtosis and skew = 0, we have 4.5σ quality.

Table B.5 Negative skew table: single tail (left-hand side of distribution) probability = 0.00135 or .135 percent

Kurtosis:	*Skew: 0*	*Skew: −0.2*	*Skew: −0.4*	*Skew: −0.6*	*−Skew: 0.8*	*Skew: −1.0*
−1.2	−1.727	−1.871	−1.895	−1.803	−1.636	—
−1.0	−1.966	−2.134	−2.169	−2.061	−1.856	—
−0.8	−2.210	−2.400	−2.454	−2.349	−2.108	−1.822
−0.6	−2.442	−2.648	−2.726	−2.646	−2.395	−2.052
−0.4	−2.653	−2.869	−2.969	−2.926	−2.699	−2.314
−0.2	−2.839	−3.060	−3.179	−3.173	−2.993	−2.608
0.0*	*−3.000**	−3.224	−3.358	−3.385	−3.259	−2.914
0.2	−3.140	−3.364	−3.510	−3.564	−3.488	−3.206
0.4	−3.261	−3.484	−3.639	−3.715	−3.681	−3.468
0.6	−3.366	−3.588	−3.749	−3.837	−3.844	−3.693
0.8	−3.458	−3.678	−3.844	−3.951	−3.981	−3.883
1.0	−3.539	−3.757	−3.925	−4.044	−4.044	−4.043
1.2	−3.611	−3.826	−3.997	−4.124	−4.145	−4.177
1.4	−3.674	−3.887	−4.060	−4.193	−4.278	−4.290
1.6	−3.731	−3.941	−4.115	−4.253	−4.351	−4.403
1.8	−3.782	−3.990	−4.164	−4.307	−4.414	−4.470
2.0	−3.828	−4.034	−4.208	−4.354	−4.468	−4.539
2.2	−3.870	−4.073	−4.247	−4.396	−4.517	−4.600
2.4	−3.908	−4.109	−4.283	−4.434	−4.560	−4.653
2.6	−3.943	−4.141	−4.315	−4.467	−4.598	−4.700
2.8	−3.975	−4.171	−4.344	−4.498	−4.632	−4.741
3.0	−4.004	−4.199	−4.371	−4.526	−4.662	−4.778
3.2	−4.031	−4.224	−4.396	−4.551	−4.690	−4.810
3.4	−4.056	−4.247	−4.418	−4.574	−4.715	−4.840
3.6	−4.079	−4.268	−4.439	−4.595	−4.738	−4.866
3.8	−4.101	−4.288	−4.458	−4.615	−4.759	−4.890
4.0	−4.121	−4.307	−4.476	−4.633	−4.778	−4.911

*When kurtosis and skew = 0, we have 3σ quality.

Table B.6 Negative skew table: single tail (right-hand side of distribution) probability = 0.99865 or 99.865 percent

Kurtosis:	*Skew: 0*	*Skew: −0.2*	*Skew: −0.4*	*Skew: −0.6*	*−Skew: 0.8*	*Skew: −1.0*
−1.2	1.727	1.496	1.230	0.975	0.747	—
−1.0	1.966	1.696	1.384	1.089	0.836	—
−0.8	2.210	1.912	1.555	1.212	0.927	0.692
−0.6	2.442	2.129	1.740	1.348	1.023	0.766
−0.4	2.653	2.335	1.930	1.496	1.125	0.841
−0.2	2.839	2.522	2.116	1.655	1.235	0.919
0.0*	*3.000**	2.689	2.289	1.817	1.356	1.000
0.2	3.140	2.836	2.447	1.976	1.485	1.086
0.4	3.261	2.966	2.589	2.127	1.619	1.178
0.6	3.366	3.079	2.716	2.265	1.754	1.277
0.8	3.458	3.179	2.829	2.396	1.887	1.381
1.0	3.539	3.267	2.930	2.512	2.004	1.491
1.2	3.611	3.346	3.020	2.617	2.122	1.602
1.4	3.674	3.41	3.100	2.713	2.243	1.713
1.6	3.731	3.479	3.173	2.799	2.345	1.815
1.8	3.782	3.535	3.238	2.877	2.439	1.916
2.0	3.828	3.586	3.298	2.949	2.525	2.023
2.2	3.870	3.633	3.352	3.013	2.604	2.115
2.4	3.908	3.675	3.401	3.073	2.676	2.202
2.6	3.943	3.714	3.446	3.127	2.743	2.283
2.8	3.975	3.750	3.487	3.177	2.805	2.358
3.0	4.004	3.782	3.525	3.223	2.861	2.428
3.2	4.031	3.813	3.561	3.265	2.914	2.493
3.4	4.056	3.841	3.593	3.305	2.962	2.553
3.6	4.079	3.867	3.624	3.341	3.007	2.610
3.8	4.101	3.891	3.652	3.375	3.050	2.662
4.0	4.121	3.914	3.678	3.407	3.089	2.711

*When kurtosis and skew = 0, we have 3σ quality.

Table B.7 Negative skew table: single tail (left-hand side of distribution) probability = 0.0000034 or .00035 percent

Kurtosis:	*Skew: 0*	*Skew: −0.2*	*Skew: −0.4*	*Skew: −0.6*	*Skew: −0.8*	*Skew: −1.0*
−1.2	−1.732	−1.877	−1.897	−1.803	−1.636	—
−1.0	−1.999	−2.177	−2.190	−2.063	−1.856	—
−0.8	−2.336	−2.557	−2.557	−2.376	−2.109	−1.822
−0.6	−2.761	−3.035	−3.021	−2.765	−2.407	−2.052
−0.4	−3.277	−3.609	−3.588	−3.252	−2.770	−2.315
−0.2	−3.866	−4.257	−4.244	−3.843	−3.221	−2.625
0.0*	*−4.500**	−4.946	−4.955	−4.520	−3.771	−2.998
0.2	−5.151	−5.648	−5.690	−5.252	−4.414	−3.455
0.4	−5.800	−6.340	−6.423	−6.007	−5.127	−4.005
0.6	−6.431	−7.011	−7.136	−6.760	−5.879	−4.642
0.8	−7.038	−7.651	−7.820	−7.494	−6.642	−5.346
1.0	−7.615	−8.257	−8.469	−8.197	−7.397	−6.089
1.2	−8.161	−8.828	−9.080	−8.865	−8.127	−6.846
1.4	−8.675	−9.365	−9.655	−9.495	−8.823	−7.597
1.6	−9.159	−9.868	−10.193	−10.087	−9.485	−8.330
1.8	−9.614	−10.340	−10.697	−10.641	−10.110	−9.031
2.0	−10.041	−10.782	−11.169	−11.161	−10.697	−9.698
2.2	−10.442	−11.196	−11.610	−11.647	−11.247	−10.331
2.4	−10.819	−11.584	−12.024	−12.102	−11.764	−10.928
2.6	−11.174	−11.949	−12.411	−12.528	−12.247	−11.490
2.8	−11.508	−12.292	−12.775	−12.928	−12.701	−12.018
3.0	−11.823	−12.615	−13.117	−13.303	−13.126	−12.514
3.2	−12.120	−12.919	−13.439	−13.656	−13.526	−12.980
3.4	−12.401	−13.205	−13.742	−13.988	−13.901	−13.418
3.6	−12.666	−13.476	−14.028	−14.300	−14.254	−13.830
3.8	−12.917	−13.732	−14.298	−14.595	−14.587	−14.217
4.0	−13.155	−13.975	−14.553	−14.873	−14.901	−14.582

*When kurtosis and skew = 0, we have 4.5σ quality.

Table B.8 Negative skew table: single tail (right-hand side of distribution) probability = 0.9999966 or 99.99966 percent

Kurtosis:	*Skew: 0*	*Skew: −0.2*	*Skew: −0.4*	*Skew: −0.6*	*−Skew: 0.8*	*Skew: −1.0*
1.2	1.732	1.496	1.230	0.975	0.747	—
1.0	1.999	1.706	1.384	1.089	0.836	—
0.8	2.336	1.963	1.562	1.212	0.927	0.692
0.6	2.761	2.285	1.773	1.349	1.023	0.766
0.4	3.277	2.686	2.034	1.505	1.125	0.841
0.2	3.866	3.166	2.359	1.690	1.236	0.919
0.0*	*4.500**	3.706	2.750	1.914	1.360	1.000
0.2	5.151	4.283	3.202	2.188	1.502	1.086
0.4	5.800	4.875	3.696	2.517	1.668	1.179
0.6	6.431	5.464	4.213	2.895	1.867	1.280
0.8	7.038	6.038	4.737	3.310	2.105	1.394
1.0	7.615	6.592	5.257	3.751	2.386	1.524
1.2	8.161	7.121	5.765	4.203	2.702	1.675
1.4	8.675	7.624	6.255	4.657	3.049	1.852
1.6	9.159	8.099	6.725	5.105	3.418	2.059
1.8	9.614	8.548	7.173	5.541	3.798	2.296
2.0	10.041	8.972	7.599	5.964	4.183	2.560
2.2	10.442	9.372	8.003	6.370	4.567	2.848
2.4	10.819	9.748	8.387	6.760	4.944	3.151
2.6	11.174	10.104	8.750	7.132	5.312	3.466
2.8	11.508	10.439	9.094	7.488	5.669	3.785
3.0	11.823	10.756	9.420	7.827	6.014	4.105
3.2	12.120	11.055	9.729	8.149	6.347	4.423
3.4	12.401	11.338	10.022	8.456	6.666	4.735
3.6	12.666	11.606	10.300	8.749	6.973	5.041
3.8	12.917	11.861	10.564	9.028	7.267	5.339
4.0	13.155	12.102	10.815	9.293	7.549	5.628

*When kurtosis and skew = 0, we have 4.5σ quality.

APPENDIX C:

The F Tables[1]

1. Reprinted with permission from Addison-Wesley Publishing Co. (*The Power of Statistical Thinking;* M. Leitnaker, R. Sanders, and C. Hild; Addison-Wesley, 1995.)

Table C.1 F table for 95 percent confidence

Denominator degrees of freedom
$F\alpha$ for $\alpha = .05$

v_2 \ v_1	Numerator Degrees of Freedom 1	2	3	4	5	6	7	8	9	10	12	15	20	24	30	40	60
1	161.4	199.5	215.7	224.6	230.2	234.0	236.8	238.9	240.5	241.9	243.9	245.9	248.0	249.1	250.1	251.1	252.2
2	18.51	19.00	19.16	19.25	19.30	19.33	19.35	19.37	19.38	19.40	19.41	19.43	19.45	19.45	19.46	19.47	19.48
3	10.13	9.55	9.28	9.12	9.01	8.94	8.89	8.85	8.81	8.79	8.74	8.70	8.66	8.64	8.62	8.59	8.57
4	7.71	6.94	6.59	6.39	6.26	6.16	6.09	6.04	6.00	5.96	5.91	5.86	5.80	5.77	5.75	5.72	5.69
5	6.61	5.79	5.41	5.19	5.05	4.95	4.88	4.82	4.77	4.74	4.68	4.62	4.56	4.53	4.50	4.46	4.43
6	5.99	5.14	4.76	4.53	4.39	4.28	4.21	4.15	4.10	4.06	4.00	3.94	3.87	3.84	3.81	3.77	3.74
7	5.59	4.74	4.35	4.12	3.97	3.87	3.79	3.73	3.68	3.64	3.57	3.51	3.44	3.41	3.38	3.34	3.30
8	5.32	4.46	4.07	3.84	3.69	3.58	3.50	3.44	3.39	3.35	3.28	3.22	3.15	3.12	3.08	3.04	3.01
9	5.12	4.26	3.86	3.63	3.48	3.37	3.29	3.23	3.18	3.14	3.07	3.01	2.94	2.90	2.86	2.83	2.79
10	4.96	4.10	3.71	3.48	3.33	3.22	3.14	3.07	3.02	2.98	2.91	2.85	2.77	2.74	2.70	2.66	2.62
11	4.84	3.98	3.59	3.36	3.20	3.09	3.01	2.95	2.90	2.85	2.79	2.72	2.65	2.61	2.57	2.53	2.49
12	4.75	3.89	3.49	3.26	3.11	3.00	2.91	2.85	2.80	2.75	2.69	2.62	2.54	2.51	2.47	2.43	2.38
13	4.67	3.81	3.41	3.18	3.03	2.92	2.83	2.77	2.71	2.67	2.60	2.53	2.46	2.42	2.38	2.34	2.30
14	4.60	3.74	3.34	3.11	2.96	2.85	2.76	2.70	2.65	2.60	2.53	2.46	2.39	2.35	2.31	2.27	2.22
15	4.54	3.68	3.29	3.06	2.90	2.79	2.71	2.64	2.59	2.54	2.48	2.40	2.33	2.29	2.25	2.20	2.15
16	4.49	3.63	3.24	3.01	2.85	2.74	2.66	2.59	2.54	2.49	2.42	2.35	2.28	2.24	2.19	2.15	2.11

17	4.45	3.59	3.20	2.96	2.81	2.70	2.61	2.55	2.49	2.45	2.38	2.31	2.23	2.19	2.15	2.10	2.06
18	4.41	3.55	3.16	2.93	2.77	2.66	2.58	2.51	2.46	2.41	2.34	2.27	2.19	2.15	2.11	2.06	2.02
19	4.38	3.52	3.13	2.90	2.74	2.63	2.54	2.48	2.42	2.38	2.31	2.23	2.16	2.11	2.07	2.03	1.98
20	4.35	3.49	3.10	2.87	2.71	2.60	2.51	2.45	2.39	2.35	2.28	2.20	2.12	2.08	2.04	1.99	1.95
21	4.32	3.47	3.07	2.84	2.68	2.57	2.49	2.42	2.37	2.32	2.25	2.18	2.10	2.05	2.01	1.96	1.92
22	4.30	3.44	3.05	2.82	2.66	2.55	2.46	2.40	2.34	2.30	2.23	2.15	2.07	2.03	1.98	1.94	1.89
23	4.28	3.42	3.03	2.80	2.64	2.53	2.44	2.37	2.32	2.27	2.20	2.13	2.05	2.01	1.96	1.91	1.86
24	4.26	3.40	3.01	2.78	2.62	2.51	2.42	2.36	2.30	2.25	2.18	2.11	2.03	1.98	1.94	1.89	1.84
25	4.24	3.39	2.99	2.76	2.60	2.49	2.40	2.34	2.28	2.24	2.16	2.09	2.01	1.96	1.92	1.87	1.82
26	4.23	3.37	2.98	2.74	2.59	2.47	2.39	2.32	2.27	2.22	2.15	2.07	1.99	1.95	1.90	1.85	1.80
27	4.21	3.35	2.96	2.73	2.57	2.46	2.37	2.31	2.25	2.20	2.13	2.06	1.97	1.93	1.88	1.84	1.79
28	4.20	3.34	2.95	2.71	2.56	2.45	2.36	2.29	2.24	2.19	2.12	2.04	1.96	1.91	1.87	1.82	1.77
29	4.18	3.33	2.93	2.70	2.55	2.43	2.35	2.28	2.22	2.18	2.10	2.03	1.94	1.90	1.85	1.81	1.75
30	4.17	3.32	2.92	2.69	2.53	2.42	2.33	2.27	2.21	2.16	2.09	2.01	1.93	1.89	1.84	1.79	1.74
40	4.08	3.23	2.84	2.61	2.45	2.34	2.25	2.18	2.12	2.08	2.00	1.92	1.84	1.79	1.74	1.69	1.64
60	4.00	3.15	2.76	2.53	2.37	2.25	2.17	2.10	2.04	1.99	1.92	1.84	1.75	1.70	1.65	1.59	1.53
120	3.92	3.07	2.68	2.45	2.29	2.17	2.09	2.02	1.96	1.91	1.83	1.75	1.66	1.61	1.55	1.50	1.43
∞	3.84	3.00	2.60	2.37	2.21	2.10	2.01	1.94	1.88	1.83	1.75	1.67	1.57	1.52	1.46	1.39	1.32

Table C.2 F table for 90 percent confidence

Denominator degrees of freedom
Fα for α = .01

v_2 \ v_1	1	2	3	4	5	6	7	8	9	10	12	15	20	24	30	40	60
	Numerator Degrees of Freedom																
1	4,052	4,999	5,403	5,625	5,764	5,859	5,928	5,982	6,022	6,056	6,106	6,157	6,209	6,235	6,261	6,287	6,313
2	98.50	99.00	99.17	99.25	99.30	99.33	99.36	99.37	99.39	99.40	99.42	99.43	99.45	99.46	99.47	99.47	99.48
3	34.12	30.82	29.46	28.17	28.24	27.91	27.67	27.49	27.35	27.23	27.05	26.87	26.69	26.60	26.50	26.41	26.32
4	21.20	18.00	16.69	15.98	15.52	15.21	14.98	14.80	14.66	14.55	14.37	14.20	14.02	13.93	13.84	13.75	13.65
5	16.26	13.27	12.06	11.39	10.97	10.67	10.46	10.29	10.16	10.05	9.89	9.72	9.55	9.47	9.38	9.29	9.20
6	13.75	10.92	9.78	9.15	8.75	8.47	8.26	8.10	7.98	7.87	7.72	7.56	7.40	7.31	7.23	7.14	7.06
7	12.25	9.55	8.45	7.85	7.46	7.19	6.99	6.84	6.72	6.62	6.47	6.31	6.16	6.07	5.99	5.91	5.82
8	11.26	8.65	7.59	7.01	6.63	6.37	6.18	6.03	5.91	5.81	5.67	5.52	5.36	5.28	5.20	5.12	5.03
9	10.56	8.02	6.99	6.42	6.06	5.80	5.61	5.47	5.35	5.26	5.11	4.96	4.81	4.73	4.65	4.57	4.48
10	10.04	7.56	6.55	5.99	5.64	5.39	5.20	5.06	4.94	4.85	4.71	4.56	4.41	4.33	4.25	4.17	4.08
11	9.65	7.21	6.22	5.67	5.32	5.07	4.89	4.74	4.63	4.54	4.40	4.25	4.10	4.02	3.94	3.86	3.78
12	9.33	6.93	5.95	5.41	5.06	4.82	4.64	4.50	4.39	4.30	4.16	4.01	3.86	3.78	3.70	3.62	3.54
13	9.07	6.70	5.74	5.21	4.86	4.62	4.44	4.30	4.19	4.10	3.96	3.82	3.66	3.59	3.51	3.43	3.34
14	8.86	6.51	5.56	5.04	4.69	4.46	4.28	4.14	4.03	3.94	3.80	3.66	3.51	3.43	3.35	3.27	3.18
15	8.68	6.36	5.42	4.89	4.56	4.32	4.14	4.00	3.89	3.80	3.67	3.52	3.37	3.29	3.21	3.13	3.05
16	8.53	6.23	5.29	4.77	4.44	4.20	4.03	3.89	3.78	3.69	3.55	3.41	3.26	3.18	3.10	3.02	2.93

17	8.40	6.11	5.18	4.67	4.34	4.10	3.93	3.79	3.68	3.59	3.46	3.31	3.16	3.08	3.00	2.92	2.83
18	8.29	6.01	5.09	4.58	4.25	4.01	3.84	3.71	3.60	3.51	3.37	3.23	3.08	3.00	2.92	2.84	2.75
19	8.18	5.93	5.01	4.50	4.17	3.94	3.77	3.63	3.52	3.43	3.30	3.15	3.00	2.92	2.84	2.76	2.67
20	8.10	5.85	4.94	4.43	4.10	3.87	3.70	3.56	3.46	3.37	3.23	3.09	2.94	2.86	2.78	2.69	2.61
21	8.02	5.78	4.87	4.37	4.04	3.81	3.64	3.51	3.40	3.31	3.17	3.03	2.88	2.80	2.72	2.64	2.55
22	7.95	5.72	4.82	4.31	3.99	3.76	3.59	3.45	3.35	3.26	3.12	2.98	2.83	2.75	2.67	2.58	2.50
23	7.88	5.66	4.76	4.26	3.94	3.71	3.54	3.41	3.30	3.21	3.07	2.93	2.78	2.70	2.62	2.54	2.45
24	7.82	5.61	4.72	4.22	3.90	3.67	3.50	3.36	3.26	3.17	3.03	2.89	2.74	2.66	2.58	2.49	2.40
25	7.77	5.57	4.68	4.18	3.85	3.63	3.46	3.32	3.22	3.13	2.99	2.85	2.70	2.62	2.54	2.45	2.36
26	7.72	5.53	4.64	4.14	3.82	3.59	3.42	3.29	3.18	3.09	2.96	2.81	2.66	2.58	2.50	2.42	2.33
27	7.68	5.49	4.60	4.11	3.78	3.56	3.39	3.26	3.15	3.06	2.93	2.78	2.63	2.55	2.47	2.38	2.29
28	7.64	5.45	4.57	4.07	3.75	3.53	3.36	3.23	3.12	3.03	2.90	2.75	2.60	2.52	2.44	2.35	2.26
29	7.60	5.42	4.54	4.04	3.73	3.50	3.33	3.20	3.09	3.00	2.87	2.73	2.57	2.49	2.41	2.33	2.23
30	7.56	5.39	4.51	4.02	3.70	3.47	3.30	3.17	3.07	2.98	2.84	2.70	2.55	2.47	2.39	2.30	2.21
40	7.31	5.18	4.31	3.83	3.51	3.29	3.12	2.99	2.89	2.80	2.66	2.52	2.37	2.29	2.20	2.11	2.02
60	7.08	4.98	4.13	3.65	3.34	3.12	2.95	2.82	2.72	2.63	2.50	2.35	2.20	2.12	2.03	1.94	1.84
120	6.85	4.79	3.95	3.48	3.17	2.96	2.79	2.66	2.56	2.47	2.34	2.19	2.03	1.95	1.86	1.76	1.66
∞	6.63	4.61	3.78	3.32	3.02	2.80	2.64	2.51	2.41	2.32	2.18	2.04	1.88	1.79	1.70	1.59	1.47

APPENDIX D:

Additional References for Tolerance Design

Statistical Issues in Setting Product Specifications; S. Read and T. Read; Hewlett-Packard Journal; June 1988

How to Perform Statistical Tolerance Analysis Vol. 11; N. D. Cox; ASQC 1986

Probability and Its Applications for Engineers; D. H. Evans; Marcel Dekker, 1992

Statistical Tolerancing: The State of the Art, Part 1. Background; D. H. Evans; Journal of Quality Technology, Vol. 6 No. 4; Oct. 1974

Statistical Tolerancing: The State of the Art, Part 2. Methods for Estimating Moments; D. H. Evans; Journal of Quality Technology, Vol. 7 No. 1; Jan. 1975

Statistical Tolerancing: The State of the Art, Part 3. Shifts and Drifts; D. H. Evans; Journal of Quality Technology, Vol. 7 No. 2; Apr. 1975

Computer Programs for the Quadrature Approximation for Statistical Tolerancing; D. H. Evans and D. R. Faulkenburg; Journal of Quality Technology, Vol. 8 No. 2; Apr. 1976

Engineering Design—A Materials and Processing Approach; G. E. Dieter; McGraw-Hill, 1991

Dimensioning and Tolerancing for Quality Production; M. F. Spotts; Prentice Hall Inc., 1983

Allocation of Tolerance to Minimize Cost of Assembly; M. F. Spotts; Journal of Engineering for Industry; Vol. 95 No. 3, Aug. 1973

Tolerance Control in Design and Manufacturing; O. R. Wade; Industrial Press; 1967

Dimensioning and Tolerancing ANSI Y 14.5M-1982; AMSE 1982

Design Issues in Mechanical Tolerance Analysis; K. W. Chase and W. H. Greenwood; Manufacturing Review, Vol. 1 No. 1, ASME Mar. 1988

A New Tolerance Analysis Method for Designers and Manufacturers; W. H. Greenwood and K. W. Chase; Transactions of the AMSE; Vol. 109; 1987

Quality, Reliability and Process Improvement 8th Ed.; N. L. Enrich; Industrial Press, 1985

Assessment and Optimization of Dimensional Tolerances; D. B. Parkinson; Computer-Aided Design; Vol. 17 No. 4; May 1985

The Optimization Problem with Optimal Tolerance Assignment and Full Acceptance; W. Michael and J. N. Siddall; Journal of Mechanical Design; Vol. 103 No. 4; Oct. 1981

Mechanism Design: Accounting for Manufacturing Tolerances and Costs in Function Generating Problems; G. H. Sutherland and B. Roth; Journal of Engineering for Industry; Vol. 97 No. 1; Feb. 1975

Minimum Exponential Cost Allocation of Sure-Fit Tolerances; D. Wilde and E. Prentice; Journal of Engineering for Industry; Vol. 97 No. 4; Nov. 1975

Some New Developments in Tolerance Design in CAD; G. Zhang and M. Porchet; DE-Vol. 65 No. 2; Advances in Design Automation; ASME 1993

Tolerancing the Components of an Assembly for Minimum Cost; J. Peters; Journal of Engineering for Industry; Vol. 92 No. 2; Aug. 1970

A Better Way to Specify Part Tolerances; S. Jakuba; Machine Design; Oct. 8, 1987

Calculation of Tolerance Based on a Minimum Cost Approach; Journal of Engineering for Industry; Vol. 94 No. 2; May 1972

Benderizing Tolerances—a simple practical probability method of handling tolerances for limit stack-ups; A. Bender; Graphic Science; Dec. 1962

Additional References for Experimental Tolerance Design:

The New Experimental Design; T. Mori; ASI 1990

A Primer on the Taguchi Method; R. Roy; Van Nostrand Reinhold; 1990

Quality Engineering Using Robust Design; M. Phadke; Prentice Hall, 1989

Total Quality Development; D. Clausing; ASME Press; 1993

Engineering Quality By Design; T. Barker; Marcel Dekker; 1990

Taguchi on Robust Technology Development; G. Taguchi; ASME; 1993

Taguchi Techniques for Quality Engineering; P. J. Ross; McGraw Hill; 1988

Optimization of Mechanical Assembly Tolerances by Incorporating Taguchi's Quality Loss Function; B. W. Cheng and S. Maghsoodloo; Journal of Manufacturing Systems; Vol. 14 No. 4; SME 1995

Robust Product Design Using Manufacturing Adjustments; K. Otto; DE-Vol. 68; Design Theory and Methodology—DTM '94; ASME 1994

A Method for Early Manufacturability Evaluation of Proposed Tolerance Plans for Thin Walled Parts; A. K. Fathailall and J. R. Dixon; DE-Vol. 68; Design Theory and Methodology: DTM '94; ASME 1994

Design of Tolerances for Quality; Chang-Xue Feng and A. Kusiak; DE–Vol. 68; Design Theory and Methodology—DTM '94; ASME 1994

Optimal Tolerancing: The Link Between Design and Manufacturing Productivity; M. Iannuzzi and E. Sandgren; DE-Vol. 68; Design Theory and Methodology—DTM '94; ASME 1994

JIS–Japanese Industrial Standard: General Tolerancing Rules for Plastic Dimensions; JIS K 7109; Japanese Standards Association, 1986

Achieving Customer Satisfaction Through Robust Design; L. B. Boza, T. J. Ciaccia, D. A. Gatenby, R. W. Muise, K. K. Ng, and G. M. Yanizeski; AT&T Journal; Jan./Feb. 1994

Robust Design by Tolerance Allocation Considering Quality & Manufacturing Cost; R. Soderberg; DE-Vol. 69-1; Advances in Design Automation; ASME 1994

Producer-Consumer Tolerances; Y. Fathi; Journal of Quality Technology; Vol. 22 No. 2; Apr. 1990

APPENDIX E:

Software Suppliers for Tolerance Design

ANOVA-TM
Advanced Systems & Designs, Inc.
1208 East Maple Street, Suite 100
Troy, Michigan 48083
Phone: 810-616-9818; Fax: 810-585-7408

Crystal Ball Monte Carlo Simulator
Decisioneering Inc.
1515 Arapahoe Street, Suite 1311
Denver, Co. 80202
Phone (303) 534-1515, (303) 534-4818, or 800-289-2550
www.decisioneering.com

Training, Books, and Software for Quality Engineering Applications
(QFD, TRIZ, Robust Design, Tolerance Design, GD&T, and SPC)
American Supplier Insititute
17333 Federal Drive, Suite 220
Allen Park, MI 48101
Phone: 313-336-8877; Fax: 313-336-3187

WinRobust
Abacus Digital
6 Lookout View Road
Fairport, NY 14450
Phone: 716-223-1368

INDEX

A

Accelerated Testing (Nelson), 41
Activities list development, 37
Additive model, 275
Adjusted *Z* transform, 63, 67–72, 393–401
Advanced Systems & Designs, 308
American Supplier Institute (ASI), 14, 266, 308, 325, 411
  circuit case study, 330–37
Analysis of means (ANOM)
  noise experiments and, 274–76
  statistical variance experiments and, 280
Analysis of variance (ANOVA), 44–45
  accounting for variation using experimental data, 307–8
  ANOVA-TM example for, 318–23
  computer-aided, 308
  defined, 306
  degrees of freedom for, 311
  error variance and pooling, 312
  error variance and replication, 312–13
  error variance and utilizing empty columns, 313
  example of, 308–10
  F-ratio (variance ratio), 306, 313–14
  sensitivity analysis, 153
  statistical variance experiments and, 280–81
  variation between parameters, 306
  variation within control factors, 306
  WinRobust example for, 314–18
ANOVA-TM, 308, 318–23, 411
ANSI Y14.5M–1982 Standard, 10
Antisynergistic, 255
Arithmetic average, 55–56
Assembly constraint, 105
Assembly gap, 102–3, 107
Asymmetric nominal-the-best, 217–18
@ *Risk,* 163

B

Barker, Thomas, 278, 282
Bathtub curve, 48
Bechwith, 31
Benchmarking: The Search for Industry Practices that Lead to Superior Performance (ASQC), 29, 31, 34
Beta distribution, 169–70
Biased sample, 57
Binomial distribution, 168
Bjorke, Oyvind, 185
Borgovini, R., 35
Boundary dimension, 107
Breipohl, A. M., 153
Build-test-fix methods, 5, 17, 248

C

Cagan, J., 189
Capability process (Cp). *See* Process capability
Cause-and-effect diagrams, 35
Central limit theorem, 63

Central tendency, 52, 55
Champy, James, 219–20
Chase, K. W., 104, 113, 115, 132, 152, 163, 187–89
Clausing, Don, 14, 29, 34
Clements, Richard Barrett, 96
Coefficient of variation (COV), 61
Cohen, Lou, 29, 34
Common cause, 204–5
Company cost-driven process, 327–28
Component manufacturing process output distribution sensitivity analysis, 151
Component-part tolerancing, 43
Component tolerance analysis, quality-loss function and multiple-, 232–37
Component tolerances, relating customer tolerances to, 230–31
Compound noise factor (CNF), 262, 270, 272–73, 275
Computer-aided ANOVA, 308
Computer-aided design (CAD), role of, 153
Computer-aided engineering (CAE), role of, 153
Computer-aided tolerancing (CAT)
  See also Crystal Ball
  advantages/disadvantages of, 194–95
  drive module for computer-aided assembly analysis, 373–79
  Monte Carlo simulations, 63, 153, 163–66
  role of, 43, 160–61
  Pro/ENGINEER, 161, 162
  software and platform options, 161–62
  sources of information, 185
Computer-Aided Tolerancing (Bjorke), 185
Concept design
  off-line quality control and, 14–17
  sensitivity analysis and, 154–58
Concept selection, subsystem, 28–39
Confounding, 253
Cost (s)
  -based optimal tolerance analysis, 187–90, 195
  evaluating aggregated low-level tolerances, 238–40
  impact of tolerances on, 11
  linking cost and functional performance, 201–2
  of quality, example of, 202–6
  tolerance design and product, 18, 20
  versus tolerance plots, 190
Cost-driven process
  company, 327–28
  manufacturing capability and, 328–29
Cost improvement parameters (CIP), 351–52, 356
Creveling, C. M., 18, 31, 34, 36, 38
Criticality analysis, 35
Critical-to-function (CTF) parameters, 347, 350–51
Crystal Ball, 153, 160, 411
  engineering analysis reports with, 184–85
  how to use, 175–81
  Monte Carlo simulations, 163–66, 170–72, 181–84, 285–92
  probability distributions, 167–70
  sensitivity analysis and, 172–81
Custer, L., 313
Customer tolerances
  defined, 209
  expressing, 224
  relating to engineering tolerances, 229
  relating to subsystem/component tolerances, 230–31
  understanding, 10

D

Data
  role of, 51–52
  sample, 51
  sample versus population, 52
Data analysis
  graphical methods of, 52–54
  quantitative methods of, 54
Degrees of freedom (DOF), 66–67
  ANOVA and, 311
  for evaluating interactions, 259
  experimental analysis with, 251
  for fractional factorial orthogonal array, 251
  for full factorial, 251

tables, 403–7
for two-level full factorial, 252–55
Deming, W. Edwards, 4
Descriptive statistics
arithmetic average, 55–56
coefficient of variation, 61
distributions, use of, 62–63
first moment about the mean, 55–56
fourth moment about the mean, 59–61
kurtosis, 59–61
second moment about the mean, 56–57
skew, 58–59
standard deviation, 57
third moment about the mean, 58–59
variance, 56–57
Design
See also Concept design; Parameter design; Tolerance design
ideal function of, 17
inspection device, 333
matching with manufacturing processes, 12
specification, 18
Designed experiments
See also Statistical variance experiment
build-test-fix approach, 248
degrees of freedom, 251–55
drive system and use of, 380–87
fractional factorial orthogonal arrays used in, 247, 249–51
full factorial, 248–55
interactions, 255–59
Designers, role of, 9
Design of experiments (DOE), 206
Deterioration characteristics in design, developing tolerances for, 241–44
Deterioration deceleration factors, 242
Deterioration noise, 36, 263–64
Dimensioning for Interchangeable Manufacturing (Fortini), 104, 113
Direct linearization method (DLM), 152
Displacement vector, 110
Distributions, use of, 62–63
DPMO (defects per million opportunities), 76
DPU (defects per unit), 77, 78
Drive-system case studies
defining tolerances for standard drive-module components, 367–69
drive module described, 365–67
drive module for computer-aided assembly analysis, 373–79
drive module for RSS assembly analysis, 371–73
drive module for worst-case assembly analysis, 369–71
drive system and use of designed experiments, 380–87
drive system described, 364
Duane Plots, 47
Dynamic root sum of squares (DRSS), 130–31, 138–39, 193–94

E

Eastman, George, 4
Eastman Kodak Co., 363
Edison, Thomas, 4
Elsayed, E. A., 231, 240
Empirical sensitivity analysis, 149, 153
Engineering arrays, 255
Engineering Methods for Robust Product Design (Fowlkes and Creveling), 18, 31, 34, 36, 38
Engineering metrics, transition to, 32
Engineering Quality by Design (Barker), 278
Engineering tolerances, relating customer tolerances to, 229
Engineers, role of, 9
Enhanced version of quality function deployment (EQFD), 14
Envelope, 102
Environmental noise, 36, 263
Errico, J. R., 292–93
Error variance
pooling and, 312
replication and, 312–13
utilizing empty columns, 313
Experimental design space, 247
Experimental error
and induced noise, 267–70
use of term, 306

Experimental Methods for Engineers (Holman), 31
Exponential distribution, 169
External noise, 36, 263

F

Failure analysis and root-cause analysis for variability, 34–35
Failure Mode, Effects, and Criticality Analysis (Borgovini, Pemberton and Rossi), 35
Failure modes, effects, and criticality analysis (FMECA), 34–35
Fault tree analysis (FTA), 35
Fault Tree Analysis Application Guide (Mahar and Wilbur), 35
First moment about the mean, 55–56
First time yield (FTY), 78
Fishbone diagrams, 35
Fisher, Ronald, 4
Ford, Henry, 4
Ford Motor Co., 204
Formulas, list of sample statistics, 82
Fortini, E. I., 104, 113
Foster, Lowell, 6
Fourth moment about the mean, 59–61
Fowlkes, W. Y., 18, 31, 34, 36, 38
Fractional factorial orthogonal arrays, 48
  degrees of freedom for, 251
  used in tolerance design, 247, 249–51
F-ratio (variance ratio), 306, 313–14
Frequency distribution table, 52–53
Freund, J. E., 313
Full factorial experiments, 248–55
  degrees of freedom for, 251
  degrees of freedom for two-level, 252–55
Functional limits, 209
Functional performance, linking cost and, 201–2

G

Gao, J., 152
Gaussian distribution, 62, 167
Geometric dimensioning and tolerancing (GD&T)
  defined, 5–6
  role of, 9, 11
Geometrics III (Foster), 6
Graphical methods of data analysis, 52–54
Greenwood, W. H., 104, 113, 115, 132, 152
Geometric distribution, 168
Grand total sum of squares (GTSS), 308–9
Greenwood, W. H., 187–89
Guide to Quality Control (Ishikawa), 35

H

Hahn, G. J., 132
Hammer, Michael, 219–20
Handbook of Statistical Methods in Manufacturing (Clements), 96
Handbook of Transducers, 31
Harry, Mikel J., 5, 88–90, 103, 128, 129, 130, 132, 147
Hauglund, L. F., 187–89
High-level tolerances, 224
Histograms, 54
Holman, J. P., 31
Hsiang, T. C., 231, 240
Hypergeometric distribution, 168

I

Industrial Revolution, impact of, 4
Inferential statistics
  adjusted Z transform, 63, 67–72
  student-t distribution, 63, 66–67
  Z transformation process, 63–66
Inherent robustness, sensitivity analysis and, 14–15
Initial tolerance design
  description of, 90–98
  process capability and, 96–98
  searching for published, 92, 93
  sources of information, 92, 93
  statistical process control and, 96–98
  three-sigma paradigm, 88–90
  through discussions with manufacturer, 93, 95
Interactions
  defined, 255–57
  degrees of freedom for evaluating, 259
  quantifying the effect of, 257–59

virtual, 257
Interchangeability, role of, 4
Intermediate tolerance, 225
Introduction to Quality Engineering (Taguchi), 225, 278
Introduction to Statistical Quality Control (Montgomery), 98
Ishikawa, K., 35
Ishikawa diagrams, 35

J
Japan, role of, 4, 5
Johnson, G. W., 31

K
Kurfess, T., 189
Kurtosis, 59–61

L
LabView Graphical Programming (Johnson), 31
Lagrange multiplier method, 189
Larger-the-better (LTB), 214–17
Leptokurtic distribution, 59, 60
Lienhard, 31
Life-cycle cost (LCC), 79
Life testing, 41
Likelihood of occurrence, 64
Linear root sum of squares, 133, 135–38, 140–45
Linear sensitivity factor, 231–32
Linear worst-case tolerance analysis, case study, 109–10
Lognormal distribution, 169
Loosli, B. G., 187–89
Lower specification limit (LSL), 204
Low-level tolerances, 225
evaluating aggregated, 238–40

M
Magleby, S. P., 152
Mahar, D., 35
Manufacturer's process capability recommendations
advantages/disadvantages of, 192–93
descriptions of, 87–99
Manufacturing capability and cost-driven process, 328–29
Manufacturing processes
matching design with, 12
parameter-robustness optimization, 88
Marangoni, 31
Mathematical sensitivity analysis, 149–50
Mating dimension, 101–2, 107
Maximum assembly gap, 103, 108, 129
Mean square deviation (MSD)
ANOVA and, 307
error, 268
factor, 268
quality-loss function, 211–12, 214, 216, 243
Mean square due to a design parameter, 306
Mean square due to a experimental error, 306
Mean square due to error, 267
Mean sum of squares, 307
Mean time between service calls (MTBSC), 29
Mean time to failure (MTTF), 29
Measurement Systems, Application and Design, 31
Mechanical Applications in Reliability Engineering (Sadlon), 30, 35
Mechanical Design Process, The (Ullman), 26
Mechanical Measurements (Bechwith, Marangoni, and Lienhard), 31
Mesokurtic distribution, 59, 60
Meyers, R., 192
Minimum assembly gap, 103, 129
Minimum gap constraint, 102, 108
Monotonic, 256
Monte Carlo simulations, 63, 153
Crystal Ball, 163–66, 170–72, 181–84, 285–92
statistical variance experiments and, 282–92
Montgomery, Douglas, 98, 192
Motorola Corp., 5, 76, 90, 102
dynamic root sum of squares (DRSS), 130–31

Motorola Corp. (*continued*)
static root sum of squares, 131–32
Multiple-component tolerance analysis, quality-loss function and, 232–37

N

Nelson, W., 41
Noise (s)
concept design and signal-to-noise ratio experimentation, 15
constructing noise diagrams and noise maps, 265–66
deterioration, 36, 263–64
experimental error and induced, 267–70
external/environmental, 36, 263
identifying for inclusion in a tolerance experiment, 264–65
parameter design and optimizing insensitivity to, 17–18
proximity, 263
subsystem noise diagrams and map construction, 35–36
surrogate, 264
types of, 262–64
unit-to-unit, 36, 263
Noise factor experiment
analysis of means for, 274–76
experiment design, 271–72
setting up and running the, 273–74
verification of predicted response, 276
Noise factors, compound, 262, 270, 272–73
Nominal assembly gap, 103, 107
Nominal mating dimension, 107
Nominal-the-best (NTB), 210–13, 229–30
Nonlinear relationships, quality-loss function and, 240–41
Nonlinear root sum of squares, 132–33, 145–47, 194
Nonlinear tolerance sensitivity analysis, 151–53
Nonlinear worst-case tolerance analysis, 104–8, 111–22
Nonnormal distribution, 61
adjusted *Z* transform and, 67–72
Normal distribution, 54, 62, 167

O

O'Conner, Patrick, 29, 35
Off-line quality control, 13
concept design, 14–17
parameter design, 17–18, 19
One-dimensional tolerance stacks, 127–30
One-factor-at-a-time sensitivity analysis, 153–54
On-target engineering, 206
Orthogonal arrays. *See* Fractional factorial orthogonal arrays

P

Parameter design
distributions and, 62–63
off-line quality control and, 17–18, 19
Parametric Technology Corp., 162
Peace, G., 36, 38
Pemberton, S., 35
Phadke, Madhve, 32, 38
Pitt, Hy, 98
Platykurtic distribution, 59, 60
Poisson distribution, 77, 168
Pooling of variances, 128, 145
Pooling sum of squares, 312
Population data, sample data versus, 52
Population statistics, 52, 165
Pox, G. E. P., 63
Practical Reliability Engineering (O'Conner), 29, 35
Probabilistic Systems Analysis (Breipohl), 153
Probability density function, 64
Probability distributions, 165–66
See also under type of
Crystal Ball and, 167–70
Probability of interference, 128
Process capability (Cp)
description of, 44–45, 72–76
dynamic root sum of squares and, 130–31
initial tolerances and, 96–98
libraries, 98–99
manufacturer's, 87–99, 192–93
quality-loss function and, 79–81
six-sigma process metrics, 76–78

Process diagrams
Crystal Ball Monte Carlo simulations, 170–72
statistical tolerance analysis for stackups and, 133–47
worst-case tolerance analysis for stackups and, 105–8
Process flow diagramming, use of, 7
Product benchmarking experimentation and evaluation, 15
Product design, 42–47
Product development and reliability
allocating reliability factor requirements, 36–37
calculating reliability gaps, 30
capability index calculations, 44–45
failure analysis and root-cause analysis for variability, 34–35
Gate 1, 39
Gate 2, 42
Gate 3, 47
new control factor variances calculated, 45
new control factor variances calculated, role of suppliers, 45–46
phase of, 26–27
preparation for, 26, 28
product reliability goal defined, 28–29
reliability growth activities list development, 37
reliability requirements allocation, 31–33
reliability signal-to-noise factor, 32–34, 38–39
set points optimized, 42
subsystem noise diagrams and map construction, 35–36
subsystem parameter optimization development, 37–38
subsystem robustness testing and optimization completed, 39–42
subsystem tolerance design, 42–44
system baseline reliability defined (MTBSC and MTTF units), 29–30
system baseline reliability defined (MTBSC or MTTF and S/N terms), 30–31
verifying improvements at system level, 46–47
verifying variance reduction, 46
Pro/ENGINEER, 161, 162
Proximity noise, 263
Pugh, Stuart, 34
concept selection process, 14, 15, 16

Q

Quadratic-loss function (QLF). *See* Quality-loss function
Quality
cost of quality, example of, 202–6
culture, 206
linking cost and functional performance, 201–2
step function and, 206–7
tolerance design and product, 18, 20
Quality control issues, development of, 4–5
Quality Engineering in Production Systems (Taguchi), 225
Quality engineering process
comparions of designs, 91
steps in, 13–14
Quality Engineering Using Robust Design (Phadke), 32, 38
Quality-function deployment (QFD), 21
Quality Function Deployment (Cohen), 29, 34
Quality-loss coefficient, 208, 209
Quality-loss economic coefficient, constructing the, 220–22
Quality-loss function (QLF), 11, 78
asymmetric nominal-the-best, 217–18
capability process (Cp) index and, 79–81
converting traditional tolerance problem into, 237–38
defined, 79
developing your own, 219–22
evaluating aggregated low-level tolerances, 238–40
example of, 210
example of tolerancing using, 229–30
high versus low level tolerances, 224–25
larger-the-better, 214–17
linear sensitivity factor, 231–32

Quality-loss function (QLF) (*continued*)
multiple-component tolerance analysis and, 232–37
nominal-the-best, 210–13, 229–30
nonlinear relationships and, 240–41
relating customer to engineering tolerances, 229
relating customer to subsystem/component tolerances, 230–31
role of, 208–9
smaller-the-better, 213–14
Taguchi tolerancing equations, 225–28
types of, 210–18
Quality problems
break-in period, 20
out-of-the-box, 20
Quantitative methods of data analysis, 54

R

Range of assembly gap, 103, 108
Rao, S. S., 29, 163
Reengineering the Corporation (Hammer and Champy), 219
Reliability Based Design (Rao), 29
Reliability bathtub curve, 48
Reliability for the Technologies, 29, 35
Reliability growth and product development. *See* Product development and reliability
Repetition, 268
Replication
defined, 268
error variance and, 312–13
Resolution V design, 257
Response surface, 192
Response Surface Methods (Meyers and Montgomery), 192
Robustness optimization, system, 39–42
Robustness test fixture, 364
Rolled throughput yield, 78
Root-cause analysis for variability, 34–35
Root sum of squares (RSS)
advantages/disadvantages of, 193–94
drive module for RSS assembly analysis, 371–73
dynamic, 130–31, 138–39, 193–94
linear, 133, 135–38, 140–45
nonlinear, 132–33, 145–47, 194
one-dimensional tolerance stacks, 127–30
process diagrams for, 133–47
role of, 43, 126–27
static, 131–32, 139–40, 194
Z transform, 127, 128–29
Ross, P., 313
Rossi, M., 35
Rule of seven, 125

S

Sadlon, R., 30, 35
Safety factor, Taguchi's economical, 225–28, 230
Sample data, 51
versus population data, 52
Sample standard deviation, 57
Sample statistics, 52, 165
Sample variance, 56
Second moment about the mean, 56–57
Sensitivity analysis
advantages/disadvantages of, 194
ANOVA, 153
component manufacturing process output distribution, 151
concept design and, 154–58
Crystal Ball and, 172–81
empirical, 149, 153
inherent robustness and, 14–15
mathematical, 149–50
nonlinear tolerance, 151–53
one-factor-at-a-time, 153–54
tolerance-range, 150–51
Shapiro, S. S., 132
Shewhart, Walter, 4
Shift vector, 131–32
Signal-to-noise ratio (S/N)
based reliability factor, 32–34, 38–39
experimentation, 15
nominal the best, 31
system baseline reliability defined (MTBSC or MTTF and S/N terms), 30–31
Six Sigma, 5
process metrics, 76–78

worst-case tolerance analysis, 102–3
Six Sigma Mechanical Design Tolerancing (Harry and Stewart), 5, 88–90, 147
Skew, 58–59
Smaller-the-better (STB), 213–14
SPC for the Rest of Us (Pitt), 98
Spearman rank correlation coefficients, 172
Special cause, 204–5
Stackups
See also Root sum of squares
one-dimensional tolerance stacks, 127–30
process diagrams for statistical tolerance analysis and, 133–47
Stackups, worst-case tolerance analysis and
linear, 109–10
nonlinear, 104–8, 111–22
process diagrams for, 105–8
Standard deviation, 57
used in statistical variance experiment, 278
Standard error of the mean, 63
Standardized normal distribution, 62
Static root sum of squares, 131–32, 139–40, 194
Statistical process control (SPC)
charts, 124–26
common versus special cause, 204–5
initial tolerance design and, 96–98
sources of information, 98
Statistical tolerance analysis. *See* Root sum of squares
Statistical Tolerance Analysis Using a Modification of Taguchi's Method (Errico and Zaino), 292
Statistical variance experiment
creating accurate output response data from three-level array experiments, 292–96
defined, 278
instructions for running three-level, 281–92
output statistcs compared, 296–99
preparing to run, 279–92
selecting right orthogonal array for, 279–81
standard deviation in, 278
worst-case, 300–304
Statistics for Experimenters (Pox), 63
Step function, 206–7
Stewart, Reigle, 5, 88–90, 103, 128, 129, 130, 132, 147
Student-*t* distribution, 63, 66–67, 392
Subsystem concept selection, 28–39
Subsystem tolerance design, 42–44
relating customer to, 230–31
Summing of variances, 127
Sum of squares, 307, 309
Surrogate noise, 264
Sustained process mean shift, 131
Synergistic, 256
System of Experimental Design (Taguchi), 225
System robustness optimization, 39–42

T

Taguchi, Genichi, 4, 10, 187, 195–96, 199–200, 231, 240, 242, 243, 278, 308
Taguchi methods, 13
See also Quality-loss function
relationship between traditional tolerancing and, 6–7
for technology robustness optimization, 15
Taguchi Methods (Peace), 36, 38
Taguchi on Robust Technology Development (Taguchi), 225, 243
Taguchi tolerancing
developing tolerances for deterioration characteristics in design, 241–44
equations, 225–28
example of tolerancing using quality-loss function, 229–30
linear sensitivity factor, 231–32
overview of, 20–23
relating customer to engineering tolerances, 229
relating customer to subsystem/component tolerances, 230–31
Technology development, defined, 14
Texas Instruments, 98, 162
Third moment about the mean, 58–59

Three-sigma paradigm, 88–90
Time based rate of change, 241
TI/TOL 3D+^{3}, 162
Tolerance/tolerancing
  See also under type of
  assembly gap, 103, 108
  defined, 10–11
  developing versus designing, 6
  feasible, 88
  high versus low level, 224–25
  historical background of, 3–6
  intermediate, 225
  loosening, 46
  relating customer to engineering, 229
  reliability bathtub curve and, 48
  Taguchi, 20–23
  typical, 88
Tolerance analysis
  See also Statistical variance experiment
  distributions and, 63
  Tolerance design
  See also Design; Designed experiments
  cost of quality, example of, 202–6
  customer tolerance, 207
  developing for deterioration characteristics in design, 241–44
  distributions and, 62–63
  fractional factorial orthogonal arrays used in, 247, 249–51
  initial, 90–98
  linking cost and functional performance, 201–2
  product cost and quality and, 18, 20
  step function, 206–7
  subsystem, 42–44
Tolerance design process
  company cost-driven process, 327–28
  manufacturing capability and cost-driven process, 328–29
  steps for, 325–27
Tolerance design process, case study
  adding noise, 338
  ANOVA and quality-loss function and process capability, 349–50
  ANOVA in, 344–49
  calculating variances and mean square deviation, 353–54
  constructing quality-loss function, 332–34
  cost improvement parameters, 351–52, 356
  critical-to-function parameters, 347, 350–51
  data entry, 340–43
  identifying costs, 352, 356
  identifying parameters for tolerancing, 330–31
  improving quality where parameter design was not done, 331
  interactions in, 343–49
  measuring response characteristics, 343
  quantifying cost of reducing parameter standard deviations, 354–55
  setting up and running the experiment, 334–40
  six-sigma attained with, 357
  system reliability improved with, 357
  two- versus three-level experiments, 337
  working with suppliers, 352
Tolerance development
  relationship between Taguchi approach and traditional, 6–7
  role of engineers and designers, 9
  steps in, 6, 7, 8
Tolerance-range sensitivity analysis, 150–51
TOLTECH, 185
Total Design (Pugh), 14, 34
Total Quality Development (Clausing), 14, 29, 34
Total sum of squares (TSS), 309
Traditional tolerance analysis
  See also under type of
  advantages/disadvantages of, 191–96
  converting, into quality-loss function problem, 237–38
  relationship between Taguchi approach and, 6–7
  role of, 85–86
Trend charts, 172
Triangular distribution, 167
TRIZ (theory of inventive problem solving), 14

U
Ullman, David, 15, 26, 28, 152, 154–58
Unigraphics 3-D, 161
Uniform distribution, 167
Unit manufacturing cost (UMC), 79
Unit-to-unit noise, 36, 263
Unit vector, 103
Upper specification limit (USL), 204

V
Variability
  output, 124
  special-cause, 124
Variance (s)
  defined, 56–57
  pooling of, 128, 145
  reduction, 5
  sample, 56
  summing of, 127
Variance adjustment factor (VAF), 327–28, 353
Variance analysis. *See* Analysis of variance (ANOVA); Statistical variance experiment
Variance ratio (F-ratio), 306, 313–14
Variation Systems Analysis, 162
Vasseur, H., 189, 190
Vector chain, 105
Vectors, 105
Virtual interactions, 257
VKC process, 189
Voice of the customer, 15, 21
VSA 3-D Tolerance Analysis Tool Kit, 162

W
Wasserman, Gary, 33
Weibull distribution, 169
Whitney, Eli, 4
Wilbur, J. W., 35
WinRobust software, 308, 411
  ANOVA example using, 314–18
Worst-case tolerance analysis, 43
  advantages/disadvantages of, 193
  case study of linear, 109
  conducting a tolerance experiment for, 300–304
  drive module for worst-case assembly analysis, 369–71
  linear, 109–10
  mating dimension, 101–2, 107
  nonlinear, 104–8, 111–22
  process diagrams for stackups and, 105–8
  six-sigma, 102–3
  when to use, 101
Worst-Case Tolerance Analysis with Nonlinear Problems (Greenwood and Chase), 104
Wu, Alan, 325

Y
Yates, Frank, 4

Z
Zaino, N. A., 292–93
Zero defects, 78
Zhang, Y.-K., 154–58
Z transform
  adjusted, 63, 67–72
  adjusted tables, 393–401
  description of, 63–66
  root sum of squares and, 127, 128–29
  tables, 389–91